OXFORD MATHEMATICAL MONOGRAPHS

Series Editors

OXFORD MATHEMATICAL MONOGRAPHS

A. Belleni-Moranti: *Applied semigroups and evolution equations*
A. M. Arthurs: *Complementary variational principles* 2nd edition
M. Rosenblum and J. Rovnyak: *Hardy classes and operator theory*
J. W. P. Hirschfeld: *Finite projective spaces of three dimensions*
A. Pressley and G. Segal: *Loop groups*
D. E. Edmunds and W. D. Evans: *Spectral theory and differential operators*
Wang Jianhua: *The theory of games*
S. Omatu and J. H. Seinfeld: *Distributed parameter systems: theory and applications*
J. Hilgert, K. H. Hofmann, and J. D. Lawson: *Lie groups, convex cones, and semigroups*
S. Dineen: *The Schwarz lemma*
S. K. Donaldson and P. B. Kronheimer: *The geometry of four-manifolds*
D. W. Robinson: *Elliptic operators and Lie groups*
A. G. Werschulz: *The computational complexity of differential and integral equations*
L. Evens: *Cohomology of groups*
G. Effinger and D. R. Hayes: *Additive number theory of polynomials*
J. W. P. Hirschfeld and J. A. Thas: *General Galois geometries*
P. N. Hoffman and J. F. Humphreys: *Projective representations of the symmetric groups*
I. Györi and G. Ladas: *The oscillation theory of delay differential equations*
J. Heinonen, T. Kilpelainen, and O. Martio: *Non-linear potential theory*
B. Amberg, S. Franciosi, and F. de Giovanni: *Products of groups*
M. E. Gurtin: *Thermomechanics of evolving phase boundaries in the plane*
I. Ionescu and M. Sofonea: *Functional and numerical methods in viscoplasticity*
N. Woodhouse: *Geometric quantization* 2nd edition
U. Grenander: *General pattern theory*
J. Faraut and A. Koranyi: *Analysis on symmetric cones*
I. G. Macdonald: *Symmetric functions and Hall polynomials* 2nd edition
B. L. R. Shawyer and B. B. Watson: *Borel's methods of summability*
M. Holschneider: *Wavelets: an analysis tool*
Jacques Thévenaz: *G-algebras and modular representation theory*
Hans-Joachim Baues: *Homotopy type and homology*
P. D. D'Eath: *Black holes: gravitational interactions*
R. Lowen: *Approach spaces: the missing link in the topology–uniformity–metric triad*
Nguyen Dinh Cong: *Topological dynamics of random dynamical systems*
J. W. P. Hirschfeld: *Projective geometries over finite fields* 2nd edition
K. Matsuzaki and M. Taniguchi: *Hyperbolic manifolds and Kleinian groups*
David E. Evans and Yasuyuki Kawahigashi: *Quantum symmetries on operator algebras*
Norbert Klingen: *Arithmetical similarities: prime decomposition and finite group theory*
Isabelle Catto, Claude Le Bris, and Pierre-Louis Lions: *The mathematical theory of thermodynamic limits: Thomas–Fermi type models*
D. McDuff and D. Salamon: *Introduction to symplectic topology* 2nd edition
William M. Goldman: *Complex hyperbolic geometry*
Charles J. Colbourn and Alexander Rosa: *Triple systems*
V. A. Kozlov, V. G. Maz'ya, and A. B. Movchan: *Asymptotic analysis of fields in multi-structures*
Gérard A. Maugin: *Non-linear waves in elastic crystals*
George Dassios and Ralph Kleinman: *Low frequency scattering*
Gerald W. Johnson and Michael L. Lapidus: *The Feynman Integral and Feynman's Operational Calculus*
W. Lay and S. Y. Slavyanov: *Special functions: a unified theory based on singularities*
A. Carbone and S. Semmes: *A graphic apology for symmetry and implicitness*

A Graphic Apology for Symmetry and Implicitness

ALESSANDRA CARBONE

Department of Mathematics and Computer Science
University of Paris XII

and

STEPHEN SEMMES

Department of Mathematics
Rice University

OXFORD
UNIVERSITY PRESS

Great Clarendon Street, Oxford OX2 6DP

Oxford University Press is a department of the University of Oxford.
It furthers the University's objective of excellence in research, scholarship,
and education by publishing worldwide in

Oxford New York

Athens Auckland Bangkok Bogotá Buenos Aires Calcutta
Cape Town Chennai Dar es Salaam Delhi Florence Hong Kong Istanbul
Karachi Kuala Lumpur Madrid Melbourne Mexico City Mumbai
Nairobi Paris São Paulo Singapore Taipei Tokyo Toronto Warsaw
with associated companies in Berlin Ibadan

Oxford is a registered trade mark of Oxford University Press
in the UK and in certain other countries

Published in the United States
by Oxford University Press Inc., New York

First published 2000

A catalogue record for this book is available from the British Library

Library of Congress Cataloging in Publication Data
(Data available)

ISBN 0 19 850729 1 (Hbk)

Typeset by the authors
Printed in Great Britain
on acid-free paper by
Biddles Ltd., Guildford and King's Lynn

PREFACE

Large or complicated objects are often described *implicitly*, through some kind of rule or pattern. Instead of listing all of the particles in an object, one might specify a recipe by which they are assembled without giving them individually. Indeed, this may be the only practical possibility, an "explicit" description being too large to be manageable or even feasible.

One of the main themes of the present book is that large or complicated objects which admit comparatively short descriptions have some kind of compensating internal symmetry. As a philosophical matter, this might be considered obvious or inevitable, but concretely it is not entirely clear what form this symmetry should take.

Classical forms of symmetry in mathematics, such as group actions or algebras of infinitesimal symmetries, are rather limited in comparison with algorithmic recursion and logical reasoning, which allow for all sorts of foldings, splicings, and substitutions. On the other hand, recursion and reasoning are so flexible that they do not themselves need to follow any simple or coherent pattern. This is connected to the incomputability of measurements of information content, such as those provided by Kolmogorov complexity and algorithmic information theory [Kol68, Cha87, Cha92, LV90, Man77].

In other words, as one allows greater freedom in the kinds of symmetries and patterns which are admitted, it becomes easier for efficient implicit descriptions for individual objects to exist, and to accommodate wider ranges of phenomena, but it also becomes more difficult to work with the classes of implicit descriptions themselves. It becomes more difficult to find compact implicit descriptions, for instance, or to determine the existence of ones of at most a given size, or to make comparisons through implicit descriptions, or to extract information from them (unless the information in question is special, and suitably compatible with the particular type of implicit description), etc. One can also think in terms of *sets* of symmetries, patterns, and implicit descriptions as objects in their own right, with their own internal structure which becomes more complex as well. Thus one has trade-offs to consider, between increasing flexibility for individual objects, at the cost of additional complications for working with collections of objects, making comparisons between them, and so on.

In this book we aim for a middle ground. We begin with questions about combinatorial models for some phenomena related to formal proofs (especially those connected to the use and elimination of "lemmas"). This leads to various geometric constructions and questions of computational complexity. We emphasize constructions which resemble propositional proofs in their recursive power, but we also look at stronger forms of recursion.

Our primary combinatorial framework deals with oriented graphs, and is similar in spirit to some basic ideas from topology and geometry. To an oriented graph G and a choice of basepoint v in G there corresponds a "visibility graph" $\mathcal{V}_+(v, G)$, with a canonical projection π from $\mathcal{V}_+(v, G)$ to G. The visibility graph is always a tree, and it plays a role somewhat analogous to that of universal covering spaces and exponential mappings in topology and geometry. (We shall expand on these themes in Chapter 4.)

One can think of the graph G as providing an "implicit description" of the visibility $\mathcal{V}_+(v, G)$, which itself can be infinite or of exponential size compared to G. When the visibility is much larger than G there is clearly a compensating degree of "symmetry", as in the principle mentioned above, even if the form of this symmetry is not the same or as intricate as ones that play such a basic role in topology and geometry.

An advantage of this more primitive type of symmetry is that it lends itself more naturally to questions of computational complexity at roughly the level of the P and NP classes [GJ79, Pap94]. (This is as opposed to towers of exponentials, for instance; note that the visibility graph, when it is finite, is never larger than the original graph by more than a (single) exponential.) In Chapter 13, we shall describe a problem about mappings between graphs and their induced mappings between visibility graphs which is NP-complete and closely related to the satisfiability problem in propositional logic. There are other problems which lie naturally in NP or co-NP, but seem to have a reasonable chance of not being complete (without being in P either), or which can be solved in polynomial time, but have the possibility to be "better" than that. (We shall discuss this further in the text.)

The computational problems about graphs considered here are typically rather different from ones most often mentioned in complexity theory (as in [GJ79, Pap94]). They often concern the possibility of "folding" one graph into another (as opposed to embeddings or the existence of special paths). One could also think of this in terms of asking about the existence of certain kinds of symmetry or patterns. Another basic issue concerns computations or analysis of "large" objects directly at an implicit level, using much smaller descriptions of the objects involved. In the main text we shall mention a number of questions of this nature concerning oriented graphs, i.e., questions which have very simple interpretations in terms of the associated visibility graphs, but for which one would like to have solutions which can work directly and efficiently at the level of the given graphs themselves.

The possibility of making computations directly at an implicit level is closely connected to the dichotomy between deterministic and nondeterministic Turing machines. It is often the case that there is a simple *nondeterministic* approach that works at the implicit level, and the problem is to know whether there is one which is deterministic and still reasonably efficient. The NP-complete problem mentioned above (from Chapter 13) fits into this context, and we shall see some other examples in Chapter 10.

The size of the visibility of an oriented graph G reflects the internal complexity of G in roughly the same manner as for universal coverings and exponential mappings of Riemannian manifolds. This form of complexity turns out to be closely connected to measuring the lengths of "chains of focal pairs" (Section 4.14) in G, and it arises naturally in several ways in formal proofs and other combinatorial structures (such as finite-state automata).

The type of complexity that one has with graphs and their visibilities is roughly like that of propositional proofs, as indicated above. We shall also consider stronger forms of recursion, analogous to the use of quantifiers in formal proofs. In terms of ordinary experience, one might strike an analogy between this and the behavior of something like a tree or other plant: there is a fairly simple overall structure, built out of leaves; the leaves themselves have a nontrivial internal structure; and so forth. A nice feature of formal logic is that it provides natural and fairly precise ways in which to think about limitations on the strength of the recursion involved, and about different levels of strength of recursion. It also does this in a manner that often lends itself to comparison with human reasoning. One of the most basic limitations to impose on a formal proof is on the depth of nesting of quantifiers which is permitted. In this book we shall try to bring out combinatorial or geometric manifestations of this kind of limited implicitness in simple and concrete ways, especially in Chapter 16. (In terms of sheer complexity, each new layer of quantification in a proof tends to correspond roughly to another level of exponentiation.)

We have tried to provide some flexibility in this book, in various ways, and concerning different potential readers in particular. In one direction might be students and researchers interested in complexity theory, formal logic, and some other topics related to theoretical computer science. Another direction could include mathematicians who have not delved into these areas too much, but who are interested in them (or are curious about them). In general, readers interested in the representation of information, complexity, formal language theory, geometry, or group theory may have use for this book, at least in part.

Different readers may be more concerned with some parts of the book than others, or may wish to pass over some parts at some times (perhaps to return to them on other occasions). We have tried to make it reasonably easy for a reader to proceed as he or she might wish, with some modularity in the text.

A large portion of the book is fairly self-contained, and involves only reasonably basic notions (like graphs or algorithms). Parts of the book deal with interactions or connections with other areas, and for these some knowledge of the other areas would often be helpful. Normally we might try to give a sketch or review of some of the relevant points, as well as some references.

A chart showing the material of the book is given in Fig. 0.1, with topics of the chapters and some groupings of them indicated. The text divides into three basic groups, with basic theory and contexts of application in the first, more on symmetry and complexity in the second, and more on symmetry and

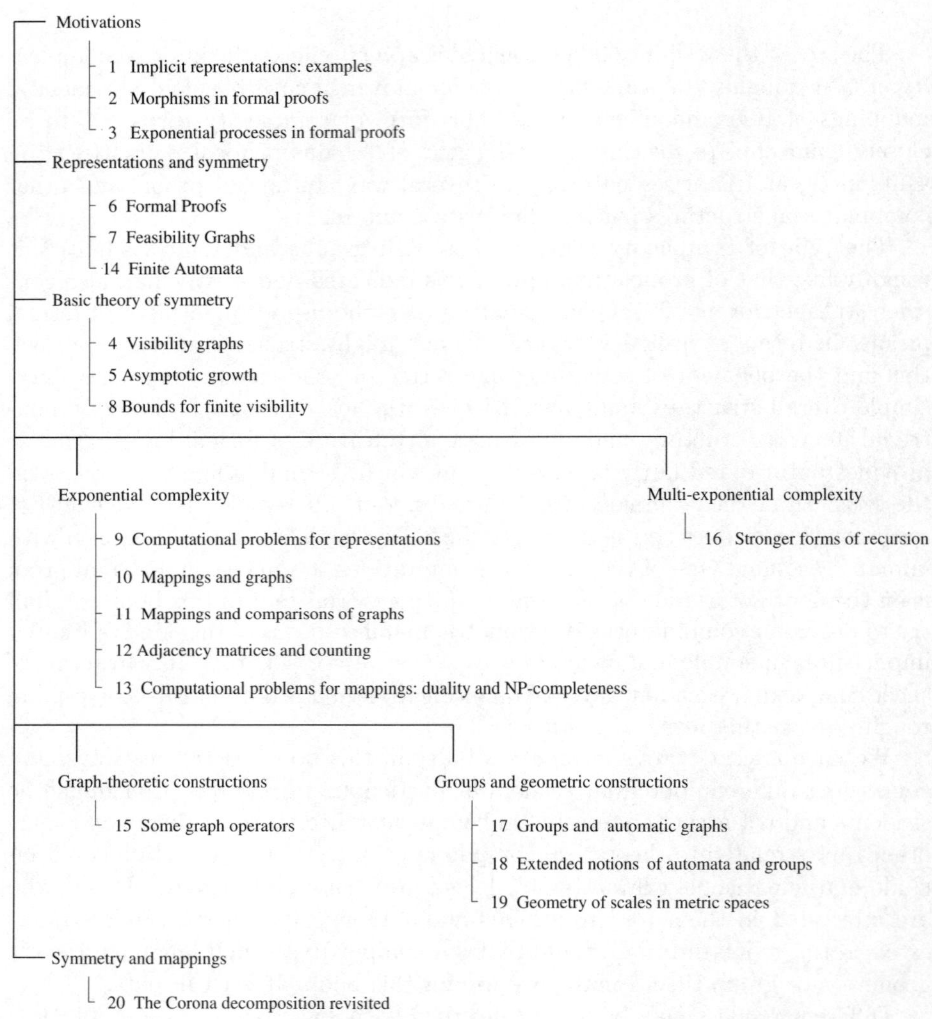

FIG. 0.1. A chart for the book

constructions in the third.

Some of the motivations related to this book have been described above, and a more detailed introduction is provided in Chapter 1. Chapter 2 concerns the ideas of morphisms and monotonicity in logic and complexity, and Chapter 3 contains some examples of propositional tautologies which help to illustrate some of the basic combinatorial phenomena. From here we are lead to the notion of visibility graphs in Chapter 4.

The passage from a graph to its visibility provides a simple geometric model for the idea of converting an implicit description into one which is completely explicit. In the context of formal proofs this type of conversion can be achieved through the elimination of *cuts*. Roughly speaking, the *cut rule* in formal logic is an extension of the more famous Modus Ponens rule (which says that if one knows A and if one knows that A implies B then one is allowed to conclude that B holds), and it provides a technical mechanism for allowing the use of *lemmas* in a formal proof. See Appendix A for a brief review of some basic elements of formal logic, including the statement of Gentzen's theorem on the elimination of cuts (in classical logic), and the notion of the *logical flow graph*, which traces the flow of atomic occurrences in a formal proof. In Chapter 6 we look at the way that the logical flow graph of a formal proof can evolve under the standard method for eliminating cuts, and we analyze it in comparison with the passage to the visibility in some detail. In particular, one can start with "short" proofs whose logical flow graphs have a lot of internal structure (as measured by the size of the visibility graph), and we shall see how the duplication of subproofs that occurs in cut elimination can simultaneously increase the size of the graph and simplify its geometry, with an effect roughly like that of the visibility. There are also circumstances in which the logical flow graph is too "microscopic" to properly reflect the inner workings of a given proof, and we shall discuss the evolution of the logical flow graph and related graphs in this type of situation too. We shall look more precisely at what happens when there are no cuts, or when the cuts are very simple.

The elimination of cuts in formal proofs is a very complicated subject, and one which is far from being fully understood. Basic references include the books [Gir87c, Tak87] of Girard and Takeuti. Beyond classical logic or Brouwer's Intuitionistic Logic (and their variants), there is also the more recent development of "Linear Logic", introduced by Girard [Gir87a, Gir89b, Gir89a, GLT89, Gir90, Gir95a].

Chapter 7 concerns the notion of *feasibility graphs*, which provides an intermediate setting between formal proofs and abstract graphs. Roughly speaking, a feasibility graph is a labelled graph in which a computation or collection of computations is represented in a certain way. The concept was motivated in part by the mathematical theory of *feasible numbers* [Par71], in which computations in arithmetic can be coded implicitly into formal proofs. For feasibility graphs, a complete accounting of the underlying computation is given by the passage to the visibility in a direct manner. While feasibility graphs are simpler than formal proofs of feasibility, they do capture some of the phenomena which can occur using cut and contraction rules in formal logic. They do not directly reflect the stronger forms of recursion associated to quantifier rules, but this can be accommodated through more elaborate geometry, as in Chapter 16. (Sections 18.3 and 18.4 are also relevant for this.)

We look at the size and rate of growth of the visibility in more detail in Chapters 5 and 8. There is a simple dichotomy between exponential and polynomial rates of growth when the visibility is infinite, but the finite case is more involved and closer to many of the themes of this book. Related questions of computability are considered in Chapters 9 and 12. In the latter, we look at the well-known method of *adjacency matrices* in the context of visibility graphs.

In Chapters 10 and 11 we discuss mappings between graphs, which provide a basic tool for making comparisons between them (and doing so implicitly vis-á-vis the associated visibility graphs), and for expressing certain kinds of symmetry. When the graphs are suitably labelled, as in the setting of feasibility graphs, one can sometimes interpret mappings as making comparisons between different calculations. Note that there is not at present a good notion of a "morphism" between formal proofs, by which to make suitable comparisons, as in other parts of mathematics. (This is discussed in Chapter 2.) This is another reason for looking at simpler combinatorial models for some of the phenomena related to formal proofs.

The ability to make comparisons through mappings is connected to a number of questions of computational complexity, as we shall see in Chapters 9, 10, and 11. In Chapter 13, we consider the "visibility mapping problem", in which one is given a pair of mappings $g : G \to K$ and $h : H \to K$ between oriented graphs, and one asks whether the induced mappings between visibilities have different images. This is analogous to the NP-complete problem of regular expression inequivalence [GJ79], and we give a similar NP-completeness result for it.

The way that mappings between graphs induce mappings between visibility graphs is analogous to induced mappings between covering spaces in topology, and it provides a different way of looking at finite automata and regular languages. That is, the assignment of letters in an alphabet to the different transitions of a finite automaton can be reformulated as a mapping between a pair of graphs, and the associated regular language corresponds to a subset of the image of the induced mapping between visibilities. This is discussed in Chapters 14 and 15, where we also point out how many standard constructions for making new automata from old ones have natural geometric interpretations which make sense more broadly for mappings between graphs.

In Chapter 17 we discuss groups and their Cayley graphs. Finitely-presented groups provide basic examples of implicit descriptions, whereby a finite set of generators and relations specifies a possibly-infinite group whose global structure may be very complicated. Even in the very special case of nilpotent groups, the large-scale geometry of the Cayley graphs is quite remarkable. For instance, there are some questions about fractal geometry for which the known examples, up until quite recently, were all or nearly all connected to the asymptotic behavior of Cayley graphs of nilpotent groups. These include questions concerning the possibility of certains forms of "calculus" on metric spaces.

For computational problems, the class of "automatic groups" introduced in [ECH$^+$92] has become quite important. This class contains many examples from geometry and topology, and the underlying concepts are very helpful practically for making certain types of computations. There is a quadratic-time algorithm for solving the word problem in automatic groups, for instance. Some of the ideas behind automatic groups apply perfectly well to graphs in general (and not just Cayley graphs), and they provide a way to make implicit descriptions of geometric structures without the presence of a group operation. We discuss some examples of this in Chapter 17, including graphs which approximate self-similar fractals.

In Chapter 19 we provide some general mechanisms by which to approximate the geometry of a given metric space by a graph. There are two basic choices, according to whether one wants to move around in the given space by paths or by scales. (For a human being this is analogous to the difference between moving around by car or by helicopter.) The two choices lead to very different ways in which an automatic structure might be used to describe a given geometry implicitly. For nilpotent groups both choices make sense, but moving around by scales seems to work better geometrically in some respects, and is more efficient in certain ways. We shall discuss this further in the main text, but for the moment let us mention that nilpotent groups cannot be automatic unless they admit an abelian subgroup of finite index [ECH+92].

Although examples of self-similar geometry are well known (see [Fal90]), it is not so clear in general how to decide what constitutes a "pattern" in geometry and what does not. In effect, one can think of notions of automatic structures as providing one way to approach this type of question, and a way that fits pretty well with a lot of basic examples. See [DS97] for a very different type of approach, closer to the traditions of geometric measure theory than algorithmic recursion. (Of course the problem of deciding what should be meant by "patterns" or "structure" occurs in many other contexts as well, even in this book.)

The notions of automatic structures mentioned above are based on finite-state automata and regular languages, and can be made more flexible by allowing more general types of operations than the ones employed by standard automata. We shall pursue these matters further in Chapter 18.

The last chapter, Chapter 20, provides an second look at the "Corona decomposition" from Chapter 8, with more emphasis on mappings and symmetry than before.

The second author was partially supported by the U.S. National Science Foundation. The authors would also like to thank the Institut des Hautes Études Scientifiques (Bures-sur-Yvette, France), where some of the work for this book was performed in 1996 and the summer of 1997, and the Courant Institute in New York University, at which the authors had the opportunity to visit and work in the spring of 1997. We would like to thank too G. David for assistance with fractal pictures, M. Gromov for numerous observations, references, and points about emphasis, L. Levin for pointing out a couple of oversights, A. Nabutovsky, P.S. Subramamian, and D. Welsh for some remarks and suggestions, and P. Schupp for some clarifications. A number of these came up during the Workshop on Geometry and Complexity at the Fields Institute (University of Toronto, May 1997), and the authors would like to thank the organizers of this workshop (A. Khovanskii and A. Nabutovsky) for the opportunity to attend. We are grateful to an anonymous reviewer for his or her suggestions as well.

Finally, we would like to mention the Paris complexity seminar of M. Gromov and A. Slissenko, which began in the Fall of 1994. The present endeavor has some of its roots in this seminar. A very positive feature of the seminar was its breadth and depth, and readiness for any starting point and any direction. This book might be considered as an attempted continuation of the spirit of the Paris seminar, or an outgrowth in printed form. Perhaps it will not be the only one!

Paris A.C.
Houston S.S.
March 2000

CONTENTS

1

INTRODUCTION

The topics of this book are largely motivated by broad themes of implicit description and internal symmetry, formal proofs and feasibility, combinatorial models, and computational complexity. In this chapter, we try to give an outline of the "big picture".

1.1 Examples of implicit descriptions

A fundamental method for making implicit descriptions is provided by the notion of an *algorithm* (typically formalized as a *Turing machine* [HU79, Pap94]). A given algorithm might describe a particular object, or instead a function when inputs are allowed. To obtain the result *explicitly* one must execute the algorithm, and in general one may not even know whether the execution of the algorithm will halt in a finite number of steps.

A more tangible example is provided by the possibility of defining a group through a *finite presentation.* One starts with a finite collection of generators $g_1, \ldots, g_n$ and a finite collection of *relations*, i.e., words which are postulated to reduce to the identity element of the group in question. One also declares as trivial any word which can be obtained from the relations (or the empty word) by taking conjugates, inverses, products, and by cancelling subwords of the form $g_i g_i^{-1}$ and $g_i^{-1} g_i$. The group is then defined by taking the set of all words and identifying any two which differ by a trivial word. (Note that "words" are allowed to include the inverses of the g_i's as well as the g_i's themselves here.)

This construction is elegant theoretically, but it can be quite frustrating in its implicitness. There are examples of such groups in which it is not possible to decide algorithmically whether a given word is trivial. (See [Man77], for instance.)

Let us consider now the concept of *regular languages*, in which one can specify a set of words through simple rules. In the previous situation we simply used all words, but the identifications that result from the relations create other difficulties (which can be quite complicated).

Fix a finite set Σ of symbols, and let Σ^* denote the set of all finite words made up of elements of Σ. One sometimes refers to Σ as an alphabet, and calls its elements letters. We include in Σ^* the *empty word*, which we denote by ϵ. A subset of Σ^* is called a *language* (over Σ).

Given a pair of languages L_1, L_2, we can define their *concatenation* $L_1 L_2$ to be the set of all words in Σ^* which can be realized as an element of L_1 followed immediately by an element of L_2. This defines a kind of "product" of languages, which permits us to define the positive powers L^i of a language L recursively through the formulae $L^i = LL^{i-1}$ when $i \geq 1$, $L^0 = \{\epsilon\}$. The *Kleene closure* of a language L is defined to be the language L^* given by

$$L^* = \bigcup_{i=0}^{\infty} L^i. \tag{1.1}$$

Regular languages are languages which can be constructed recursively in the following way. The empty language is considered to be regular, as is the language which consists only of the empty word, and the languages that contain only a single word given by one symbol in Σ (a word of length 1). Finite concatenations and unions of regular languages are considered regular, as are Kleene closures of regular languages. (The language consisting of only the empty word is thus also covered as the Kleene closure of the empty language.)

This defines completely the set of regular languages over the alphabet Σ. An individual regular language can be described succinctly through a *regular expression*, as follows. A regular expression is itself a word in which one can use letters from the alphabet Σ and some additional symbols. The symbol $\emptyset$ is permitted as a regular expression, and denotes the empty language. The symbol ϵ is a regular expression, and it corresponds to the language consisting of only the empty word in Σ^*. (Note that ϵ denotes the empty word in Σ^*, but is nonempty as a regular expression.) Each individual letter in Σ is a regular expression, which corresponds to the language consisting of only that letter. If r and s are regular expressions which denote the languages R and S, then $(r + s)$ is a regular expression which denotes the language $R \cup S$, (rs) is a regular expression which denotes the concatenation RS, and (r^*) is a regular expression which denotes the Kleene closure R^*. See [HU79] for more information about regular expressions. There are standard conventions for leaving out parentheses when this would not cause trouble. In this regard, one might accept certain laws for regular expressions, such as commutativity for sums, associativity for sums and products, and distributivity for expanding products of regular expressions involving sums. These laws are compatible with the languages being described, i.e., $(r + s)t$ and $rt + st$ represent the same languages, and so on.

Through the use of regular expressions one can specify *infinite* sets of words through *finite* expressions. These sets have very special internal structure. Every finite set of words is trivially regular, but even for finite sets the efficiency of such a representation can reflect interesting information about the set.

We can also use words and regular expressions to represent geometric objects. Let us illustrate this idea with an example. Given any equilateral triangle, we can split it into four smaller equilateral triangles as in Fig. 1.1. Each of the new triangles is one-half the size of the original one, in terms of sidelength. We label the new triangles with the letters U (upper), L (left), R (right), and M (middle), as shown in Fig. 1.1.

Let us fix now a particular triangle T_0 in which to work. We can interpret words over U, L, R, and M as subtriangles of T_0, by repeating the recipe above. The first letter specifies one of four subtriangles of T_0, the next letter specifies

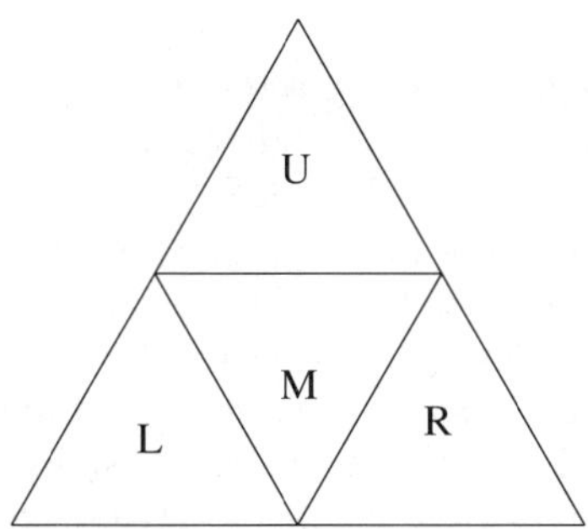

FIG. 1.1. Subdivision of triangles

FIG. 1.2. The Sierpinski gasket (through a number of levels)

a subtriangle of that one, and so forth. A word of length n represents a triangle in T_0 with sidelength equal to 2^{-n} times the sidelength of T_0.

A regular expression in U, L, R, and M represents a collection of words which we can interpret now as a collection of triangles, or more simply as a *subset* of T_0 by passing to the union. A basic example is provided by

$$(U + L + R)^k, \tag{1.2}$$

which corresponds to the kth level of a standard construction of the *Sierpinski gasket*, a well-known fractal set. (See Fig. 1.2.)

All words in U, L, and R of length k belong to this language, and each one of them represents a triangle of size 2^{-k} (times the size of the original triangle T_0) in Figure 1.2. For instance, the uppermost triangle of this size in the picture is represented by the word $UU \cdots U$, with k occurrences of the letter U.

For large values of k, there are really two types of symmetry involved here, and two levels of implicitness. To see this, consider regular expressions of the form

$$A_1 A_2 \cdots A_k, \tag{1.3}$$

where each A_i represents a sum of three of the four letters U, L, R, and M. Our example (1.2) is itself an implicit representation of a regular expression of this form. In general, one cannot compress a regular expression of the form (1.3) in this way. On the other hand, (1.3) is already rather special for regular expressions which represent sets of 3^k words of length k.

If E is a regular expression of the form (1.3), then the corresponding set S of words over U, L, R, and M enjoys a lot of special structure. If w is any word of length j, $j < k$, and $S_j(w)$ denotes the set of words v of length $k - j$ such that wv lies in S, then $S_j(w)$ is either empty, or looks the same independently of w (and is in fact represented by the regular expression $A_{j+1}A_{j+2}\cdots A_k$). For the example in (1.2), the different $S_j(w)$'s are closely related even for different values of j. For generic expressions of the form (1.3), one does not have this additional symmetry. The sets that they describe do have substantial symmetry, though, compared to general sets of 3^k words of length k.

Let Z denote the subset of T_0 which is associated to a regular expression E as above (of the form (1.3)). That is, E defines a set S of words over U, L, R, and M, and we take Z to be the union of the triangles determined by these words. The preceding discussion about the $S_j(\cdot)$'s corresponds to geometric statements about the structure of Z. In this regard, imagine fixing a word w of length j as before, and let $\tau(w)$ denote the corresponding subtriangle of T_0. This provides us with a "snapshot" of Z, namely the part of Z which lies in the interior of $\tau(w)$. This snapshot is determined by $S_j(w)$ in a simple way, and all of these snapshots will look the same when E is of the form (1.3), and the snapshots are nonempty. In the special case of (1.2), snapshots associated to different j's will also be a lot alike. This corresponds to the usual self-similarity property for the actual Sierpinski gasket, i.e., the set that arises from the analogous *infinite* construction.

These examples illustrate ways in which one mathematical object might represent another, and how the conciseness of such a representation may reflect internal symmetry of the (typically larger) object being represented.

1.2 Formal proofs and cut elimination

Let us look now at the idea of implicit description in a more general way, using the language of mathematical logic as a formalization of reasoning.

The power to make lemmas is a crucial feature of ordinary reasoning. Once we prove a lemma we can use it over and over again, and we can even make recursions in which the lemma is applied to data obtained from a previous application. (The latter possibility occurs through the use of quantifiers.)

The use of lemmas is closely related to the idea of symmetry. Short proofs of "hard" statements are typically obtained by finding subtle patterns or relationships and then coding them into a small number of lemmas which might be used many times, at least implicitly. Without the lemmas (or something analogous), one would verify pieces of information separately, bit by bit, and they could be more independent of each other.

The idea of "lemmas" can be formalized through the *cut rule* in the sequent calculus [Gir87c, Tak87, CS97]. The cut rule is an extension of the more famous rule of *Modus Ponens*, which says that if A is true, and A implies B, then B must also hold. Together with the *contraction* rule one can prove a single fact once and use it many times. A precise description of the cut and contraction rules in the sequent calculus is provided in Section A.1 in Appendix A, and we shall see some concrete examples in Chapters 3, 4, and 16.

By contrast, proofs without cuts are much more rigid and explicit. For instance, proofs without cuts satisfy the "subformula property", which means that every logical formula that occurs in the course of the proof also occurs within the final conclusion. (Imagine how ordinary mathematics would be with this restriction.)

It may be surprising, then, that there is a procedure for eliminating cuts from a given proof which always works in some general situations, such as ordinary propositional and predicate logic. This is Gentzen's famous "cut elimination" theorem [Gen34, Gen69, Gir87c, Tak87, CS97]. See also Theorem A.1 in Appendix A for a more precise statement, and Chapter 6 for a discussion of some of the main ingredients in the standard method.

The elimination of cuts does not come for free, however, in the sense that it may lead to large expansion in the size of the given proof. Indeed, there are numerous examples where the shortest proof with cuts is much shorter than any proof without cuts [Ore82, Ore93, Sta74, Sta78, Sta79, Tse68]. Typically, one might expect exponential expansion for propositional proofs and nonelementary expansion in predicate logic. (One can think in terms of having the possibility for a new exponential in expansion for each layer of nested quantifiers.) In the context of arithmetic, the elimination of cuts can lead to infinite proofs, and this comes from mathematical induction.

It is easy to see informally how the prohibition against lemmas can lead to much larger proofs in ordinary mathematical activity. The "John–Nirenberg theorem" from real analysis provides a nice illustration of this; see [CS97]. The examples in Chapters 3, 4, and 16 provide further illustrations of this phenomenon.

Still, the precise nature of cut elimination remains mysterious in general, and it provides one of the motivating themes behind this book.

1.3 Feasibility

In order to connect formal proofs to particular mathematical objects, one can use the idea of *feasibility*. In the original context of arithmetic, one introduces a unary predicate F for which the intended meaning of $F(x)$ is that "x is a feasible number". One typically assumes that 0 is feasible, and that the property of feasibility is preserved by sums, products, and the successor function (addition by 1). The idea is that a formal proof of the feasibility of some term should encode an actual construction of the term, at least implicitly; thus one does not allow induction to be applied to formulae which contain an occurrence of F, since that would lead to proofs of $\forall x F(x)$ which do not really contain information.

The concept of feasibility arose from philosophical questions about the concrete meaning of very large numbers, and the mathematical treatment comes from [Par71]. See also [Dra85].

For our purposes, a basic point is that a formal proof of the feasibility of some term t can be very small compared to the actual size of t. (Compare with [Car00].) We would like to say that for this to happen there must be a substantial amount of compensating "symmetry" or special structure in the term which is being represented. This principle is quite reasonable, but it is not clear how to articulate it precisely, let alone prove it rigorously. One of the aims of this book is to explore a variety of contexts in which similar matters arise.

The idea of feasibility can easily be applied to other mathematical structures, as discussed in [CS]. For instance, one can apply it to regular expressions, and show that there are short proofs of the feasibility of expressions of the form (1.2) for large values of j, using arguments like some of the ones found in [Car00]. This would not work for generic expressions of the form (1.3), but reflects the special structure in (1.2). One can also apply the idea of feasibility to finitely-generated groups, or to rational numbers, for instance. In the latter case, one can code the action of $SL(2, \mathbf{Z})$ on $\mathbf{Q}$ by projective transformations, to make short proofs of feasibility in an interesting dynamical way.

In each situation, a formal proof provides an implicit description of the object in question, with the level of implicitness related to the nature of the proofs and the logical formulae being used. Roughly speaking, one obtains "one level" of implicitness through cuts and contractions, even without quantifiers, while additional layers of quantification correspond to additional levels of implicitness.

In the context of arithmetic, for instance, there are examples of formal proofs of feasibility due to Solovay which allow for non-elementary growth in the terms. These proofs have many layers of nested quantifiers, and the underlying *logical flow graphs* (the definition of which is given in Appendix A) have a large number of oriented *cycles*, which are themselves nested in an interesting way. See [Car00, Cara] for more information, and also Chapter 16 below.

The minimal length of a formal proof of feasibility of a given object provides a measurement of information content in a similar spirit as algorithmic information theory [Kol68, Cha87, Cha92, LV90]. For formal proofs one can limit the level of implicitness by restricting the kinds of formulae or rules that are allowed. One can think in terms of different measurements of information content adapted to different forms of reasoning.

In this book we shall be typically concentrate on levels of implicitness which correspond to the use of cuts and contractions, but no quantifier rules. This is closer to the dichotomy between polynomial and exponential complexity (as opposed to superexponential growth). It also lends itself well to combinatorial and geometric situations. The next level of implicitness, corresponding to a single layer of quantification, is less flexible, but it also lends itself fairly well to geometric expression. We already saw a natural version of this in Section 1.1. With stronger forms of recursion come more refined patterns, which may be less accessible to the human eye.

See Chapter 16 for more discussion about different levels of implicitness, and their combinatorial and geometric interpretations.

1.4 Combinatorial models

One of our broad goals has been to find ways to model various combinatorial aspects of formal proofs, models which are more *geometric* than *syntactical.* One such model is presented in [Carc], in connection with *Craig interpolation*, in the context of proofs without cuts. In Chapter 7 we shall present a notion of *feasibility graphs*, which is loosely based on the idea of formal proofs of feasibility (without quantifier rules).

The idea of feasibility graphs is roughly the following. One starts with some set of objects and operations defined on them. One can then use oriented graphs to describe constructions over the set by permitting the edges and "focusing branch points" (where there is more than one incoming edge) to represent designated operations on the set. The "defocusing branch points" (where there is more than one outgoing edge) represent duplications, much like the ones in formal proofs which can occur through the use of contractions. More precisely, duplications which take place after other operations are performed correspond to the use of both contractions and cuts in formal proofs. The constructions described by feasibility graphs can be coded back into formal proofs of feasibility (without quantifier rules) in a simple manner. Only special proofs of feasibility are represented by such graphs, but some of the basic exponential effects are captured in this model.

Stronger forms of recursion can be obtained using feasibility graphs to describe the construction of other feasibility graphs. We shall discuss this in Chapter 16.

For a feasibility graph, the conversion from implicit description to explicit construction is achieved simply through the elimination of defocusing branch points. This is accomplished by the *visibility* of the underlying feasibility graph, which represents the way that a graph "looks" from the perspective of a given vertex. (See Chapter 4.) This leads to the problem of estimating the size of the visibility in terms of the geometry of the underlying graph. We shall address this problem in Chapters 4, 5, and 8.

Oriented graphs are quite interesting by themselves, without additional labellings, as for feasibility structures. One can also go in a different type of direction — starting from graphs — more along the lines of topology and spaces of higher dimension, like polyhedra or manifolds.

In topology, there are some basic constructions, like those of homotopy and homology groups, which are obtained from a given topological space. The spaces can be quite a bit more complicated than graphs, and there are also more complicated types of symmetries involved, e.g., involving homeomorphisms. A basic point is to look at structures or objects which are independent of specific combinatorial or other realizations of an underlying space, which amounts to respecting certain symmetries. Constructions like those of homotopy and homology groups do this, and this is a key reason for looking at them.

These two general directions, of richer topology or feasibility graphs, are somewhat different from each other, in terms of kinds of symmetry involved, and constraints, and for flexibility in making constructions, using various kinds of operations, etc. In this book, we often try to strike balances between considerations like symmetry and flexibility.

A basic theme will be to make comparisons between graphs through mappings from one to the other, and through induced mappings between associated visibility graphs. See Chapters 10 and 11. This is somewhat analogous to some activities in topology. In the context of feasibility graphs, such mappings can induce correspondences between the underlying computations, as we discuss in Section 11.5. We shall also consider expressions of internal symmetry in graphs, in the way that one graph can be "folded" into another. Again, this has some analogues in topology and geometry, and there are some differences too.

1.5 Formal proofs and algorithmic complexity

Given a propositional formula (in formal logic), can one determine whether or not it is a propositional tautology through a polynomial-time algorithm?

Here "propositional" means the version of logic without quantifiers. Thus $(x \wedge y) \vee \neg z$ is an example of a propositional formula, while $\forall x \exists y G(x, y)$ is not. A propositional formula is a *tautology* if it is "true" for all possible assignments of truth values to the variables. The formula $(x \wedge y) \vee \neg z$ is not a tautology, but $\neg(x \wedge \neg x)$ and $(p \wedge q \supset r) \supset (p \supset \neg q \vee r)$ are tautologies.

The question stated above turns out to be equivalent to the famous P = NP problem. Indeed, the set of non-tautologies is NP-complete, because of the celebrated Cook-Levin theorem on the NP-completeness of the "satisfiability" problem. (See [Pap94].) In the context of oriented graphs, this leads to the NP-completeness of the "visibility mapping problem" (Section 13.1).

Does every propositional tautology admit a proof of polynomial size?

This is not known, and the existence of a propositional proof system in which every tautology has a proof of polynomial size (compared to the size of the tautology) is equivalent to the NP = co-NP question in complexity theory [CR79]. (We shall return to this in Section 1.8.)

It is known that proofs of polynomial size do *not* always exist when one uses standard systems but without the cut rule (or a suitable variant thereof) [Tse68, Ore82, Sta78, Hak85]. People have looked at some natural examples of families of tautologies for which polynomial-sized proofs with cuts might not exist, and in some cases this has resulted in the discovery of polynomial-sized proofs which are far from obvious. Related references include [Ajt88, Ajt90, Bus87, Rii93, Rii97, BIK$^+$96, BIK$^+$92, Pud91].

Thus one is lead to broad questions about the lengths of proofs, and the concept of "difficulty" for the existence of short proofs. The idea of feasibility provides a way to incorporate basic mathematical objects (like numbers, words in a language, elements of a group, etc.) into a similar framework. One would like to say that the existence of short proofs connected to large or complicated objects is related to some kind of concrete symmetry in those objects.

In the setting of oriented graphs, it is natural to ask when a given tree admits a representation as the visibility of a much smaller graph. This is roughly similar to asking when a given (large) proof without cuts can be "folded" into a much smaller one with cuts. One can consider analogous questions for feasibility graphs. We shall return to themes like these later in the book, starting in Section 9.2.

1.6 The role of symmetry

It is perhaps natural to think that negative answers to questions in complexity (like the standard ones related to NP and co-NP) can be difficult to establish precisely because counterexamples are necessarily hard to describe. In other words, the families that one can write down should have to have special structure, precisely because they can be written down.

In dealing with large objects, or infinite families of objects, one has the unavoidable problem that human beings are limited to some kind of implicit description. One simply cannot start listing objects of size 10^{10} explicitly. One can have huge databases to investigate, but for human beings it can be necessary or strongly desirable to have some kind of intermediate processing.

Thus one constantly looks for new mechanisms by which to make complicated objects comprehensible. One might say that one looks for new and useful forms of symmetry and recursion.

At the same time, one might like to have some coherence to symmetries and patterns that one employs. For a human being it can be hard to keep track of a large number of disconnected forms of structure. There are inevitable trade-offs between coherence of patterns and flexibility or range of patterns. (One might think in terms of "patterns" in the class of patterns.)

As usual, we often try to aim here for intermediate situations in which one has the possibility to move in either direction. This is one reason for limiting the strength of the implicitness involved.

1.7 Partial symmetries

It is sometimes important to be flexible about the nature of symmetry. One can have situations in which nice fragments are glued together in an irregular manner, or in which overlapping structures are difficult to separate from each other.

Consider the case of acoustic signals, i.e., sounds that we hear in everyday life. Human beings learn to accomplish many feats which remain inaccessible computationally. One can use Fourier analysis to decompose a signal into pure waveforms, which can lead to simple "explanations" for apparently complicated signals, but this is not the whole story by any means. In real life one typically has sounds that start and stop, overlapping each other in tricky ways, or with complicated recurrences. Imagine a person speaking, and then imagine two people speaking at the same time, in an airplane.

To change a sound by affecting the speed in modest but non-constant ways can have little effect for a human being, and be much more subtle computationally.

Similar considerations apply to images. One can easily have objects of different texture overlapping each other. The local structure of a particular object can be very complicated in its own right, with subtle variations from place to place.

Pictures can look different at different scales. Imagine a chain-link fence for instance. To a fly it has enormous holes which one can go through, while to a person it is more like a wall. For a bird it might be more like a periodic lattice. Or imagine a bookshelf, in which individual books melt into blurry horizontal continua as one steps backward, away from the wall.

In reasoning we also have fragmentations and irregular borders. We have many basic patterns and tools which can be distinct, but interact in complicated ways.

There are powerful mathematical tools for analyzing certain types of patterns. The matter is much more complicated when the patterns are allowed modest distortions and irregularities. With algorithms and Turing machines we have enormous flexibility in the notion of "patterns", but it can be hard to know which ones to choose.

Some references related to signal processing are [Mar82, Bar96, Coi91, R$^+$92, Dau93, MS95].

1.8 Computational complexity

Let us take a moment to review some notions and facts pertaining to computational complexity. Basic references include [GJ79, HU79, vL90a, Pap94]. For simplicity, we shall restrict ourselves to "decision" problems, i.e., problems for which the answer should be "yes" or "no" (as opposed to "17", or "yellow").

Originally one began with the distinction between *algorithmic decidability* and *undecidability*. There was the additional nuance that a set could be *recursively enumerable* — generated by a computer program — but not *decidable*, because of the non-existence of a computer program to generate the complement of the set. Thus, for instance, the tautologies in classical predicate logic, or the trivial words in a finitely-presented group, are always recursively enumerable, but they need *not* be algorithmically decidable. These are basic results.

Algorithmic decidability is far from practical feasibility in the amount of space or time that might be required. Thus one looks at more restricted classes of algorithms, such as the class P of *polynomial-time algorithms*, which answer a given question in an amount of time that is bounded by a polynomial of the size of the input. This is a useful concept, and one which fits nicely with a lot of basic situations, but one should keep in mind that n^3 (or n^2, etc.) complexity can be problematic in numerical analysis, and that physically-meaningful quantities (like Planck's constant, or the number of particles in the solar system) have modest logarithms.

There are many natural problems which are not known to be in P but do lie in the class NP, which means that they can be solved in polynomial time by a *nondeterministic* Turing machine. Roughly speaking, this is equivalent to saying that for each occurrence with an answer of "yes", there is an "effective witness", i.e., a justification which can be given in polynomial size and verified in polynomial time. For this concept the phrase "succinct certificate" is also used. See p181f of [Pap94] for a more precise formulation of this.

As an example, let us consider the travelling salesman problem. For this one starts with a graph in which the edges are labelled by positive numbers to measure distance, and one specifies a particular size $k > 0$. The problem is to decide whether it is possible to visit all of the vertices in the graph exactly once through a path of total length at most k. When the answer is "yes", then there is a path, and that gives an effective witness. (There are some variants of this, e.g., with paths which go through the vertices at least once, rather than exactly once.)

To answer "yes" or "no" directly is apparently much more difficult. In particular, it is not known whether one can solve the problem in polynomial time. One can make naive searches of paths, but this involves exponential complexity.

It is also not known whether the NP condition should imply the existence of effective witnesses for answers of "no". In the context of the travelling salesman problem, for instance, it is not at all clear what such an effective witness should be. We shall say more about this in a moment.

Another example is the *satisfiability* problem, in which one is given a Boolean expression f in some number of variables, and asked whether f ever attains the value 1. An affirmative answer always has an effective witness, i.e., a choice of truth assignments for the Boolean variables for which the value of f is 1. The validity of such a witness can be verified easily, but it is apparently much more difficult to find such a satisfying truth assignment, or even to detect its existence.

The famous P=NP question asks if every NP problem has a polynomial-time solution, and this is not known. It is known that many particular examples of NP problems are actually *NP-complete*, which means that a polynomial-time algorithm for the given problem implies the existence of polynomial-time algorithms for all NP problems. The travelling-salesman and satisfiability problems mentioned above are both NP-complete. (See [GJ79, HU79, Pap94].)

The asymmetry between answers of "yes" and "no" in the definition of the NP class suggests that one also consider the class co-NP of problems whose "complement" (with the roles of "yes" and "no" reversed) lies in the NP class. It is not known whether co-NP = NP. This would follow if P = NP were known to be true, but a priori one could have co-NP = NP without P = NP. (This possibility does occur when one makes computations relative to certain *oracles*. See [Pap94], especially p351.)

The *validity* of Boolean expressions (without quantifiers) provides an important example of a co-NP problem. Recall that a Boolean expression is said to be *valid* if it takes only the value 1. If we write SAT^0 for the (essentially equivalent) version of the satisfiability problem in which the role of the truth value 1 is played by 0 instead, then the validity problem is the complement of SAT^0. Because SAT^0 is NP-complete, it follows that its complement, the validity problem, is *co-NP-complete.* See p220 of [Pap94], and also p142 there for some related remarks about complements of computational problems and languages.

One does not know whether the validity problem (or any other co-NP complete problem) lies in NP, but there is a natural way to try to make effective witnesses for positive answers to validity, through formal proofs. Every valid formula admits a (propositional) proof of at most *exponential* size, but it is not known whether valid formulae always admit proofs of *polynomial size.* As in Section 1.5, the NP = co-NP question is equivalent to the existence of *some* proof system in which valid formulae always admit proofs of polynomial size [CR79].

Another interesting class is given by the intersection of NP and co-NP. This is the class of problems for which there are effective witnesses for both "yes" and "no" answers. It is not known whether every such problem lies in the class P, however. A priori, one would still have to perform exponential searches through sets of possible effective witnesses. It is also possible that P is equal to the intersection of NP and co-NP, even though P is not equal to NP. (These possibilities do occur relative to certain *oracles*, as explained on p351 of [Pap94].)

It turns out that the "primality" problem lies in the intersection of NP and co-NP. In this problem, one tries to determine whether a given number is prime, with the caveats that the number should be given in binary (or some other base strictly larger than 1), and that measurements of complexity should be expressed as functions of the size of the binary (or other) representation. These are standard conditions for many problems. There is a simple effective witness for a number not to be prime, namely, a realization of it as a product of two smaller numbers. The effective witnesses for primality to hold are more subtle; see [Pra75], and p222 of [Pap94]. It is not known whether the primality problem lies in P, but it is known that there are polynomial-time *randomized* algorithms for deciding primality, and that primality lies in P if the Riemann hypothesis holds. (See 11.5.7 on p273 of [Pap94].)

So far we have concentrated on the amount of *time* needed to perform a computation. One can also look at the amount of *space* required. More precisely, one counts only the space needed for actual computational work, and not the read-only tape on which the input is provided, or the write-only tape on which the answer is given. (See p35 of [Pap94].)

Two classes of particular interest are L and NL, in which the amount of space used is bounded by a constant times the *logarithm* of the size of the input string. For L one must use a deterministic algorithm, while NL allows nondeterministic Turing machines. Thus L is contained in NL, but it is not known if the inclusion is proper. It is known that both L and NL are contained in P, but it is not known whether one has equality here either. (See p148-9 of [Pap94].)

A logarithmic bound on space is a strong condition. In practice, one needs a logarithmic amount of space merely to do very simple things, such as describe a particular object under consideration (like a vertex in a graph), or track a counter. Keep in mind that integers can be represented in logarithmic space, through binary representations.

As an example, consider the set of binary strings $\{0^i 1^i : i \in \mathbf{Z}_+\}$. The problem of deciding when a given binary string lies in this set is in the class L.

The "reachability" problem is an important example of an element of NL. Given an oriented graph G and a particular pair of vertices v, w in G, this problem asks whether there is a path in G which goes from v to w. This turns out to be NL-complete (Theorem 16.2 on p398 of [Pap94]), so that a deterministic solution of it in logarithmic space would imply a similar solution for any other problem in NL.

Let us mention that *regular languages* (as in Section 1.1) are precisely the ones that are recognized by Turing machines which never use more than a bounded amount of space. (See p55 of [Pap94].) Note that the language $\{0^i 1^i : i \in \mathbf{Z}_+\}$ is not regular.

It turns out that the class co-NL of problems whose complements lie in NL is the same as NL itself, i.e., NL = co-NL. There is a general result of this nature for complexity classes defined in terms of nondeterministic space, as a consequence of a theorem of Immerman and Szelepscényi. See p151ff of [Pap94].

The classes L and NL are also closely related to ones used to measure the complexity of *parallel* computation (Theorem 16.1 on p395 of [Pap94]). Namely, the parallel computation class NC_1 is contained in L, while NL is contained in NC_2. There is a hierarchy of NC classes related to parallel computation, all of which are contained in P, and for which strict inclusion in P remains unknown. The hierarchy would collapse if $\mathrm{P} = \mathrm{NC}_1$, and this is closely connected to the question of whether it is possible to represent Boolean *circuits* in terms of Boolean *expressions* with only polynomial expansion in size. The latter can be seen as a concrete problem about the dichotomy between implicit descriptions and explicit constructions, and we shall say more about it in Sections 4.12 and 7.11.

There are other natural complexity classes which are expected to be larger than P. One is PSPACE, the class of problems which can be solved by a Turing machine which uses only polynomial space. For this class it turns out that the deterministic and nondeterministic versions are equivalent, and indeed the passage from nondeterministic to deterministic Turing machines entails only a *quadratic* expansion in the amount of *space* required. This is a consequence of a well-known theorem of Savitch; see p150 of [Pap94]. There are numerous problems which are known to be PSPACE-complete (Chapter 19 of [Pap94]), including analogues of standard NP-complete problems in which the underlying objects are *infinite* but *periodic* (p483ff in [Pap94]).

Let us record the chain of inclusions

$$\mathrm{L} \subseteq \mathrm{NL} \subseteq \mathrm{P} \subseteq \mathrm{NP} \subseteq \mathrm{PSPACE}\ (= \mathrm{NPSPACE}) \tag{1.4}$$

from p148 of [Pap94]. Here NPSPACE denotes the nondeterministic version of PSPACE, which is also sometimes called NSPACE. Note that the inclusion L $\subseteq$ PSPACE is known to be strict, as mentioned on p149 of [Pap94]. This implies that at least one of the inclusions in the chain (1.4) is strict. No particular one of these inclusions is known to be strict, but they are all expected to be.

Another interesting class consists of the "succinct problems" discussed in [Pap94], beginning on p492. For this one tries to make computations concerning large objects which are defined only implicitly. For instance, one considers graphs of exponential size defined implicitly though much smaller Boolean circuits which determine the pairs of vertices that are connected by edges. One can then ask whether a given graph of this type admits a Hamiltonian cycle, for instance. (A *Hamiltonian* cycle is one which passes through every vertex in the graph exactly once.) This problem turns out to be complete for the nondeterministic complexity class NEXP, in which one allows running times which are exponentials of polynomials in the size of the input.

We shall encounter numerous computational questions in the course of this book. In many cases there will be a natural "effective witness" whose verification appears to be much simpler than the direct solution of the original problem. Sometimes the problems will fit naturally into the NP class, and sometimes not (i.e., they might be in a smaller class).

2

MORPHISMS IN LOGIC AND COMPLEXITY

The idea of "morphisms" is central to many areas of mathematics. There are homomorphisms between groups, algebras, semigroups, etc., linear mappings between vector spaces, continuous mappings between topological spaces, algebraic mappings between varieties, etc. In each case, one can use classes of morphisms to express comparisons and interactions between objects of interest.

One might like to make similar comparisons in contexts of formal proofs or complexity theory. We shall discuss some ways in which this can arise in this chapter.

2.1 Morphisms and formal proofs

No general notion of a "morphism" between formal proofs currently exists, but what might it entail? What are some examples of prospective morphisms?

Let us start with the special case of "embeddings", for which there are obvious examples coming from *subproofs* of a given proof. That is, one can take everything in a given proof up to a certain stage, and then stop, to get a smaller proof of an intermediate statement. (An example is shown on the left side of Fig. 2.1.)

There is a more general kind of "embedding" of one proof into another, which comes from the notion of an "inner proof" (introduced in [Car97]). For this one is not required to keep *everything* in the original proof up to a certain stage, but

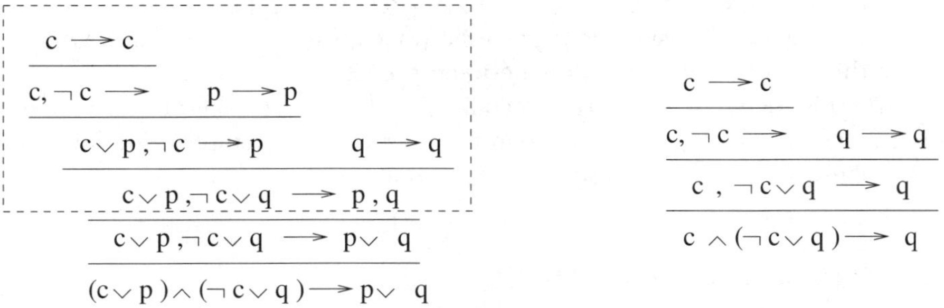

$$\begin{array}{c}
c \longrightarrow c \\ \hline
c, \neg c \longrightarrow \qquad p \longrightarrow p \\ \hline
c \vee p, \neg c \longrightarrow p \qquad q \longrightarrow q \\ \hline
c \vee p, \neg c \vee q \longrightarrow p, q \\ \hline
c \vee p, \neg c \vee q \longrightarrow p \vee q \\ \hline
(c \vee p) \wedge (\neg c \vee q) \longrightarrow p \vee q
\end{array}
\qquad\qquad
\begin{array}{c}
c \longrightarrow c \\ \hline
c, \neg c \longrightarrow \qquad q \longrightarrow q \\ \hline
c, \neg c \vee q \longrightarrow q \\ \hline
c \wedge (\neg c \vee q) \longrightarrow q
\end{array}$$

FIG. 2.1. A formal proof, a subproof, and an inner proof. The diagram on the left shows a formal proof, in "sequent calculus" (which is reviewed Section A.1 in Appendix A). The portion of the diagram surrounded by the square is a subproof of this proof. The diagram on the right gives an example of an inner proof. The inner proof is obtained from the original one by keeping everything that does not involve p.

only certain pieces of information which are chosen in accordance with suitable consistency conditions. See Fig. 2.1 for an example.

What about other kinds of morphisms between proofs? A basic situation that occurs in ordinary mathematics is the compression of a proof through more efficient use of "lemmas". That is, one can try to find a general lemma which incorporates several different parts of a given proof, in such a way that the original proof can be "folded" into a smaller one. The possibility for doing this is often facilitated by allowing for less precision in the theorem in question; for instance, instead of giving explicit solutions to some system of equations or inequalities, one might consider statements about their existence.

These phenomena have counterparts in the setting of formal logic. One sees this clearly in the subject of "cut elimination", in which proofs are "unfolded" by replacing *each application* of general lemmas with explicit computations. We shall return to this in Chapter 6, with particular emphasis on "duplication of subproofs" ((6.4) and (6.5)), which provides the main step in the unfolding of lemmas. In this book we shall also look at combinatorial models for this kind of duplication, especially in the context of graphs. There the idea of morphisms is easier to manage, through various classes of mappings between graphs.

Another basic question about formal proofs is the following: what does it mean for two proofs to be "identical", in essence? Is there a notion that one might define? In practice, it is easy to see sometimes that a pair of proofs are essentially the same, except for some kind of rearranging. With cut elimination, different choices in the process can lead to proofs which are different in some overall apparent shape, but which may be essentially the same nonetheless. These relations between proofs may not follow something as simple and exact as mappings or isomorphisms in other contexts.

Concerning foldings and unfoldings connected to proofs, a related matter is the use and role of notation, and names for objects and notions that one might employ in general. One might introduce a notion, use it repeatedly, and this can lead to various compressions and patterns (or capturing of patterns). One can look at this in terms of morphisms between proofs too.

With this, one can also have "recodings" which affect the appearance of a proof, without changing the main content. This can be a complication for trying to say when two proofs are essentially the same, as above.

2.2 Morphisms and monotonicity

Many standard problems in computational complexity theory (as in [GJ79, Pap94]) have natural "monotonicity" properties. Let us mention two examples.

The first problem is that of 3-coloring: given a graph G, when is it possible to assign to each of the vertices in G one of three "colors", so that no two adjacent vertices have the same color? This problem turns out to be NP-complete [Pap94]. It enjoys the monotonicity property that if a graph G admits a 3-coloring, then this remains true for any subgraph of G.

In the "clique" problem, one is given a graph G and a number k, and one is asked to decide whether G contains a k-*clique*, i.e., a set of k vertices so that any two of these vertices are connected by an edge in G. This problem is also NP-complete, and it has the monotonicity property that if a graph G contains a k-clique, then this is also true for any graph which contains G as a subgraph. A second monotonicity property is that the existence of a k-clique for a fixed value of k implies the existence of a k-clique for all smaller values of k.

In both cases, there is a monotonicity property with respect to embeddings of graphs, although the monotonicity properties go in opposite directions. Of course there are similar monotonicity properties for other types of computational problems.

Let us restrict ourselves for the moment to graphs which do not contain edges that are "loops", i.e., with both endpoints at a single vertex. It is easy to see that whether loops are allowed or not is not important for the clique or 3-coloring problems. (Loops do not affect the existence of cliques, and they automatically rule out colorings.)

Let G_1 and G_2 be a pair of such graphs, and suppose that there is a *mapping* between them. We shall discuss mappings between graphs more carefully in Chapter 10, but for the moment let us simply say that to have a mapping from G_1 into G_2 there should be a mapping from vertices in G_1 to vertices in G_2, and a mapping from edges in G_1 to edges in G_2, and that these two mappings should satisfy the usual compatibility conditions. (That is, if a vertex v in G_1 is an endpoint of an edge e, then the same should be true for the images of v and e in G_2.) If there is a mapping from G_1 into G_2, then the existence of a 3-coloring on G_2 implies the same for G_1, and the existence of a clique of size k in G_1 implies the same for G_2. In other words, a 3-coloring on G_2 can be "pulled back" to G_1, while k-cliques are pushed forward from G_1 to G_2. For both situations, we are using the assumption that there are no edges which are loops, to know that adjacent vertices are never mapped to the same vertex.

This type of monotonicity property is often described in the context of graphs in terms of collapsing vertices together. Indeed, every mapping between graphs can be realized (up to isomorphic equivalence) through the operations of collapsing vertices together, identifying pairs of edges which have the same endpoints, and adding edges and vertices. This is well known, and easy to verify. (Note that the possibility of multiple edges between a fixed pair of vertices does not play a role in the 3-coloring and clique problems.)

In Chapters 10, 11, and 13, we shall encounter a number of other features or questions about graphs, with various monotonicity and symmetry properties connected to mappings between them.

2.3 Combinatorial "proof systems"

Given a graph G, one can demonstrate the existence of 3-coloring in an effective way by producing the coloring and showing that adjacent vertices never have the same color. How might one "prove" that a graph G does *not* admit a 3-coloring? (Compare with Sections 1.5 and 1.8.)

A method for doing this has been given by Hajós [Haj61]. (See also [Ber91, Pud98].) For this discussion, let us again restrict ourselves to graphs which do not contain edges with both endpoints at the same vertex, and also require that our graphs do not have multiple edges between any fixed pair of vertices. In the "Hajós calculus", one begins with any graph which is a complete graph on four vertices, i.e., any graph which has exactly four vertices and one edge between any pair of distinct vertices. If a graph G has already been constructed, then one is permitted to add vertices and edges to it freely. One is also allowed to collapse any pair of non-adjacent vertices to a single point. In this case multiple edges could appear, and these should be reduced to single edges in order to maintain compatibility with the requirements mentioned above.

Finally, one is allowed to combine two graphs G_1, G_2 (with disjoint sets of edges and vertices) through the following operation of "joining". Suppose that e_1, e_2 are edges in G_1, G_2, respectively, and let a_i, b_i be the endpoints of e_i in G_i, $i = 1, 2$. A new graph is obtained by taking the union of G_1 and G_2, identifying the vertices a_1 and a_2 together, removing the edges e_1 and e_2, and adding a new edge between b_1 and b_2.

It is easy to see that the complete graph on four vertices does not admit a 3-coloring, and that these three rules for making new graphs from old ones preserve this property. Hajós proved that these rules are also complete, in the sense that every graph which does not admit a 3-coloring can be derived in this manner.

However, it is not clear that if G is a graph which is not 3-colorable, then there should exist a derivation in the Hajós calculus whose total size is bounded by a fixed polynomial in the size of G. If this were true, then the set of non-3-colorable graphs would satisfy the NP property, and one would be able to conclude that NP = co-NP, since the 3-coloring problem is NP-complete. Similar matters came up in Section 1.8, in connection with validity of propositional formulae, and formal proofs as a means to show that a formula is valid. (See also p219f of [Pap94].)

The example of the Hajós calculus illustrates how the concept of a "proof system" can be meaningful in purely combinatorial or even geometric terms, and not just in the sense of formal logic. Some other mathematical examples are discussed in [Pud98].

3

EXPONENTIAL PROCESSES AND FORMAL PROOFS

In this chapter, we look at some basic examples of formal proofs in propositional logic. In particular, we shall indicate some simple ways in which exponential complexity can arise.

3.1 Preliminaries

In this discussion of formal proofs, we shall use *sequent calculus*, as reviewed in Appendix A. Recall that a *sequent* is an expression of the form

$$A_1, A_2, \ldots, A_m \to B_1, B_2, \ldots, B_n, \tag{3.1}$$

where the A_i's and the B_j's are themselves *logical formulae*. Thus a sequent contains *collections* of formulae, or, more properly, *multisets* of formulae. This simply means that we count repeated occurrences of the same formula separately, but we do not care about the ordering of the formulae.

The sequent above is interpreted as meaning that if all of the A_i's are true, then at least one of the B_j's should be true as well. Let us emphasize, though, that a sequent is not a formula, nor is the sequent arrow $\to$ a logical connective.

Sequents of the form $A \to A$ are treated as axioms, as are the ones of the form

$$\Gamma, A \to A, \Delta, \tag{3.2}$$

where Γ and Δ are arbitrary multisets of formulae. A formal proof consists of a tree of sequents in which axioms are combined into more complicated sequents through certain rules of inference. See Appendix A for a more precise review (including a list of rules of inference).

Here are some basic examples. Let $p_1, p_2, \ldots, p_m$ be propositional variables. The sequent

$$\to p_1, p_2, \ldots, p_m, \neg p_1 \wedge \neg p_2 \wedge \cdots \wedge \neg p_m \tag{3.3}$$

says that at least one of the p_i's is true, or that they are all false. It is easy to give a formal proof of this in about m steps. One starts with the axioms

$$p_i \to p_i \tag{3.4}$$

for $i = 1, 2, \ldots, m$, and one combines them one-by-one using rules of inference concerning negations and conjunctions. Specifically, one employs the following general rules:

$$\neg : right \qquad \frac{A, \Gamma \to \Delta}{\Gamma \to \Delta, \neg A}$$

$$\wedge : right \qquad \frac{\Gamma_1 \to \Delta_1, A \quad \Gamma_2 \to \Delta_2, B}{\Gamma_{1,2} \to \Delta_{1,2}, A \wedge B}$$

(See Section A.1 for more information.)

We can get a slightly more interesting example as follows. Write S_j for the formula

$$\neg p_1 \wedge \cdots \wedge \neg p_{j-1} \wedge p_j. \tag{3.5}$$

We interpret S_1 to be p_1. Consider now the sequent

$$\to S_1, S_2, \ldots, S_n, \neg p_1 \wedge \neg p_2 \wedge \cdots \wedge \neg p_n. \tag{3.6}$$

We can prove this in $O(n^2)$ steps, by combining proofs of (3.3) for $m \leq n$. To see this, let us start with

$$\to p_1, p_2, \ldots, p_n, \neg p_1 \wedge \neg p_2 \wedge \cdots \wedge \neg p_n, \tag{3.7}$$

from (3.3), and then combine it with its version for $n - 1$,

$$\to p_1, p_2, \ldots, p_{n-1}, \neg p_1 \wedge \neg p_2 \wedge \cdots \wedge \neg p_{n-1}, \tag{3.8}$$

to get

$$\to p_1, p_2, \ldots, p_{n-1}, S_n, \neg p_1 \wedge \neg p_2 \wedge \cdots \wedge \neg p_n. \tag{3.9}$$

Specifically, we used the $\wedge$: *right* rule to combine the p_n from (3.7) with the formula $\neg p_1 \wedge \neg p_2 \wedge \cdots \wedge \neg p_{n-1}$ from (3.8) to get the copy of S_n in (3.9). After applying the $\wedge$: *right* rule we get two copies of p_i for $1 \leq i \leq n - 1$, but these can be reduced to one copy of each through the *contraction* rule, which is given as follows:

$$\frac{\Gamma \to \Delta, A, A}{\Gamma \to \Delta, A} \tag{3.10}$$

(There is a similar rule for contracting formulae on the *left* side of the sequent.) This gives (3.9), and we can use the same method to systematically convert all of the p_i's into S_i's, thereby obtaining a proof of (3.6) in the end. It is not hard to verify that the number of steps in this proof is $O(n^2)$.

This proof of (3.6) is rather different from one that a human being might typically make. A human being might proceed as follows: if p_1 is true, then we are finished, because S_1 is true; if not, p_2 might be true, in which case S_2 is true; if p_2 is also not true, but p_3 is, then S_3 is true, etc. Thus at least one of the S_j's is true, or $\neg p_1 \wedge \neg p_2 \wedge \cdots \wedge \neg p_n$ is, and this corresponds exactly to the standard interpretation of (3.6).

This type of informal argument corresponds roughly to the following formal proof. To handle the "transitions" from the the jth level to the $(j+1)$th, we would like to use the sequent

$$\neg p_1 \wedge \neg p_2 \wedge \cdots \wedge \neg p_j \to S_{j+1}, \neg p_1 \wedge \neg p_2 \wedge \cdots \wedge \neg p_{j+1}. \tag{3.11}$$

This can be proved by combining two copies of the axiom

$$\neg p_1 \wedge \neg p_2 \wedge \cdots \wedge \neg p_j \to \neg p_1 \wedge \neg p_2 \wedge \cdots \wedge \neg p_j \tag{3.12}$$

with

$$\to p_{j+1}, \neg p_{j+1} \tag{3.13}$$

(which itself comes from the axiom $p_{j+1} \to p_{j+1}$ using the $\neg$: *right* rule mentioned above). More precisely, we apply the $\wedge$: *right* rule twice to attach $\neg p_1 \wedge \neg p_2 \wedge \cdots \wedge \neg p_j$ to each of p_{j+1} and $\neg p_{j+1}$. This gives (3.11), but with two copies of $\neg p_1 \wedge \neg p_2 \wedge \cdots \wedge \neg p_j$ on the left side of the sequent arrow instead of just one. This duplication can be fixed using the contraction rule. Once we have (3.11) for each j, we can get (3.6) as follows. We start with

$$\to S_1, \neg p_1, \tag{3.14}$$

which comes from the axiom $p_1 \to p_1$ using $\neg$: *right*, since $S_1 = p_1$. We combine this and (3.11) with $j = 1$ using the *cut rule* (see Section A.1) to get

$$\to S_1, S_2, \neg p_1 \wedge \neg p_2. \tag{3.15}$$

In this application of the cut rule, the occurrences of $\neg p_1$ on the left side of (3.11) (when $j = 1$) and on the right side of (3.14) are removed, and the other formula occurrences in the two sequents are kept, and included into a single sequent. We then combine (3.15) and (3.11) with $j = 2$ using the cut rule again to get

$$\to S_1, S_2, S_3, \neg p_1 \wedge \neg p_2 \wedge \neg p_3. \tag{3.16}$$

Proceeding in this manner, we can get (3.6) for any value of n, using only a *linear* number of steps.

Note the difference between the first example (3.3) and the second example (3.6) in the role of *ordering.* For a human being the ordering of the p_i's is quite important in the second example, but not in the first. We are cheating here slightly, because we have not been careful about the definition of logical formulae and the role of parentheses (for which the ordering of the p_i's plays a role), but still there is a significant issue involved.

Next we look at examples in which there is a nontrivial amount of *branching.*

3.2 A process of branching

Now suppose that we have propositional variables a_i, b_i, $i \geq 1$. We shall use the symbol $\supset$ for the connective of implication. (Remember that $\to$ is being used for the sequent arrow.) Let Γ_n denote the set of $2n$ formulae given by

$$\{a_i \supset (a_{i+1} \vee b_{i+1}),\ b_i \supset (a_{i+1} \vee b_{i+1}) \ : \ 1 \leq i \leq n\}, \tag{3.17}$$

and consider the sequent

$$a_1 \vee b_1,\ \Gamma_n \to a_{n+1}, b_{n+1}. \tag{3.18}$$

It is easy to see that this sequent is valid, as follows. The hypotheses tell us that either a_1 or b_1 is true, and that in either case we may conclude that one of a_2 and b_2 is true. We can repeat the process until we reach the desired conclusion, that one of a_{n+1} and b_{n+1} is true. It is not hard to formalize this into a proof with $O(n+1)$ steps.

Let us be careful and do the exercise explicitly. We construct the proofs recursively, as follows. For the $n = 0$ case we want to prove that

$$a_1 \vee b_1 \to a_1, b_1. \tag{3.19}$$

This is easy to do, by combining the axioms $a_1 \to a_1$ and $b_1 \to b_1$ using the $\vee : left$ rule (analogous to the $\wedge : right$ rule before). Suppose now that we have a proof of (3.18) for some value of n, and let us transform it into a proof for $n+1$. As above, we can prove

$$a_{n+2} \vee b_{n+2} \to a_{n+2}, b_{n+2} \tag{3.20}$$

by combining the axioms $a_{n+2} \to a_{n+2}$ and $b_{n+2} \to b_{n+2}$ using the $\vee : left$ rule. We then combine this with a proof of (3.18) to obtain a proof for

$$a_1 \vee b_1,\ \Gamma_n,\ a_{n+1} \supset (a_{n+2} \vee b_{n+2}) \to b_{n+1},\ a_{n+2},\ b_{n+2} \tag{3.21}$$

using the $\supset$: $left$ rule. (This is analogous to the $\vee : left$ rule, except that the occurrence of a_{n+1} on the right side of (3.18) ends up on the left side of (3.21).) We can do the same thing to replace the b_{n+1} on the right side of (3.21) with

$$b_{n+1} \supset (a_{n+2} \vee b_{n+2}) \tag{3.22}$$

on the left, and adding new occurrences of a_{n+2}, b_{n+2} to the right side of (3.21) in the process. In the end we obtain

$$a_1 \vee b_1,\ \Gamma_{n+1} \to a_{n+2},\ b_{n+2},\ a_{n+2},\ b_{n+2}. \tag{3.23}$$

From here we can apply two contractions on the right to obtain

$$a_1 \vee b_1,\ \Gamma_{n+1} \to a_{n+2},\ b_{n+2}. \tag{3.24}$$

This is the same as (3.18), with n replaced by $n+1$.

In this recursive construction, we used the previous proof of (3.18) only once (as a subproof), and we added a constant number of new steps to it. This leads to a linear bound (in n) on the number of steps in the proof as a whole. For the size of the proof (total number of symbols), one gets a quadratic bound.

There is a kind of exponential process which underlies this proof, indeed, a process which is analogous to stochastic or Markov processes. This is easier to understand at the "human" level of reasoning. We begin with the knowledge that one of a_1 and b_1 is true, and at each step we learn that one of a_j, b_j is true, but we never know which. By the time that we arrive to the conclusion of a_{n+1} or b_{n+1}, we have treated an exponential number of possibilities, at least implicitly. Still, we managed to make a short proof (without *cuts*) by organizing it properly.

Let us formalize this idea as follows. Let $\mathcal{B}_{2n+2}$ denote the set of all Boolean sequences of length $2n + 2$, which represent the possible truth values of the variables a_i, b_i, $1 \leq i \leq n + 1$. Let $\mathcal{H}(n)$ be the subset of $\mathcal{B}_{2n+2}$ of sequences where $a_1 \vee b_1$ and the formulae in Γ_n are all true. How does this subset behave?

To measure complexity of subsets in $\mathcal{B}_{2n+2}$, let us use the notion of *cells*. A *cell* is a subset of $\mathcal{B}_{2n+2}$ which can be defined by assigning specific truth values to some of the variables and leaving the rest free. Thus a subset of $\mathcal{B}_{2n+2}$ with only one element is always a cell, and corresponds to a complete assignment of truth values to the variables. Every subset of $\mathcal{B}_{2n+2}$ is a finite union of cells, and we can measure the complexity of a given subset of $\mathcal{B}_{2n+2}$ in terms of the minimum number of cells needed to represent it as a union of cells. This is related to notions of *entropy* and *information*, as in [Ash65, LM95, Sin76, Sin94].

Lemma 3.1 *Notation and assumptions as above. If $\{C_\alpha\}$ is any collection of cells in $\mathcal{B}_{2n+2}$ whose union is equal to $\mathcal{H}(n)$, then there must be at least 2^{n+1} different cells among the C_α's.*

Proof Suppose that we have a collection of cells $\{C_\alpha\}$ whose union is $\mathcal{H}(n)$. Let K denote the subset of $\mathcal{B}_{2n+2}$ so that for each j we have that

$$\text{either} \quad a_j = 1 \text{ and } b_j = 0, \quad \text{or} \quad a_j = 0 \text{ and } b_j = 1. \tag{3.25}$$

Notice that $K \subseteq \mathcal{H}(n)$. We claim that no two distinct elements of K can lie in the same C_α.

Indeed, let C be any cell in $\mathcal{B}_{2n+2}$ which is contained in $\mathcal{H}(n)$. Thus C determines a truth assignment for some number of the variables, and the inclusion in $\mathcal{H}(n)$ ensures that C must *assign* the value 1 to at least one of a_j and b_j for every j. In particular, C cannot leave both a_j and b_j free for any choice of j. This implies that an element of K which lies in C must be completely determined by these assignments, and is therefore completely determined by C itself.

In other words, $C \cap K$ consists of at most a single element for each cell C which is contained in $\mathcal{H}(n)$. Any realization of $\mathcal{H}(n)$ as a union of cells must therefore involve at least 2^{n+1} different cells, since there are 2^{n+1} different elements of K. This proves the lemma. □

Thus the hypotheses of the sequent (3.18) actually contain an exponential amount of "information", in a certain sense. One can see a kind of branching process taking place inside of the proof of (3.18); this is closely related to the use of the *contraction* rule in the passage from (3.23) to (3.24). Although contractions were also used in both approaches to the second example (3.6) in Section 3.1, we did not have the same kind of branching or exponential effects there as we do here. For that matter, such effects do not occur in the first example in Section 3.1 either (for which the contraction rule was not used). (In the second example, there is a kind of *nilpotency*, which kept exponential branching from happening.)

3.3 A stronger process of branching

We shall now describe a more complicated example, in which the exponential effects of branching are connected to the use of the cut rule. This example is based on a result of Statman [Sta78, Bus88], who showed that the sequents in question have very simple proofs with cuts which are of polynomial size, but for which all cut-free proofs are of exponential size. Similar results were obtained by Orevkov [Ore82, Ore93], and earlier work in this direction was accomplished by Tseitin [Tse68].

Let c_i and d_i be propositional variables, $i \geq 1$, and define formulae A_i, B_i, and F_i in the following manner. We first set

$$F_k = \bigwedge_{j=1}^{k} (c_j \vee d_j) \tag{3.26}$$

for any $k \geq 1$, and then $A_1 = c_1$, $B_1 = d_1$ and

$$A_{i+1} = F_i \supset c_{i+1}, \quad B_{i+1} = F_i \supset d_{i+1} \tag{3.27}$$

when $i \geq 1$.

Consider the sequent

$$A_1 \vee B_1, A_2 \vee B_2, \ldots, A_n \vee B_n \rightarrow c_n, d_n. \tag{3.28}$$

This example is similar to the one in the previous section, in terms of human reasoning. In this regard, assume that the formulae on the left side of the sequent (3.28) are true, and imagine trying to prove that at least one of c_n and d_n is true. Under these "hypotheses", one knows that at least one of c_1 and d_1 is true, and one can argue that at least one of c_j and d_j is true for each j. For this one must carry along more information from the previous variables c_i, d_i than before, in order to make each new step, but the basic idea is much the same.

Although the pattern of reasoning is similar to the one before, the formalization of the argument is different. To handle the more complicated transitions which occur here, one uses the cut rule, in much the same manner as in the second method for proving (3.6) in Section 3.1.

Let us be more precise. The basic building block in the proof is given by the sequent

$$F_i,\, A_{i+1} \vee B_{i+1} \to F_{i+1}. \tag{3.29}$$

We shall first explain how to prove this in a bounded number of steps, and then we shall combine a series of these proofs using cuts to get (3.28).

It is not hard to derive

$$F_i,\, A_{i+1} \to c_{i+1} \tag{3.30}$$

from $F_i \to F_i$ and $c_{i+1} \to c_{i+1}$ using the $\supset$: *left* rule. Similarly, we have that

$$F_i,\, B_{i+1} \to d_{i+1}. \tag{3.31}$$

We can combine the two using $\vee$: *left* to get

$$F_i,\, A_{i+1} \vee B_{i+1} \to c_{i+1},\, d_{i+1}. \tag{3.32}$$

This employs also a contraction on the left side, to reduce two copies of F_i into a single one. We can further reduce this to

$$F_i,\, A_{i+1} \vee B_{i+1} \to c_{i+1} \vee d_{i+1}. \tag{3.33}$$

using the $\vee$: *right* rule.

We can combine (3.33) with $F_i \to F_i$ again using $\wedge$: *right* to obtain

$$F_i, F_i,\, A_{i+1} \vee B_{i+1} \to F_i \wedge (c_{i+1} \vee d_{i+1}). \tag{3.34}$$

This leads to (3.29), because we can contract the two copies of F_i on the left side of (3.34) into one copy, and because we can rewrite the right-hand side of (3.34) using the fact that

$$F_{i+1} = F_i \wedge (c_{i+1} \vee d_{i+1}). \tag{3.35}$$

(Strictly speaking, we should have been more precise about parentheses in the definition of F_j, but this does not cause trouble.)

Once one has (3.29) for each i, one can combine these sequents using the *cut rule* to get (3.28). To be more precise, one uses (3.32) instead of (3.29) at the last step, and one observes that F_1 is the same as $A_1 \vee B_1$ for the starting point.

Notice that the basic building blocks (3.29), (3.32) were each derived in a bounded number of steps. They were constructed directly from axioms, not through induction. If one insists that they be derived from axioms whose distinguished occurrences are atomic formulae (i.e., formulae without connectives), then one would use a linear number of steps to get $F_i \to F_i$. In the end, we get a proof of (3.28) for which the total number of steps is either linear or quadratic in n, depending on whether or not we insist on starting from axioms with atomic main formulae.

This proof uses the cut rule in a strong way. It can be shown that all proofs that do not use cuts are necessarily of *exponential* size [Sta78]. (See [Bus88] for a simplification of a proof of Takeuti of Statman's result.) The rough idea is that any cut-free proof must use the $\vee : left$ rule many times, and that it must "maintain" the branches in the proof-tree that arise from the $\vee : left$ rule.

Let us illustrate how exponential expansion arises naturally in this example by describing a simple recipe for making a proof that does not use cuts. For this it is a bit more convenient to work with the sequent

$$A_1 \vee B_1, A_2 \vee B_2, \ldots, A_n \vee B_n \to F_n \tag{3.36}$$

instead of (3.28). Let us first show how a cut-free proof of (3.36) for an arbitrary choice of n leads to a cut-free proof of (3.28) for $n+1$, and then explain how to make a cut-free proof of (3.36).

Given a proof of (3.36), we can combine it with the axiom $c_{n+1} \to c_{n+1}$ using the $\supset: left$ rule to obtain

$$A_1 \vee B_1, A_2 \vee B_2, \ldots, A_n \vee B_n, A_{n+1} \to c_{n+1}. \tag{3.37}$$

This follows from the definition (3.27) of A_{n+1}. Similarly we can derive

$$A_1 \vee B_1, A_2 \vee B_2, \ldots, A_n \vee B_n, B_{n+1} \to d_{n+1} \tag{3.38}$$

from (3.36) using the axiom $d_{n+1} \to d_{n+1}$. Applying the $\vee : left$ rule to (3.37) and (3.38) yields

$$A_1 \vee B_1, A_2 \vee B_2, \ldots, A_{n+1} \vee B_{n+1} \to c_{n+1}, d_{n+1}, \tag{3.39}$$

at least if we also use contractions on the left to get rid of the extra copies of $A_j \vee B_j$, $j = 1, 2, \ldots, n$. This sequent is the same as (3.28), but with n replaced by $n+1$. We conclude that a cut-free proof of (3.36) for a given choice of n leads to a cut-free proof of (3.28) for $n+1$. (Note that the $n = 1$ case of (3.28) can be obtained directly from the axioms $c_1 \to c_1$ and $d_1 \to d_1$ and the $\vee : left$ rule.)

Now let us explain how one can make a cut-free proof of (3.36) for each value of n, recursively. The $n = 1$ case is already an axiom, because $A_1 \vee B_1$ is the same as F_1. Thus we suppose that we have a cut-free proof of (3.36) for some choice of $n \geq 1$, and we try to use it to make a cut-free proof for $n+1$. To do this we use the proof that we already constructed for (3.39). More precisely, we first convert (3.39) into

$$A_1 \vee B_1, A_2 \vee B_2, \ldots, A_{n+1} \vee B_{n+1} \to c_{n+1} \vee d_{n+1} \tag{3.40}$$

using the $\vee : right$ rule. Then we combine (3.40) with (3.36) using the $\wedge : right$ to obtain a proof of

$$A_1 \vee B_1, A_2 \vee B_2, \ldots, A_{n+1} \vee B_{n+1} \to F_n \wedge (c_{n+1} \vee d_{n+1}). \tag{3.41}$$

For this step we also employ contractions on the left-hand side to get rid of the duplicate copies of $A_j \vee B_j$, $1 \leq j \leq n$. This yields (3.36) for $n+1$, because (3.41) is the same as (3.36) for $n+1$, by (3.35).

These constructions provide cut-free proofs of (3.36) and (3.28) for all $n \geq 1$. Unlike the earlier proofs with cuts, the number of steps in these proofs grows exponentially with n, because we needed *three* copies of (3.36) at level n to make our proof of (3.36) at level $n + 1$. (Note that this did not happen in the proof described in Section 3.2.)

A version of this duplication also occurs implicitly in the earlier proof with cuts. This can be traced back to the use of contractions in the derivations of (3.32) and (3.29). Let us think about what would happen if we did not use contractions in either of those derivations. In that case we would get a proof

$$F_i,\, F_i,\, F_i,\, A_{i+1} \vee B_{i+1} \rightarrow F_{i+1} \tag{3.42}$$

for any choice of i, and this proof would use neither cuts nor contractions. In the previous situation we had only one F_i on the left-hand side, and we were able to exchange it for a single copy of F_{i-1} using a cut. If we start now from (3.42), then we need to make three cuts to exchange the three copies of F_i for copies of F_{i-1}, and we would end up with 9 copies of F_{i-1} on the left side. We would then have to make 9 cuts in order to replace these 9 copies of F_{i-1} with 27 copies of F_{i-2}, and so forth. In the end we would get a sequent which has 3^i copies of F_1, and we would stop there, since F_1 is the same as $A_1 \vee B_1$.

To eliminate contractions completely from the proof, one should replace (3.32) with

$$F_i,\, F_i,\, A_{i+1} \vee B_{i+1} \rightarrow c_{i+1},\, d_{i+1}. \tag{3.43}$$

In the end, one would obtain a proof without contractions of a sequent which is the same as (3.28) except for many extra copies of the $A_j \vee B_j$'s, which are produced from the multiple cuts mentioned above. To get rid of these redundant formulae, one would use contractions. This would not be so bad, because the contractions would all occur *below* the cuts.

The construction just described provides a concrete illustration of the "duplication of subproofs" which is used in the standard method of cut elimination to simplify cuts over contractions. We shall describe the general method in some detail in Chapter 6.

Note that this same method could be applied to the second proof of (3.6) in Section 3.1, to simplify the cuts over the contractions there. Again this would lead to exponential expansion in the proof, even though there is a reasonably short proof without cuts in this situation, namely the first proof of (3.6) described in Section 3.1. If in this first proof we insisted that contractions not be applied until after all of the applications of the logical rules (namely, $\wedge : right$ and $\neg : right$), then we would again be lead to the same kind of duplications and exponential expansion as before.

Instead of *eliminating* cuts from a proof, one can look as well for constructions of an intermediate nature, with some simplification of the structure of a proof, and less cost in expansion. Some results along these lines are presented in [Carb, Car99]. In particular, this concerns the possibility of oriented *cycles* in the logical flow graph of a proof, and transformations of proofs which can take such cycles apart. The results in [Carb, Car99] apply to both propositional and predicate logic, with corresponding levels of complexity, compared to the complexity of cut elimination (which can be much larger, in both cases).

3.4 Comparisons

The examples described in Sections 3.2 and 3.3 are very similar in the kind of exponential activity through branching which occurs. We shall see another example of this in Section 4.8, in the context of feasible numbers.

There are some important differences in the way that this exponential activity is represented in these three examples. For the first example we had short proofs without cuts, but not for the second. In the examples related to feasible numbers, one can see the exponential activity of the proof clearly in the underlying "logical flow graph" (whose definition is recalled in Section A.3 in Appendix A), while this does not work for the proofs from Section 3.3. We shall discuss this further in various parts of Chapter 6 (including Section 6.15).

Although these examples differ in their *representation* of exponential activity, the activity itself is very similar in all three cases. Let us look again at the proof in Section 3.2 and the one with cuts in Section 3.3, for instance. The proof in Section 3.2 did not use cuts, but the binary rule $\supset$: $left$ played a somewhat similar role as the cuts did in Section 3.3. Indeed, if we go back and look at the proof in Section 3.2, we see that the $\supset$: $left$ rule and the contraction rule interacted with each other in a very similar manner as the cut and contraction rules did in Section 3.3. One can "disentangle" the $\supset$: $left$ rule from contractions in the proof in Section 3.2, so that the contractions do not occur until *after* all the applications of the $\supset$: $left$ rule, in the same manner as described near the end of Section 3.3. This leads to the same kind of duplication of subproofs and exponential expansion as in Section 3.3.

This transformation of the proof in Section 3.2 has the effect of making it more explicit, in much the same way as for the proof in Section 3.3. Let us look at the proof in Section 3.2 in terms of "paths", as in a Markov chain. Imagine that the propositional variables a_i, b_i represent different locations in a map, or different states in some kind of physical system or computing machine. The "hypotheses" of the sequent (3.18) can be interpreted as saying that we can start at at least one of a_1 or b_1, and that when we reach any a_i or b_i, we can proceed next to at least one of a_{i+1} or b_{i+1}, without telling us which one. We can interpret the conclusion of (3.18) as saying that we can always reach at least one of a_{n+1} or b_{n+1} in this manner. The proof given in Section 3.2 verifies this in a kind of implicit way, while a "transformed" proof — with the contractions being applied *after* the $\supset$: $left$ rules — actually checks each individual trajectory starting from a_1 or b_1, to see that it eventually arrives at one of a_{n+1} or b_{n+1}.

The sequent (3.28) in Section 3.3 can be interpreted analogously. In this case the propositional variables are called c_i and d_i, but again one can think of paths which start at either c_1 or d_1 and proceed through the c_i's and d_i's to eventually reach c_n or d_n. As before, every sequence $\alpha_1, \alpha_2, \ldots, \alpha_n$ with $\alpha_i \in \{c_i, d_i\}$ for all i is accepted, but now the validation of the transition from α_j to α_{j+1} involves the α_i's with $i < j$ too, which was not the case before. In terms of a formal proof, this more extended validation is accommodated through the use of the cut rule.

The phenomena of branching and duplication indicated above are quite basic in formal proofs, and they provide one of the main motivations behind the analysis of graphs and their visibilities given below, beginning in Chapter 4. We shall pursue this analysis both at a purely geometric level, and with additional combinatorial structure, as in the notion of *feasibility graphs* (Chapter 7).

3.5 The pigeon-hole principle

Let us mention one more family of explicit propositional tautologies, based on the well-known *pigeon-hole principle.*

Let $p_{i,j}$ be propositional variables, $i, j \geq 1$, and consider the sequent

$$\Gamma_n \to \Delta_n, \tag{3.44}$$

where

$$\Gamma_n = \Big\{ \bigvee_{j=1}^{n} p_{i,j} : i = 1, 2, \ldots, n+1 \Big\} \tag{3.45}$$

and

$$\Delta_n = \{p_{l,k} \wedge p_{m,k} : 1 \leq l < m \leq n+1, 1 \leq k \leq n\}. \tag{3.46}$$

This is a valid sequent, and in fact it is a coding of the pigeon-hole principle. To see this, think of $p_{i,j}$ as representing the statement that the ith pigeon is contained in the jth box. Then the hypotheses of (3.44) become the assertion that each of $n+1$ pigeons lies within at least one of n boxes, while the conclusion says that at least two of these pigeons lie in the same box.

It is easy to write down a proof of the pigeon-hole principle in ordinary mathematics, using induction on n. Proofs of modest size of (3.44) in propositional logic are much more subtle, but propositional proofs of polynomial size have been found by Buss [Bus87]. Proofs without cuts require exponential size [Hak85].

This example is quite different from the previous ones, in the nature of its underlying *symmetry.* For (3.44), any particular propositional variable $p_{i,j}$ has essentially the same role as any other one, and indeed one could apply arbitrary permutations to the i's in $\{1, 2, \ldots n+1\}$ or to the j's in $\{1, 2, \ldots n\}$ without really changing the conceptual content of (3.44) as a whole. (As usual, there are technical points about parentheses here, but we shall ignore this for the present.) This was not the case for the examples in Sections 3.2 and 3.3, in which there was an important *ordering* of the underlying variables. The ordering of propositional variables also played an important role in the second example (3.6) in Section 3.1, but not in the first one (3.3).

3.6 Proofs, sets, and cells

Let $p_1, \ldots, p_n$ be a collection of propositional variables, and let $\mathcal{B}_n$ denote the set of all possible truth assignments for these variables. We shall think of each element of $\mathcal{B}_n$ as simply being a binary string of length n.

Every logical formula in the variables $p_1, \ldots, p_n$ defines a subset of $\mathcal{B}_n$, namely the set of truth assignments for which the formula takes the value 1. We shall sometimes use the same letter to denote the formula and the underlying set.

Testing the validity of a sequent (in propositional logic) is essentially the same as testing whether the intersection of a given collection of subsets of $\mathcal{B}_n$ is empty or not. To be precise, suppose that we are given a sequent

$$D_1, D_2, \ldots, D_k \to E_1, E_2, \ldots, E_r \tag{3.47}$$

in which the formulae D_i, E_j are made up out of the propositional variables $p_1, \ldots, p_n$ only. Then the validity of this sequent is equivalent to

$$\Big(\bigcap_{i=1}^{k} D_i\Big) \cap \Big(\bigcap_{j=1}^{r} E_j^c\Big) = \emptyset, \tag{3.48}$$

where E_j^c denotes the complement of the subset of $\mathcal{B}_n$ determined by E_j. This follows from the usual soundness and completeness theorems for propositional logic.

The "soundness of propositional logic" simply means that (3.48) holds for every provable sequent. This is not hard to show, and indeed each rule of inference has a natural interpretation at the level of sets, and one can check that they all preserve (3.48). For axioms one has (3.48) automatically, because one of the D_i's is necessarily the same as one of the E_j's.

Let us forget about formal proofs for the moment, and try to look at complexity issues related to (3.48) directly. For this purpose it will be convenient to use some concepts from Section 3.2.

Recall that a subset of $\mathcal{B}_n$ is called a *cell* if it is defined by specifying the truth values of some of the propositional variables while leaving the others free. This is the same as saying that the set corresponds to a logical formula which consists only of a conjunction of propositional variables and their negations. Every subset of $\mathcal{B}_n$ is a finite union of cells, because every singleton is a cell, but the minimal number of cells needed to realize a given set can be exponentially large, as in Lemma 3.1.

Definition 3.2 *If A is a subset of $\mathcal{B}_n$, then the* complexity *of A will be used to mean the smallest number m such that A is the union of m cells. The empty set is interpreted as having complexity equal to 0. The complexity of A will be denoted as* $\mathrm{com}(A)$.

Lemma 3.3 *If A and B are subsets of $\mathcal{B}_n$, then*

$$\mathrm{com}(A \cup B) \leq \mathrm{com}(A) + \mathrm{com}(B) \tag{3.49}$$
$$\mathrm{com}(A \cap B) \leq \mathrm{com}(A) \cdot \mathrm{com}(B). \tag{3.50}$$

Proof This is an easy consequence of the definitions, using DeMorgan's laws for the intersection of a union for the second inequality. □

The multiplicative bound for intersections leads to the possibility of exponential growth in the complexity of intersections of sets which have small complexity. We saw this before, in Lemma 3.1, and one can make other examples as well.

The realization of sets as unions of cells gives a way to deal with questions of nonemptiness, i.e., for intersections of sets which are unions of cells. In other words, one can reduce to cases of intersections of cells, for which the problem is much simpler. One then has the issue of the number of these cases, and this is bounded as in Lemma 3.3. However, it is easy to have exponential complexity, as with the examples mentioned above. (See also the remarks at the end of the section.)

Let us look at two more families of examples, coming from the propositional versions of the pigeon-hole principle (Section 3.6). For this we use propositional variables $p_{i,j}$, with $1 \leq i \leq n+1$ and $1 \leq j \leq n$, and we denote by $\mathcal{B}$ the set of all truth assignments for these variables. Although these truth assignments are no longer arranged as single sequences, this is not a problem for the notion of cells, or the complexity of subsets of $\mathcal{B}$. Consider the formulae

$$A_i = \bigvee_{j=1}^{n} p_{i,j}, \qquad 1 \leq i \leq n+1. \tag{3.51}$$

These are the formulae which appeared on the left side of the sequent (3.44). Writing A_i also for the corresponding subset of $\mathcal{B}$, we have that $\mathrm{com}(A_i) \leq n$ for all i, by the definitions.

Lemma 3.4 *The complexity of the set*

$$\bigcap_{i=1}^{n+1} A_i \tag{3.52}$$

is equal to n^{n+1}.

Proof Let A denote the set given in (3.52). The fact that the complexity of A is $\leq n^{n+1}$ follows easily from Lemma 3.3. Thus we only need to show that $\mathrm{com}(A) \geq n^{n+1}$.

To do this, we argue in the same way as for Lemma 3.1. Let K be the set of truth assignments in $\mathcal{B}$ which assign the value 1 to $p_{i,j}$ for exactly *one* choice of j for each integer i with $1 \leq i \leq n+1$. This choice of j is allowed to depend on i. It is easy to check that K is contained in A and has n^{n+1} elements, since j runs through the range $1 \leq j \leq n$.

Let C be any cell in $\mathcal{B}$ which is contained in A, and suppose that τ is an element of K which lies in C. We claim that τ is uniquely determined by C. To show this, we begin by fixing an arbitrary choice of i_0 with $1 \leq i_0 \leq n+1$, and we let j_0 be the unique integer such that $1 \leq j_0 \leq n$ and $p_{i_0,j_0} = 1$ for the truth assignment τ.

Since C is a cell, it is defined by a partial assignment of truth values to the $p_{i,j}$'s. For each $j \neq j_0$, this truth assignment must leave $p_{i_0,j}$ free or set it equal to 0, since τ lies in C. It must also set p_{i_0,j_0} to be 1 or leave it free, for exactly the same reason. In fact, C cannot leave p_{i_0,j_0} free, because it has to assign the value 1 to $p_{i_0,j}$ for at least one value of j (since $C \subseteq A$), and we have already eliminated all of the other choices of j.

Thus we conclude that there is exactly one choice of j such that C assigns the value 1 to $p_{i_0,j}$, namely j_0. This implies that τ is determined uniquely by C, since we can do this for each i_0, $1 \leq i_0 \leq n+1$.

We can rephrase this by saying that if C is any cell contained in A, then C cannot contain more than one element of K. This means that A cannot be expressed as the union of fewer than m cells, where m is the number of elements in K, which is n^{n+1}. Thus the complexity of A is at least n^{n+1}, and the lemma follows. □

This lemma says that the subset of $\mathcal{B}$ which corresponds to the left-hand side of the sequent (3.44) has large complexity in n. Let us consider now the right-hand side. Set

$$B_{l,m,k} = (\neg p_{l,k}) \vee (\neg p_{m,k}), \tag{3.53}$$

where $1 \leq l < m \leq n+1$ and $1 \leq k \leq n$. These formulae correspond to the *negations* of the ones on the right side of (3.44), and the validity of (3.44) is equivalent to the statement that

$$\Big(\bigcap_{i=1}^{n+1} A_i\Big) \cap \Big(\bigcap_{k=1}^{n} \bigcap_{m=1}^{n+1} \bigcap_{l=1}^{m-1} B_{l,m,k}\Big) = \emptyset, \tag{3.54}$$

as in (3.48).

Lemma 3.5 *The complexity of the set*

$$\bigcap_{k=1}^{n} \bigcap_{m=1}^{n+1} \bigcap_{l=1}^{m-1} B_{l,m,k} \tag{3.55}$$

is at least $(n+1)^n$.

This case is a bit different from the previous one, because a single $p_{i,j}$ occurs in more than one $B_{l,m,k}$.

Proof This can be proved in almost the same manner as before. In this case, we take H be the set of truth assignments τ in $\mathcal{B}$ such that for each j there is an $i = i(j)$ such that τ assigns the value 1 to $p_{i(j),j}$ and the value 0 to $p_{i,j}$ when $i \neq i(j)$. It is easy to see that H is contained in the intersection (3.55), and that H has $(n+1)^n$ elements. (As usual, the j's are chosen from the range $1 \leq j \leq n$, while the i's run through the range $1 \leq i \leq n+1$.)

Let C be any cell which is contained in the intersection (3.55) and contains some element τ of H, and let us show that C determines τ uniquely. Let $i(j)$ be as above, so that τ assigns the value 1 to $p_{i(j),j}$ and the value 0 to $p_{i,j}$ when $i \neq i(j)$. Fix j for the moment, $1 \leq j \leq n$.

Because τ is contained in C, we have that C must either assign the value 1 to $p_{i(j),j}$ or leave it free. Let us check that this implies that C must assign the value 0 to $p_{i,j}$ whenever $i \neq i(j)$. Assume first that $i < i(j)$. Since C is contained in (3.55), it is contained in $B_{i,i(j),j}$ in particular, and so either $p_{i,j}$ or $p_{i(j),j}$ must take the value 0 for any given truth assignment in C. This implies that C must assign the value 0 to $p_{i,j}$, since we know that it either assigns the value 1 to $p_{i(j),j}$ or leaves it free. The same argument works when $i > i(j)$; the only difference is that we write $B_{i(j),i,j}$ in that case, instead of $B_{i,i(j),j}$.

This shows that C determines $i(j)$ uniquely, and this works for each choice of j. We conclude that τ itself is determined by C, so that C cannot contain more than one element of H. From this it follows that (3.55) cannot be expressed as the union of fewer than $(n+1)^n$ cells, since H has $(n+1)^n$ elements (and is contained in (3.55)). This implies that the complexity of (3.55) is at least $(n+1)^n$, as desired. □

In view of Lemmas 3.4 and 3.5, it is perhaps not too surprising that the propositional version of the pigeon-hole principle (3.44) does not admit cut-free proofs of polynomial size in n. After all, the validity of the sequent is equivalent to the emptiness of the intersection of the sets (3.52) and (3.55), and we know that each of these sets must have large complexity. It is natural to think that a cut-free proof should have to check the emptiness of the intersection in some direct way.

Unfortunately, this idea does not seem to work in such a simple manner. Let us describe now a kind of "counterexample" based on the example in Section 3.2. For this it will be more convenient to work with propositional variables a_i, b_i, c_i, d_i, with $1 \leq i \leq n+1$, say. The idea is that c_i and d_i should be the same as the negations of a_i and b_i, respectively, but this will not be imposed directly.

Let Θ_n and Δ_n be the collections of formulae given as follows:

$$\Theta_n = \{c_i \vee a_{i+1} \vee b_{i+1},\ d_i \vee a_{i+1} \vee b_{i+1} : 1 \leq i \leq n\}, \tag{3.56}$$

$$\Delta_n = \{a_i \wedge c_i,\ b_i \wedge d_i : 1 \leq i \leq n\} \tag{3.57}$$

Consider the sequent

$$a_1 \vee b_1,\ \Theta_n \to a_{n+1},\ b_{n+1},\ \Delta_n. \tag{3.58}$$

If we think of c_i and d_i as being the negations of a_i and b_i, then this sequent is essentially the same as the one in (3.18). In any case, one can construct a proof of modest size of this sequent, without cuts, in practically the same manner as before. Let us sketch the argument. We start with the $n = 0$ case, in which Θ_n and Δ_n are interpreted as being empty, and (3.58) reduces to

$$a_1 \vee b_1 \to a_1, b_1. \tag{3.59}$$

This can be proved by combining the axioms $a_1 \to a_1$ and $b_1 \to b_1$ using the $\vee : left$ rule.

Now suppose that we have constructed a proof of (3.58) for some choice of n, and we want to obtain a proof for $n+1$. The first step is to convert (3.58) into

$$a_1 \vee b_1, \Theta_n, c_{n+1}, d_{n+1} \to \Delta_{n+1}. \tag{3.60}$$

This is easy to accomplish, by combining (3.58) with the axioms $c_{n+1} \to c_{n+1}$ and $d_{n+1} \to d_{n+1}$ using the $\wedge : right$ rule. More precisely, we combine c_{n+1} with a_{n+1} on the right to get $a_{n+1} \wedge c_{n+1}$, and similarly we combine d_{n+1} with b_{n+1} to get $b_{n+1} \wedge d_{n+1}$ on the right side of the sequent. These formulae are then combined with Δ_n to get Δ_{n+1} on the right side of the sequent. This leaves us with occurrences of c_{n+1} and d_{n+1} on the left side of the sequent, as in (3.60).

Just as in Section 3.2, we can pass from (3.60) to

$$a_1 \vee b_1, \Theta_{n+1} \to a_{n+2}, b_{n+2}, a_{n+2}, b_{n+2}, \Delta_{n+1}. \tag{3.61}$$

To do this, we first make a proof of

$$a_{n+2} \vee b_{n+2} \to a_{n+2}, b_{n+2}, \tag{3.62}$$

by combining the axioms $a_{n+2} \to a_{n+2}$ and $b_{n+2} \to b_{n+2}$ using the $\vee : left$ rule. We then combine (3.60) with (3.62) twice, using the $\vee : left$ rule twice, to convert the occurrences of c_{n+1} and d_{n+1} on the left side of (3.60) into $c_{n+1} \vee a_{n+2} \vee b_{n+2}$ and $d_{n+1} \vee a_{n+2} \vee b_{n+2}$ (still on the left side). We absorb these new formulae into Θ_n to get Θ_{n+1} on the left side, and in the end we obtain (3.61), with the new occurrences of a_{n+2} and b_{n+2} on the right side of the sequent (compared to (3.60)). Note that we only use (3.60) and its supporting proof once here, while we use two copies of (3.62) and its supporting proof.

Once we have (3.61), we can easily obtain the analogue of (3.58) with n replaced by $n+1$, by contracting the duplicate occurrences of a_{n+2} and b_{n+2} on the right side of the sequent. Thus we can always pass from a proof of (3.58) to a proof of its analogue for $n+1$, so that in the end we get a proof for each n. These proofs are cut-free and require only a linear number of steps as a function of n, as one can easily verify. (This is because there are only a bounded number of steps needed in going from n to $n+1$ for any n, or in the $n = 0$ case.)

On the other hand, the sets involved in this sequent have complexity of exponential size. To be more precise, let X_n denote the *intersection* of the sets of truth assignments which correspond to the formulae $a_1 \vee b_1$ and the elements of Θ_n, and let Y_n denote the *union* of the sets of truth assignments corresponding to the formulae a_{n+1}, b_{n+1}, and the elements of Δ_n. The validity of (3.58) is equivalent to

$$X_n \backslash Y_n^c = \emptyset, \tag{3.63}$$

just as in (3.48). One can show that the complexity of each of X_n and Y_n^c is of exponential size as a function of n, using arguments very similar to the ones for Lemmas 3.1, 3.4, and 3.5.

Thus the situation for (3.58) is seemingly somewhat similar to that of the pigeon-hole principle. We have a family of propositional tautologies, in which the formulae on the left sides of the sequents are all disjunctions of propositional variables, and in which the formulae on the right side are all conjunctions of propositional variables (or individual propositional variables). The set represented by the left side of the sequent has complexity of exponential size, as before, as does the complement of the set represented by the right side of the sequent. The lower bounds on complexity are somewhat stronger in the present case than for the pigeon-hole principle, because we had $n(n+1)$ propositional variables in the latter case, as compared to $4n+2$ variables as in (3.58).

Still, we have short cut-free proofs in this case, and not for the pigeon-hole principle. This shows that we cannot get large lower bounds for the sizes of cut-free proofs just using the notion of complexity of sets that we have been considering, even under very strong restrictions on the formulae which appear in the sequent.

It might be reasonable to hope for results of this nature if one also assumes a lot of "symmetry" in the given sequent. Our short cut-free proofs for (3.58) relied heavily on the fact that there was a clear "ordering" of the propositional variables, and this kind of structure in the endsequent will be disrupted by sufficiently strong forms of symmetry, as we have for the propositional version of the pigeon-hole principle. In that case, every $p_{i,j}$ has essentially the same role as any other $p_{l,k}$, and in fact one can permute the first and second indices freely in the problem without really changing the information in the hypotheses or conclusion. In particular, there is no preferred place to *start*, as there was for (3.58). (Similar points were mentioned before, near the end of Section 3.5.)

It is not clear how to make a proof system which adapts itself automatically to the symmetries of a given problem in a good way.

Notice, incidentally, that the first example (3.3) given in Section 3.1 also had a lot of symmetry, but that the relevant sets of truth assignments in that case were of low complexity.

One can also think about proofs like these in geometric terms. The validity of (3.58) is equivalent to (3.63), and one can think of "proving" the emptiness of the set on the left side of (3.63) by systematically cutting up the space $\mathcal{B}$ of truth assignments, and showing that the set is restricted to smaller and smaller portions of it, until there is no place left in $\mathcal{B}$ for the set to be. The proof of (3.58) described above could easily be given a geometric formulation of this type.

Let us end this section with the following remark. Let $\mathcal{B}_n$ be the set of binary sequences of length n, and suppose that we have some collection $\{A_1, A_2, \ldots, A_n\}$ of subsets of $\mathcal{B}_n$. As above, one can ask whether the intersection of all of the A_i's is empty or not. This makes sense as a computational problem, at least if we specify how the A_i's are represented. Let us suppose that the A_i's are given as finite unions of cells. Then this problem is just a reformulation of the complement of the "satisfiability" problem for certain types of Boolean expressions. (Compare with Section 1.8.) In fact, one has co-NP completeness in the special case where each A_i is a union of 3 cells, and each cell is defined by only one assignment of a truth value (in the binary sequences). This follows from the corresponding result for satisfiability ("3SAT"), as on p183 of [Pap94].

It is rather nice to think in the geometric manner of sets and cells. In particular, this fits well with symmetry in geometric terms.

4

GRAPHS AND THEIR VISIBILITIES

For the record, a *graph* is a (nonempty) set of *vertices* together with a collection of *edges*. Each edge has two endpoints, which are vertices in the graph. We allow edges whose endpoints are the same vertex, and we also allow multiple edges with the same endpoints.

We shall typically restrict ourselves to *oriented* graphs, in which the edges flow *from* one endpoint *to* the other.

Let us adopt the convention that the word "graph" automatically means "finite graph" in this book unless the contrary is explicitly stated. This will not be the case for the terms "visibility graph" or "tree", which are permitted to be infinite. We shall normally ask our trees to be locally finite, however, which means that they have only finitely many edges attached to any give vertex. These conventions are generally not too significant, but they can be convenient for avoiding technicalities, and they accommodate the main situations of concern.

4.1 Optical graphs

For the beginning of this book we shall restrict ourselves to *optical graphs*, for convenience of exposition. By an optical graph we simply mean an oriented graph with the property that each vertex u has at most *three* edges attached to it, for which there are never more than *two* edges oriented *away* from u or more than two oriented *towards* u.

This assumption is not at all crucial, but it helps to simplify some of the writing, and basic phenomena in which we shall be interested occur already in this case. We shall dispense with this assumption officially in Section 8.6, where we explain how the assertions made up to that point can be extended to the general situation.

The word "optical" refers to the idea of "looking" through the graph, following rays of light. We shall pursue this with the notion of the "visibility graph" in Section 4.2.

A vertex u in an optical graph G is called a *branch point* if it has two edges attached to it which are both pointing towards u or both oriented away from u. In the first case we say that u is a *focusing* branch point, and in the second case we call u a *defocusing* branch point. We shall be interested in the exponential divergences which may result from the branching within an optical graph.

Optical graphs often come with additional combinatorial data that reflect some other construction or process. For instance, formal proofs have underlying *logical flow graphs* which trace the flow of occurrences of formulae in the proof. This notion was introduced by Buss [Bus91], and a related graph associated to proofs was introduced earlier by Girard [Gir87a]. For our purposes, it is better to use a variant of the notion of Buss, in which we restrict ourselves to *atomic* occurrences of formulae, as in [Car97]. Logical flows graphs also carry natural orientations, as in [Car97]. It is easy to check that logical flow graphs are optical graphs, with branch points coming from the contraction rule for formal proofs. See Section A.3 for definitions and additional information. One can also use optical graphs to encode computations in a given mathematical context more directly, as in the notion of *feasibility graphs* discussed in Chapter 7. In these settings, the idea of "rays of light" in an oriented graph corresponds roughly to fixing some piece of information and asking where it came from, or how it will be used later. The answers to these questions are affected strongly by the arrangement of the branch points, and we shall look at this in some detail.

4.2 The definition of the "visibility"

Let G be an optical graph, and let v be a vertex in G. We want to define the *visibility* of G from v. This will be a rooted tree which represents the way that G looks from the perspective of v. For simplicity of exposition, we shall confine ourselves to what one sees from v in the directions of *positive* orientation, but one could just as well look at the negative orientations, or both simultaneously.

Let us set some terminology. By a *path* in G we mean any (finite) ordered succession of adjacent vertices together with a succession of edges connecting them. In some cases, there may be more than one edge connecting a given pair of vertices, so that the choices of the edges would matter. A path may traverse one or more edges whose endpoints are a single vertex, in which case the vertex would be repeated (and is considered to be adjacent to itself). *Degenerate paths* are included as paths, with only one vertex and no edges. An *oriented* path is one in which the succession of vertices and edges respects the orientation, i.e., the edge from the jth vertex to the $(j+1)$th vertex should be oriented from the jth to the $(j+1)$th vertex. We might also use the phrase "positively-oriented path", with "negatively-oriented paths" defined analogously (going in the direction opposite to that of the orientations).

By the *length* of a path we mean the number of edges that it traverses, counted with multiplicities. That is, we count an edge each time that it is traversed, irrespective of whether it is traversed more than once.

The *visibility* $\mathcal{V}_+(v, G)$ of G from v is a graph whose vertices are themselves the oriented paths in G which start at v (and may end anywhere in G). We include the degenerate path that consists of v alone, without any edges or other vertices. This degenerate path represents a vertex in $\mathcal{V}_+(v, G)$ which we call the *basepoint* of $\mathcal{V}_+(v, G)$. Two vertices p_1, p_2 in $\mathcal{V}_+(v, G)$ are connected by an edge $\mathcal{V}_+(v, G)$ oriented from p_1 to p_2 exactly when the corresponding paths in G have the property that the path associated to p_2 is obtained from the path associated to p_1 by adding an edge in G at the end of it. We attach one such edge in $\mathcal{V}_+(v, G)$ from p_1 to p_2, and these are the only edges that we attach. (See Fig. 4.1 for some examples.)

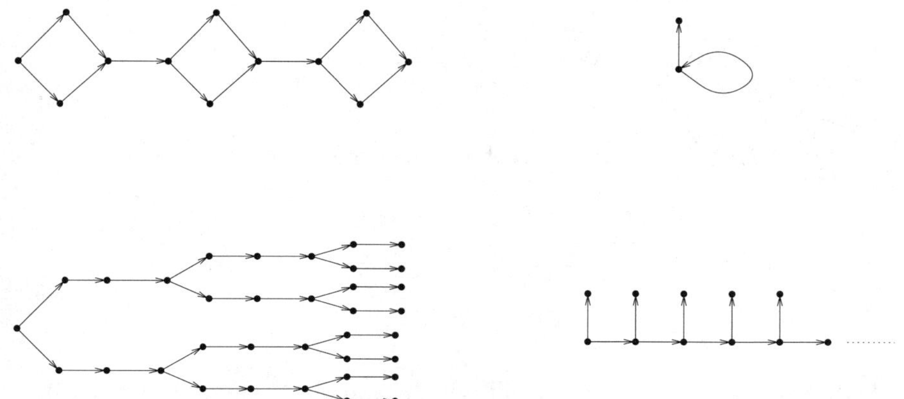

FIG. 4.1. In this picture, there are two examples of optical graphs (above), and associated visibility graphs (below). In the example on the left, the vertex v at which the visibility is taken is the one on the far left, and in the example on the right, it is the lower vertex. In the example on the left, one sees repeated branching in the visibility graph, which leads to exponential expansion in general. The visibility graph is finite in this case. In the example on the right, the visibility graph is infinite, and grows at a linear rate.

This defines $\mathcal{V}_+(v, G)$ as an oriented graph, which may be infinite. Note that $\mathcal{V}_+(v, G)$ is always *locally finite*, however; the number of edges going into any vertex is never more than 1, and the number of edges coming out of any vertex is never more than the maximum of the number of edges coming out of any vertex in G. This is easy to see from the definitions. We shall discuss some basic properties of $\mathcal{V}_+(v, G)$ in the next sections, after looking at some more examples.

One can also define a version of the visibility graph based on negatively-oriented paths, rather than positively-oriented paths. This version is more convenient in some situations. The two cases of positively and negatively-oriented paths are easily seen to be equivalent to each other, in the sense that one can reverse the orientation on G to pass from one to the other. This visibility graph based on negatively-oriented paths will sometimes be called the *negative visibility* or *negatively-oriented visibility*, and denoted by $\mathcal{V}_-(v, G)$.

4.3 Some examples

Consider a graph G like the one pictured in Fig. 4.2, and let v be the vertex on the far left side of the picture. The passage from G to the visibility $\mathcal{V}_+(v, G)$ represents a kind of branching process, in which the duplications that are implicit in G are made explicit. One can show that the visibility $\mathcal{V}_+(v, G)$ is of finite but exponential size compared to the size of G, and we shall discuss this kind of phenomenon in more detail later in the chapter. In the special case of three "steps" for the graph, the visibility graph is shown in Fig. 4.1.

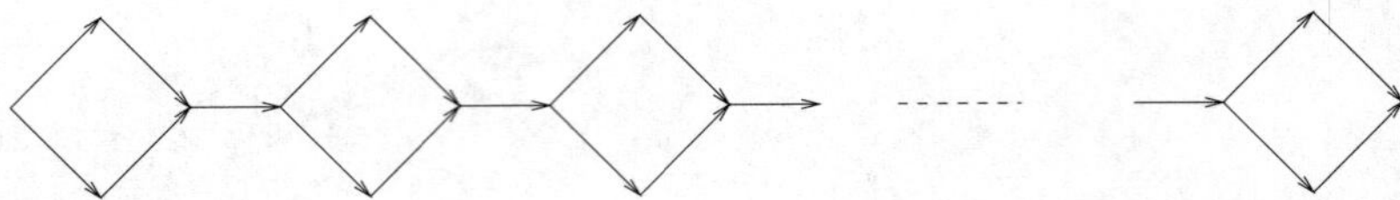

FIG. 4.2. A graph whose visibility is of finite but exponential size

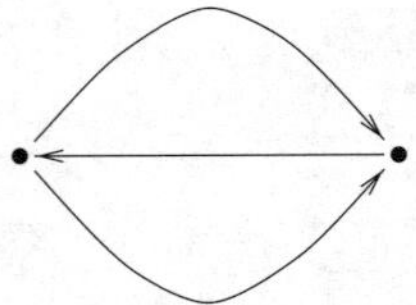

FIG. 4.3. A graph whose visibility is the standard (infinite) binary tree

For the graph shown in Fig. 4.3, the visibility is *infinite*. This is due to the presence of *oriented cycles*, around which oriented paths can go as many times as they like. In this case, the visibility graph is the standard binary tree, and in particular it grows at an *exponential* rate. If there were only one loop, then the visibility would be an infinite ray, with only *linear* growth. Instead we could choose our graph to have several loops "in sequence", and then the visibility would have polynomial growth of higher degree determined by the number of loops. We shall discuss these possibilities in more detail in the next chapter.

There is a certain sense in which the two examples in Fig. 4.2 and Fig. 4.3 are based on the same pattern. In the first example, this pattern is repeated a finite number of times, while in the second it occurs only once but feeds back into itself, which leads implicitly to an infinite process.

Next we consider the graph H given in Fig. 4.4. All of the branch points on the left side of Fig. 4.4 are *defocusing*, while on the right side they are all *focusing*. For simplicity we assume that the transition between the two occurs in a uniform way, after some number k of steps away from p. The same number of steps from the interface to the vertex q is then also given by k.

The total number N of vertices in this graph H is roughly proportional to 2^k. (That is, N is bounded from above and below by constant multiples of 2^k). This is easy to check, by summing a geometric series. The number of oriented paths from p to q is also roughly proportional to 2^k. Indeed, every oriented path from p to the interface between the defocusing and focusing branch points can be continued to q in a unique way, and there are about 2^k different ways to cross the interface. Every oriented path from p to q has to cross the interface somewhere, and thus the total number of such paths is about 2^k.

The number of oriented paths from p to any vertex z in H is never more than about 2^k. Indeed, for z in the right side of the picture the paths to z are again determined by the way that they cross the interface, and there are never more than about 2^k ways of doing this. The number of ways of crossing the interface is much smaller when z is closer to the interface than to q, and for z in the left

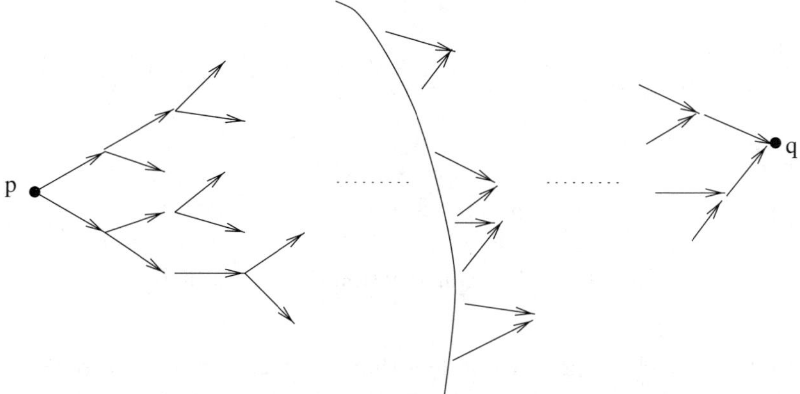

FIG. 4.4. The graph H

side of the picture there is only one oriented path from p to z.

These considerations imply that the number of vertices in the visibility graph $\mathcal{V}_+(p, H)$ is $O(N^2)$. A more careful accounting shows that $\mathcal{V}_+(p, H)$ has about $N \log N$ vertices. However, one can make graphs in which the visibility is of quadratic size (compared to the underlying graph) by adding a "tail" to the end of H.

More precisely, imagine that we enlarge H by adding a string of vertices $r_1, \ldots, r_L$ and oriented edges $e_1, \ldots, e_L$, where e_1 goes from q to r_1 and e_i goes from r_{i-1} to r_i when $i > 1$. Let H^* denote the graph that results, with no other additional edges or vertices. Then the number of distinct oriented paths in H^* that go from p to r_i is exactly the same as the number of oriented paths in H which go from p to q. If z is a vertex which lies already in H, then the number of oriented paths in H^* which go from p to z is the same as in H.

For simplicity, let us suppose that L is at least as large as the number N of vertices in H. Then the total number of vertices in H^* is $N + L$, and hence lies between L and $2L$, while the number of vertices in the visibility $\mathcal{V}_+(p, H^*)$ is roughly proportional to $N \cdot L$. This is easy to show, using the observations above (concerning the number of oriented paths in H^* which begin at p and end at a given vertex). If we take L to be equal to N, then the number of vertices in the visibility $\mathcal{V}_+(p, H^*)$ is roughly proportional to the square of the number of vertices in H^* itself. If we take L to be approximately N^α, $\alpha \geq 1$, then the total number of vertices in $\mathcal{V}_+(p, H^*)$ is roughly proportional to $N^{\alpha+1}$, which is approximately the same as the number of vertices in H^* raised to the power $(\alpha + 1)/\alpha$.

Every real number s in the interval $(1, 2]$ can be realized as $(\alpha + 1)/\alpha$ for some $\alpha \geq 1$. Thus the preceding construction shows that for any such s we can find families of graphs for which the size of the visibility is roughly proportional to the size of the graph raised to the sth power. One could get more complicated functions (than powers) through suitable choices of L, and one could obtain

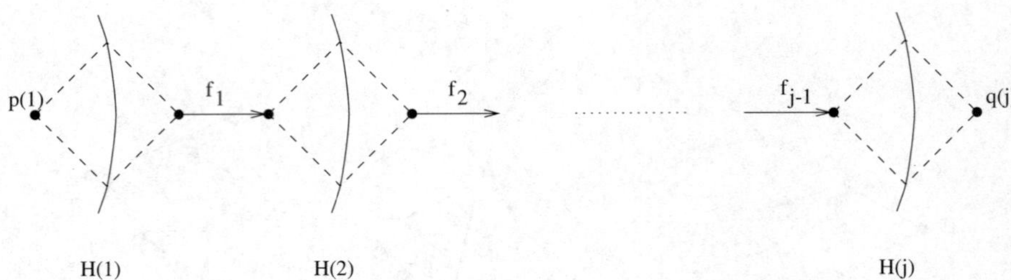

FIG. 4.5. The construction of the graph H_j

similar effects by choosing L to be less than N (e.g., a fractional power of N). However, in this type of construction the size of the visibility is never more than *quadratic* in the size of the underlying graph.

To go beyond quadratic growth one can proceed as follows. Let $j \geq 2$ be given, and let us define a new optical graph H_j as follows. We begin by taking j identical but disjoint copies of H, which we denote by $H(i)$, $1 \leq i \leq j$. Let $p(i)$ and $q(i)$ denote the vertices in $H(i)$ which are analogous to the vertices p and q in H. We define H_j to be the graph obtained by taking the union of the $H(i)$'s, $1 \leq i \leq j$, together with oriented edges f_i that go from $q(i)$ to $p(i+1)$ for $i = 1, 2, \ldots, j-1$. (See Fig. 4.5.) For convenience we can define H_j when $j = 1$ by simply taking H itself.

The number of vertices in H_j is equal to $j \cdot N$, but the number of oriented paths in H_j which go from $p(1)$ to $q(j)$ is roughly proportional to N^j, by the same considerations as above. (The constants which are implicit in this estimate are allowed to depend on j, but not on N.) One can also add a string of L vertices to H_j beginning at $q(j)$ to obtain a graph H_j^* which is analogous to H^* above. This gives one greater flexibility in making graphs for which the size of the visibility is approximately prescribed as a function of the size of the underlying graph. For instance, if we take L to be N^α, $\alpha \geq 1$, as before, then the total number of vertices in the visibility of H_j^* (starting at $p(1)$) will be roughly comparable to $N^{j+\alpha}$, while the size of H_j^* will be roughly comparable to N^α. (Again we are not being precise about the dependence on j here.)

We can rephrase this by saying that the number of vertices in the visibility of H_j^* is roughly comparable to the number of vertices in H_j^* raised to power s, where $s = (j + \alpha)/\alpha$. We can get any $s \geq 1$ that we want here, by choosing $j \in \mathbf{Z}_+$ and $\alpha \geq 1$ appropriately. We could also obtain functions which are not simply powers, by choosing L in other ways.

For fixed j and any choice of L, the size of the visibility of H_j^* is never be more than a constant multiple of the size of H_j^* raised to the $(j+1)$th power. This is because the number of vertices in H_j^* is $jN + L$, while the number of vertices in the visibility $\mathcal{V}_+(p(1), H_j^*)$ is $O(N^j(N + L))$, as one can check.

To obtain rates of growth which are greater than polynomial, one should allow j to vary as well. For instance, one can take the graph G discussed at the beginning of this section (Fig. 4.2) and attach a chain of L vertices to the end of it, in the same manner as for H^* and H_j^*. In fact, one can view this construction as a special case of the story of H_j^*, in which the parameter k controlling the depth of H is chosen to be 1. By choosing L appropriately, one can obtain almost any rate of growth in the size of the visibility compared to the size of the underlying graph, so long as this rate of growth is less than the one for the graph G itself. For these choices of G, the visibility is of exponential size compared to the size of the underlying graph. In general, the size of the visibility can never be more than exponential in the size of the underlying graph (at least when the visibility is finite), as we shall discuss in Section 4.7.

This second type of construction — taking the graph G from the beginning of the section and adding a tail to it — can also be used to give examples where the size of the visibility is roughly a (specified) power of the size of the graph. However, the H_j^*'s do this more efficiently, in terms of the number of *alternations* between defocusing and focusing branch points in the underlying graph. For instance, one only needs a single transition from defocusing to focusing vertices in H^* to get quadratic growth in the visibility, while in the second type of construction the number of such alternations would have to grow logarithmically in the size of the graph.

In Chapter 8 we shall give more precise bounds for the visibility when it is finite, bounds which take this type of alternation into account. These bounds will fit very well with the kind of polynomial growth that we have seen here for the H_j^*'s.

4.4 Visibility and depth

To clarify the relationship between the size of an optical graph K and its visibility, it is helpful to consider also the notion of "depth", as in the following definition.

Definition 4.1 *Let K be an oriented graph, and let v be a vertex in K. We define the* depth *of K beginning at v to be the length of the longest oriented path in K which starts at v. (This may be infinite when there are cycles present, or when K is allowed to be infinite itself.)*

It is not hard to show that the size of the visibility of an optical graph enjoys an exponential upper bound in the depth of the given graph. (See Lemma 4.9 in Section 4.7 below.) This kind of exponential expansion can easily occur, as one can see from the examples in the previous section. We can make a more "direct" example as follows. Let Λ be the optical graph that consists of only the left side of the graph H in Section 4.3 (pictured in Fig. 4.4). That is, we keep all of the defocusing branch points but none of the focusing ones, to get a binary tree of depth k starting from the root p, where p is the same vertex as in H. The total number of vertices in Λ is given by

$$\sum_{l=0}^{k} 2^l = 2^{k+1} - 1. \tag{4.1}$$

The visibility $\mathcal{V}_+(p, \Lambda)$ has exactly the same number of vertices as Λ does, because there is exactly one oriented path in Λ which begins at p and ends at any prescribed vertex z in Λ. (In fact Λ and its visibility are isomorphic as graphs, as one can easily check, using the "canonical projection" described in Section 4.5.)

Notice that the graph Λ and its visibility $\mathcal{V}_+(p, \Lambda)$ are as large as they can be, given that Λ is an optical graph with depth k starting at p. This is not hard to see.

The graphs G and H_j from Section 4.3 are similar to Λ in that they also have an exponential gap between the depth of the graph and the size of the visibility, but in these examples we see different phenomena in terms of the sizes of the graphs themselves. For the graph G (pictured in Fig. 4.2), for instance, the size of the graph is approximately the same as its depth, and it is exponentially smaller than the size of the visibility, as mentioned in Section 4.3. For the H_j's we have an exponential gap between the size of the graph and the depth, and then only a polynomial difference between the size of the graph and the size of the visibility.

Let us look now at the possible behavior of these gaps in a more general way, starting with the following observation.

Lemma 4.2 *If K is an optical graph and v is a vertex in K, then the depth of K starting from v is the same as the depth of the visibility $\mathcal{V}_+(v, K)$ starting at the basepoint.*

Proof This is not hard to check. There are two main points. The first is that any oriented path in the visibility $\mathcal{V}_+(v, K)$ of length ℓ which begins at the basepoint can be "projected down" to an oriented path in K of length ℓ which begins at v. This can be derived from the definitions, but it is a little easier to understand in terms of the "canonical projection" from $\mathcal{V}_+(v, K)$ to K defined in Section 4.5. Conversely, an oriented path in K which begins at v can be "lifted" in a natural way to an oriented path in $\mathcal{V}_+(v, K)$ which begins at the basepoint and has the same length as the original path. This is also not difficult to show, and we shall discuss it in more detail in Section 4.6. Once one has these two facts, the equality of the depths of K and the visibility (starting from v and the usual basepoint, respectively) follows immediately. □

Thus, in making comparisons between the size of a graph or its visibility and the depth, we do not need to worry about the possible distinction between the depth of the graph and the depth of the visibility.

Let us continue to suppose that we have an optical graph K and a vertex v in K, as in the lemma. For simplicity, we make another assumption now, which is that every vertex in K can be reached by an oriented path which begins at v. This was true for all of the examples in Section 4.3, and it ensures that the visibility $\mathcal{V}_+(v, K)$ is at least as large as K itself. We can always reduce to this case anyway, by throwing away any part of K which cannot be reached by an oriented path that begins at v.

Let us also ask that K contain no nontrivial oriented cycles. This was true for all of the examples in Section 4.3, except the second one (pictured in Fig. 4.3), and it implies that the depth of K is finite. In fact, the absence of nontrivial oriented cycles implies that no oriented path in K can pass through the same vertex twice, and hence that the depth of K must be less than the number of vertices of K.

To summarize, the preceding assumptions imply that

$$\text{depth of } \mathcal{V}_+(v, K) = \text{depth of } K \leq \text{ size of } K \leq \text{ size of } \mathcal{V}_+(v, K). \tag{4.2}$$

We know too that the visibility $\mathcal{V}_+(v, K)$ must be finite and of at most exponential size compared to the depth of K in this case, as in Lemmas 4.8 and 4.9 in Section 4.7 below.

In particular, we cannot have an exponential gap between both the depth of K and the size of K on the one hand and between the size of K and the size of the visibility on the other; if we did, then the size of the visibility would be too large compared to the depth. This fits very well with the examples in Section 4.3, from which we see that it is easy to make the graph be large compared to the depth, or small compared to the visibility, even if we cannot do too much of both at the same time. (Imagine varying the parameters j and k in the definition of H_j, for instance, including the possibility of taking j large. Remember that the graph G pictured in Fig. 4.2 amounts to the same thing as an H_j with $k = 1$.)

The relationship between the size of a graph G, its depth, and the size of its visibility reflect the nature and extent of the "activity" which occurs within G, in a way which is similar to the notion of *entropy* from information theory and dynamical systems (as in [Ash65, LM95, Mañ87, Sin76, Sin94]). We shall discuss some of these analogies further in Section 4.13.

4.5 The canonical projection

Let G be an optical graph, as before, and fix a vertex v in G.

There is a canonical projection $\pi : \mathcal{V}_+(v, G) \to G$ which sends the basepoint of the visibility $\mathcal{V}_+(v, G)$ to v. We define this mapping as follows. Each vertex p in $\mathcal{V}_+(v, G)$ represents a path in G which begins at v, and we take $\pi(p)$ to be the vertex in G which is the endpoint of this path. Now suppose that we are given an edge ϵ in $\mathcal{V}_+(v, G)$, and we want to associate to it an edge $e = \pi(\epsilon)$ in G. If ϵ is the edge that goes from the vertex p_1 to the vertex p_2 in $\mathcal{V}_+(v, G)$, then the path in G represented by p_2 is obtained from the path in G represented by p_1 by the addition of a single edge, and we take that to be our edge e.

This defines π as a map from vertices in $\mathcal{V}_+(v, G)$ to vertices in G, and from edges in $\mathcal{V}_+(v, G)$ to edges in G, with the obvious compatibility condition between the two. This mapping also respects the orientations on the two graphs, by construction.

Let us record a few basic facts.

Lemma 4.3 *If p is a vertex in $\mathcal{V}_+(v, G)$, then p has exactly one edge going into it, unless p happens to be the basepoint, in which case it has no edges going into it.*

Proof This is easy to check from the definitions. Indeed, p represents an oriented path in G which begins at v, and this path either traverses no edges, and so represents the basepoint in $\mathcal{V}_+(v, G)$, or it traverses at least one edge, in which case we can consider the path in G which agrees with the one associated to p except that it omits the last step. This new path determines a vertex p' in $\mathcal{V}_+(v, G)$ such that there is an edge in $\mathcal{V}_+(v, G)$ that goes from p' to p, and it is easy to see that p' is uniquely determined by this property. This proves the lemma. □

Lemma 4.4 *If p is a vertex in $\mathcal{V}_+(v, G)$, then p has the same number of outgoing edges in $\mathcal{V}_+(v, G)$ as $\pi(p)$ has in G.*

Proof This is an immediate consequence of the definitions. □

From these observations we see that $\mathcal{V}_+(v, G)$ is an optical graph, since G is, and that $\mathcal{V}_+(v, G)$ has no focusing branch points. The defocusing branch points in $\mathcal{V}_+(v, G)$ correspond exactly to those of G under the canonical projection π.

One can think of the passage to the visibility as being the universal procedure for eliminating the focusing branch points in a graph while leaving the defocusing branch points alone. This is quite standard, but it is useful to name it explicitly, and to study it as a mathematical process in its own right.

4.6 Basic properties of the visibility

Let G, v, $\mathcal{V}_+(v, G)$, etc., be as in the preceding section. We want to describe now a canonical way for lifting paths from G to $\mathcal{V}_+(v, G)$.

Let p be an oriented path in G which begins at v and has length n. Given $j = 0, 1, 2, \ldots, n$, let p_j denote the initial subpath of p of length j, i.e., the path in G which begins at v and follows p for exactly the next j steps before stopping. Thus p_0 is the degenerate path at v, while p_n is the same as p itself.

Each of these paths p_j determines a vertex in $\mathcal{V}_+(v, G)$, by definition of the visibility. For each $0 \leq j < n$ we also have that there is an edge in $\mathcal{V}_+(v, G)$ that goes from the vertex associated to p_j to the vertex associated to p_{j+1}, again by definition of the visibility. Thus in fact we get an oriented path in the visibility $\mathcal{V}_+(v, G)$, which we denote by $\lambda(p)$. This path begins at the basepoint of $\mathcal{V}_+(v, G)$, and its projection back down to G via $\pi : \mathcal{V}_+(v, G) \to G$ gives p back again. This is easy to derive from the definitions, and one can also check that $\lambda(p)$ is determined uniquely by these two properties. (That is, we have specified the starting point of $\lambda(p)$ in $\mathcal{V}_+(v, G)$, and one can show that each successive step is determined uniquely by the requirement that $\lambda(p)$ project back down to p in G. This follows from the definitions.)

We call $\lambda(p)$ the *canonical lifting of the path p to the visibility $\mathcal{V}_+(v, G)$*. It is a simple analogue of a standard construction for covering surfaces in topology. (We shall discuss this analogy further in Section 4.11.)

Note that the length of $\lambda(p)$ is the same as the length of p itself.

It is easy to have pairs p_1, p_2 of oriented paths in a graph G which begin at v and cross each other several times. This cannot happen with the liftings $\lambda(p_1)$, $\lambda(p_2)$ to the visibility; they cannot cross each other several times, but must remain disjoint as soon as they split apart a single time. Roughly speaking, this is because $\lambda(p_1)$, $\lambda(p_2)$ do not only track the succession of vertices in p_1, p_2, but also their entire histories in p_1, p_2.

Similarly, although an oriented path p in G can cross itself many times, the canonical liftings are always *simple paths*, i.e., they never pass through the same vertex twice. Again this is because each vertex in the canonical lifting corresponds to a whole initial subpath of p, and the initial subpaths of p are distinct by virtue of having different lengths, for instance.

The following proposition provides an alternative characterization of the lifting $\lambda(p)$.

Proposition 4.5 *For each vertex p in $\mathcal{V}_+(v, G)$, there is a unique oriented path in $\mathcal{V}_+(v, G)$ from the basepoint to p.*

Proof Let p be given, so that p corresponds exactly to an oriented path in G which begins at v. Its lifting $\lambda(p)$ to a path in $\mathcal{V}_+(v, G)$ is oriented and goes from the basepoint to p (now viewed as a vertex in $\mathcal{V}_+(v, G)$), and this gives the existence part of the proposition.

As for uniqueness, suppose that α is another oriented path in $\mathcal{V}_+(v, G)$ which begins at the basepoint and ends at p. Let m be the length of α, and let q_i, $0 \leq i \leq m$, be the ith vertex in $\mathcal{V}_+(v, G)$ which appears in the path α. Thus q_0 is the basepoint of $\mathcal{V}_+(v, G)$, and q_m is p itself.

Since α is an oriented path in $\mathcal{V}_+(v, G)$, we know that there is an edge in $\mathcal{V}_+(v, G)$ which goes from q_i to q_{i+1} for each $i = 0, 1, \ldots, m-1$. (Of course the edge is allowed to depend on i.)

Each q_i represents an oriented path in G which begins at v. The preceding condition of adjacency implies that the path represented by q_{i+1} is obtained from the path represented by q_i by adding exactly one more step at the end. Of course q_0 represents the degenerate path at v which traverses no edges.

By using this fact repeatedly, we obtain that each q_i is actually an initial subpath of q_m, and that it is the initial subpath of length i. In particular, q_m itself has length m. We already know that q_m is the same as p, and from these observations one can easily check that α must in fact be $\lambda(p)$, as desired. This proves the proposition. □

Corollary 4.6 *If α is an oriented path in $\mathcal{V}_+(v, G)$ from the basepoint to a vertex p, then $\pi : \mathcal{V}_+(v, G) \to G$ maps α to the path in G represented by p.*

Proof This can be derived from the proof, or from the fact that α must coincide with $\lambda(p)$, by uniqueness. □

Given a vertex p in $\mathcal{V}_+(v, G)$, we can define its *distance to the basepoint* as the number of edges in the unique oriented path from the basepoint to p. This is the same as the *length* of the path in G represented by p, since this path in G has the same length as its canonical lifting $\lambda(p)$ in $\mathcal{V}_+(v, G)$.

If q is another vertex in $\mathcal{V}_+(v, G)$, and if there is an edge in $\mathcal{V}_+(v, G)$ which goes from p to q, then the distance from q to the basepoint is exactly 1 greater than the distance from p to the basepoint. This is easy to check.

Corollary 4.7 *$\mathcal{V}_+(v, G)$ is a tree.*

Proof Since $\mathcal{V}_+(v, G)$ is connected, we only have to show that there are no nontrivial unoriented (finite) paths which begin and end at the same vertex, but otherwise do not pass through the same vertex more than once.

Suppose to the contrary that there is such a path L in $\mathcal{V}_+(v, G)$. Let p be a vertex in L *furthest* from the basepoint (with respect to the distance to the basepoint mentioned above). This point exists, since L is finite. This choice of p ensures that L cannot traverse an edge in $\mathcal{V}_+(v, G)$ which flows out of p; for if L did traverse such an edge, then the other endpoint q of that edge would lie in L and be further from the basepoint than p is, as in the observation just prior to the statement of the corollary.

On the other hand, there is at most one edge in $\mathcal{V}_+(v, G)$ which flows into p, by Lemma 4.3. As a result, L should cross this edge twice (since L begins and ends at the same vertex), and pass through the vertex at the other end of the edge twice as well. This gives a contradiction, and the corollary follows. $\square$

Using Proposition 4.5, it is not hard to see that the visibility of $\mathcal{V}_+(v, G)$ starting from the basepoint is isomorphic to $\mathcal{V}_+(v, G)$ itself in a natural way. That is, we know from Section 4.5 that $\mathcal{V}_+(v, G)$ is an optical graph in its own right, so that we can define its visibility in the same way as before. (The possibility of $\mathcal{V}_+(v, G)$ being infinite does not cause any real problems here.) This is not very interesting, since we do not get anything new, but one should take note of the possibility.

4.7 The size of the visibility

Let G, v, and $\mathcal{V}_+(v, G)$ be as before.

Lemma 4.8 *$\mathcal{V}_+(v, G)$ is a finite graph for every vertex v in G if and only if there are no (nontrivial) oriented cycles in G. If we fix a vertex v in G, then $\mathcal{V}_+(v, G)$ is finite if and only if there is no nontrivial oriented cycle in G which can be reached by an oriented path that starts at v.*

For us an *oriented cycle* means an oriented path which begins and ends at the same vertex.

Proof We may as well prove only the second part, of which the first is a consequence.

Suppose first that G has a nontrivial oriented cycle which is accessible by an oriented path starting from v. Then there are infinitely many distinct oriented paths in G beginning at v, because one can go from v to the cycle and then traverse the cycle as many times as one please.

Conversely, suppose that G contains no nontrivial oriented cycle which can be reached from v. From this it follows that if p is any oriented path in G which begins at v, then p cannot go through any vertex more than once. This is easy to verify, and it implies that the length of p is strictly less than the total number of vertices in G.

This universal bound on the length of these paths implies that there are only finitely many of them. From here we obtain the finiteness of the visibility, as desired. □

In the next lemma we give more precise bounds when the visibility is finite.

Lemma 4.9 *If G contains no oriented paths starting at v and having length greater than k, then $\mathcal{V}_+(v, G)$ contains at most 2^{k+1} vertices. (This holds in particular when k is 1 less than the total number of vertices in G and there is no nontrivial oriented cycle in G which is accessible from v by an oriented path, as in the proof of the preceding lemma.)*

Proof Let S_j denote the set of vertices in $\mathcal{V}_+(v, G)$ which can be reached from the basepoint by an oriented path of length j, $j \geq 0$. We want to estimate the number N_j of elements of S_j. Notice that S_0 consists of only the basepoint, so that $N_0 = 1$. In general we have

$$N_{j+1} \leq 2 \cdot N_j \tag{4.3}$$

for all $j \geq 0$. Indeed, the definition of S_j ensures that for each element p of S_{j+1} there is a $q \in S_j$ such that there is an edge in $\mathcal{V}_+(v, G)$ that goes from q to p. There can be at most two p's corresponding to any given q, since $\mathcal{V}_+(v, G)$ is an optical graph, and (4.3) follows from this.

Thus we have that

$$N_j \leq 2^j \quad \text{for all } j, \tag{4.4}$$

and $\bigcup_{j=0}^{k} S_j$ has at most 2^{k+1} elements. (This works for any optical graph G.)

We can also describe S_j as the set of vertices in $\mathcal{V}_+(v, G)$ which represent oriented paths in G which begin at v and have length equal to j. This reformulation follows from the remarks just prior to the statement of Corollary 4.7. If k is chosen as in the statement of the lemma, then every oriented path in G which begins at v has length at most k. This means that every vertex in $\mathcal{V}_+(v, G)$ lies in some S_j with $j \leq k$, and hence $\mathcal{V}_+(v, G)$ contains at most 2^{k+1} vertices. This proves the lemma. □

Corollary 4.10 *If $\mathcal{V}_+(v, G)$ has only finitely many vertices, then it has at most 2^n vertices, where n is the number of vertices in G.*

Proof Indeed, if $\mathcal{V}_+(v, G)$ has only finitely many vertices, then there are no oriented paths in G which begin at v and which can reach a nontrivial oriented cycle in G, because of Lemma 4.8. This permits us to apply Lemma 4.9 to obtain the desired bound. □

Corollary 4.11 *If G is an optical graph with at most n vertices, and if the visibility $\mathcal{V}_+(v, G)$ has more than 2^n vertices, then $\mathcal{V}_+(v, G)$ is infinite, and G contains a nontrivial oriented cycle which can be reached by an oriented path beginning at v.*

Proof This follows from the previous corollary and its proof. □

To summarize a bit, there are some exponential upper bounds for the size and growth of the visibility that we always have.

We have seen in Section 4.3 how the visibility can be far from exponentially-large as compared to the underlying graph. From these examples one is lead to the notion of "long chains of focal pairs" (Definition 4.17), which can be used to detect exponential growth in the visibility. More precisely, the presence of a long chain of focal pairs leads to a lower bound on the size of the visibility (Proposition 4.18), and there are roughly similar upper bounds in terms of the length of the longest chain of focal pairs given in Chapter 8. For infinite visibilities there is a similar test for exponential versus polynomial growth, in terms of the oriented cycles in the graph, and we shall explain this in Chapter 5.

Before we get to that, let us pause briefly to look at some ways in which visibility graphs arise in connection with some other mathematical structures.

4.8 Formal proofs and logical flow graphs

Logical flow graphs of formal proofs are always optical graphs, to which the notion of visibility can be applied. How does the visibility behave in relation to the underlying proof?

For this discussion we use the *sequent calculus* (Section A.1 in Appendix A) for making formal proofs, and the construction of the *logical flow graph* is reviewed in Section A.3.

Of course one normally expects proofs without cuts to be simpler in their structure than proofs with cuts, and that is the case here. For a proof without cuts, the logical flow graph contains no nontrivial oriented cycles [Car97], and so the visibility graphs are all finite, as in Lemma 4.8. In fact the visibility of the logical flow graph of a cut-free proof is never more than *quadratic* in size as compared to the size of the logical flow graph itself. We shall discuss this further in Section 6.12. This kind of quadratic growth can certainly occur, and indeed the graphs H and H^* mentioned in Section 4.3 reflect well the kind of structure that can arise in the logical flow graph of a proof without cuts.

For proofs with cuts, one can have exponential expansion of the visibility, coming from the interaction between cuts and contractions. The basic idea is captured well by the first example pictured in Section 4.3.

The notion of *feasible numbers* provides a nice setting in which this kind of picture can be seen concretely. For instance, one can make a formal proof of the feasibility of 2^{2^n} in $O(n)$ steps which employs cuts but not quantifier rules, and for which the logical flow graph is practically the same as the graph shown in Fig. 4.2 in Section 4.3. Let us sketch this example without going too far into technical details about formal proofs and feasible numbers, and refer to [Car00] for more precise information, as well as other examples which exhibit more intricate structure of *cycling* in their logical flow graphs.

In the study of feasible numbers, one works in the setting of ordinary arithmetic, but one adds to the usual language a new predicate $F(\cdot)$, for which the intended meaning of $F(x)$ is that "x is feasible". One assumes certain special axioms and rules, which have the effect of saying that $F(0)$ is true (so that 0 is a feasible number), and that feasibility is preserved by sums, products, and the successor function (addition by 1).

With these rules, one can easily prove that $F(n)$ is true for every nonnegative integer n. One simply begins with $F(0)$ and then applies the rule for the successor function n times. This leads to a proof of $F(n)$ in $O(n)$ lines.

One can do better than this, using the rule for products. For instance, one can make a proof of $F(2^n)$ in $O(n)$ lines by first making n different proofs of $F(2)$, and then combining them using the rule for multiplications. In terms of formal logic this proof is pretty trivial, because it does not use cuts or contractions. In other words, one might say that this kind of proof is completely explicit.

By using cuts and contractions one can make a proof of $F(2^{2^n})$ in $O(n)$ steps. The main point is to give a proof of

$$F(2^{2^j}) \to F(2^{2^{j+1}}) \tag{4.5}$$

for any integer j, using only a few steps (independently of j). This is easy to do. One starts with two copies of the axiom

$$F(2^{2^j}) \to F(2^{2^j}), \tag{4.6}$$

and then combines them using the rule for multiplications to get

$$F(2^{2^j}), F(2^{2^j}) \to F(2^{2^{j+1}}) \tag{4.7}$$

(since the product of 2^{2^j} with itself is $2^{2^{j+1}}$). One then applies a contraction on the left-hand side to get (4.5). To get a proof of $F(2^{2^n})$, one strings together proofs of (4.5) for $j = 0, 1, 2, \ldots, n-1$ using cuts to get a proof of

$$F(2) \to F(2^{2^n}) \tag{4.8}$$

in $O(n)$ steps, and then one combines this with a proof of $F(2)$ again using a cut.

In this proof we used cuts and contractions to amplify greatly the effect of a single proof of $F(2)$. Remember that in the earlier proof without cuts or contractions we had n separate proofs of $F(2)$ which were combined to make a single proof of $F(2^n)$. In other words, we only got out as much material as we put in, while here we used cuts and contractions to make certain "duplications".

If we did *not* use contractions as above, but instead took (4.7) as our basic building block instead of (4.5), then we could still build a proof of $F(2^{2^n})$ using cuts, but the number of cuts would double with each stage of the construction. In the end we would need 2^n proofs of $F(2)$ in order to build a single proof of $F(2^{2^n})$, and we would be back to the same kind of "explicit construction" as before.

In terms of the logical flow graph, the contractions correspond to *defocusing* branch points, while each use of a multiplication rule lead to a *focusing* branch point. The use of cuts permits one to alternate between the two systematically, and this would not be possible otherwise.

It is precisely this kind of alternation which leads to exponential effects in the visibility, as we shall see later on, beginning in Section 4.14.

If one takes the final occurrence of $F(2^{2^n})$ in the proof and asks "where did this come from?", then one is lead naturally to look at the analogue of the visibility of the logical flow graph, but with the orientations reversed. That is, the logical flow graph traces the way that the formulae are used in the proof, and in this case the duplications which occur in the proof correspond exactly to the splitting of defocusing branch points which is accomplished through the negatively-oriented visibility.

Indeed, the structure of the logical flow graph is the same in essence as that of the first graph pictured in Section 4.3. One can even forget about formal proofs and simply think about having an oriented graph in which numbers are attached to vertices, with some rules given to specify the transitions from point to point. In this case one could think of the focusing vertices as representing multiplications of the numbers attached to the preceding vertices, while a defocusing vertex would be interpreted as a kind of "duplicator" for the number given there.

This idea will be pursued more formally in Chapter 7, through the notion of *feasibility graphs.* For these structures the visibility always has a simple and direct interpretation as providing an explicit rendering of an implicit construction.

There are more elaborate proofs of feasibility of certain large numbers, using quantifiers to increase more dramatically the gap between the number of steps in the proof and the size of the number whose feasibility is being established. The logical flow graphs of these proofs are also much more complicated, with nested layers of cycling, as described in [Car00].

Conversely, it is shown in [Car00] that cycles in the logical flow graph are *necessary* under conditions like these. Specifically, there is a constant $c > 0$ so that any proof of $\to F(m)$ with no more than $c \cdot \log\log m$ lines must contain a cycle. This should be compared with the existence of proofs of $\to F(2^{2^n})$ in $O(n)$ steps which do not have cycles, as above. In particular, this would not work for $\to F(2^{2^{2^n}})$.

In Chapter 16, we explain how similar phenomena of nesting and recursion can be captured combinatorially through the use of feasibility graphs which themselves describe the construction of other feasibility graphs.

4.9 Comparison with L-systems

L-systems, or *Lindenmayer systems*, are certain systems for producing words over an alphabet which have been used in modelling in biology [PL96, RS80]. The simplest of these are called *D0L-systems*, the definition of which we shall review now. The "D" here stands for "deterministic", while the 0 indicates that the type of substitutions used will be context-free (0 symbols of context). A brief discussion of these systems can also be found in [HU79], beginning on p390.

Let Σ be a nonempty finite set of letters, and let Σ^* denote the set of words over Σ. Suppose that to each letter in Σ we associate a word in Σ^*. If a is a letter in Σ, let us write $h(a)$ for the word to which it is associated. It is easy to extend h to a mapping from Σ^* to itself, in such a way that the extension is a *homomorphism*, i.e., so that $h(uw) = h(u)h(w)$ for all words $u, w \in \Sigma^*$, where uw denotes the concatenation of u and w (and similarly for $h(u)h(w)$). In more explicit terms, if we also use h to denote this extension, then h sends the empty word to itself, and it maps $a_1 a_2 \cdots a_k$ to $h(a_1)h(a_2)\cdots h(a_k)$ whenever $a_1, a_2, \ldots, a_k$ are elements of Σ.

Let ω be a word in Σ^*, which is often called the *axiom* or *starting word* of the system. A D0L-system is exactly given by the combination of an alphabet Σ, a collection of assignments of a word in Σ^* to each letter in Σ (or, equivalently, a homomorphism h from Σ^* to itself), and a word ω in Σ^* like this. From this data, one gets the sequence of words

$$\omega,\ h(\omega),\ h^2(\omega),\ h^3(\omega), \ldots, \tag{4.9}$$

and these are the words *generated by the D0L-system*.

One of the reviewers of this book pointed out that some of the statements given here concerning the size and growth of visibility graphs are closely related to the theory of growth functions for L-systems. For a D0L-system as above, the *growth function* would be the function $f(n)$ defined by taking $f(n)$ to be the length of $h^n(\omega)$ for each nonnegative integer n. (When $n = 0$, we interpret h^0 as being the identity mapping on Σ^*.) Thus the growth function is a mapping from the set of nonnegative integers to itself.

There is a simple and precise correspondence between D0L-systems and oriented graphs and their visibilities that one can make, which we now describe. Suppose that one has a D0L-system as above. We want to define an oriented graph G associated to it. For the set of vertices of G, we take the set Σ. If a and b are elements of G, then we attach an edge in G going from a to b for each occurrence of b in $h(a)$. We do this for all choices of a and b, including the case where $b = a$, and these are all of the edges that we attach. If $h(a)$ is the empty word, then there are no edges in G for which a is the initial vertex.

This graph encodes the same information as in the *growth matrix* associated to a D0L-system, as on p31 of [RS80]. More precisely, the growth matrix is the same as the adjacency matrix of the graph. We shall return to adjacency matrices in general in Chapter 12.

Lemma 4.12 *Under the conditions just given, if a is an element of Σ, and if n is a nonnegative integer, then the length of $h^n(a)$ is equal to the number of oriented paths in G which begin at a and have length exactly equal to n.*

Proof This is not hard to check from the definitions. There is a more precise statement, which makes the verification simpler; namely, given $a \in \Sigma$ and a nonnegative integer n, there is a one-to-one correspondence between occurrences of a letter $b \in \Sigma$ in $h^n(a)$ and oriented paths in G which go from a to b and have length equal to n. To establish this, one can use induction on n. If $n = 0$, then $h^n(a) = a$ automatically, and $h^n(a)$ has length 1. On the other hand, there is exactly 1 oriented path in G starting at a with length 0, namely the degenerate path at a which traverses no edges (and ends at a). If the statement has been established for some value of n, then one can derive the corresponding statement for $n + 1$ by looking at the correspondence between the words $h(c)$, $c \in \Sigma$, and the way that edges are attached to G, and using the fact that paths of length $n + 1$ are obtained from paths of length n by adding exactly one more step at the end. This derivation for $n + 1$ is easy to verify. □

If the starting word ω of the D0L-system is of the form $a_1 a_2 \cdots a_k$, then we have that

$$h^n(\omega) = h^n(a_1)h^n(a_2)\cdots h^n(a_k) \tag{4.10}$$

for all nonnegative integers n. This follows from the fact that $h : \Sigma^* \to \Sigma^*$ is a homomorphism. If $|w|$ denotes the length of a word w, then we get that

$$|h^n(\omega)| = \sum_{i=1}^{k} |h^n(a_i)|. \tag{4.11}$$

Lemma 4.12 can be applied to relate this to numbers of oriented paths in G.

Now let us go in the other direction. If G is any oriented graph, then one can define a D0L-system which is associated to G in this way. Specifically, one takes Σ to be the set of vertices of G, and to each element a in Σ one assigns a word $h(a)$ in Σ^* such that the number of occurrences of a letter $b \in \Sigma$ in $h(a)$ is the same as the number of edges in G that go from a to b.

The particular choice of word $h(a)$ that one uses here does not matter, as long as the number of occurrences is correct. Thus there can be multiple D0L-systems which are compatible with G in this way. This reflects a well-known observation that growth functions for D0L-systems do not depend on the ordering of letters in the words $h(a)$, $a \in \Sigma$.

The choice of starting word ω in the D0L-system does not play a role in the correspondence with an oriented graph G, so that one can change ω without affecting G.

Remark 4.13 Oriented graphs and their visibilities are also closely related to finite-state automata and the languages that they recognize. We shall discuss this in Chapter 14, which includes a review of the basic notions. For the moment, let us mention a few points which are related to the present topics.

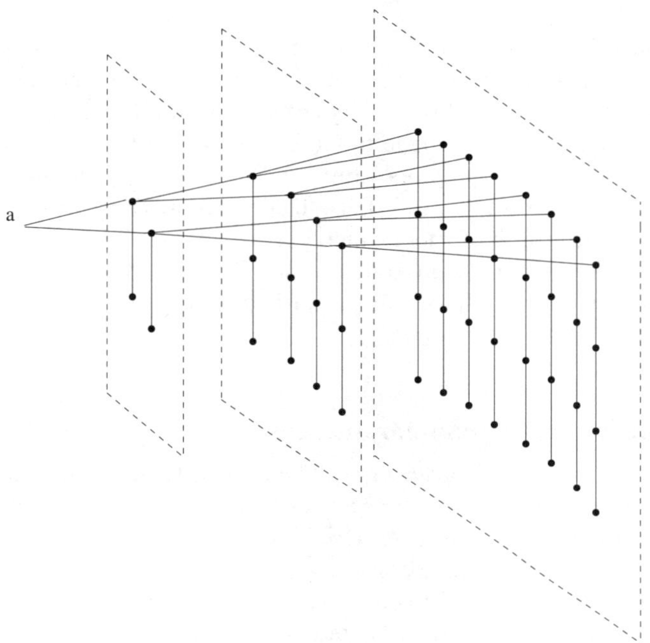

FIG. 4.6. This picture shows a visibility graph and "fronts" in it, as correspond to words generated by a D0L-system associated to the underlying oriented graph (as in Lemma 4.12 and its proof). Each vertex in the visibility graph comes from an oriented path in the original graph, which is indicated in the "sheets" above. The sheets shown correspond to the words $h(a)$, $h^2(a)$, and $h^3(a)$, in the notation of this section.

With both finite automata and D0L-systems, there are associated families of words, and oriented graphs which are related to the system. However, the ways that these fit together are somewhat different. In an automaton, the words correspond to *paths* in the graph. In a D0L-system, the words being generated can be viewed as corresponding to collections of paths, as in Lemma 4.12 and its proof. Suppose that the starting word in the D0L-system consists of a single letter, and let us think of collections of oriented paths in the graph starting at the vertex associated to this letter as being collections of vertices in the corresponding visibility graph. Then the words generated by the D0L-system give exactly a family of successive "fronts" in the visibility graph, or parallel ridges, starting from the basepoint of the visibility. See Fig. 4.6.

To put it another way, automata and D0L-systems can both be viewed in terms of oriented graphs, paths in them, and their associated visibility graphs, but they use these graphs and paths in different ways, with different "slices" of them.

With this correspondence between D0L-systems and oriented graphs, one can move back and forth between the two, concerning growth functions for D0L-systems and growth in the visibility graph in particular. In this way, some of the observations mentioned in this book concerning visibility graphs can be seen as versions or relatives of known results for D0L-systems, as in [RS80]. Specific instances of this come up in Chapters 5 and 12. Similar issues arise in relation to automata and regular languages, and this will come up in Chapters 5 and 12 too. Alternatively, one can think of the geometry as providing different perspectives for some of the known results, and giving other elements as well.

See [RS80] for (very interesting) mathematical results and questions pertaining to L-systems.

4.10 "Visibility" in Riemannian manifolds

The idea of the visibility of an optical graph has a well-known cousin in Riemannian geometry, with geodesic rays as "rays of light". The extent to which these rays of light are tangled up in a given manifold provides a way to measure the "internal activity" of the manifold. We shall review some basic examples and concepts related to these topics in this section. Some textbooks for Riemannian manifolds are [Boo75, Spi79] and, at a somewhat more advanced level, [KN69]. The lectures in [Gro94] provide a concise and concrete introduction to many important aspects of Riemannian geometry, and additional information can be found in [Bes78, CE75].

Roughly speaking, a Riemannian manifold is a space in which there are local measurements of length, volume, and so forth, just as in Euclidean space. The geometry can be quite different from that of a Euclidean space, because of curvature. Very near any given point, the space will look approximately like a Euclidean space, to a certain degree of precision, but the way that this happens can change as one moves in the space, and this can affect the overall geometry.

In a Riemannian manifold, there is a special class of curves, called *geodesics*. These curves play the role of straight lines in Euclidean spaces. They are characterized by the property that they give a path of shortest length between a pair of points on the curve, at least when the two points are close enough together. We shall say a bit more about this in a moment. Geodesics can also be described by an explicit second-order ordinary differential equation. In some situations, a geodesic curve gives a path of shortest length between any two points on the curve, whether or not they are close together. This is true in Euclidean spaces, for instance. In other situations it is not true, because of bumps in the space, or the way that a geodesic might wrap around the space, and so on. We shall see examples of this in a moment.

A basic example of a Riemannian manifold is provided by the standard (round) 2-dimensional sphere, for which the "complete" geodesics are given by the great circles. More precisely, the "Riemannian metric" on the sphere is the one that it inherits from being a surface in ordinary 3-dimensional Euclidean space $\mathbf{R}^3$, with the induced measurements of length and area.

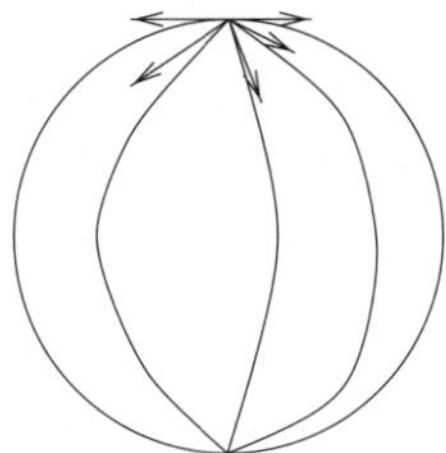

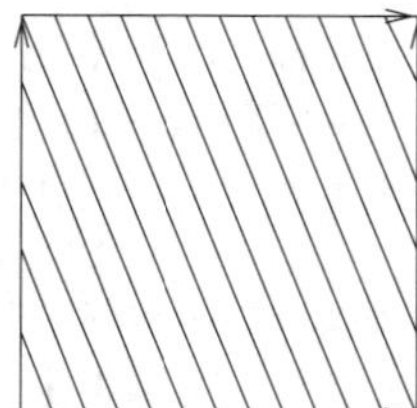

FIG. 4.7. Geodesic rays and the way that they can wrap around a space, in the 2-dimensional round sphere (on the left), and a diagram for a 2-dimensional (flat) torus (on the right)

Given a pair of points p, q in the sphere which are not antipodal to each other, there is a unique great circle C that goes through them. This circle C defines two geodesic arcs between p and q, a shorter arc and a longer arc. The shorter arc is the path of minimal length between p and q. The longer arc is an example of a curve which is a geodesic, and so locally gives paths of shortest length between two points, but which is not a path of smallest length between its endpoints p and q. If p and q are antipodal with respect to each other, then there is a continuous family of great circles passing through them, and all of the resulting geodesic arcs have the same length and are of minimal length.

In general, we would like to consider the family of all geodesic rays which emanate from a given point in a Riemannian manifold. This reflects the way that the manifold "looks" from the given point, and from the inside of the manifold. Let M be a fixed compact Riemannian manifold (without boundary), like a sphere or a torus (to which we shall return in a moment). Fix also a point x in M, which will serve as a basepoint. One can look around in M in all directions, from the point x. In this connection, let $T_x M$ denote the vector space of all tangent vectors to M at x, which is called the *tangent space* to M at x. This gives the set of all directions in which one can go, starting at x. More precisely, the unit vectors in the tangent space give the directions, and the lengths of these vectors can be used to say how fast (or far) one would go in those directions.

Given a unit vector u in $T_x M$, there is a unique unit-speed geodesic ray $\gamma_u : [0, \infty) \to M$ in M which begins at x ($\gamma_u(0) = x$) and whose initial velocity vector is u ($\gamma_u'(0) = u$). This is a consequence of existence and uniqueness theorems for ordinary differential equations. Some illustrations related to this are given in Fig. 4.7, and we shall say more about these pictures as we proceed.

The *exponential mapping* from $T_x M$ to M is defined as follows. Given a vector v in $T_x M$, we can write it as $t \cdot u$, where u is a unit vector in $T_x M$, and t is a nonnegative real number. By definition, the exponential mapping takes v and associates to it the point $\gamma_u(t)$ in M. In other words, $\gamma_u(t)$ is the point in M that we "see" when we look in the direction of u at distance t. On the standard 2-dimensional sphere, for instance, one can "see" the point antipodal to x in all directions u. This is illustrated in the left side of Fig. 4.7. If one continues the same amount of length, and in any direction, then one can see x itself. If one goes further, then one gets to the antipodal point again, and then x again, etc. Other points in the sphere are also repeated, but it is only for the antipodal point and x itself that one sees the point in every direction (starting from x). All of this works for any choice of basepoint x in the standard (round) 2-sphere. For that matter, the whole picture can be rotated to adjust for any other basepoint. A rotation will not change anything in the geometry.

Geodesic rays typically wrap around a given manifold in more complicated ways than in the standard 2-sphere. As another example, let us look at the (standard) 2-dimensional flat torus. This can be realized as the Cartesian product of two copies of the unit circle $\mathbf{S}^1$ in $\mathbf{R}^2$. One can also start with a square, as in the picture on the right-hand side of Fig. 4.7, and glue the edges together in a certain way. Specifically, one can glue the top edge to the bottom edge in a "parallel" and even manner (as indicated by the arrows on the edges in the picture), and one can glue the left and right sides of the square to each other similarly. Instead of starting with a square, one could use other parallelograms, and this would lead to spaces which are similar but not quite the same. (They would be the same *topologically*, but the geometry would be changed.)

One might think of a torus topologically in terms of the surface around a donut, or an inner tube, inside of $\mathbf{R}^3$. This is the same topologically as the flat torus that we are considering here, but they are not quite the same geometrically. Our torus is *flat*, which means that around every point, the torus looks locally exactly like the standard 2-dimensional Euclidean space, with exactly the same geometry. This is not true for the surface of a donut, or an inner tube; one cannot flatten them out, even locally, without changing the geometry. In our case, the local equivalence with the standard 2-dimensional Euclidean geometry comes out nicely if one thinks of producing the torus by taking a standard square, as in Fig. 4.7, and identifying the edges of it as before. In doing this, one can keep the Euclidean geometry (locally) from the square. This is also compatible with the gluing along the edges.

Imagine that one chooses a basepoint x in the torus, as in the discussion of the exponential mapping. Which point one takes does not really matter, because one can slide points around to get to any other choice of basepoint, without changing the geometry. Consider the space of tangent vectors to the torus at x, as before. The unit vectors correspond to points in a circle in the tangent space (because the torus is 2-dimensional), and one can describe them by an angle θ. For each unit vector, one gets a geodesic ray in the torus, as before.

Our torus is flat, and so looks locally like ordinary 2-dimensional Euclidean space. In such a local model, a geodesic arc is a straight line segment. In particular, this is what happens if we represent the torus as a square with its sides identified, as in Fig. 4.7. A geodesic arc is a straight line segment, and when it reaches one of the sides, it does not have to end there, but can continue on the opposite side, following the identification of points in the two sides. This can take place repeatedly, and an example is shown in Fig. 4.7.

If one initializes the angles in the right way, then the angle θ of a unit tangent vector u in our tangent space will be the same as the angle of the associated geodesic in the square, which will be represented by parallel line segments, as in Fig. 4.7. When the angle θ is a rational multiple of π, the geodesic is periodic, and wraps around a closed curve in the torus. For the picture in the square, this means that one can start somewhere on one of the segments, follow it along, crossing the identifications in the edges whenever they come up, and eventually return to the place where one began. The whole geodesic ray goes along finitely many segments, over and over again. When the angle is an irrational multiple of π, the geodesic never closes up, and in fact the geodesic as a whole gives a dense subset of the torus. (The latter reduces to a well-known elementary fact in number theory.) In the picture with the square, one would have infinitely many parallel line segments, which never come back to the place where they start. For rational multiples of π, the geodesic rays become more and more complicated as the denominators in the rational numbers increase, in the sense that more and more segments are needed to represent the whole geodesic ray. They can approximately fill up the torus, as in Fig. 4.7, even if they do not become dense. These phenomena are quite different from the case of the sphere, where all of the geodesics are periodic, and have the same period.

Note that there are other Riemannian manifolds, besides round spheres, for which every geodesic is periodic. See [Bes78] for more information.

How can we measure the complexity of the "wrapping" of geodesic rays in a given manifold? One approach is to count the number of times that geodesic rays emanating from x go through a given point $y \in M$. More precisely, define $N_t(y)$ by

$$N_t(y) = \text{the number of vectors in } T_xM \text{ of length less than } t \quad (4.12)$$
$$\text{which are mapped to } y \text{ by the exponential mapping.}$$

A single geodesic ray may go through a single point many times, as in the periodic examples before, and this method counts each crossing separately. Set

$$N_t = \text{the average of } N_t(y) \text{ over } y \in M. \quad (4.13)$$

In other words, N_t is the integral of $N_t(y)$ over $y \in M$, divided by the total volume of M. The volume element on M used for defining the integral and for determining the volume of M comes from the Riemannian metric on M. The rate of growth of N_t as $t \to \infty$ provides one way to measure the overall complexity of the exponential mapping.

For the standard n-dimensional sphere, N_t grows *linearly* in t. This is not hard to see from the earlier discussion, with the simple periodicity which took place. More precisely, $N_t(y)$ is infinite when y is equal to x or the point antipodal to x and t is at least the circumference of the sphere (or half the circumference when y is antipodal to x). For all other points y, $N_t(y)$ is finite, and grows at uniform linear rate.

In general, $N_t(y)$ may be infinite sometimes, but one can show that N_t will remain finite for all t. This implies that $N_t(y)$ will not be infinite on a set of positive measure in M for any t. To see that these assertions hold, one can write the integral of $N_t(y)$ over M as an integral of a smooth function over a ball in the tangent space T_xM, through a change of variables, using the exponential mapping. The function $N_t(y)$ counts the multiplicities for the change of variables, and does not appear in the integral of the domain, but reflects how often the domain wraps around points in the image. One has to include a Jacobian in the in the integral on the domain, but this remains bounded on each ball. In this way, the rate of growth of N_t can be expressed in terms of the Jacobian of the exponential mapping, and each N_t can be seen to be finite.

For flat n-dimensional tori, like the 2-dimensional version before, N_t grows like $O(t^n)$. We shall say more about this below. In general, N_t can grow at an *exponential* rate (and no more), and this occurs for manifolds of negative curvature. This happens in particular for the *hyperbolic* metrics (which have constant negative curvature) that always exist on closed 2-dimensional surfaces with at least two handles. Note that the torus has one handle, and a sphere has no handles. Also, standard spheres have *positive* curvature (which is constant), while flat tori have curvature equal to 0. An excellent treatment of the meaning of the *sign* of curvature can be found in [Gro94].

The exponential mapping from T_xM to M is somewhat similar to the canonical projection from the visibility of a given oriented graph G back down to G itself. For this comparison, it is helpful to reformulate $N_t(y)$ as follows:

$$N_t(y) = \text{the number of distinct geodesic arcs in } M \text{ which go from } x \text{ to } y \text{ and have length less than } t. \tag{4.14}$$

This reformulation uses the existence and uniqueness theorem for second-order differential equations, which gives a one-to-one correspondence between geodesic rays starting at x and the possible initial conditions at x, which are given by tangent vectors at x.

At any rate, $N_t(y)$ is roughly analogous to counting the number of oriented paths which go from a given vertex in a graph G to another one (and with length at most a given number), and the size of these numbers provides a measure of the internal activity of the given graph G. We shall see another version of this in Section 4.11.

Let us come back for a moment to the example of the 2-dimensional flat torus from above, and the rate of growth of the quantity N_t. Let us use the realization of the torus from a square, as in Fig. 4.7, with identifications between the sides. Imagine "unwrapping" this picture, in the following manner. Instead of identifying the top and bottom edge, for instance, imagine that one places another copy of the square above the one that we have, with the bottom edge of the new copy lining up evenly with the top edge of the original square. When one tries to go across the top edge of the original square, as with a geodesic arc, one can simply continue into the new square above, instead of following the identifications and going to the bottom of the original square, as before. One can do this for all four sides of the square, adding new squares in all four directions. One can then repeat this along the new edges which are created. One can do the whole process inside of a standard 2-dimensional Euclidean plane, in which the original square sits. By doing this repeatedly, the whole plane will be filled up by the square and copies of it, in a kind of tessellation. One can think of points in the squares as corresponding to points in the torus in the same manner as for the original square, except that now points in the torus will be represented in all of the squares (and hence will be represented infinitely many times).

If one does this, then a geodesic ray in the torus can be unwrapped to a ray in a straight line in the ordinary Euclidean plane. It will not pass through any square more than once, but, on the other hand, the squares represent repeated copies of the single square before. The total behavior of the geodesic ray in the torus now corresponds to the combination of what the straight ray in the Euclidean plane does in all of these squares.

Using this description, one can show that the quantity N_t is equal to a constant times t^2 in this case. This is because one can transfer the question to an analogous one on the Euclidean plane, where the result is more easily seen. More precisely, if $N_t(y)$ is defined as before, and if $\widetilde{N}_t(z)$ is defined in an analogous manner for the Euclidean plane, then the integral of $N_t(y)$ over the torus will be the same as the integral of $\widetilde{N}_t(z)$ over the whole plane. This is because they count essentially the same events, just rearranged in a certain way. With $N_t(y)$, a single ray might go through y several times, but in the Euclidean plane this does not happen. However, the ray may pass through points in different squares which correspond to the same point in the torus, and this is the counterpart of what can occur in the torus.

Inside the Euclidean plane, $\widetilde{N}_t(z)$ is equal to 1 at points z at distance less than t from the basepoint, and it is equal to 0 otherwise. This follows from unwinding the definitions. Thus the integral of $\widetilde{N}_t(z)$ over the plane is equal to the area of a disk of radius t, which is πt^2. The integral of $N_t(y)$ over the torus is equal to the same quantity, and N_t is this integral divided by the area of the torus. The area of the torus is the same as the area of the original square.

The same argument applies to flat tori more generally, and in all dimensions. One gets that N_t is equal to a constant times t^n, where n is the dimension.

There are versions of this which apply to hyperbolic geometry as well. Instead of a square as a basic building block, one has regions with more sides, and they fill up a hyperbolic space instead of a Euclidean space. One can make computations like the ones above, and get exponential behavior for N_t (as t tends to infinity), because of the exponential rate of growth of volume in hyperbolic space. The exponential growth can also be viewed in another way, concerning the manner in which the copies of the basic building blocks fit together, with more branching involved than with squares in the previous case.

To fit with the previous discussion, one could start with a compact Riemannian manifold which is hyperbolic, which means that it has constant curvature

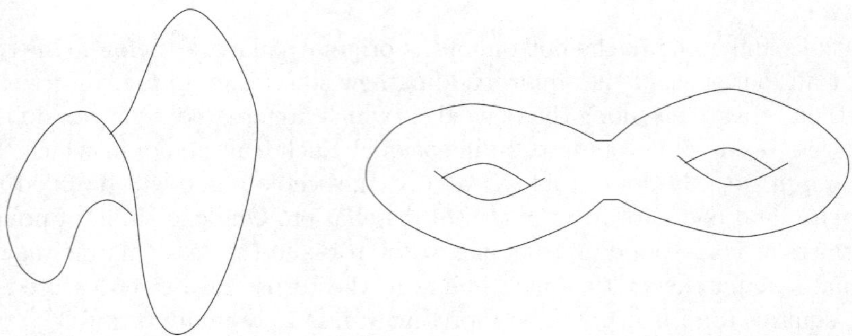

FIG. 4.8. The left side shows how a negatively-curved 2-dimensional surface would look inside $\mathbf{R}^3$, and the right side shows a two-handled torus. The two-handled torus admits Riemannian metrics with negative curvature, and even constant negative curvature, but this does not come from an embedding into $\mathbf{R}^3$, as in the picture.

equal to -1. This would play the role of the flat torus before. Similar considerations apply (at least in part) more generally to compact manifolds whose curvature is negative, but not necessarily constant. For these there is still the same kind of behavior of exponential growth, even if one might not get as precise information, or formulae. We shall say more related to these matters in a moment, and in Section 4.11.

To get a compact Riemannian manifold with negative curvature in dimension 2, one can start with an ordinary torus and add a handle to it. Such a two-handled torus is shown in the right side of Fig. 4.8. This gives the right *topology* for the manifold, but not the right *geometry*. In order to have negative curvature, the surface should look locally like the picture on the left side of Fig. 4.8.

In the saddle-shaped surface on the left side of Fig. 4.8, one can take slices of the surface with 2-dimensional planes and get curves roughly as shown in Fig. 4.9. There are some natural orthogonal axes to take, around any point in the surface, where this kind of picture is optimized in a certain way at that point; these are axes in the directions associated to the two principal curvatures of the surface at the point, together with an axis normal to the surface at the point. The negative curvature of the surface is manifested in the way that one of the curves in our slices points up, while the other one points down. For a surface of positive curvature, like a piece of a standard 2-dimensional sphere, the two slices would point in the same general direction. The direction in which the slices are going does not matter, and one could simply rotate the whole picture anyway. The more important matter is whether the slices both point in the same general direction, or in opposite directions.

It turns out that there are metrics on the two-handled torus which have constant negative curvature. This is not so obvious from looking at embeddings into $\mathbf{R}^3$, and indeed metrics of negative curvature on the two-handled torus, or on

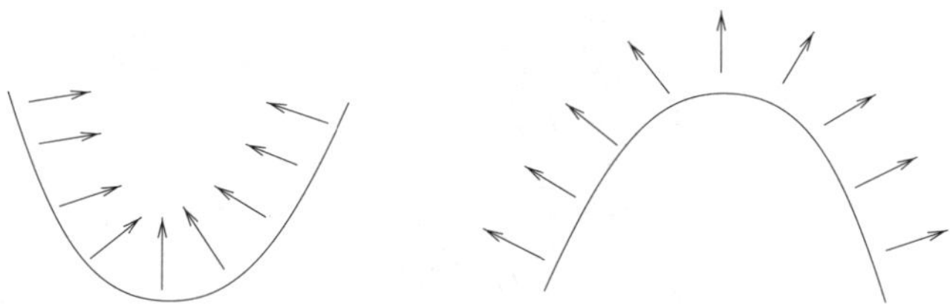

FIG. 4.9. In this picture, imagine the curves as slices of the surface on the left side of Fig. 4.8, in different directions.

any other compact surface (without boundary), cannot arise from an embedding into $\mathbf{R}^3$. Here is a classical argument for proving this. Fix a point p in $\mathbf{R}^3$, and let q be a point on the surface whose distance from p is maximal. Such a point always exists, if the surface is compact. On the other hand, around q, the surface cannot look locally like the one on the left side of Fig. 4.8. This is because the distance from q to p is maximal among all points on the surface. If the surface looks locally like the one on the left side of Fig. 4.8, then there would be a way to move along the surface in a direction away from p, so that the distance to p would become larger. To put it another way, with this choice of q, the surface would have to lie in a closed ball whose boundary sphere passes through q, namely the ball centered at p and with radius $|q - p|$, and this is incompatible with negative curvature of the surface at q. This is analogous to the second-derivative test for maxima and minima in calculus, i.e., the matrix of second derivatives of a function should be nonpositive at a maximum (or local maximum) of the function.

This is similar to the situation earlier, with the flat torus. The *topology* of the torus can be realized by a surface in $\mathbf{R}^3$ in a standard way, as the boundary of a donut, or as an inner tube, but this does not give a metric of curvature 0.

At any rate, metrics of negative curvature on the two-handled torus do exist, and this would continue to work if one added more handles. These metrics can be realized by embeddings into Euclidean spaces of higher dimension. With the local picture on the left side of Fig. 4.8, one can already see important features about the way that geodesics behave in a Riemannian manifold of negative curvature. Namely, they spread apart from each other, at least locally; there may be crossings globally, as on a two-handled torus, from the geodesics wrapping around the manifold. This is quite different from what happens in a space of positive curvature, like a round sphere, where they tend to come back together, or from spaces of curvature 0, like the flat torus, or ordinary Euclidean spaces. In spaces with curvature 0, the geodesics keep apart (at least locally, again, without including crossings that might occur globally, as on a torus), but there is not the kind of increasing spreading that takes place in spaces of negative curvature. The increasing spreading leads to exponential growth in quantities like N_t.

To make another comparison between the case of a flat torus and what happens for surfaces of negative curvature, let us consider the picture shown in Fig.

FIG. 4.10. Repeating squares in a different way, with exponential growth, rather than tiling a 2-dimensional Euclidean plane

4.10. Imagine that each of the boxes in the picture is an isometric copy of a single standard square, and that all of the angles are of 90°.

The repeating of the squares here is similar to what we did before, in "unwrapping" the flat torus. In that situation we also repeated copies of a single square, but we did this inside of a plane. This was suited to the circumstances, with the unwrapping of geodesics into straight lines in the plane in particular. Here we allow more repeating of the squares, roughly speaking, with more squares going around a corner of a square, i.e., 6 instead of 4. This leads to exponential expansion which is similar to what occurs in the case of negative curvature. In fact, one can consider this as a different kind of negative curvature, which is manifested in the increased total angle around the corners.

The unwrapping of the flat torus to get a Euclidean plane that we did before can be seen as an instance of a general construction, which we discuss in the next section. This construction is called the *universal covering* of a topological space, and it is also somewhat similar to the visibility of a graph. It applies without a choice of a Riemannian structure on a manifold, but if one has such a structure, then it leads in a natural way to a Riemannian structure on the universal covering as well.

The analogy between exponential mappings and the visibility works in a slightly different way in the context of *Lorentzian* geometry instead of Riemannian geometry. Roughly speaking, Lorentzian manifolds provide a model for the geometry of *space-time*, in the same way that Riemannian manifolds provide a model for the geometry of space. Mathematically this means that the field of quadratic forms used to make measurements are negative in a "time-like" direction and positive in the complementary number of "space-like" directions (rather than being positive in all directions, as with a Riemannian metric). This leads to natural local partial-orderings on the manifold, corresponding to paths which are future-oriented and time-like, i.e., which do not try to go backwards in time or faster than the speed of light. Some references for Lorentzian geometry include [HE73, O'N83].

4.11 Universal covering spaces

Let M be a compact topological space, and fix a basepoint $x \in M$. For simplicity, we shall restrict our attention to topological spaces which are manifolds, although this is not important (or can be weakened) for many of the basic concepts and facts. To be a manifold means that every point in the space has a neighborhood which is homeomorphic to an open set in an ordinary Euclidean space. Spheres and tori provide basic examples. The requirement that M be compact is not needed for the basic notions, but it will be important for some other parts.

To define the *universal covering space* of M (with basepoint x), one begins by taking $\mathcal{P}_x(M)$ to be the space of all continuous paths p in M which begin at x. This is a huge infinite-dimensional space, but we can reduce its size by taking a quotient of it. To do this, we define a relation $\sim$ on $\mathcal{P}_x(M)$ by saying that $p_1 \sim p_2$ when p_1 and p_2 end at the same point in M, and admit a continuous deformation from one to the other which keeps both endpoints fixed. In other words, p_1 and p_2 should be *homotopically equivalent*. See Fig. 4.11 for some examples. This defines an equivalence relation on $\mathcal{P}_x(M)$, as one can show, and the universal covering $\widetilde{M}$ of M (based at x) is defined to be the set of equivalence classes in $\mathcal{P}_x(M)$ corresponding to this equivalence relation. See Fig. 4.12 for the case where M is a circle.

Let us assume that M is connected, so that every pair of points in M can be connected by an arc. In this case, the universal covering spaces associated to different choices of the basepoint x will be isomorphic to each other, in a way that has natural properties. This also ensures that the universal covering involves all of M. Otherwise, one only deals with a component of M (containing the basepoint).

The universal covering space is similar to the visibility of a graph, except that we use *all* paths in the space, at least initially. In general, something like the restriction to oriented paths does not make sense, without additional structure. In the context of Riemannian manifolds, as in the previous section, we dealt with this in a different way, by restricting ourselves to *geodesic* arcs.

The universal covering space, like the visibility of a graph, reflects some of the internal activity in a given space. We shall discuss this further in a moment, after

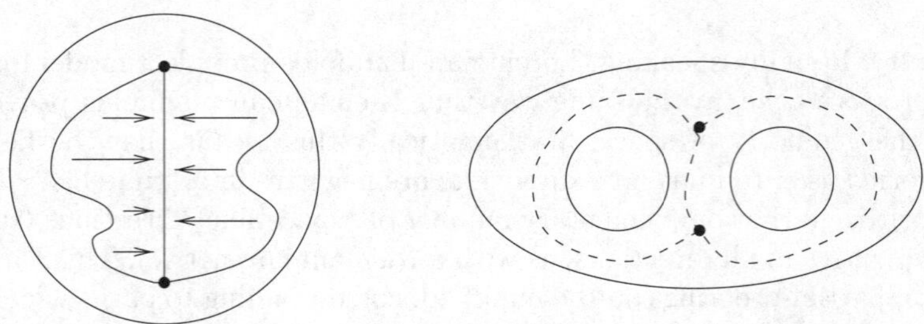

FIG. 4.11. In these pictures, one has three paths between a pair of points in a region in the plane. In the picture on the left, the region is a disk, and each of the three paths can be deformed to each other (while keeping the endpoints fixed and staying in the disk). In the picture on the right, the region has two holes, and none of the paths can be deformed to the other (inside the region), because they cannot get around the holes.

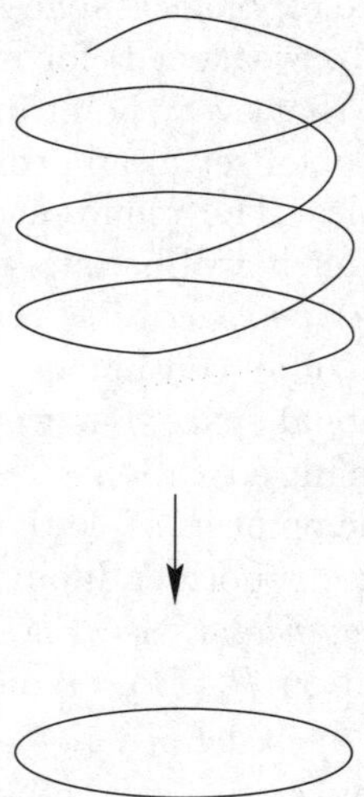

FIG. 4.12. The universal covering of a circle (shown with the canonical projection back down to the circle)

describing some examples, and reviewing some more about the basic notions. See [AS60, Mas91] for more information and details about covering spaces.

To understand better the equivalence relation defined above, it is helpful to think about the case where M is a circle. In this situation, the existence of a continuous deformation between two paths with the endpoints held fixed turns out to depend on (and only on) whether the two paths wrap around the circle the same number of times. In counting the number of times a path wraps around the circle, one counts both positive wrappings and negative wrappings, and one counts them positively and negatively, respectively. In other words, if a path goes all the way around once in the positive orientation, and then backs up, and goes all the way back around once in the negative orientation, then the total wrapping that would be counted is 0. That this total "wrapping number" (or "winding number") determines exactly when there is a continuous deformation between two paths is a basic fact from topology.

By contrast, any pair of paths in the 2-sphere with the same endpoints can be continuously deformed into each other, while keeping the endpoints fixed. This is true for spheres of any dimension larger than 1, and is another fact from topology. Basically, one can reduce to the case where the curves avoid a single point; then they lie in a region which is homeomorphic to a plane of the the same dimension, where the deformation can be made in a simple way (by taking linear combinations of the paths). It is the first step that causes trouble when the dimension is 1.

For these spaces, the passage to the universal covering does not really bring anything new, as it does for the circle, for which the universal covering is homeomorphic to a line. One can look at the latter in terms of the picture in Fig. 4.12, and unwinding the infinite spiral in the top part of the picture into a line.

A 2-dimensional torus is the same topologically as the Cartesian product of two circles. In this case, paths can wrap around the two circular directions independently of each other, and one can keep track of two different winding numbers accordingly. The universal covering space can be identified with a plane, and this can be derived from the corresponding statement for a circle. This also ends up being essentially the same as the "unwrapping" of a 2-dimensional torus discussed in Section 4.10.

A basic feature of the universal covering space $\widetilde{M}$, which is implicitly used in the preceding paragraphs, is that it comes with a natural topology. This topology can be described as follows. Let α be a point in $\widetilde{M}$. By definition, α is an equivalence class of paths in M which begin at our basepoint x and end at some other point in M. Let p be one of those paths. (It does not matter which path representing α that one chooses.) A point β in $\widetilde{M}$ will be considered "close" to α if it can be represented by a path q in M which consists of the path p together with a small arc added at the end. This is the same as saying that any path q' which represents β can be deformed to a path of this form. One can make this more formal, by defining a system of neighborhoods around α in $\widetilde{M}$. Specifically, if y is the endpoint of p, and if U is a neighborhood of y in M, then one can define a neighborhood V of α in $\widetilde{M}$ associated to U by taking the set of all points β in $\widetilde{M}$ which can be represented by a path q which consists of the path p together with a path which starts at the endpoint y of p and remains inside of U.

This defines a topology for $\widetilde{M}$. There is also a *canonical projection* $\Pi : \widetilde{M} \to M$, defined as follows. Let α be a point in $\widetilde{M}$, and let p be a path in M which represents α. If the endpoint of p is $y \in M$, then one sets $\Pi(\alpha) = y$. This endpoint y is the same for all paths in the equivalence class specified by α, by the definition of the equivalence relation.

It is not hard to see that $\Pi : \widetilde{M} \to M$ is continuous with respect to the topology on $\widetilde{M}$ described above. It is even a local homeomorphism, which means that for each point α in $\widetilde{M}$ there is a neighborhood V of α in $\widetilde{M}$ which is mapped homeomorphically onto an open subset of M by Π. To get this, one can choose V so that it is associated to an open set U in M containing $\Pi(\alpha)$ in the same manner as before, where U is homeomorphic to a standard open ball. We are assuming that M is a manifold, which means exactly that every point in M has a neighborhood which is homeomorphic to an open ball in a Euclidean space of a fixed dimension. If one chooses V in this way, then the restriction of Π to V will be a homeomorphism onto U. This is not hard to show, using the fact that all paths between a given pair of points in an open ball (and hence in U) can be continuously deformed into each other, while keeping the endpoints fixed. This is similar to (and, in fact, equivalent to) the analogous statement for curves in a plane, which came up in the discussion of spheres earlier.

As an example, let us consider the case where M is a circle. We can think of M as being the same as the unit interval $[0, 1]$ with its endpoints identified with each other, or, equivalently, as being the same as the real numbers modulo 1. In this case, the canonical projection from $\widetilde{M}$ to M is topologically equivalent to the standard quotient mapping from $\mathbf{R}$ to $\mathbf{R}/\mathbf{Z}$, i.e., to the mapping which takes a real number and keeps only its fractional part.

One can do something similar for a 2-dimensional torus, in which the projection from $\widetilde{M}$ to M is topologically equivalent to the standard quotient mapping from $\mathbf{R}^2$ to $\mathbf{R}^2/\mathbf{Z}^2$. As before, this is very similar to the discussion of "unwrapping" a torus in Section 4.10.

Now let us look at a natural way to talk about the "rate of growth" in a universal covering space. Roughly speaking, one can do this by looking at the number of "sheets" in $\widetilde{M}$ which lie within a ball of a given radius. To make this precise, we shall assume for simplicity that M is a smooth manifold equipped with a Riemannian metric, but this is not really needed, and one could work with much less. Remember that we are assuming that M is compact, which will be important in some of the steps below.

Under these conditions, one can define a Riemannian structure on the universal covering space $\widetilde{M}$, through the requirement that the canonical projection $\Pi : \widetilde{M} \to M$ preserve the Riemannian structure. In other words, local measurements of length and volume would be the same in $\widetilde{M}$ and M, through the mapping Π. One can do this because Riemannian metrics are defined in a purely local way. Using this Riemannian structure on $\widetilde{M}$, one can define the distance between two points in $\widetilde{M}$ to be the minimal length of curves between them. If M is a flat torus, for instance, this leads to the usual Euclidean geometry on a plane (as the universal covering space).

Fix a basepoint α in $\widetilde{M}$, and let $B_{\widetilde{M}}(\alpha, t)$ be the open ball in $\widetilde{M}$ with center α and radius t. This ball in $\widetilde{M}$ is defined in terms of the distance function on $\widetilde{M}$ just mentioned. Let $V(t)$ denote the volume of this ball, as defined through the

Riemannian metric. The growth of $V(t)$ as $t \to \infty$ provides a measurement of the complexity of the universal covering of M, and reflects some of the internal structure in M.

The rate of growth of $V(t)$ does not depend too strongly on the choice of basepoint α, nor on the choice of Riemannian metric on M. It is not difficult to write down inequalities along these lines. To within certain bounds, $V(t)$ is determined by the topology of M, through the fundamental group of M. The *fundamental group* of M can be defined as follows. One begins by taking the set of equivalence classes β of curves in M which both begin and end at our basepoint x in M. For these equivalence classes of loops, there is a natural group operation, which comes from combining two loops into a single curve by tracing out one loop and then the other. The inverse operation on the set of equivalence classes comes from taking a loop and replacing it with the one which is the same except that the parameterization goes backward. It is not too hard to show that these operations are well-defined on the set of equivalence classes of loops based at x, and that they define a group.

For a compact manifold, the fundamental group is always finitely generated.

If β is an equivalence class of paths in M from the basepoint x to some other point $z \in M$, then one can get an equivalence class β' of loops in M which is reasonably close to β. Namely, one can add a curve in M from z to x to the ends of the curves in the equivalence class β, and take β' to be the equivalence class which contains the curves that result from this. These curves will all lie in the same equivalence class, as one can check. A key point is that the curves in the equivalence class β might wrap around M a lot, but one can take the path from z to x which is added to them to be quite direct. One can take this path to have length which is as small as possible, and less than or equal to the diameter of M, in particular. Note that the diameter is finite, since M is compact. Because of this, β' will be reasonably close to β.

Using considerations like these, one can show that $V(t)$ behaves roughly like a combinatorial "volume function" for the fundamental group, to which we shall return in Chapter 17. The distinction between polynomial and exponential growth of the volume function depends only on the fundamental group, for instance.

This volume function $V(t)$ is closely related to the measurements concerning the exponential mapping that were discussed in Section 4.10. Before we look at this, let us make some normalizing assumptions. Remember that x is our basepoint in M, for the definition of the universal covering space. We shall also use x as the point at which one bases the exponential mapping into M, as in Section 4.10. There is a natural choice of basepoint in $\widetilde{M}$ corresponding to x, namely, the element of $\widetilde{M}$ that represents the equivalence class of paths in M which begin and end at x and which are equivalent to the constant path, i.e., the path that does not move. From now on, let us assume that the basepoint α in $\widetilde{M}$ which is used in the definition of $V(t)$ above is this point in $\widetilde{M}$ that corresponds to the constant path at x. Note that $\Pi(\alpha) = x$.

Let $N_t(z)$ and N_t be as in (4.12) and (4.13) in Section 4.10. Under the normalizations mentioned above, we have that

$$N_t \cdot \mathrm{Vol}(M) \geq V(t) \qquad \text{for all } t > 0, \tag{4.15}$$

where $\mathrm{Vol}(M)$ denotes the volume of M with respect to our Riemannian metric. The volume is finite, since M is compact. Because N_t is defined to be the average of $N_t(z)$ over M, $N_t \cdot \mathrm{Vol}(M)$ is the same as the integral of $N_t(z)$ over M.

Let us sketch a proof of (4.15). We begin with the following result. Let z be an element of M, and let β be an equivalence class of paths in M which go from x to z. This is the same as saying that β is an element of $\widetilde{M}$ such that $\Pi(\beta) = z$. The result is that this class of curves includes at least one geodesic from x to z. This is a well-known theorem, and it is proved by finding a path in the equivalence class which has minimal length, among all paths in the equivalence class. Such a minimizer is automatically a geodesic, but one has to be careful about why the minimizer exists. In part this uses our assumption that M be compact.

Thus, every point β in $\widetilde{M}$ leads to at least one geodesic in M. The next main point is that if β lies in $B_{\widetilde{M}}(\alpha, t)$, then this geodesic can be taken to have length less than t. Indeed, if β lies in $B_{\widetilde{M}}(\alpha, t)$, then it means that the distance from α to β in $\widetilde{M}$ is less than t. This is the same as saying that there is a curve in $\widetilde{M}$ which goes from α to β and has length less than t. We can project this curve to one in M using Π, and this new curve will begin at x, end at z, and have length less that t. In fact, the length of the new curve in M is the same as the length of the original curve in $\widetilde{M}$; this is because of the way that we chose the Riemannian structure on $\widetilde{M}$, so that it would be preserved under the mapping $\Pi : \widetilde{M} \to M$.

This new curve in M goes from x to z, and we would like to say that it lies in the equivalence class of curves in M which is determined by β. One way to prove this is as follows. Instead of considering this statement just for the curve in $\widetilde{M}$ from α to β and its projection into M, one can consider the analogous statements for each of the initial subpaths of this curve as well. It is easy to check that the analogous statements hold for the initial subpaths which stay close to the initial point α. Basically this is because one can work in a small neighborhood which is homeomorphic to a ball, where it is easy to make deformations between paths. If one knows that the statement holds for some given initial subpath, then one can show that the statement continues to hold for initial subpaths which go a bit further. This uses arguments similar to ones employed for the very beginning of the curve; one can work mostly in a small neighborhood of the endpoint of the given initial subpath, with small curves being added to that initial subpath in that neighborhood. One can choose this neighborhood to be homeomorphic to a ball, so that it is easy to make deformations inside of it.

In this fashion, one can show that the statement in which we are interested works for initial subpaths near the beginning of our curve in $\widetilde{M}$, and that once it works for some initial subpath, it keeps working for a while longer, if one has not already reached the end. Using these observations, one can show that the statement works for all of the initial subpaths of our curve, including the original curve itself.

To summarize, we get that if β is a point in $\widetilde{M}$ which lies in $B_{\widetilde{M}}(\alpha, t)$, then there is a curve in $\widetilde{M}$ that goes from α to β and has length less than t, and the projection of this curve in M gives a curve which goes from x to $\Pi(\beta)$, has length less than t, and lies in the equivalence class of curves in M defined by β. As indicated in the previous step in this discussion, this equivalence class β also contains a curve from x to $\Pi(\beta)$ which is a geodesic, and whose length is as small as possible. Hence the length is less than t. In short, each $\beta \in B_{\widetilde{M}}(\alpha, t)$ leads to at least one geodesic in M which starts at x and has length less than t. Different choices of β lead to different geodesics, because the geodesic lies in the equivalence class determined by β, by construction.

Let us reformulate this as follows. Given a point z in M, define $P_t(z)$ by

$$P_t(z) = \text{the number of } \beta\text{'s in } B_{\widetilde{M}}(\alpha, t) \text{ such that } \Pi(\beta) = z. \tag{4.16}$$

Then

$$N_t(z) \geq P_t(z) \qquad \text{for all } z \in M \text{ and } t > 0. \tag{4.17}$$

Remember that $N_t(z)$ counts the number of geodesics in M which begin at x, end at z, and have length less than t, as in (4.14) in Section 4.10. The discussion above implies that there is at least one such geodesic associated to each $\beta \in B_{\widetilde{M}}(\alpha, t)$ which satisfies $\Pi(\beta) = z$. This gives (4.17).

To get (4.15), one integrates (4.17) over $z \in M$. The integral of $P_t(z)$ gives $V(t)$; this is a kind of "change of variables", using the mapping $\Pi : \widetilde{M} \to M$. Locally, Π preserves measure between $\widetilde{M}$ and M, because of the way that we chose the Riemannian structure on $\widetilde{M}$. Globally, one should take the multiplicities into account, since different points in $\widetilde{M}$ can be mapped to the same point in M, and $P_t(z)$ exactly counts the relevant multiplicities.

If we define P_t to be the average of $P_t(z)$ over M, just as N_t is the average of $N_t(z)$ over M, then we can write the integral inequality as

$$N_t \geq P_t \qquad \text{for all } t > 0. \tag{4.18}$$

At any rate, this completes our sketch of the proof of (4.15). We shall say more about what this inequality means a little later, but first let us mention a refinement of it. If the Riemannian metric on M has nonpositive curvature (everywhere on M), then

$$N_t \cdot \text{Vol}(M) = V(t) \qquad \text{for all } t > 0. \tag{4.19}$$

That is, we have equality in (4.15), and, in fact, in the other inequalities above as well.

Indeed, if M has nonpositive curvature, then it is a basic theorem in Riemannian geometry that for each point z in M and each equivalence class of curves in M that go from x to z there is a *unique* geodesic in that class. We shall say a bit more about this below. This does not work for arbitrary Riemannian manifolds, as in the case of standard spheres.

Once one has this uniqueness statement, one can basically reverse all of the previous inequalities, and get equalities. The main point is to obtain that

$$N_t(z) = P_t(z) \qquad \text{for all } z \in M \text{ and } t > 0, \tag{4.20}$$

as a strengthening of (4.17). To be precise, one argues as follows. Let $z \in M$ and $t > 0$ be given, and suppose that we have a geodesic path p from x to z in M of length less than t. Let β denote the point in $\widetilde{M}$ which gives the equivalence class of paths in M that contains p. The uniqueness assertion tells us that p is the only geodesic in this equivalence class. The main remaining point is that β actually lies in $B_{\widetilde{M}}(\alpha, t)$. In other words, we want to say that there is a path in $\widetilde{M}$ which goes from α to β and has length less than t. To obtain this, one would like to "lift" the path p in M to one in $\widetilde{M}$ from α to β, where the lifted path has the property that its projection to M by Π is the same as p. With our definition of $\widetilde{M}$, one can get this lifting in an automatic manner, and this would work for any path in M which begins at x. Namely, each initial subpath of p is a curve in M starting at x which is contained in some equivalence class of curves, and hence each initial subpath of p leads to a point in $\widetilde{M}$. The family of these subpaths then gives rise to a curve of points in $\widetilde{M}$, and this is the curve that we want. This lifted curve in $\widetilde{M}$ has the same length as the original curve in M, because of the way that we chose the Riemannian metric on $\widetilde{M}$.

This shows that β does lie in $B_{\widetilde{M}}(\alpha, t)$ under the conditions above. Thus every geodesic in M that begins at x and has length less than t leads to a point β in $B_{\widetilde{M}}(\alpha, t)$, and the property of uniqueness implies that different geodesics are associated to different β's. From this one obtains that $N_t(z) \leq P_t(z)$, which is what we needed to establish (4.20), since we already have (4.17).

Given (4.20), one can obtain (4.19) in the same way that (4.15) was before, from (4.17). Similarly,

$$N_t = P_t \qquad \text{for all } t > 0. \tag{4.21}$$

This follows from (4.20), since N_t and P_t are the averages of $N_t(z)$ and $P_t(z)$ over M, by definition.

The uniqueness of geodesics in the equivalence classes when the curvature on M is nonpositive is often given through the following result of Hadamard and Cartan: if M has nonpositive curvature at every point, then the tangent space $T_x M$ and the exponential mapping from $T_x M$ to M give a topologically-equivalent realization of the universal covering of M and the mapping Π from it to M. Note that one should be careful about the geometry on the universal covering space; $T_x M$ is a vector space and comes with a natural flat geometry, and this will not be the same as the geometry that we have chosen on the universal covering space, unless the original Riemannian metric on M has curvature equal to 0 everywhere. There are some natural relationships between the geometry on $T_x M$ and the one on the universal covering space, though.

Let us look at some examples. If M is a standard sphere $\mathbf{S}^n$, with its "round" metric, and if $n \geq 2$, then $\mathbf{S}^n$ is simply-connected, i.e., all paths with the same endpoints can be continuously deformed to each other. This was mentioned earlier in the section. In this case, $\widetilde{M}$ is essentially the same as M. In particular, the volume function $V(t)$ remains bounded for all t, with $V(t) \leq \mathrm{Vol}(M)$ for all t. By contrast, $N(t)$ grows linearly in t as $t \to \infty$, as indicated in Section 4.10.

Now suppose that M is the 2-dimensional flat torus described in Section 4.10. The curvature is equal to 0 everywhere in this case, so that the formula (4.19) applies. In effect, the discussion in Section 4.10 gave a derivation of this formula for this concrete example. The constructions in this section can be viewed as giving general versions of the same basic procedures. In Section 4.10, we also obtained (in effect) that the universal covering of the flat torus, with a Riemannian metric inherited from the flat torus as above, is equivalent to a 2-dimensional Euclidean plane (with its usual geometry). This lead to the fact that N_t is equal to a constant times t^2, since the volume function is equal to a constant times t^2.

Similar statements hold for flat tori in general, and for arbitrary dimensions. The case of $n = 1$ corresponds to a circle, for which the universal covering space is equivalent to the real line. In n dimensions, N_t and the volume function $V(t)$ are equal to constant multiples of t^n.

Now let us consider the case where M is a two-handled torus. As a topological space (or a smooth manifold), the two-handled torus can be given as in the picture on the right-hand side of Fig. 4.8 in Section 4.10. This realization is as a 2-dimensional surface embedded in $\mathbf{R}^3$.

As in Section 4.10, there are Riemannian metrics on this manifold which have negative curvature, and even constant negative curvature, but they do not arise from embeddings into $\mathbf{R}^3$. When the curvature is constant, there is a concrete realization of the universal covering space of the manifold, as 2-dimensional hyperbolic space. One can do many of the same things in this situation as for the flat torus in Section 4.10; in particular, there are concrete and simple descriptions of the geodesics the 2-dimensional hyperbolic space, and explicit formulae for the volume of balls, which lead to an explicit formula for N_t, just as in the case of the flat torus. The geometry is different in this situation, but it can still be given explicitly.

Let us take a slightly different view of this, using the universal covering. Suppose that we start from a metric that comes from an embedding of the two-handled torus in $\mathbf{R}^3$ as in the right side of Fig. 4.8. That is, the metric is induced from the one in $\mathbf{R}^3$, with the corresponding notions of length and area.

This metric does not have negative curvature at all points, but it is easy to think about in concrete visual terms. One can also look at the way that paths wrap around in this surface, with the possibility of wrapping around the two handles independently of each other.

A basic point is that the volume function $V(t)$ grows exponentially in t as $t \to \infty$. This comes down to the topology, and the way that curves can wrap around the surface. There is substantial *noncommutativity* in this, which can be expressed in terms of the fundamental group of the surface, and this leads to exponential growth in $V(t)$. In an n-dimensional torus, this kind of noncommutativity does not take place, and the fundamental group is isomorphic to $\mathbf{Z}^n$.

It is nice to look at this concretely for a metric coming from an embedding of the manifold as in Fig. 4.8. On the other hand, and as indicated earlier in the section, a change in the Riemannian metric on the manifold does not change the behavior of $V(t)$ too strongly, since the manifold is compact. Given any two Riemannian metrics on this space (or any other compact smooth manifold), each is bounded by a constant times the other, because the space is compact. This leads to some bounds between the corresponding $V(t)$'s. In particular, exponential growth for one implies exponential growth for the other.

In other words, setting aside the issue of special metrics on the two-handled torus, one can see the presence of exponential effects in simpler terms. For arbitrary Riemannian metrics on the two-handled torus, whether or not they have negative curvature everywhere, notice that one has exponential growth for the number of geodesics, because of the lower bounds like (4.15).

Let us come back for a moment to the function $P_t(z)$, defined in (4.16). This function is not quite constant in z, but the dependence on $z \in M$ is pretty mild. To this end, let z' be another point in M, and let q be a path in M which goes from z to z'. If p is any path in M which goes from x to z, then one can convert it into a path that goes from x to z' by adding q to the end of it. If two paths from x to z lie in the same equivalence class (so that there is a continuous deformation between them, while keeping the endpoints fixed), then the same is true after one adds q to the ends of the paths.

Each point $\beta \in \widetilde{M}$ corresponds to an equivalence class of curves in M, and when $\Pi(\beta) = z$, these curves go from x to z. The operation just described leads to a way to convert points $\beta \in \widetilde{M}$ such that $\Pi(\beta) = z$ to points $\beta' \in \widetilde{M}$ such that $\Pi(\beta') = z'$.

This operation can be reversed. Specifically, let q' denote the path in M which goes from z' to z, and which is the same as q, except that it goes backwards. Using q', one gets an operation which takes points $\beta' \in \widetilde{M}$ such that $\Pi(\beta') = z'$ and converts them into points $\beta \in \widetilde{M}$ such that $\Pi(\beta) = z$, by adding q' to the ends of paths as above. This operation is the inverse of the previous one. To see this, let p be a path in M that goes from x to z again, and consider the path p_1 that one gets if one first adds q to the end of p, and then adds q' to the end of that path. This new path p_1 is not literally the same as p, but they do lie in the same equivalence class. More recisely, one can make a deformation from p to p_1, through paths which go from x to z, by looking at paths of the following type. One starts with the path p, and then adds to the end of it a path that follows q part of the way towards z', and then turns around and comes back to z exactly the same way that it left, along q. This family of paths gives a continuous deformation between p and p_1 in which the endpoints x and z are preserved.

Thus, when one adds q to a path from x to z, and then adds q' to that path, one does not change the equivalence class of the original path. A similar statement applies to paths from x to z', and the operation on them which adds q' to the end of the path, and then q. These statements imply that the transformations from $\{\beta \in \widetilde{M} : \Pi(\beta) = z\}$ to $\{\beta' \in \widetilde{M} : \Pi(\beta) = z'\}$ and back again, coming from these operations on paths, are inverses of each other.

These transformations are useful for looking at the counting function $P_t(z)$ and its dependence on z, but there is an extra ingredient involved. Suppose that we start with a point $\beta \in \widetilde{M}$ such that $\Pi(\beta) = z$, and transform it in this way to a point $\beta' \in \widetilde{M}$ such that $\Pi(\beta') = z'$. What happens to the distance from these points to the basepoint α in $\widetilde{M}$? Basically, the addition of the path q that takes place in this transformation leads to a path in $\widetilde{M}$ that goes from β to β', and which is a kind of lifting of q. This is similar to the lifting of paths which takes place in the sketch of the proof of (4.20) above. (For (4.20), one was assuming that the manifold had nonpositive curvature, but this part of the argument did not use that, and works in general.) The situation is slightly different here, but one can make a lifting of q to a path from β to β' in $\widetilde{M}$ in a manner analogous to the one before, and the length of the lifted path will be the same as the length of q in M. The latter is also similar to what happened before, and it comes from the way that we chose the Riemannian metric for $\widetilde{M}$.

For the present purposes, one may as well choose q so that its length is as small as possible, among all paths in M that go from z to z'. In particular, one can choose it so that its length is bounded by the diameter of M. The diameter of M is finite, since M is compact, and this gives an upper bound for the length of q which does not depend on z, z', or t.

Thus, given $\beta \in \widetilde{M}$ with $\Pi(\beta) = z$, one can make this kind of transformation to get a point $\beta' \in \widetilde{M}$ such that $\Pi(\beta') = z'$ and the distance from β to β' in $\widetilde{M}$ is bounded by the diameter of M. The distance from β' to α is therefore bounded by the sum of the distance from β to α and the diameter of M, and there is a similar inequality with the roles of β and β' reversed.

Using this, one can make comparisons between $P_t(z)$ and analogous quantities for z'. At first one would make shifts in t by a bounded amount, to compensate for the change in z, and the resulting changes in distance to α in $\widetilde{M}$. One can make other estimates from these, which avoid this, but which can be less precise in other ways (i.e., with constant factors). At any rate, the basic point is that quantities like $P_t(z)$ are pretty stable. Because of this, the average behavior for $P_t(z)$ in z, which is reflected in the volume function $V(t)$, is approximately the same as the behavior of $P_t(z)$ for individual z's.

The universal covering space and the canonical projection from it to the original space are similar in spirit to the visibility of an oriented graph and the canonical projection from it to the original graph. This analogy provides a helpful guide for some of our later discussions about mappings between graphs and their role in making comparisons between different structures, as in Chapters 10 and 11.

Universal covering spaces are often associated to much higher levels of computational complexity than visibility graphs, however. Indeed, the universal covering of a given manifold is closely connected to its fundamental group, which can be an arbitrary finitely-presented group, at least when the manifold has dimension ≥ 4. There are finitely-presented groups for which the word problem (of recognizing when a given word represents the identity element of the group) is algorithmically unsolvable, as in [Man77]. In terms of geometry, this corresponds to taking a pair of paths in M with the same endpoints, and asking whether they are homotopically equivalent, i.e., whether they determine the same element in the universal covering space. (One can just as well work with spaces which are finite polyhedra and paths which are piecewise-linear to make these geometric problems purely combinatorial ones.) By contrast, visibility graphs tend to be associated to computational problems with at most a (linear) exponential degree of complexity. In Chapter 13, for instance, we shall describe an NP-complete problem about mappings between oriented graphs and their induced mappings between the associated visibility graphs.

There are some natural analogies between the kind of unwinding of topological spaces which occurs with universal covering spaces and the kind of unwinding of formal proofs which occurs with cut elimination. In the case of propositional proofs, the level of complexity entailed is on the order of one exponential, while for proofs with quantifiers, the complexity can be much stronger (nonelementary). The propositional case also has a number of aspects which are like graphs and their visibilities. We shall discuss these matters further in Chapter 6.

One feature of cut elimination for formal proofs is that it unwinds oriented cycles in the logical flow graph in an appropriate sense. There is some extra structure involved in this, and one might say that the elimination of cuts forces cycles to be like straight lines; on the other hand, there are different processes that one can use, given in [Car99], in which cycles are turned into "spirals". This is not as strong as what one gets from cut elimination, but the amount of complexity which is needed is much less in general. In particular, the amount of expansion is elementary, with double-exponential increase in the number of lines (as compared to the nonelementary expansion that occurs with cut elimination). See [Car99] for more information.

In general, in working with formal proofs and structures related to them, one might keep in mind the way that different geometric aspects can correspond or be connected to different types of complexity.

4.12 Boolean circuits and expressions

By a *Boolean function* we mean a function f of some collection of (Boolean) variables $x_1, \ldots, x_n$, where both the function and the variables take values in $\{0, 1\}$. Every Boolean function can be represented as a *Boolean expression*, which means a combination of the Boolean variables using the connectives $\wedge, \vee, \neg$ of conjunction, disjunction, and negation. This is very well known.

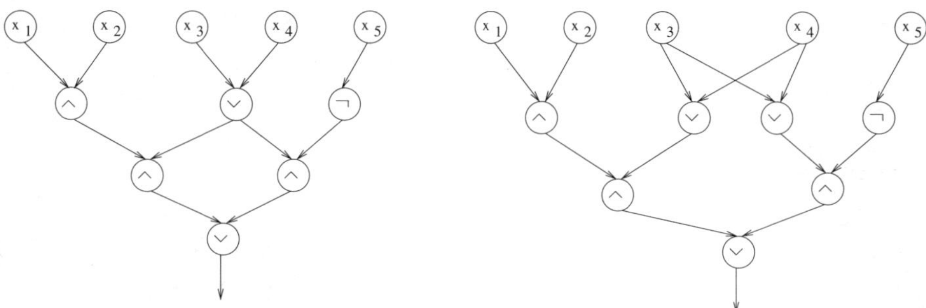

FIG. 4.13. A pair of Boolean circuits, both of which represent the expression $((x_1 \wedge x_2) \wedge (x_3 \vee x_4)) \vee ((x_3 \vee x_4) \wedge \neg x_5)$. In the circuit on the left, there are two edges coming out of one of the $\vee$-vertices, and this corresponds to the duplication of $x_3 \vee x_4$ in the expression being represented. On the right-hand side, multiple outgoing edges only occur at the input vertices, and this makes it necessary to duplicate the $\vee$-vertex which had two outgoing edges before.

There is another way to represent Boolean functions, through *Boolean circuits*, which can be described as follows. (See Fig. 4.13 for an example.) One starts with an oriented graph G, which is free of nontrivial oriented cycles. (This includes simple loops.) Each vertex in the graph should be marked with a label which is either a Boolean variable x_i, a designation of 1 ("true") or 0 ("false"), or a connective $\wedge, \vee, \neg$. If a vertex is marked with a Boolean variable, or with "true" or "false", then there should be no incoming edges at that vertex. Let us call these vertices *input vertices*. If the vertex is labelled with $\neg$ then it should have exactly one edge going into it, while vertices marked with either $\wedge$ or $\vee$ should have two incoming edges. Let us call a vertex with no outgoing edges an *output vertex*.

Such a circuit can represent a Boolean function, or, more generally, a mapping from $\{0,1\}^n$ to $\{0,1\}^m$, where n is the number of Boolean variables used in the input vertices, and m is the number of output vertices. Indeed, an assignment of values to the Boolean variables leads to assignments at all of the other vertices, simply by following the rules of the circuit one step at a time (applying a connective $\wedge, \vee, \neg$ exactly when one reaches a node so marked). One can do this in a consistent manner because of the (assumed) absence of oriented cycles in the underlying graph G. (For instance, as in the definition on p80 of [Pap94], one can label the vertices of the circuit by integers in such a way that the existence of an edge from the ith vertex to the jth vertex implies that $j > i$.)

Here is a classical example, or rather a family of examples. For each positive integer n, let $PARITY_n : \{0,1\}^n \to \{0,1\}$ be the Boolean function of n variables which takes the value 1 when an odd number of the variables are equal to 1, and is equal to 0 otherwise. This can be written as

$$PARITY_n(x_1, x_2, x_3, x_4, \ldots, x_n) = (\cdots(((x_1 \oplus x_2) \oplus x_3) \oplus x_4) \oplus \cdots \oplus x_n), \quad (4.22)$$

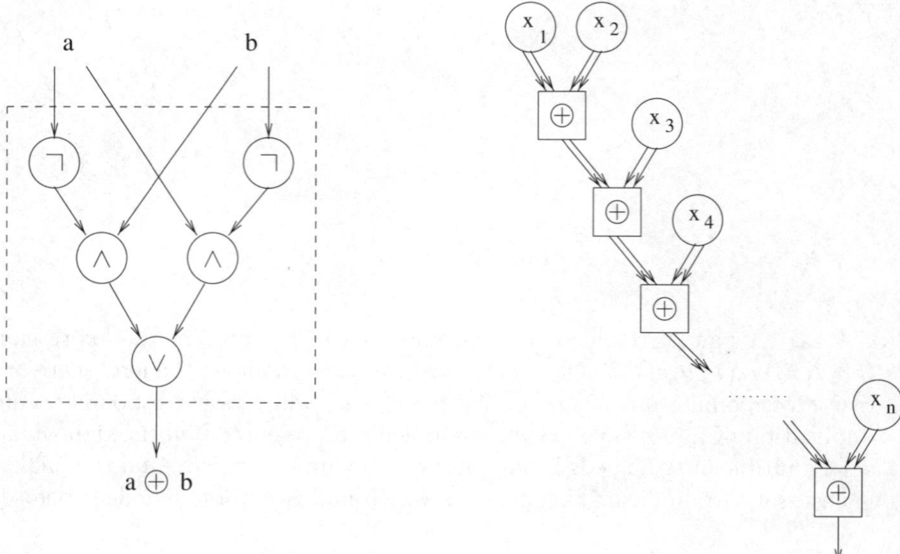

FIG. 4.14. This picture shows how one can make a Boolean circuit for $PARITY_n$ whose size is linear in n, following the formula (4.22). The diagram on the left gives a circuit for computing $a \oplus b$, and copies of this can be combined to provide a circuit for $PARITY_n$, as in the diagram on the right. The variables are all used twice as inputs to their corresponding $\oplus$-boxes, and the outputs of these boxes are used twice as inputs into the boxes that come immediately afterward (except for the last box, where there is no box after it). These double uses of variables and outputs as inputs in the boxes are indicated by the double arrows in the diagram on the right, and they appear more explicitly in the diagram of a single box and its inputs on the left.

where $a \oplus b$ gives the sum of a and b modulo 2, which is the same as the "exclusive or" operation. In terms of $\neg$, $\wedge$, and $\vee$, $a \oplus b$ can be written as

$$a \oplus b = (\neg a \wedge b) \vee (a \wedge \neg b). \tag{4.23}$$

One can expand (4.22) using (4.23) to get a Boolean expression for $PARITY_n$ in the variables x_i, $1 \leq i \leq n$, i.e., an expression using $\neg$, $\wedge$, and $\vee$, but not $\oplus$. If one does this, then the resulting expression will have size on the order of 2^n, as one can check. (This is not to say that this is a representation as a Boolean expression of approximately minimal size, however.) On the other hand, one can make a Boolean circuit that represents $PARITY_n$, following the formula in (4.22) in a natural way, and which is of linear size in n. This is indicated in Fig. 4.14.

Here is another way in which to represent $PARITY_n$. Let us assume for simplicity that $n = 2^k$ for some integer k. One can always reduce to this case, by choosing k as small as possible so that $n \leq 2^k$. The condition $n \leq 2^k$ implies that one can obtain $PARITY_n$ from $PARITY_{2^k}$, simply by setting $2^k - n$ of the arguments of $PARITY_{2^k}$ to be 0. By taking k as small as possible, we get that $2^k < 2n$, so that n and 2^k are of essentially the same size.

To compute $PARITY_{2^k}$, one can do the following. First put the 2^k variables into 2^{k-1} disjoint pairs, and apply the operation $\oplus$ to each of them. This gives 2^{k-1} outputs, which one can put into 2^{k-2} pairs (when $k \geq 2$). Again one can apply $\oplus$ to each of these pairs, to get 2^{k-3} outputs (when $k \geq 3$). This process can be repeated, until there is only one output, which is exactly the value of $PARITY_{2^k}$. This computation is essentially the same as the one before, but arranged differently. For $k = 3$, we can write this as

$$((x_1 \oplus x_2) \oplus (x_3 \oplus x_4)) \oplus ((x_5 \oplus x_6) \oplus (x_7 \oplus x_8)). \tag{4.24}$$

As in the previous case, one can realize this way of making the computation in terms of a Boolean circuit. The circuit for $a \oplus b$ on the left side of Fig. 4.14 would again be employed as a basic building block. Now, the overall structure of the circuit would look like a binary tree, with these building blocks at the junctures.

One can convert this circuit, or the expression using $\oplus$, into a Boolean expression of modest size. Although we are still combining 2^k variables, we are doing it in such a way that there are only k "levels". This is reflected in the depth of the graph underlying the Boolean circuit, which is on the order of k.

In general, a Boolean expression is practically the same as a Boolean circuit in which no vertex has more than one outgoing edge. There is a simple procedure for converting an arbitrary Boolean circuit (with a single output vertex) into a circuit where every vertex has at most one outgoing edge, in effect by taking the negatively-oriented visibility based at the output vertex and interpreting it as a Boolean circuit in a straightforward manner. (Compare with p396 of [Pap94], for instance.) This passage to the visibility can lead to exponential expansion in the size of the circuit. The circuit for $PARITY_n$ indicated in Fig. 4.14 gives a nice example of this. The same basic phenomenon also occurs in the example in Fig. 4.13, but in a more limited way. The second way of realizing the parity function in terms of $\oplus$ behaves quite differently; there is much less expansion connected to the negatively-oriented visibility graph, and this is related to the much smaller depth of the underlying graph.

Although all Boolean functions can be represented by Boolean circuits, it is well known that most Boolean functions of n variables need Boolean circuits of exponential size to represent them. More precisely, most such functions need circuits whose size is at least on the order of $2^n/n$. See p82, 83, and 86 of [Pap94]. This is proved by a counting argument, with elementary upper bounds for the total number of circuits with at most a certain number of gates, and comparison of this with the total number of Boolean functions. A consequence of this is that most Boolean functions of n variables can only be represented by Boolean expressions of the same exponential size as well, since Boolean expressions give rise to Boolean circuits in a simple way.

If a given Boolean function can be represented by a circuit of a certain size, what might one be able to say about the size of the smallest Boolean expression which represents the same function? As above, there is a straightforward way to convert any Boolean circuit into a Boolean expression, and in general this method can lead to exponential expansion in the size. This does not mean that there are not *other* Boolean expressions for which the increase in size is much smaller, however (and which might be obtained in a very different way).

Note that any Boolean function of n variables can be represented by a Boolean expression of size $O(n^2 2^n)$. See p79 of [Pap94]. As above, for most Boolean functions, any circuit which represents them has size at least on the order of $2^n/n$. Thus, for most functions, the smallest possible size of a Boolean expression which represents it cannot be too much larger than the size of any circuit which represents it.

This broad issue is not understood very well, and indeed it is not known whether every Boolean circuit of size m can be represented by a Boolean expression of polynomial size in m. This is related to the complexity question $\mathrm{P} = \mathrm{NC}_1$?, concerning the possibility that polynomial-time algorithms always admit efficient parallel representations. (See p386 of [Pap94].) To be more precise, this type of complexity question involves *families* of Boolean circuits, rather than individual circuits, and families that are *uniform*, in the sense that they can be produced by algorithms satisfying certain conditions. Given a uniform family of Boolean circuits, one would then ask about uniform families of Boolean expressions which represent the same Boolean functions, and the increase in size that might be needed for this. Nonuniform versions of this are of concern too, and this brings one back to problems about individual circuits.

These and other topics about Boolean circuits are somewhat analogous to matters related to the lengths of propositional proofs and the use of cuts. In both situations there are standard methods that involve some kind of duplication process which can lead to exponential growth. There are ways of using "building blocks" in both situations which are similar to each other; for Boolean circuits, this is illustrated by the examples in Fig. 4.14. In the setting of formal proofs, it is known that exponential expansion may be unavoidable for propositional proofs without cuts (as in the example of Statman [Sta78] mentioned in Section 3.3), but the precise mechanisms behind this kind of behavior remain unclear. It is not known if there are propositional tautologies for which proofs with cuts have to be of exponential or otherwise super-polynomial size, compared to the size of the tautology.

With formal proofs, one seems to be missing (so far, anyway) rough counting arguments like the one for Boolean circuits mentioned above. On the other hand, in the context of Boolean functions, there seem to be difficulties in furnishing reasonably-explicit examples with complexity properties like the ones that are known to occur for many instances. Compare with p83 of [Pap94].

Note that the idea of "families" (including uniform families) comes up naturally for tautologies and formal proofs, as well as for circuits. A number of examples of this have arisen earlier in this book.

Some good references concerning results related to the analogy between propositional proofs and Boolean circuits are [Kra95, Pud98].

We shall return to some related themes concerning Boolean circuits in Section 7.11.

4.13 Combinatorial dynamical systems

Let G be an optical graph, or just an oriented graph, for that matter, and suppose that to each vertex v in G there is associated a set of points $S(v)$. Suppose also that to each oriented edge e in G going from a vertex u in G to another vertex w there is associated a mapping $\phi_e : S(u) \to S(w)$. This defines a kind of "combinatorial dynamical system", in which every oriented path in G determines a mapping between two sets of points, i.e., the compositions of the mappings associated to the edges traversed by the path.

For instance, the sets $S(v)$ might all be the same set X, so that all the mappings involved send X into itself. The graph G then has the effect of specifying ways in which the ϕ_e's can be composed. Alternatively, the $S(v)$'s might be proper subsets of a larger set X, and one might be interested in keeping track of the way that points move between different parts of X.

As another basic scenario, imagine that one starts with a mapping $\psi : X \to X$ which is not injective, e.g., something like a polynomial mapping. In studying the inverse images of points in X under ψ, one might choose the ϕ_e's so that they represent branches of the inverse of ψ, defined on various subsets of X.

Here is a "model" situation. Given an oriented graph G and a vertex v in G, let $S(v)$ denote the set of all oriented paths in G which end at v. One might wish to impose the additional conditions, e.g., that the paths be as long as possible, including infinite paths when G has nontrivial oriented cycles. Given an edge e going from a vertex u in G to a vertex w, there is an obvious mapping $\phi_e : S(u) \to S(w)$, in which one simply takes an oriented path which ends at u and adds e to it to get a path which ends at w.

We shall mention some variations of this in a moment. In particular, in some contexts it is natural to have mappings ϕ_e which are defined only "partially" on a set $S(u)$, i.e., only on a subset of $S(u)$. This leads to moderately different kinds of situations. One might also consider having probabilities associated to transitions from u to other vertices, along the different edges e.

Before proceeding, let us pause for some definitions.

Definition 4.14 (Maximal paths and marked paths) *Let G be an oriented graph, and let t be an oriented path in G, which we allow to be infinite in either direction (in which case G should have at least one nontrivial oriented cycle). We call t a* maximal path *if it cannot be extended further in either direction as an oriented path in G. (This is considered to be automatically true when t is infinite in both directions, but otherwise one has to check whether the initial vertex of* *t has a predecessor in G, or whether the final vertex of t has a successor in G, when these initial or final vertices exist.)*

By a marked path *we mean a path t in which a particular occurrence of a vertex v in it has been specified. (If there are nontrivial oriented cycles in G, then t might pass through a given vertex v many times, which is the reason for specifying the particular* occurrence *of v in t. Otherwise, if there are no oriented cycles, then a choice of v determines the occurrence of v on the path.)*

One can think of a marked path as really being a combination of two paths, a path which ends at the particular vertex v, and another which begins at that particular vertex. If there is a possibility of doubly-infinite paths which are periodic, then one may prefer to be more precise and think of the vertices of the paths as being labelled by integers in a specific way, and then think of a marked path as being one in which a particular integer has been specified.

Here is a second model situation. Let G be an oriented graph, and let v be any vertex in G. Define $S(v)$ to be the set of marked oriented paths in G which are maximal and for which the marked vertex is v. Given an edge e going from a vertex u to a vertex w, there is a natural partially-defined mapping from a subset of $S(u)$ into $S(w)$ which is given as follows. Let t be an element of $S(u)$, so that t represents a maximal marked path for which the marked vertex is u. Denote by f the edge in G that t traverses immediately after it passes the marked occurrence of u. If f is equal to our given edge e, then we take t to be in the domain of ϕ_e, and we define $t' = \phi_e(t)$ to be the marked path which is the same as t as a path, and which is marked at the occurrence of w in t that is immediately after the occurrence of u in t which was marked originally.

In this second model situation, each set $S(v)$ is the disjoint union of the domains of the mappings ϕ_e corresponding to oriented edges e which begin at v, at least if there are any such edges. This follows easily from the maximality of the paths. (For this one only needs to know that the paths are maximal in the "forward" direction.)

One might compare this with the classical "Bernoulli shifts", which are mappings of the following sort. One starts by taking a set Σ of "symbols", and considering the space X of doubly-infinite sequences with values in Σ. On this space there is a natural "shift mapping", which simply slides the values of a given sequence one step.

In our context, we are using paths in graphs in place of sequences of symbols, but the effect is similar. For instance, one could take G to be the infinite graph whose set of vertices is given by the set $\mathbf{Z}$ of all integers, and for which we attach exactly one (oriented) edge from the jth vertex to the $(j+1)$th vertex for each element of Σ. In this case, a maximal oriented path in G is the same in essence as a doubly-infinite sequence with values in Σ. One could also take a graph G with one vertex, and an edge from that vertex to itself for each element of Σ. One would still have doubly-infinite oriented paths in G, going around the individual loops at the one vertex over and over again. There would be some modest differences, in that paths would not come directly with an indexing by integers.

There are more precise versions of shifts, in which the same basic shift mapping is used, but the collection of sequences in the domain is restricted. Sometimes one might describe the collection of sequences to be used in terms of paths in an oriented graph, as in [LM95].

Instead of shifts based on doubly-infinite sequences, one might use singly-infinite sequences, indexed by the set of nonnegative integers. If one uses a backward shift mapping, then this mapping is still well defined on the set of sequences with values in Σ. The shift mapping would not be one-to-one in this situation, as it was in the case of doubly-infinite sequences. (A forward shift mapping would not be well defined on the set of singly-infinite sequences.)

More generally, there are Bernoulli shifts based on any group or semigroup Γ. Instead of sequences, one would look at mappings from Γ into Σ. This amounts to the same thing as doubly-infinite sequences in the case where Γ is the group of integers, or to singly-infinite sequences if Γ consists of the nonnegative integers. For any group or semigroup Γ, one can define shift mappings, using translations in the group. In the case of the integers, the usual (backward) shift mapping corresponds to translations by 1, and all other shifts can be obtained from this one by compositions, including inverses when one is using all of the integers. Groups and semigroups in general need not have a single generator like this, but the collection of all of the shifts makes sense in the same way. See Chapter 2 of [CP93] for more on Bernoulli shifts associated to arbitrary groups.

It may be that Γ has a finite set of generators, in which case one can look at the shifts associated to them. All other shifts coming from translations in Γ will be compositions of these (together with their inverses, if Γ is a group). For another perspective on shifts related to finitely-generated groups and their Cayley graphs, see [Gro87], beginning on p236. We shall come back to finitely-generated groups and Cayley graphs in Chapter 17.

Shift mappings provide important examples of dynamical systems, and they are also used to make models for other ones. See [CP93, LM95, Mañ87, Sin76, Sin94] for more information. The study of these models and comparisons with other dynamical systems is closely connected to notions of *entropy*, in which (roughly speaking) one measures the number of different transitions between some states within a certain number of steps. More precisely, a limit is involved, as the number of steps becomes large. See [LM95, Mañ87, Sin76, Sin94]. Notions of entropy in the context of dynamical systems are closely related to earlier ones in information theory, as in [Ash65]. In our setting, when there are no oriented cycles, one might look at the number of oriented paths going from some vertices to others (with probabilities of transitions taken into account if appropriate).

In thinking about a notion like that of "combinatorial dynamical systems", one might keep in mind other types of situations besides the sort of groups and semigroups of mappings ordinarily considered in dynamical systems. We saw examples of this in Chapter 3 (starting in Section 3.2), in connection with propositional logic. That is, one can try to interpret the "reasoning" which underlies the verification of a given propositional tautology as a kind of dynamical process, in which various cases are considered, and transitions are made from one context to another, using the information in the hypothesis of the statement. Notions of entropy could be very useful here too, for measuring the amount of information underlying a proof, as we discussed in Chapter 3. Just as "positive entropy" reflects a kind of exponential complexity in a dynamical system, one would like to have measurements for formal proofs which are "large" in the case of short proofs with cuts when all of the cut-free proofs are necessarily much larger.

Note that the Kolmogorov of "Kolmogorov complexity theory" also played a fundamental role in the use of entropy in probability theory and dynamical systems. See [Mañ87, Sin76, Sin94], for instance. Both types of ideas can be seen as providing measurements of information content, but in very different ways. Roughly speaking, entropy deals only with certain kinds of "countings" (or distributions of multiplicities), rather than more precise descriptions of objects, as in Kolmogorov complexity and algorithmic information theory.

In dealing with combinatorial objects, like formal proofs, it is natural to look for intermediate measurements of information content, measurements which can be like entropy in their simplicity and manageability, but which can be more sensitive to the given structure at hand as well. Graphs and their visibilities provide a kind of laboratory in which to work, and one in which the basic objects can often be adapted to more elaborate kinds of structure. We shall return to some of these matters elsewhere in the book.

4.14 Exponential expansion

In this section, we describe a simple criterion for exponential complexity of visibility graphs, in terms of the presence of a "long chain of focal pairs". We shall discuss the necessity of this criterion in Chapter 8.

We begin with a more primitive concept.

Definition 4.15 (Focal pairs) *Let G be an optical graph. By a* focal pair *we mean an ordered pair (u, w) of vertices in G for which there is a pair of distinct oriented paths in G from u to w. We also require that these paths arrive at w along different edges flowing into w.*

In particular, w should be a *focusing* branch point under these conditions. The pair of paths may agree for some time after leaving u, but eventually they have to split apart at a *defocusing* branch point. Thus a focal pair always involves at least one focusing and one defocusing branch point.

The requirement that the paths arrive at w along different edges is convenient but not serious. It has to happen on the way to w anyway (unless there are cycles present, which we discuss below). Note that w may very well *not* be the first focusing branch point after u, as there may be many others with branches coming in from other parts of the graph (not necessarily related to oriented paths from u to w). See Fig. 4.15.

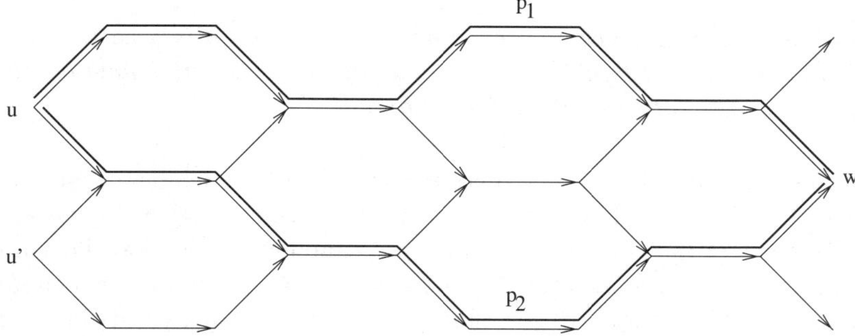

FIG. 4.15. An example of a focal pair (with a number of others being present as well)

There is a second way to have more than one oriented path from u to w, by going around a cycle some number of times before stopping at w. For the moment we shall concentrate on finite effects that do not require the presence of oriented cycles, leaving for Chapter 5 the matter of asymptotic behavior of infinite visibilities, for which oriented cycles play a crucial role.

Let us note the following.

Lemma 4.16 *Let G be an optical graph, and v be a vertex in G. Suppose that there is no vertex w in G such that (v, w) is a focal pair, and no nontrivial oriented cycle in G which passes through v. Then the canonical projection $\pi : \mathcal{V}_+(v, G) \to G$ (from Section 4.5) is one-to-one. In particular, the size of $\mathcal{V}_+(v, G)$ is less than or equal to the size of G.*

This is not hard to verify, from the definitions (and it is similar to the remarks above).

Definition 4.17 (Chains of focal pairs) *A* chain of focal pairs *in an optical graph is a finite sequence of focal pairs $\{(u_i, w_i)\}_{i=1}^n$ such that $u_{i+1} = w_i$ for each $i = 1, 2, \ldots, n-1$. We call n the* length *of the chain, and u_1 the* starting point *of the chain.*

In the graph depicted in Fig. 4.15, one can see several chains of focal pairs of length 2, while the graph H in Fig. 4.4 (in Section 4.3) has many focal pairs, but no chains of length 2. The graph in Fig. 4.2 in Section 4.3 has a long chain of focal pairs in a very simple way.

The idea of long chains of focal pairs is very natural in the context of formal proofs. One can think of having a chain of "facts", each of which is used twice in order to derive the next. By the end of the proof, the first fact is used an exponential number of times, at least implicitly. This type of phenomenon relies heavily on the use of cuts and contractions, as we have seen in Chapter 3 and Section 4.8.

Proposition 4.18 *Suppose that G is an optical graph, v is a vertex in G, and that there is a chain of focal pairs in G starting at v and with length n. Then the visibility $\mathcal{V}_+(v, G)$ contains at least 2^n different vertices.*

Proof Let $\{(u_i, w_i)\}_{i=1}^n$ be a chain of focal pairs in G which begins at v and has length n. It suffices to show that there are 2^n distinct vertices in the visibility $\mathcal{V}_+(v, G)$ which project down to w_n under the canonical projection. This amounts to saying that there are at least 2^n different oriented paths in G which go from $v = u_1$ to w_n. This is easy to see, since there are at least two distinct oriented paths α_i and β_i going from u_i to w_i for each $i = 1, 2, \ldots, n$, and there are 2^n different ways to combine the α_i's and β_i's to get paths from u_1 to w_n.

One should be a little careful about the assertion that these 2^n different combinations are truly distinct as paths in G. This is the case if we require that $u_i \neq u_j$ when $i \neq j$, and that no α_i or β_i contain a u_j except at the endpoints. If one of these conditions fails to hold, then it means that there is a nontrivial oriented cycle which passes through some u_i, and hence infinitely many oriented paths from $v = u_1$ to w_n. (In this regard, see also Chapter 5.) □

Chains of focal pairs imply the presence of alternations between defocusing and focusing branch points, but the latter is not sufficient to guarantee large growth in the visibility compared the underlying graph. For example, let us start with a graph L which is "linear", i.e., which consists of a sequence of vertices $v_1, v_2, \ldots, v_n$, and exactly one edge from v_i to v_{i+1} for $1 \leq i \leq n-1$, and no other edges. Suppose that we add to this graph a (disjoint) collection of n vertices, $u_1, u_2, \ldots, u_n$, and an edge from v_i to u_i when i is odd, and an edge from u_i to v_i when i is even, and no other edges. This will give an optical graph such that v_i is a defocusing branch point when i is odd and $i < n$, and v_i is a focusing branch point when i is even. This leads to numerous alternations between defocusing and focusing branch points along the v_i's, but this graph has no focal pairs or nontrivial oriented cycles.

There has to be a focal pair or a nontrivial oriented cycle in a graph G in order for the visibility to be larger than G, as in Lemma 4.16. Theorem 8.9 in Section 8.4 will show that the presence of a long chain of focal pairs is necessary in order for the visibility to be of large size compared to the original graph (when the visibility is finite). The proof will use a "stopping-time argument", in order to choose a piece of the visibility which has simple behavior with respect to the canonical projection, and which is about as large as possible. The remaining portions of the visibility will turn out to be visibility graphs in their own right, to which one can repeat the stopping-time argument. This will give rise to a decomposition of the original visibility graph into a tree of simpler pieces. The precise choice of these simpler pieces will involve focal pairs, in such a way that one can bound the depth of the tree of simpler pieces in terms of the length of the longest chain of focal pairs. This will lead to a bound on the size of the visibility graph, because we shall also have an estimate for the sizes of the simpler pieces.

The idea that cycles and other types of cyclic structures in the logical flow graphs of proofs is related to complexity issues is a recurring theme in [Car97, Car00, Cara]. In particular, there are results in [Car00] concerning the *necessity* of cycles for certain kinds of complexity in proofs in arithmetic, and examples to show that this can happen. The present discussion should be seen in this context.

5

ASYMPTOTIC GROWTH OF INFINITE VISIBILITIES

In this chapter, we study the size of the visibility in terms of the rate of its growth when it is infinite. The statements that we provide are very similar to results given in Section 1.3 of [ECH$^+$92], concerning the dichotomy between exponential and polynomial growth for regular languages in terms of their representations through regular expressions. Note that regular languages can be represented by "finite automata", using a process which is very close to taking the visibility of an optical graph. We shall discuss this further in Chapter 14.

As in Section 4.9, the statements described here are also closely related to known results about growth functions for L-systems. This includes the basic dichotomy between polynomial and exponential growth. See [RS80].

Matters involving the rate of growth of the visibility when it is infinite will be considered again in Chapter 12, in a different way.

5.1 Introduction

As usual, we let G be an optical graph, and we fix a vertex v in G. Let A_j denote the number of *vertices* in the visibility graph $\mathcal{V}_+(v, G)$ which can be reached by an oriented path starting from the basepoint in $\mathcal{V}_+(v, G)$ which traverses at most j edges. This is the same as the number of oriented *paths* in G beginning at v which traverse at most j edges (as in the observations in Section 4.6). If N_i is as defined in Section 4.7, then we have that

$$A_j = \sum_{i=0}^{j} N_i. \tag{5.1}$$

We shall assume that the visibility $\mathcal{V}_+(v, G)$ is infinite, and we shall investigate the asymptotic behavior of the A_j's as $j \to \infty$. It is easy to see that

$$A_j \geq j + 1 \quad \text{for all } j \geq 0 \tag{5.2}$$

as soon as the visibility $\mathcal{V}_+(v, G)$ is infinite. (That is, there has to be at least one path of any given length.) We also know from Section 4.7 that the A_j's grow at most exponentially.

Let us use the phrase "oriented loop in G" to mean a nontrivial oriented path in G whose initial and final vertices are the same, but for which no other vertex is repeated. In other words, an oriented loop is the same as a nontrivial oriented cycle which does not cross itself.

If G is an optical graph such that the visibility $\mathcal{V}_+(v, G)$ is infinite, then there must be at least one oriented loop in G which can be reached by an oriented path that begins at v. Indeed, we know from Lemma 4.9) that there has to be a nontrivial oriented cycle in G which is accessible by an oriented path which starts at v. The remaining point is that one can always find a loop "within" any nontrivial oriented cycle, and this is a standard fact which can be verified directly.

When we say that an oriented loop or cycle is accessible by an oriented path beginning at v, we actually mean (strictly speaking) that there is a single vertex on the loop or cycle which can be reached by such a path. This is the same as saying that every vertex on the loop or cycle is accessible by an oriented path from the basepoint.

If we speak of two oriented loops as being "the same", then we permit ourselves to adjust the initial and endpoints as necessary. In other words, we do not really care about the artificial selection of a basepoint, but only the circular ordering of the edges. One can check that two oriented loops are the same in this sense if their corresponding sets of vertices and edges coincide, i.e., the circular ordering of the edges is determined by the set of edges. (This is not true for oriented cycles in general.) If we say that two loops are different or distinct, then we mean that they are not "the same" in this broader sense.

Notice that two oriented loops might have the same vertices but be distinct because they do not have the same edges. The two loops might pass through a pair of adjacent vertices that are connected by two different edges, thus giving two different ways to make the transition from one vertex to the other. If all of the edges traversed by one oriented loop L_1 are also traversed by another oriented loop L_2 in the same graph, then L_1 and L_2 must be be actually the same (in the sense above). This is not hard to verify.

5.2 When loops meet

Let us now give a criterion for exponential growth of the A_j's.

Proposition 5.1 (Exponential lower bounds) *Suppose that G is an optical graph which contains a pair of distinct oriented loops which have a vertex in common and which are both accessible by an oriented path starting from the vertex v in G. Then the A_j's grow exponentially, i.e., there is a real number $r > 1$ so that*

$$r^j \leq A_j \leq 2^{j+1} \quad \text{for all } j \geq 0. \tag{5.3}$$

Note that the example pictured in Fig. 4.3 in Section 4.3 satisfies the hypotheses of this result.

Proposition 5.1 can be seen as an "asymptotic" version of Proposition 4.18 (concerning exponential expansion of the visibility when it is finite, in terms of the presence of chains of focal pairs). One can check, for instance, that the hypotheses of Proposition 5.1 are equivalent to asking that there be a focal pair (u, w) of vertices in G such that there is an oriented path from v to u and an

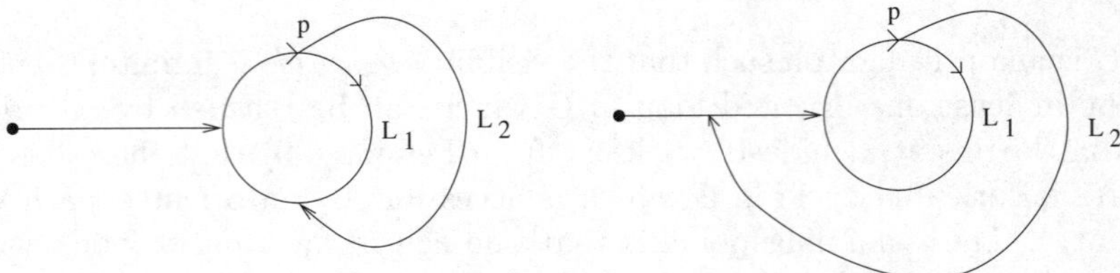

FIG. 5.1. Examples of loops that meet

oriented path from w to u (so that the pair (u, w) can "feed back into itself"). This is not difficult; the point is that the "focal pair" condition ensures that there are oriented paths from u to w which arrive at w along distinct edges, and one can make these into cycles if there is another oriented path from w back to u. If these cycles are not loops already, then one can extract loops from them which contain w and arrive at w through different edges. Conversely, if a pair of distinct oriented loops have a vertex w in common, one can follow the loops "backwards" (in the direction of the negative orientation) until one arrives at a vertex u at which they diverge. This leads to a focal pair, with $w = u$.

One can also think of the hypotheses of Proposition 5.1 as asking for the existence of arbitrarily long chains of focal pairs in G starting from v, in which in fact the same pair is repeated over and over again. Note that (finite!) optical graphs cannot contain chains of focal pairs of arbitrarily long length without having a pair of distinct loops which meet, as in the lemma. This is not hard to check (since some of the u's in the chain would have to be repeated).

Note that the converse of Proposition 5.1 is also true: exponential growth implies the existence of cycles which meet. See Corollary 5.12 below. Also, the gap between the upper and lower bounds in 5.3 is necessary, in the sense that one can make examples in which the rate of exponential expansion is as slow as one wants, by taking the loops to be sufficiently large. (Of course, one could analyze this further to get more precise bounds.)

Proof The upper bound follows from (4.4). The proof of the lower bound will be like that of Proposition 4.18, but to be precise it will be convenient to use a different kind of language.

Let L_1 and L_2 be the two loops promised in the hypotheses, and fix an oriented path α in G from v to a common vertex p of L_1 and L_2. Some examples are pictured in Fig. 5.1.

Let k be the maximum of the lengths of α, L_1, and L_2. Given any finite string σ of 1's and 2's, we can get an oriented path in G which begins at v by following α up to p, and then going around L_1 and L_2 over and over again, switching from one to the other in accordance to the code provided by the string σ. (That is, one starts with L_1 if the first entry in σ equals 1, and otherwise one starts with L_2, etc.) One can check that distinct words lead to distinct paths, since our loops L_1 and L_2 are themselves different from each other. In particular, each contains an edge not traversed by the other, by the remark made at the end of Section 5.1.

By looking at strings of length n, for instance, we get 2^n distinct paths of length at most $(n+1)\,k$. This allows one occurrence of k for the initial path α, and another for each successive tour around one of the loops L_1 or L_2. Thus

$$A_{(n+1)\,k} \geq 2^n. \tag{5.4}$$

From here the lower bound in (5.3) follows easily, with a suitable choice of $r > 1$ which can be computed from k. Strictly speaking, to get (5.3) from (5.4), we should also use (5.2) to handle the small values of j. □

5.3 When loops do not meet

In this section we shall make the assumption that

$$\begin{array}{l}\text{If two distinct oriented loops in } G \text{ are accessible}\\ \text{from the basepoint } v \text{ by an oriented path,}\\ \text{then they have disjoint sets of edges and vertices.}\end{array} \tag{5.5}$$

That is, we assume that the assumption in Proposition 5.1 does *not* hold. (Note that two loops with disjoint sets of vertices automatically have disjoint sets of edges.) We are going to show that the A_j's grow at a polynomial rate in this case, with the degree of the expansion determined in a simple way by the geometric configuration of the loops.

Our analysis will be much like the proof of Proposition 5.1. We shall look at the way in which paths can loop around cycles, but we shall want to be a little more careful this time in understanding how paths can go from one cycle to another. We shall need to derive both upper and lower bounds this time; for Proposition 5.1 we simply used the exponential upper bounds that are always true, given in (4.4).

We begin with some lemmas about the possible interactions between oriented cycles, loops, and paths under the assumption that (5.5) holds.

Lemma 5.2 (From cycles to loops) *Assume that (5.5) holds. If Γ is a non-trivial oriented cycle in G such that there is an oriented path from the basepoint v to Γ, then in fact there is a loop L in G such that Γ and L pass through exactly the same sets of vertices and edges. (In fact, Γ will simply go around L some number of times.)*

Remember that in general the succession of edges and vertices in an oriented cycle is not determined simply by the sets of edges and vertices, as is true in the case of loops.

Proof The main point is that every cycle can be "decomposed" into a collection of loops, and that (5.5) implies that the relevant loops must be disjoint or the same. To make a precise proof we argue by induction, i.e., we assume that the statement is true for cycles which traverse at most k edges (counting multiple occurrences separately), and then try to prove it when they traverse at most

$k+1$ edges. We can take the "base case" to be $k = 0$, which is trivial, since there are no nontrivial cycles which traverse no edges.

Thus we assume that (5.5) holds, and we let Γ be any nontrivial oriented cycle which is accessible from the basepoint v and which traverses at most $k+1$ edges. If this cycle does not cross itself, then it is already a loop, and there is nothing to do. If it does cross itself, then we can realize it as two shorter cycles Γ_1, Γ_2 which are spliced together (like wires). Note that Γ_1 and Γ_2 must have a vertex in common. Each of Γ_1 and Γ_2 can be accessed by an oriented path from v, since Γ can be so accessed. Thus we may apply our induction hypothesis to conclude that each Γ_i simply goes around a loop L_i, $i = 1, 2$, some number of times. In general, it would be possible for these two loops L_1 and L_2 to be distinct, but under the assumption (5.5) they have to be the same loop. This is because L_1 and L_2 must have a vertex in common, since Γ_1 and Γ_2 do.

Using the fact that L_1 and L_2 must be the same loop, it is not hard to check that Γ itself simply goes around $L_1 = L_2$ some number of times. (That is, nothing strange can happen in the splicing of Γ_1 and Γ_2. This uses the fact that there is only one way to go around a loop, because of the orientations.) □

Lemma 5.3 (Connections between vertices and loops) *Assume that the condition (5.5) holds, and let L be an oriented loop in G which is accessible by an oriented path in G from the basepoint v. Suppose that p is a vertex in G for which there is an oriented path σ from p to a vertex in L, and another oriented path τ from a (possibly different) vertex in L back to p. Then p lies in L, and the vertices and edges in σ and τ are all contained in L.*

Proof Let L, p, σ, and τ be as in the lemma. We can build an oriented cycle Γ in G in the following manner. We start by following τ from some vertex a in L to p. Then we follow σ from p to some vertex b in L. If $a = b$, then this defines a cycle Γ. If not, then we continue along an oriented arc of L to go from b back to a, to get an oriented cycle Γ.

There is an oriented path from the basepoint v to Γ because of the corresponding assumption for L. From Lemma 5.2 we conclude that there is some oriented loop L' in G such that Γ and L' pass through the same sets of vertices and edges. Thus L' is accessible from v also, and it contains a vertex in common with L. This implies that $L = L'$, by (5.5). Since σ and τ are contained in Γ we obtain that all of their edges and vertices are contained in L, as desired. □

Given two oriented loops L_1 and L_2 in G, let us say that L_2 *follows* L_1 if there is an *oriented* path from some vertex in L_1 to some vertex in L_2.

Corollary 5.4 (Ordering loops) *If (5.5) holds, then one cannot have two distinct oriented loops L_1 and L_2 in G such that each is accessible by an oriented path from v and each follows the other.*

Proof This is a straightforward consequence of Lemma 5.3. □

Let $\mathcal{L}$ denote the collection of oriented loops in G which are accessible by an oriented path from v. Under the assumption that (5.5) holds, the relation that

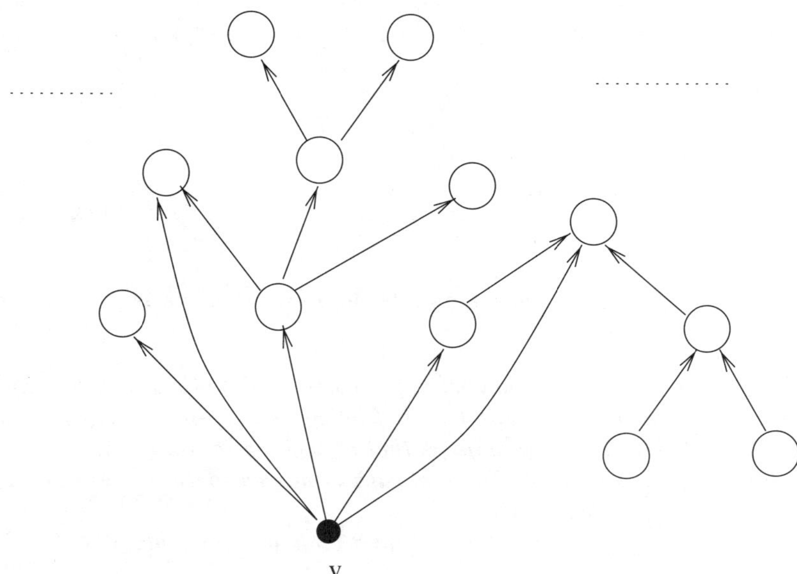

FIG. 5.2. A possible configuration for the loops in G

says that one loop follows another is a partial ordering on $\mathcal{L}$; it is automatically transitive, and Lemma 5.3 says that if each of two loops follows the other, then they are the same.

See Fig. 5.2 for a picture of the way in which the loops in G might be configured. Keep in mind that $\mathcal{L}$ contains only the loops in G which can be reached by v, but that there might be other loops in G as well, as in the picture. (Note that our assumption (5.5) does permit these other loops to intersect each other.)

The asymptotic behavior of the visibility in this case is largely determined by $\mathcal{L}$ as a partially ordered set. To make this precise, we first analyze how paths which begin at v can interact with loops. In the next lemma, we show that any path α which interacts with several loops can be decomposed into simple paths α_i (i.e., paths which never pass through a vertex more than once) and possibly multiple occurrences of loops. This decomposition is illustrated in Fig. 5.3, where the path α should be thought of as starting at v, ending at w, and cycling around the intervening loops L_i possibly many times along the way. Note that a path α_i in the bottom portion of Fig. 5.3 can go around part of the corresponding loop L_i, but not all the way around L_i, since α_i is supposed to be a simple path. We want to separate the number of complete tours around the loops from the rest, and we have to include the partial trips as being part of "the rest".

Lemma 5.5 (Decompositions of paths) *Suppose that (5.5) holds, and let α be a (finite) oriented path in G which begins at v. Then there is a finite sequence of distinct loops $L_1, L_2, \ldots, L_k$ in $\mathcal{L}$, a sequence of positive integers $m_1, m_2, \ldots, m_k$, a sequence of vertices p_j in G for $j = 1, 2, \ldots, k$, and a sequence of oriented paths*

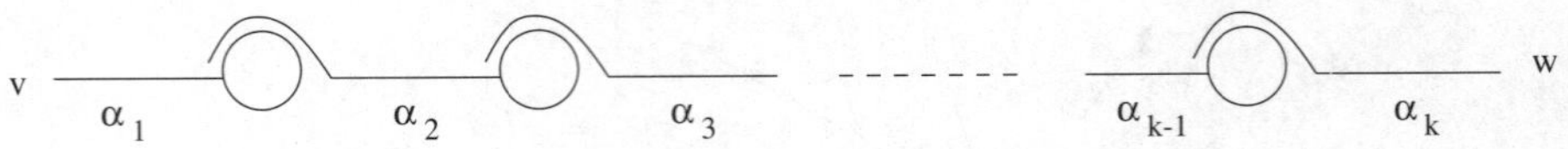

FIG. 5.3. Diagram of decompositions of paths

α_i in G, $0 \leq i \leq k$, with the following properties. (We allow k to be zero here, in which case there are no loops L_j, multiplicities m_j, or vertices p_j, and the content of the following conclusions is that α does not cross itself.)

(a) The α_i's are subpaths of α, and each is nondegenerate (traverses at least one edge), except possibly for α_k.

(b) α_j begins at p_j when $1 \leq j \leq k$ and ends at p_{j+1} when $0 \leq j < k$, and α_0 begins at v.

(c) α begins with α_0 and follows it all the way up to p_1, then goes all the way around L_1 exactly m_1 times, then follows α_1 from p_1 to p_2, then goes all the way around L_2 exactly m_2 times, and so forth until it reaches L_k, goes around it exactly m_k times, and then ends by following α_k. In particular, p_i lies in L_i for each i.

(d) Each α_i never passes through a vertex more than once.

(e) No vertex in α_i lies in L_{i+1}, except for the endpoint p_{i+1}, for all $0 \leq i < k$. (However, α_i might well spend some time in L_i before leaving it; we only know that α_i will not go all the way around L_i, because of (d).)

(f) p_j is not contained in α_r when $r > j+1$, and p_j is not contained in L_r when $r > j$.

(g) L_j follows L_{j-1} when $2 \leq j \leq k$.

(h) If α_i and α_j have a vertex q in common, $i < j$, then $j = i+1$ and $q = p_j$.

(i) If α_r has any vertices in common with L_j, then either $r = j$ or $r = j+1$.

Note that the analogue of (f) for $r < j$ is also true, and is contained implicitly in (h) and (i). In fact, (f) itself is implicitly contained in (h) and (i), but it is slightly convenient to state it separately in this manner.

Proof The existence of a decomposition with these properties is fairly straightforward to check. Let α be given as above. If α does not cross itself, then we take $k = 0$ and $\alpha_0 = \alpha$, and we are finished. Otherwise, α does cross itself, and we let p_1 be the vertex that occurs in α first and is repeated in α. Let α_0 be the subpath of α from v to the *first* occurrence of p_1 in α. Then we let β_1 be the subpath of α which goes from the first occurrence of p_1 to the last occurrence of p_1 in α. This defines a cycle, and in fact there must be an oriented loop $L_1 \in \mathcal{L}$ so that β_1 simply goes all the way around L_1 some positive number of times.

This follows from Lemma 5.2. We denote by m_1 be the number of times that β_1 goes around L_1.

Notice that α_0 does not cross itself, by construction. The vertices in α_0 which precede p_1 cannot lie in L_1 either. For if such a vertex q preceding p_1 did lie in L_1, then it would be repeated in α itself between the first and second occurrences of p_1, as α traverses L_1 for the first time. In this case, q would be a vertex which is repeated in α and which has an occurrence in α before the first occurrence of p_1 in α, in contradiction to the way that p_1 was chosen. Thus no vertex in α_0 besides the endpoint lies in L_1.

Let γ_1 be the part of α which begins at the last occurrence of p_1 and continues to the end. There is no other occurrence of p_1 in γ_1 after the starting point, by construction. It may be that the last occurrence of p_1 in α was the endpoint of α, in which case γ_1 is a degenerate path (which traverses no edges).

If γ_1 does not cross itself, then we stop here, take $k = 1$ and $\alpha_1 = \gamma_1$. In particular, this is what we do when γ_1 is degenerate. Suppose instead that γ_1 does cross itself. Let p_2 be the first vertex in γ_1 which is repeated. Note that p_2 is necessarily distinct from p_1, since γ_1 does not contain any occurrences of p_1 past the initial point. Let α_2 denote the part of γ_1 which goes from p_1 to the first occurrence of p_2, and let β_2 denote the part of γ_1 which goes from the first occurrence of p_2 to the last occurrence of p_2. Using Lemma 5.2, we conclude that there is an oriented loop $L_2 \in \mathcal{L}$ such that β_2 simply goes around L_2 some number of times. Let m_2 be the number of times that β_2 goes around L_2. Note that p_1 does not lie in L_2 by construction, since γ_1 does not pass through p_1 again after the initial occurrence.

As before, we have that α_2 does not cross itself, and the endpoint p_2 of α_2 is the only vertex in α_2 which lies in L_2.

Now we take γ_2 to be the part of γ_1 that begins at the last occurrence of p_2. Notice that γ_2 contains no occurrences of p_1, and only one occurrence of p_2, at the starting endpoint. We repeat the process until we have to stop, which happens when we reach a stage where γ_k does not cross itself (and may be degenerate). This will happen in a finite number of steps, since α is a finite path. Note that each γ_j is a subpath of α which goes from the last occurrence of p_j in α to the end of α, and that each γ_j is a proper subpath of the preceding γ_{j-1} (when $j \geq 2$). Each γ_j contains only one occurrence of p_j, at the beginning, and no occurrences of p_i for $i < j$.

In this way, we get sequences of loops L_j, $1 \leq j \leq k$, multiplicities m_j, vertices $p_j \in L_j$, and subpaths α_i of α. They satisfy (a), (b), (c), and (d) in Lemma 5.5 by construction. We also get (e) from the construction, as in the cases above. Part (f) also follows from the construction; more specifically, it follows from the fact that each γ_j contains only one occurrence of p_j, at the beginning, and no occurrences of p_i for $i < j$, as mentioned above.

Next we observe that the L_j's are *distinct*. This comes from (f), which says that $p_j \in L_j$ does not lie in L_r when $r > j$. Our assumption (5.5) then implies that the L_j's have pairwise disjoint sets of edges and vertices.

Part (g) follows from (b), which ensures that α_{j-1} is an oriented path from L_{j-1} to L_j.

Let us prove (h). Suppose that α_i and α_j have a vertex q in common, and that $i < j$. Let δ denote the subpath of α which goes between these two occurrences of q. From (c) we know that δ traverses the loops L_{i+1} and L_j a positive number of times, since α itself does this between the times that it finishes crossing the subpath α_i and it begins to cross α_j.

Since δ begins and ends at the same vertex q, it defines an oriented cycle in G. It is automatically accessible by an oriented path from v, namely a subpath of α in this case, and it is nontrivial because it goes around L_{i+1} and L_j a positive number of times. From Lemma 5.2 we conclude that this cycle simply goes around a single oriented loop L some positive number of times. This loop L must contain all of the edges and vertices of L_{i+1} and L_j, by construction, and hence it must be the same as both L_{i+1} and L_j, because of (5.5). In particular, L_{i+1} and L_j must be the same as each other, which implies that $j = i + 1$, because of the distinctness of the loops observed above.

Since L is the same as L_{i+1} we also obtain that q itself lies in L_{i+1}. From here we obtain that $q = p_{i+1}$, by (e). This gives (h).

We are left with proving (i). Suppose to the contrary that there is a vertex q in α_r which also lies in L_j, and that $r \neq j, j-1$. Let ξ be a subpath of α which connects this occurrence of q in α_r with an occurrence of q when α is going around L_j (between the subpaths α_{j-1} and α_j). This defines an oriented cycle which is accessible by an oriented path beginning at v. This cycle is nontrivial (crosses at least one edge), because $r \neq j, j-1$. Thus there is an oriented loop M in G which ξ traverses completely some (positive) number of times, as in Lemma 5.2. Since we also know that M and L_j contain a vertex in common (namely, q), we may conclude that M and L_j are the same loop, by our assumption (5.5).

On the other hand, our assumption that $r \neq j, j-1$ ensures that there is an $i \neq j$ such that ξ traverses the loop L_i. Specifically, one can take $i = j - 1$ when $r < j - 1$ and $i = j + 1$ when $r > j$; this is easy to check, using part (c) of the lemma. From here we obtain that M must be the same loop as L_i, and hence that L_i and L_j must be the same loop. This contradicts our earlier observation about the distinctness of the loops L_s. Thus (i) is established, and the lemma follows. □

This structural result for oriented paths in G which begin at v uses the assumption (5.5) in a nontrivial way. Without (5.5), our path α might start to go around some loop L, make a detour around another loop L' before going all the way around L, and then continue with L afterwards. This cannot happen when (5.5) holds, which ensures that α cannot return to a loop L once it has departed from L, as in Lemma 5.3. The next result gives another version of the same fact.

Lemma 5.6 (Multiplicities of loops) *Suppose that G satisfies (5.5) and that α is an oriented path in G which begins at v. Let $L_1, L_2, \ldots, L_k \in \mathcal{L}$ and*

$m_1, m_2, \ldots, m_k$ be the oriented loops and their multiplicities in α provided by Lemma 5.5, and let α_i, $0 \leq i \leq k$ be the subpaths of α given in Lemma 5.5. If L is an oriented loop in G which has at least one vertex in common with α, then either L is one of the L_j's, or it is not among the L_j's, and there is exactly one α_i which intersects L. In the latter situation, α will not traverse L completely, because of part (d) of Lemma 5.5.

In other words, Lemma 5.5 says α goes around each L_j a certain number of times, and Lemma 5.6 emphasizes the fact that Lemma 5.5 accounted for all complete tours around oriented loops in G that were made by α.

Proof If L is not one of the loops L_j, then L shares no vertices with any of the L_j's, because of our assumption (5.5). Thus we assume instead that L shares a vertex with both α_i and α_j for some $0 \leq i < j \leq k$. Then L_{i+1} follows L, because there is a subpath of α_i which goes from a vertex in L to p_{i+1} in L_{i+1}. Similarly, L follows L_{i+1}, because we can take a subpath of α to go from p_{i+1} to a vertex w that lies in both L and α_j. Corollary 5.4 then implies that $L = L_{i+1}$, so that we are back to the first possibility. This proves the lemma. □

In order to estimate the growth of the A_j's, we want to associate a kind of "code" to oriented paths in G which begin at v. Let α be such a path, and let us define a function f_α on $\mathcal{L}$ in the following manner. We apply Lemma 5.5 to get a collection of loops $L_1, L_2, \ldots, L_k \in \mathcal{L}$ and multiplicities $m_1, m_2, \ldots, m_k$. We set $f_\alpha(L_j) = m_j$, and we take $f_\alpha(L) = 0$ for all other loops $L \in \mathcal{L}$. This function represents our "code" for α.

Lemma 5.7 (Coding the lengths of paths) *Let α be an oriented path in G which begins at v. If G satisfies (5.5), and f_α is defined as above, then*

$$\sum_{L \in \mathcal{L}} \operatorname{length} L \cdot f_\alpha(L) \leq \operatorname{length} \alpha < \#G + \sum_{L \in \mathcal{L}} \operatorname{length} L \cdot f_\alpha(L), \tag{5.6}$$

Here $\#G$ denotes the number of vertices in G, and, as usual, "length" means the number of edges traversed.

Proof Apply Lemma 5.5 to get collections of loops $L_1, L_2, \ldots, L_k \in \mathcal{L}$, multiplicities $m_1, m_2, \ldots, m_k$, and paths $\alpha_1, \alpha_2, \ldots, \alpha_k$ as before. From Lemma 5.5 (c) we get that

$$\operatorname{length} \alpha = \sum_{j=1}^{k} m_j \cdot \operatorname{length} L_j + \sum_{i=0}^{k} \operatorname{length} \alpha_i. \tag{5.7}$$

Note that each α_i may also go around part of L_i, but we still have an equality here. That is, the m_j's count the number of complete tours that α makes around the L_j's, but this does not preclude the possibility of partial tours in the α_i's. Since

$$\sum_{j=1}^{k} m_j \cdot \operatorname{length} L_j = \sum_{L \in \mathcal{L}} \operatorname{length} L \cdot f_\alpha(L), \tag{5.8}$$

by definitions, we need only show that

$$0 \leq \sum_{j=0}^{k} \operatorname{length} \alpha_j < \#G. \tag{5.9}$$

Of course only the second inequality is nontrivial.

As in Lemma 5.5 (b), the endpoint of α_i is the starting point of α_{i+1}, and so we can combine the α_i's to get a single path η. One can think of η as being the same as α, but with all the loops removed. In particular, it does not go through any vertex twice, because of Lemma 5.5 (d) and (h). The total number of vertices in η is no greater than $\#G$, and so the length of η is strictly less than $\#G$. We also have that

$$\operatorname{length} \eta = \sum_{j=0}^{k} \operatorname{length} \alpha_j, \tag{5.10}$$

by definitions. Thus the bound on the length of η implies (5.9), which is what we wanted. This proves the lemma. □

Next we estimate how many times a single "code" $f : \mathcal{L} \to \mathbf{Z}_+ \cup \{0\}$ can arise as f_α for an oriented path α in G.

Lemma 5.8 (Bounded multiplicities for the coding of paths) *If G satisfies (5.5), then there is a constant C_0, depending only on G, with the following property: if $f : \mathcal{L} \to \mathbf{Z}_+ \cup \{0\}$ is given, then there are at most C_0 oriented paths α in G which begin at v and for which $f_\alpha = f$.*

Proof Let f be given, and suppose that α is an oriented path in G which begins at v. The function f_α is defined using the structural analysis of Lemma 5.5, as described just before the statement of Lemma 5.7. Once f_α is known, α is itself uniquely determined by the knowledge of the subpaths α_i mentioned in Lemma 5.5. (Note that the number $k+1$ of α_i's is determined by the number of loops in $\mathcal{L}$ at which $f_\alpha = f$ takes a nonzero value.) Because these subpaths α_i are *simple* (Lemma 5.5 (d)), it is easy to see that the number of different choices for them is bounded by a constant that depends on G but not on α. This proves the lemma. □

The simple method of this lemma does not give a good bound for the constant C_0, and indeed C_0 could easily be of exponential size compared to the size of G. For example, G could contain a subgraph which looks like the graph pictured in Fig. 4.2 in Section 4.3 and which lies between our initial vertex v and a loop L. This would lead to exponentially many possibilities just for the subpath α_0 of

α (as defined in Lemma 5.5). The same thing could happen between a pair of loops L, L', or for several pairs of loops along a single path.

To control better the constant C_0, one can begin with the following observation.

Lemma 5.9 *Suppose that G satisfies (5.5), and let α be an oriented path in G which begins at v. Let f_α be defined as before, and let η denote the concatenation of the subpaths α_i of α that are provided by Lemma 5.5. (This is the same as the definition of η used in the proof of Lemma 5.7.) Then η is an oriented path in G which begins at v and which is simple (i.e., does not cross itself), and α is uniquely determined by η and f_α.*

In other words, in the proof of Lemma 5.8 we used the fact that α can be recovered from the knowledge of f_α and the subpaths α_i, and now we are improving this slightly to say that one really only needs to know the concatenation of the α_i's, and not the listing of the individual pieces.

Proof The fact that η is an oriented path that begins at v and does not pass through any vertex twice follows from Lemma 5.5, especially (c), (d), and (h). In order to recover α from η and f_α, it suffices to be able to recover the subpaths α_i of α from the knowledge of η and f_α. The α_i's are all subpaths of η, but one has to be able to figure out when one α_i stops and the next one begins.

This is not hard to do. The first main point is that the collection of loops L_j, $1 \leq j \leq k$ from Lemma 5.5 is determined by the function f_α. That is, the L_j's are simply the elements of $\mathcal{L}$ at which f_α takes a nonzero value. This determines the L_j's as an *unordered* collection of loops, but we can easily recover $L_1, \ldots, L_k$ as an ordered sequence using the partial ordering on $\mathcal{L}$. That is, we know that L_{i+1} follows L_i for each $i < k$, and this is enough to recapture the linear ordering of the L_j's. This uses Corollary 5.4 as well, i.e., the fact that there cannot be a pair of distinct oriented loops L and L' in G which are accessible by an oriented path starting from v, and such that each follows the other in the sense that we defined before.

To recover α from η we can now proceed as follows. If we start at v and follow η until the first moment when it reaches L_1, then we get exactly the subpath α_0. This is easy to derive from Lemma 5.5 (or from the definitions within its proof). Similarly, we can recover α_1 by starting where α_0 left off and continuing in η until the first moment at which we arrive to a vertex in L_2, etc.

Thus we can recover all of the α_i's from η and the knowledge of f_α, and the lemma follows. □

Using Lemma 5.9, we can say that the constant C_0 in Lemma 5.8 is controlled by the number of simple oriented paths η in G which begin at v and which pass only through the loops $L \in \mathcal{L}$ for which $f(L) \neq 0$, where $f : \mathcal{L} \to \mathbf{Z}_+ \cup \{0\}$ is as in Lemma 5.8. This can be analyzed further through the methods of Chapter 8. We shall say more about this later, in Remark 5.11 and Section 5.4.

To proceed with the analysis of the growth of the visibility of G starting from v, it will be helpful to introduce some additional terminology. We call a sequence

$L_1, L_2, \ldots, L_k$ of distinct loops in $\mathcal{L}$ a *chain of loops* if L_j follows L_{j-1} when $2 \le j \le k$. We call k the *length* of the chain. We define the *depth* of $\mathcal{L}$ to be the largest integer d for which there exists a chain in $\mathcal{L}$ of length d. The depth of $\mathcal{L}$ is at least 1 as soon as it is nonempty, which happens exactly when the visibility $\mathcal{V}_+(v, G)$ is infinite. It is easy to build examples of optical graphs for which the depth of $\mathcal{L}$ is any preassigned positive integer. (Compare with Fig. 5.2.) For a graph G (which satisfies (5.5)) of a given size, the depth cannot be too large, though; the total number of loops in G is bounded by the number of vertices in G, for instance, because the loops define pairwise disjoint collections of vertices in G, under the assumption (5.5).

The following is our basic upper bound for the A_j's when G satisfies (5.5). Recall that A_j is defined in (5.1) in Section 5.1.

Proposition 5.10 (Polynomial upper bounds for the A_j's) *Suppose that the optical graph G satisfies (5.5), and that $\mathcal{L}$ has depth d. Then there is a constant C_1 (depending on G only) such that*

$$A_j \le C_1\,(j^d + 1) \quad \textit{for all } j \ge 0. \tag{5.11}$$

This quantity C_1 includes the constant C_0 from Lemma 5.8 as a factor, but it has other components as well. It will be clear from the proof of the proposition how one could try to analyze these other components more precisely. (See also Remark 5.11.)

Proof We shall use the definition of A_j as the number of distinct oriented paths in G which begin at v and traverse at most j edges. We shall bound this number using our coding in terms of the functions f_α.

Let $\mathcal{F}_j$ denote the collection of functions $f : \mathcal{L} \to \mathbf{Z}_+ \cup \{0\}$ such that

$$\sum_{L \in \mathcal{L}} \operatorname{length} L \cdot f(L) \le j \tag{5.12}$$

and such that $f(L) = 0$ except when L lies in some chain $\mathcal{C}$ of loops in $\mathcal{L}$. This chain of loops is allowed to depend on f. If α is an oriented path in G which begins at v and traverses at most j edges, then the corresponding function f_α lies in $\mathcal{F}_j$. Indeed, $f = f_\alpha$ satisfies (5.12) in this case, because of Lemma 5.7, and it takes nonzero values only on the loops L_j, $1 \le j \le k$, given by Lemma 5.5. These loops forms a *chain* in $\mathcal{L}$, because of part (g) of Lemma 5.5. Thus f_α lies in $\mathcal{F}_j$, as desired.

Let B_j denote the number of elements of $\mathcal{F}_j$. From the preceding observation and Lemma 5.8 we obtain that

$$A_j \le C_0\, B_j \quad \text{for all } j \ge 0, \tag{5.13}$$

where C_0 is as in Lemma 5.8. Thus it suffices to bound B_j.

Given a chain $\mathcal{C} = \{L_1\}_{i=1}^k$ of loops in $\mathcal{L}$, let $\mathcal{F}_j(\mathcal{C})$ denote the collection of functions $f \in \mathcal{F}_j$ such that $f(L) = 0$ when L is not in the chain, and $f(L_i) > 0$

for each L_i in the chain. Let $B_j(\mathcal{C})$ be the number of elements of $\mathcal{F}_j(\mathcal{C})$. It suffices to get a bound on $B_j(\mathcal{C})$ for each chain $\mathcal{C}$, since the number of such chains is finite (because G is finite). We allow the empty chain here, which accounts for the case where $f(L) = 0$ for all L. For this chain, there is exactly one element of $\mathcal{F}_j(\mathcal{C})$ (the zero function), and so we may restrict our attention to nonempty chains for the estimation of $B_j(\mathcal{C})$.

Fix a chain $\mathcal{C} = \{L_i\}_{i=1}^k$ of loops in $\mathcal{L}$. By hypothesis, $k \leq d$. An element of $\mathcal{F}_j(\mathcal{C})$ is described completely by a collection of k nonnegative integers which represent $f(L_i)$ and which are constrained by (5.12). This constraint implies that each of these k integers is $\leq j$, since each L_i has length at least 1. There are j^k ways to choose k-tuples of positive integers which are each at most j, and this implies that the number of elements of $\mathcal{F}_j(\mathcal{C})$ is bounded by j^d, since $k \leq d$. Of course this bound is rather crude, and could be improved by using the constraint (5.12) more efficiently. Thus we have a bound of the correct type for each $B_j(\mathcal{C})$, and this implies a similar bound for the B_j's, since there are only finitely many chains $\mathcal{C}$. From here we can derive (5.11) from (5.13), and the proposition follows. □

Remark 5.11 To get better estimates in Proposition 5.10, one can reorganize the preceding arguments somewhat, in the following manner. Let α be an oriented path in G which begins at v, and let η be the simple path associated to it, as in Lemma 5.9. If $\mathcal{C}$ is the chain of all loops in $\mathcal{L}$ through which η passes, then the function f_α is necessarily supported on $\mathcal{C}$, i.e., $f_\alpha(L) = 0$ for any loop L not in $\mathcal{C}$. Note that $f_\alpha(L)$ may be equal to 0 for some loops L in $\mathcal{C}$, when $\mathcal{C}$ is chosen in this way (depending only on η, and not α). Let $\mathcal{F}_j'(\mathcal{C})$ be defined in the same way that $\mathcal{F}_j(\mathcal{C})$ was above, except that we allow $f(L)$ to be 0 when L is a loop in the chain $\mathcal{C}$. Then the number of α's of at most a given length is bounded by the total number of simple oriented paths η in G which begin at v, times the maximum of the numbers of elements of $\mathcal{F}_j'(\mathcal{C})$ for chains $\mathcal{C}$ in $\mathcal{L}$. This uses Lemma 5.9. In other words, instead of counting the chains $\mathcal{C}$ in $\mathcal{L}$ separately, as we did before, this information is now incorporated into the total number of simple oriented paths η in G which begin at v. This has the effect of combining the estimation for the number of chains with that of the constant C_0 from Lemma 5.8.

Corollary 5.12 (The converse to Proposition 5.1) *Suppose that G is an optical graph, v is a vertex in G, and the A_j's grow faster than any polynomial in j. Then G satisfies the hypothesis of Proposition 5.1, i.e., G contains a pair of distinct oriented loops which have a vertex in common and which are accessible by an oriented path starting from v. In particular, the A_j's grow at an exponential rate, as in Proposition 5.1.*

Proof This is an immediate consequence of Proposition 5.10. That is, super-polynomial growth of the A_j's implies the failure of (5.5), since the depth d of $\mathcal{L}$ is finite (and bounded by the number of vertices in G, for instance). This is the same as saying that G contains a pair of distinct loops with the required properties. □

Proposition 5.13 (Polynomial lower bounds for the A_j's) *Suppose that the optical graph G satisfies (5.5), and that $\mathcal{L}$ has depth equal to $d \geq 1$. Then there exists a constant $C_2 > 1$ such that*

$$A_j \geq C_2^{-1}\,(j+1)^d \qquad \textit{for all } j \geq 0. \tag{5.14}$$

Again, the basic nature of the constant C_2 will be pretty clear from the proof.

Proof To prove this, we shall essentially just "reverse" the process by which our coding $\alpha \to f_\alpha$ was defined before.

Fix a chain $L_1, \ldots, L_d$ of distinct loops in $\mathcal{L}$, whose existence follows from the assumption that $\mathcal{L}$ has depth d. By definition of $\mathcal{L}$, we have that there is an oriented path α_0 in G from v to a vertex p_1 in L_1. Since the L_i's form a chain, there is an oriented path α_1 from p_1 to a vertex p_2 in L_2. By repeating this process $d-2$ times, we get a collection of oriented paths α_i in G, $0 \leq i \leq d-1$, which start at a vertex p_i in L_i when $i \geq 1$, and end at a vertex p_{i+1} in L_{i+1} for each i. These paths α_i and vertices p_j should be considered as fixed for the rest of the argument. For the sake of efficiency, let us require that p_{i+1} be the first point in L_{i+1} which is reached by α_i, $0 \leq i \leq d-1$, and that α_i does not go all the way around L_i when $1 \leq i \leq d-1$. These are easy to arrange.

Let $f : \mathcal{L} \to \mathbf{Z}_+ \cup \{0\}$ be a function which vanishes at all loops not in our chain $L_1, \ldots, L_d$. To f we can associate an oriented path ϕ_f in G as follows. We start by following α_0 from v to p_1. We then go all the way around the loop L_1 exactly $f(L_1)$ times. We then continue with α_1 from p_1 to p_2, and then go all the way around L_2 exactly $f(L_2)$ times. We repeat this process until we arrive at p_d by α_{d-1} and go around L_d exactly $f(L_d)$ times.

Distinct functions f lead to distinct paths ϕ_f. This follows from the fact that the L_i's are all different from each other. We also have that

$$\operatorname{length}\phi_f = \sum_{i=0}^{d-1} \operatorname{length}\alpha_i + \sum_{i=1}^{d} \operatorname{length} L_i \cdot f(L_i), \tag{5.15}$$

because of the definition of ϕ_f.

Since distinct choices of f yield distinct paths ϕ_f, we may conclude that A_j is bounded from below by the number of f's such that $\operatorname{length}\phi_f \leq j$. It is easy to see that the number of these f's is bounded from below by a constant times $(j+1)^d$ when j is large enough, because we are free to choose the values of f at the loops L_i as we like, and because there are d of these loops.

Thus we get (5.14) for sufficiently large j, and for j small we can use (5.2). This completes the proof of the proposition. □

One can be more precise about this construction, as in the next lemma. Let η be the path in G obtained by combining the α_i's, $0 \leq i \leq d-1$, from the proof of Proposition 5.13. This is the same as taking $\eta = \phi_f$, where $f(L) = 0$ for all L.

Lemma 5.14 *η is a simple path in G.*

Proof Suppose to the contrary that η crosses itself. This means that there is a subpath δ of η which defines a nontrivial oriented cycle in G. This cycle is accessible by v, since η begins at v. Lemma 5.2 implies that there is a loop L in G such that δ simply goes around L some number of times.

Suppose first that L is not equal to any of the loops L_i in our chain. Because of our assumption (5.5), this means that L is disjoint from all of the L_i's. In particular, it does not contain any of the p_i's. This implies that the subpath δ is wholly contained in some α_j, since it does not go through the endpoints of any α_i, except possibly for the initial endpoint of α_0.

In this case, we have that L_{j+1} follows L, and that L follows L_j if $j \geq 1$. This uses subpaths of α_j to make connections from a vertex in L to p_{j+1} in L_{j+1}, and from p_j in L_j to a vertex in L when $j \geq 1$. Because of this property, we can add L to our chain of loops in $\mathcal{L}$ and get a larger chain, of size $d+1$. This contradicts the definition of d as the depth of $\mathcal{L}$, i.e., as the largest length of such a chain. Instead of using this, we could also simply have chosen the α_i's from the start so that they did not cross themselves.

Now suppose that L is equal to a loop L_j in our chain. We chose the α_i's so that α_i would not go all the way around L_i, and this implies that δ cannot be a subpath of α_j. On the other hand, we chose the α_i's so that they did not reach the corresponding loops L_{i+1} until the final endpoint, and this implies that δ does not have any vertices in common with α_{j-1}, except for its final endpoint. If δ goes through part of α_i when $0 \leq i < j-1$, then we would have that L_{i+1} follows L_j, because a subpath of α_i would give an oriented path from a vertex in δ, which lies in L_j, to the vertex p_{i+1} in L_{i+1}. This contradicts the assumption that $L_1, \ldots, L_d$ be a chain of distinct loops in $\mathcal{L}$, since $i+1 < j$, i.e., the ordering goes the wrong way. Similarly, one can check that δ cannot go through any vertices in α_i when $i > j$, because L_j would then follow L_i.

Thus we get a contradiction in all cases. This proves Lemma 5.14. □

From the lemma we get that

$$\sum_{i=0}^{d-1} \operatorname{length} \alpha_i = \operatorname{length} \eta < \#G, \tag{5.16}$$

where $\#G$ denotes the number of vertices in G. This gives more information about the condition $\operatorname{length} \phi_f \leq j$, as in the proof of Proposition 5.13.

Also, the path ϕ_f corresponding to a given function $f : \mathcal{L} \to \mathbf{Z}_+ \cup \{0\}$ has the same relationship to the α_i's and L_i's as in Lemma 5.5. This is not hard to show. (Compare with Lemma 5.6.) In particular, the original function f is in fact the same as the coding function f_α that we chose before (just prior to the statement of Lemma 5.7), with α taken to be ϕ_f. In other words, the method of the proof of Proposition 5.13 really is the "reverse" of the earlier one.

5.4 Summary and remarks

Theorem 5.15 *Let G be an optical graph and let v be a vertex in G. Then one of the following is true. (Recall that the A_j's were defined in (5.1) in Section 5.1.)*

(i) *(Finiteness) $A_j \leq 2^n$ for all $j \geq 0$, where n is the number of vertices in G.*

(ii) *(Polynomial growth) There is a positive integer d and a constant C (depending only on G) such that*

$$C^{-1}\,(j+1)^d \leq A_j \leq C\,(j+1)^d \quad \textit{for all } j \geq 0. \tag{5.17}$$

(iii) *(Exponential growth) There is a real number $r > 1$ so that*

$$r^j \leq A_j \leq 2^{j+1} \quad \textit{for all } j \geq 0. \tag{5.18}$$

Proof If the visibility $\mathcal{V}_+(v, G)$ is finite, then it has at most 2^n vertices, where n is the number of vertices in G, by Corollary 4.10 in Section 4.7. In this case, we have the first possibility. If the visibility is infinite, then there must be a nontrivial oriented cycle in G which is accessible by an oriented path from v, by Lemma 4.9. As observed at the beginning of this section, this implies that there is an oriented *loop* in G which is accessible by an oriented path from v. If there are two distinct loops like this which share a vertex in common, then we have exponential growth as in (iii), by Proposition 5.1. If not, then (5.5) is satisfied, and we have polynomial bounds as in (ii) by Propositions 5.10 and 5.13. This proves the theorem. □

The behavior of the A_j's can also be analyzed algebraically, through the use of *adjacency matrices.* This will be discussed in Chapter 12, and in particular one can show that the A_j's can always be described in terms of finite combinations of polynomials and (complex) exponentials. See Section 12.2.

Note that the basic "decomposition" used in Section 5.3 has a version when there are distinct loops which intersect. In general, one can define an equivalence relation on vertices in G by saying that a pair of vertices are equivalent if there is an oriented path from each one to the other. It is easy to see that this is an equivalence relation. We can then partition the set of vertices in G into equivalence classes, many of which may be "trivial", in the sense that they consist only of a single vertex, and with no edge attached as a loop. The "nontrivial" equivalence classes play the same role that the loops did in Section 5.3, and we can decompose oriented paths in G into subpaths which are contained in nontrivial equivalence classes, and subpaths which go between them. In the present setting, the behavior of a path inside an equivalence class need not be as simple as it was in Section 5.3, where the main point was merely to count the number of tours around a given loop. Still, this kind of decomposition can be useful for separating the "finite" effects between equivalence classes from the unbounded repetitions which can occur within equivalence classes. Keep in mind that these "finite" effects (*between* equivalence classes) may provide the main contribution to the A_j's for modest values of j (compared to the size of G), even if the nontrivial equivalence classes predominate eventually, for j large.

One can also think of this in terms of deforming G by collapsing the vertices in a single equivalence class, and eliminating the edges which go between vertices in the same equivalence class. This leads to an oriented graph G_c, which reflects some of the finite effects in the estimates of the A_j's in a natural way. For instance, one can check that the path η described in Lemma 5.9 is determined uniquely by its projection into G_c (in the case where distinct loops do not intersect). Thus the counting of these paths η in G is controlled by the counting of general oriented paths in G_c. We shall discuss this graph G_c further in Section 8.10. (Compare also with Remark 5.11.)

5.5 Asymptotic geometry

Instead of looking merely at the rate of *growth* of the visibility, we can also look at its asymptotic *geometry*.

Fix an optical graph G and a vertex v in G, and suppose that the visibility $\mathcal{V}_+(v, G)$ is *infinite*. Let $\mathcal{A}$ denote the set of infinite oriented paths in G which begin at v. This is equivalent to looking at the infinite oriented paths in $\mathcal{V}_+(v, G)$ which begin at the basepoint.

This defines $\mathcal{A}$ as a set, but in fact it has additional structure, coming from a special class of subsets that one might call *cells*. Given a vertex s in the $\mathcal{V}_+(v, G)$, define the *cell* $\mathcal{C}(s)$ to be the subset of $\mathcal{A}$ of infinite oriented paths in G which include the path represented by s as an initial subpath. This is the same in essence as the set of infinite oriented paths in the visibility which begin at the basepoint and pass through s.

This system of subsets of $\mathcal{A}$ enjoys very simple nesting properties. If s' represents an initial subpath of s, then $\mathcal{C}(s') \supseteq \mathcal{C}(s)$. This is immediate from the definitions. If s and s' represent paths such that neither is an initial subpath of the other — so that they must diverge at some vertex in G — then the cells $\mathcal{C}(s')$ and $\mathcal{C}(s)$ are *disjoint*.

We can decide to use this system of cells as a basis for a topology of $\mathcal{A}$. That is, each cell is considered to be open, as is any union of cells, and these are all of the open subsets of $\mathcal{A}$. This defines a topological space. In fact, it is totally disconnected, because the complement of each cell can be realized as a finite union of cells (this is not hard to verify), and is therefore open.

One can think of infinite paths in G as being described by infinite sequences of vertices and edges, in such a way that $\mathcal{A}$ can be identified with a subset of an infinite Cartesian product of finite sets (the sets of vertices and edges in G). (It is enough to simply use the edges here, and the vertices will be determined by them.) If one does this, then the topology just defined on $\mathcal{A}$ is the same as one inherited from using a classical product topology on the infinite Cartesian product space. In effect, $\mathcal{A}$ is realized topologically as a subset of a Cantor set.

What do these spaces look like? Here are a couple of basic observations.

Proposition 5.16 *Notation and assumptions as above. When the visibility has exponential growth, one can find a subset of $\mathcal{A}$ which is homeomorphic to a standard Cantor set.*

Proof This can be derived from the same kind of coding argument as in the proof of Proposition 5.1. Recall that the usual Cantor set is homeomorphic to the countably-infinite product of the discrete spaces $\{1, 2\}$, using the standard product topology. The elements of this product space can be seen simply as sequences of the form $\{x_j\}_{j=1}^{\infty}$, in which each each x_j takes either the value 1 or 2. From each such infinite sequence we can get an infinite oriented path in G, as in the proof of Proposition 5.1. This defines an embedding of the Cantor set into $\mathcal{A}$, and it is not too hard to show that it is actually a homeomorphism onto its image, using the definitions of the topologies involved. □

Proposition 5.17 *Notation and assumptions as above. Suppose that the visibility has polynomial growth, and let $\mathcal{L}$ denote the (finite) collection of oriented loops in G which are accessible from the basepoint v by an oriented path, as defined before just after Corollary 5.4. Then there is a natural mapping $\phi : \mathcal{A} \to \mathcal{L}$ with the following properties.*

(a) *Suppose that L_1, L_2 are elements of $\mathcal{L}$, and that $\{s_j\}$ is a sequence of elements of $\mathcal{A}$ such that $\phi(s_j) = L_1$ for all j. If $\{s_j\}$ converges to an element s of $\mathcal{A}$ with $\phi(s) = L_2$, then L_1 follows L_2 in the sense defined just before Corollary 5.4.*

(b) *Conversely, if L_1, L_2 are elements of $\mathcal{L}$ such that L_1 follows L_2, then one can find a sequence $\{s_j\}$ in $\mathcal{A}$ with the properties described in* (a).

(c) *If $\{s_j\}$, and s are as in (a), but now with $L_2 = L_1$, then $s_j = s$ for all sufficiently large j.*

Proof Notice first that if t is any element of $\mathcal{A}$, then t represents an infinite oriented path in G which, after a finite initial subpath, simply wraps around a single loop $L = L(t)$ in $\mathcal{L}$ infinitely often. This is not difficult to verify, using Lemma 5.5. (Remember also Corollary 5.4, which implies that once an oriented path α leaves a loop L' to go to a different loop L'', α cannot go back to L' ever again.) We define $\phi(t)$ to be exactly this "terminal" loop L.

Consider now part (a). In order for a sequence $\{s_j\}$ to converge to s, where $\phi(s) = L_2$, we must have that the s_j's themselves wrap around L_2 as j gets large, with the number of tours around L_2 going to infinity as $j \to \infty$. This is not hard to check from the definitions, using also the fact that s eventually just goes around L_2 over and over again. On the other hand, each s_j must end up in L_1 eventually, since $\phi(s_j) = L_1$ for all j. This implies that L_1 must follow L_2, by the definition of "following" for loops in G.

Conversely, if it happens that L_1 follows L_2, then it is easy to find sequences of this type. One can choose s_j so that it starts at v, follows a fixed oriented path β from v to L_2, goes around L_2 at least j times, and then proceeds to L_1 along an oriented path, where it spins around for the rest of its time. It is easy to see that these paths s_j converge as $j \to \infty$ to the path s which follows β from v to L_2, and then wraps around L_2 forever. Thus $\phi(s) = L_2$, and (b) follows.

Suppose now that we are back in the situation of (a), but with $L_1 = L_2$. Let u represent a (finite) initial subpath of s which goes from v to L_2. The part of s which comes after u must simply go around L_2 forever; if it were to ever leave L_2, it would not be able to come back, because of Lemma 5.3. We can use u to make a cell $\mathcal{C}(u)$, and then the definition of our topology on $\mathcal{A}$ implies that each s_j contains u as an initial subpath when j is sufficiently large. Since we have assumed now that the s_j's all have $L_2 = L_1$ as their terminal loop, the same argument as for s implies that after such an s_j traverses u, it can only spin around L_2 over and over again (without ever leaving L_2). This implies that $s_j = s$ for sufficiently large j, as desired. □

Corollary 5.18 *If the visibility $\mathcal{V}_+(v, G)$ has exponential growth, then $\mathcal{A}$ is uncountable, with the cardinality of the continuum. If $\mathcal{V}_+(v, G)$ has polynomial growth, then it is at most countable. In this case, if $\mathcal{L}$ is the set of oriented loops in G which are accessible from the basepoint v by an oriented path (as usual), then $\mathcal{A}$ is finite exactly when the depth of $\mathcal{L}$ is* 1, *and it is countably infinite when the depth of $\mathcal{L}$ is strictly greater than* 1.

Recall that the *depth* of $\mathcal{L}$ is the length of the longest chain of elements of $\mathcal{L}$, as defined just before Proposition 5.10.

Proof It is easy to see that $\mathcal{A}$ can never have cardinality greater than that of the continuum, since each element can be represented by a sequence of edges from G (which is a finite set). When the visibility has exponential growth, the cardinality is equal to that of the continuum, because $\mathcal{A}$ contains a subset which is in one-to-one correspondence with a Cantor set, as in Proposition 5.16. Now suppose that the visibility is of polynomial growth, so that we are in the situation of Section 5.3. Every element of t of $\mathcal{A}$ follows a finite initial subpath α and then simply wraps around the terminal loop $L = L(t)$ of t infinitely often, as observed at the beginning of the proof of Proposition 5.17. From this it is clear that $\mathcal{A}$ contains at most countably many elements, since there are at most countably many finite paths in G. If $\mathcal{L}$ has depth 1, then there are only finitely-many ways for an infinite path $t \in \mathcal{A}$ to reach its terminal loop $L(t)$, because t cannot pass through any other loop in G besides $L(t)$ (since $\mathcal{L}$ has depth 1), and therefore cannot pass through any vertex more than once before reaching $L(t)$. If $\mathcal{L}$ has depth at least 2, then there are distinct loops L_1 and L_2 in $\mathcal{L}$ with L_1 following L_2, and one can get infinitely many elements of $\mathcal{A}$ by taking paths which wrap around L_2 an arbitrary (but finite) number of times, and then proceed to L_1, which they go around forever. (This is analogous to the proof of (b) in Proposition 5.17.) This completes the proof of Corollary 5.18. □

In the case where the visibility has polynomial growth, the depth of $\mathcal{L}$ can also be described as follows. Define $\mathcal{A}^j$ recursively by setting $\mathcal{A}^0 = \mathcal{A}$ and taking

$\mathcal{A}^j$ to be the set of limit points in $\mathcal{A}^{j-1}$ when $j \geq 1$, with respect to the topology that we defined before. If d is the depth of $\mathcal{L}$, then

$$\mathcal{A}^d = \emptyset \quad \text{and} \quad \mathcal{A}^{d-1} \neq \emptyset. \tag{5.19}$$

More precisely, $\mathcal{A}^j$ consists of the elements t of $\mathcal{A}$ whose terminal loop $L(t)$ can be realized as the beginning of a chain of loops $L_1, L_2, \ldots, L_{j+1}$ in $\mathcal{L}$. This is not hard to show, using Proposition 5.17, and arguments like the ones above.

This discussion of asymptotic geometry of the visibility is analogous to (but much simpler than) well-known constructions discussed in [Gro87, Pan89a, GP91] for asymptotic geometry at infinity of negatively curved groups and manifolds. Note that Cantor sets also arise as the spaces at infinity of free groups, but in general groups with relations can have "connectedness" at infinity. For negatively-curved manifolds (and their fundamental groups) one gets topological spheres as the spaces at infinity, with dimension one less than the manifold with which one started. Although the *topologies* of these spheres are the standard ones, their *geometries* can be very different (when the manifolds in question have *variable* negative curvature, as in the case of complex hyperbolic spaces).

Some topics related to these came up in Sections 4.10 and 4.11.

What about *geometry* for $\mathcal{A}$? Can we define a natural notion of *distance* on $\mathcal{A}$, and not just a topology for it? Indeed, our cells in $\mathcal{A}$ have more combinatorial structure than we have used. One way to define a distance between elements of $\mathcal{A}$ is as follows. Given $u, t \in \mathcal{A}$, let $\mathcal{C}(s)$ be the *smallest* cell which contains them both. This amounts to taking s to be the largest common initial subpath of u and t. We then define the distance between u and t to be 2^{-l}, where l is the length of the path s in G. This is very much analogous to (but simpler than) constructions in [Gro87, Pan89a]. One can just as well use a^{-l} for some fixed $a > 1$ instead of 2, and this is the same as changing the distance above by a power.

The metric $d(\cdot, \cdot)$ on $\mathcal{A}$ described above is actually an *ultrametric*, which means that

$$d(x, z) \leq \max\{d(x, y), d(y, z)\} \tag{5.20}$$

for all points $x, y, z \in \mathcal{A}$. This can be derived from the nesting properties of cells, i.e., if two cells intersect, then one must be contained in the other. This property of being an ultrametric is very strong, and reflects the disconnected nature of the topology. In particular, balls in ultrametric spaces are always both open and closed, as one can check.

The diameter of $\mathcal{A}$ with respect to this metric is at most 1, by construction. One can also show that $\mathcal{A}$ is compact. This is not hard, and it is similar to the compactness of infinite Cartesian products of finite sets, when given the usual product topology. As mentioned earlier, $\mathcal{A}$ can be viewed as a subset of such a product, and that provides a natural way to look at the compactness of $\mathcal{A}$. (I.e., one can verify that $\mathcal{A}$ is closed as a subset of this infinite Cartesian product.)

6

GEOMETRIC ASPECTS OF CUT ELIMINATION

Imagine reading a proof in a mathematical text. It may have several lemmas which interact with each other in a tricky way. In order to see better what is happening within the proof, one might try to unwind the lemmas to make explicit each basic step in the argument. This makes sense informally, in terms of our everyday experience, but it can also be treated more formally through mathematical logic. (This general theme occurs repeatedly in the writings of Kreisel [Kre77, Kre81b, Kre81a].)

Each lemma can reflect a subtle process in its own right. The interaction between lemmas may lead to complex systems which enjoy efficient representation in a proof. The unwinding of the lemmas then generates a proof whose local structure might be quite simple, but whose large-scale behavior can involve complicated patterns, patterns that reflect the fact that the proof could be compressed through the introduction of lemmas. In general, it is not easy to recognize when a proof can be made substantially smaller through the introduction of lemmas.

To give precise meaning to these ideas, we can consider formal proofs in classical logic. Specifically, we shall use *sequent calculus* (reviewed in Appendix A), for which the notion of *lemmas* is captured by the *cut rule.* Roughly speaking, the cut rule permits one to say that if A implies B and B implies C, then A implies C directly. (In fact it is somewhat more general than this.)

To "unwind the lemmas" in a formal proof in the sequent calculus, we can try to eliminate the cuts. There is a fundamental method for doing this, originally proposed by Gentzen in the 1930's, and developed further in several directions since then. In this chapter, we shall look at the combinatorics of cut elimination with particular emphasis on geometric effects, as seen through the *logical flow graph* (Section A.3) of a formal proof.

In recent years much work related to cut elimination and its complexity has been done in the context of *linear logic.* Some references include [Gir87a, Gir89a, Gir90, Gir95a, Gir95b, DJS97].

6.1 Preliminary remarks

Imagine that one has a finite presentation of a group G, and a word w over the generators of G. How might one prove the triviality of w?

Let us be more precise. A finite presentation of G consists of a finite set of generators $g_1, \ldots, g_n$ together with a finite collection of words over the g_i's and their inverses called *relations.* For simplicity, let us assume that the set of relations includes the inverses of all of its elements, and the empty word. Each of the relations is supposed to represent the identity element of G. This implies that arbitrary products of conjugates of relations also represent the identity element, as well as words that can be obtained from these through the cancellation of subwords of the form $g_i^{-1} g_i$ or $g_i g_i^{-1}$. In order to have a presentation for G, it should also be true that every trivial word arises in this manner. Alternatively, one can think of G as being given by the quotient of the free group with generators $g_1, \ldots, g_n$ by the normal subgroup generated by the relations.

So how might one prove the triviality of a given word w? The most direct approach would be to produce an explicit product of conjugates of relations from which w can be derived through cancellations. One can also make proofs which are less explicit, through lemmas that encode general recipes for the construction of trivial words. These lemmas might be used many times, and in no particular ordering, so that by the end, one may not have a clear idea of how to write an explicit product of conjugates of relations, even if one knows that this is possible in principle.

This is a familiar scenario in mathematics. In general, infinite processes might lead to proofs which are nonconstructive. Even in purely "finite" contexts, the explicit rendering of the implicit constructions given by formal proofs can be very subtle. The elimination of cuts provides exactly a way in which to do this.

Proofs with cuts can often be much shorter than proofs without, as in [Ore82, Ore93, Sta74, Sta78, Sta79, Tse68]. It is natural to expect this phenomenon to be connected to the presence of some kind of *symmetry* in the underlying language or objects. In other words, if it is possible to make a much shorter proof with cuts than without, then the "lemmas" being used ought to capture some fundamental rules or patterns in the underlying objects or structure. These patterns should then become visible when one tries to eliminate the cuts.

To understand the kinds of symmetries which might be captured by lemmas, one can look at patterns produced by cut elimination itself. In general, the elimination of cuts can lead to enormous expansion in the underlying proof, and in the standard procedures the main force behind this expansion comes from the *duplication of subproofs* involved in the simplification of a cut over a contraction. Roughly speaking, the duplication of subproofs corresponds to the fact that one can use a single lemma many times, even though it is proved only once; in a direct proof, one should give a separate proof for each application of the lemma, using particular data at hand. In the context of finitely-presented groups, for instance, one might prove a lemma that says that the square of any trivial word is trivial, and in the elimination of cuts the proof of this general lemma would be repeated for each application.

We shall describe the duplication of subproofs in cut elimination more precisely in the next section. It is already quite interesting in its simplest combinatorial form, and we shall analyze its possible effects on the underlying "logical flow graph" (Section A.3) in some detail.

For various reasons connected to complexity and automatic deduction of theorems, it would be interesting to be able to determine when a proof Π can be "compressed" through the introduction of cuts. This is a very difficult problem. One can look at it in geometric terms, as being analogous to asking when a given graph can be "folded" into a much smaller one. For formal proofs the problem is much more difficult than in purely geometric contexts. Even if one knows that the given proof Π was obtained from one with cuts through a standard method of cut elimination, there is no clear way to recover the symmetries in the larger proof. One of the reasons for this is that in the duplication of subproofs some information is lost, and there is no clear way to "guess" what has been lost in order to go backwards in the construction. (See [Car97] for information about which properties of a proof after cut elimination can be traced back to the original proof. In particular, one can look at this in connection with the notion of "inner proofs", mentioned in Section 2.1.)

Let us now proceed to a more detailed discussion of Gentzen's method of cut elimination and its effect on the geometry of the underlying logical flow graphs.

6.2 The process of cut elimination

For most of the rest of this chapter we shall need to assume that the reader has some familiarity with sequent calculus, especially the cut and contraction rules. A brief review is given in Section A.1 in Appendix A. The definition of the *logical flow graph* will also be needed, and it can be found in Section A.3 in Appendix A. Remember that some simple examples of formal proofs in the sequent calculus were described in Chapter 3.

We shall mostly concentrate on *topological* properties of logical flow graphs of proofs, such as the existence of oriented cycles and the behavior of oriented paths. Note that logical flow graphs automatically have a natural orientation (Section A.3), and that they are always *optical graphs*, as in Section 4.1. We shall often look at logical flow graphs in terms of their visibility graphs, and we shall therefore be concerned with chains of focal pairs (Definition 4.17) as well.

The presence of cuts and contractions in a formal proof is important logically and for the patterns which can appear in the logical flow graph. For instance, cuts and contractions are both needed in order to have oriented cycles in a logical flow graph [Car97, Cara]. One of the main goals of this chapter is to see what kind of geometric patterns arise naturally from a formal proof under Gentzen's method of cut elimination.

We shall restrict ourselves here to classical logic (as opposed to other calculi), although much of the discussion could be applied to other contexts. Detailed treatments of Gentzen's method are provided by [Gir87c, Tak87], and a general introduction is given in [CS97]. See Section A.2 for more information.

We should emphasize that the approach to eliminating cuts described here is not the only one that is available. In particular, there are important differences between the classical methods of cut elimination and the ones in linear logic. See [Car97] for a method of cut elimination in *propositional* logic in which one has more control over the topology of the underlying logical flow graphs.

In order to illustrate the basic idea, imagine that we have a proof Π which uses a cut over a formula A, and that we want to eliminate this cut. In general, we cannot do this in a single step, but instead we have to look at the way that A was built up inside the proof. The idea is to push the cut up higher and higher in the proof, until we get close enough to the axioms that we can eliminate it directly.

During the process of cut elimination, one often *increases* the total number of cuts, but the point is that one is able to reduce their *complexity* in a certain way. One has to be slightly careful about the precise measurement of the complexity of the cuts, in that one should take into account both the *structural complexity* of the cut formula (i.e., the number of logical connectives), as well as the extent to which the contraction rule was used in the history of the formula within the proof. A key feature of the process is that one adds new cuts only *above* the line of the proof at which one is working at a given moment. One typically starts with cuts which are as far down in the proof as possible, so that there is no danger of accidentally increasing the complexity of a cut which might otherwise occur below, or anything like that. This is helpful for showing that the process of cut elimination will actually end in a finite number of steps.

To carry out this procedure, one distinguishes cases depending on the structure of the cut formula, and whether it came from a contraction. There are particular recipes for dealing with each case, as described in [Gir87c, Tak87, CS97]. For our purposes at the moment, the specific nature of many of these recipes is not important, because they do not change the "topological" features of the logical flow graphs. That is, we shall not worry about having to add vertices or to shrink or extend edges, as is sometimes needed to accommodate the addition or removal of formulae in the process of cut elimination. Instead we shall focus on operations which can lead to the breaking of cycles or splitting of paths.

Specifically, the operations in the usual method of cut elimination that deal with the removal of logical connectives from the cut formula are not important for the *topology* of the logical flow graphs, but the duplication of subproofs that one employs to push a cut above a contraction does change the structure of the logical flow graph in a substantial way. This is discussed in [Car97, CS97], and we shall return to it in a moment.

Through the repeated application of these operations, one eventually reduces to situations in which there is a cut over a formula which comes directly from an axiom, either as a distinguished occurrence or as a weak occurrence. Consider first the situation where the cut formula comes from a distinguished occurrence in an axiom, as in the following.

$$\frac{\Gamma_1, A \to A, \Delta_1 \quad \overset{\Pi_*}{A, \Gamma_2 \to \Delta_2}}{\Gamma_1, A, \Gamma_2 \to \Delta_1, \Delta_2} \tag{6.1}$$

In this case we can remove the axiom from the proof and simply add the weak occurrences in Γ_1 and Δ_1 to the subproof Π_* without trouble, thereby obtaining a new proof of the sequent $\Gamma_1, A, \Gamma_2 \to \Delta_1, \Delta_2$, in which the last cut has been eliminated. The *topology* of the logical flow graph is not altered in this step; paths are shrunk or extended, but that is all.

Suppose instead that we have a cut over a formula which comes from a weak occurrence in an axiom, as in the following situation.

$$\frac{\Gamma_1, A \to A, \Delta_1, C \qquad \begin{array}{c}\Pi_0\\ C, \Gamma_2 \to \Delta_2\end{array}}{\Gamma_1, A, \Gamma_2 \to A, \Delta_1, \Delta_2} \tag{6.2}$$

To eliminate the cut, one can simply eliminate the subproof Π_0, take out the (weak) occurrence of C in the axiom, and add Γ_2 and Δ_2 to the axiom as weak occurrences. In other words, the sequent

$$\Gamma_1, A, \Gamma_2 \to A, \Delta_1, \Delta_2 \tag{6.3}$$

is itself an axiom already. By doing this one removes a possibly large part of the logical flow graph, and this can easily change the topology of the part that remains in a strong way. It can lead to the breaking of cycles in the proof as a whole, or to the breaking of connections between different formula occurrences in the proof as a whole.

Let us now consider the case of contractions. The following diagram shows the basic problem.

$$\frac{\begin{array}{c}\Pi_1\\ \Gamma_1 \to \Delta_1, A\end{array} \qquad \dfrac{\begin{array}{c}\Pi_2\\ A^1, A^2, \Gamma_2 \to \Delta_2\end{array}}{A, \Gamma_2 \to \Delta_2}}{\Gamma_1, \Gamma_2 \to \Delta_1, \Delta_2} \tag{6.4}$$

That is, A^1 and A^2 denote two occurrences of the same formula A, and they are contracted into a single occurrence before the cut is applied. The contraction could just as well be on the left, and this would be treated in the same way. To push the cut above the contraction, one duplicates the subproof Π_1, and uses the cut rule twice, as indicated below.

$$\begin{array}{c}\dfrac{\begin{array}{c}\Pi_1\\ \Gamma_1 \to \Delta_1, A\end{array} \qquad \dfrac{\begin{array}{c}\Pi_1\\ \Gamma_1 \to \Delta_1, A\end{array} \quad \begin{array}{c}\Pi_2\\ A^1, A^2, \Gamma_2 \to \Delta_2\end{array}}{A^2, \Gamma_1, \Gamma_2 \to \Delta_1, \Delta_2}}{\Gamma_1, \Gamma_1, \Gamma_2 \to \Delta_1, \Delta_1, \Delta_2} \\ \vdots \ \textit{contractions} \\ \Gamma_1, \Gamma_2 \to \Delta_1, \Delta_2\end{array} \tag{6.5}$$

This case can be more intricate topologically. Again vertices that were connected in the original proof can become disconnected by this operation, even though we are not throwing away anything in the graph. We can also break cycles in this operation without disconnecting vertices. In fact there are several different geometric phenomena which can result from this operation, which we discuss further in the next sections.

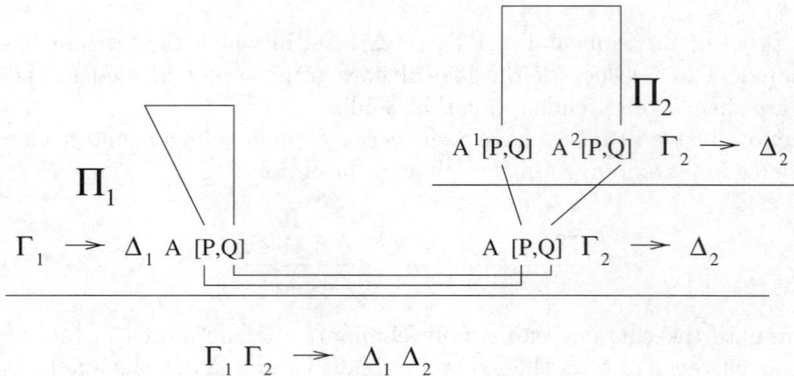

FIG. 6.1. A cycle before the duplication of subproofs

6.3 A first scenario, and the breaking of cycles

Let us describe a general situation in which the breaking of paths and cycles can occur under the duplication of subproofs given in (6.4) and (6.5) above. Let P and Q be a pair of atomic occurrences lying inside the occurrences A^1 and A^2 in (6.4), respectively. To be more precise, we think of P as lying in A^1 and Q as lying in A^2, although each has a counterpart inside the other, since A^1 and A^2 are identical as logical formulae (as they must be in order to apply the contraction in (6.4)). We assume that P and Q do not occupy the same position in A^1 and A^2, but that they do represent the same atomic formula, so that they have the possibility of being connected to each other in the logical flow graph of the proof as a whole. For instance, A^1 and A^2 might be of the form $S \vee \neg S$, where P and Q correspond to the two different occurrences of S.

Imagine first that there is a path from P to Q in the part of the logical flow graph that comes from Π_2, above the contraction. As in Fig. 6.1, the initial configuration (6.4) might also contain a path starting at Q and going down through the contraction and the cut, up into Π_1, and then down again and back through the cut and contraction a second time to arrive at P inside A^1. After the transformation from (6.4) to (6.5), this cannot happen. The path starting from Q and going into the (lower) copy of Π_1 will not have the opportunity to go back through A^1, but can only go back into Π_2 through A^2, as shown in Fig. 6.2. In this way, an oriented cycle can be broken.

Similarly, if the first path from Q to P in Π_2 did not exist, then we would not have a cycle, but we could still have a connection from Q to P through Π_1 as before. This connection would again be broken in the passage from (6.4) to (6.5). (See [Car97, Cara] for more information about these phenomena and their role in the structure of formal proofs.)

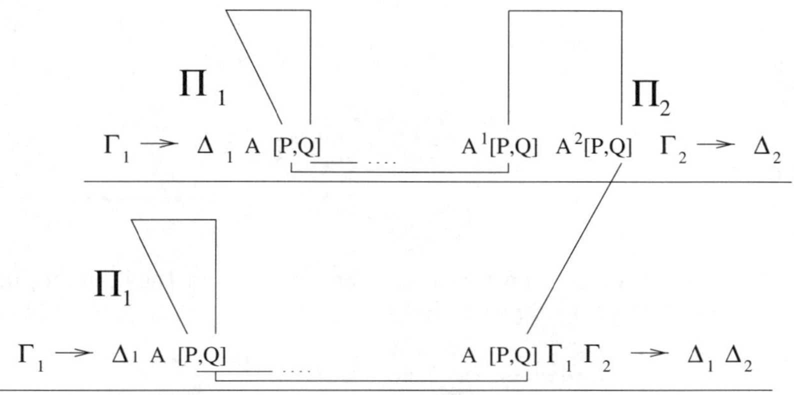

FIG. 6.2. Breaking the cycle

6.4 A second scenario, and the breaking of focal pairs

Let us consider now another kind of reduction of connectedness, namely, the breaking of focal pairs (Definition 4.15), and the reduction in length of chains of focal pairs (Definition 4.17). These phenomena can easily occur in the cancellation of subproofs (as in the transformation from (6.2) to (6.3)), and so we shall concentrate on the duplication of subproofs.

Thus we suppose again that we are in the situation of (6.4) and (6.5). Imagine that there is a node in Π_2 from which a pair of oriented paths emerges and goes through Π_2 until the two paths reach A^1 and A^2, respectively. From there the paths will proceed down through the contraction and across the cut into Π_1. Let us assume that the two paths either converge together at the contraction of A^1 and A^2 (Fig. 6.3), or later in the subproof Π_1 (Fig. 6.4). In both cases, we assume that the paths end in weak occurrences in the axioms in Π_1. In either situation, the convergence of the two paths would be broken in the passage from (6.4) to (6.5). In particular, we would lose the focal pair that we had in the logical flow graph of the original proof.

We shall discuss this situation a bit further in Section 6.7, after discussing a different kind of effect on focal pairs which can occur in the transition from (6.4) to (6.5).

6.5 A third scenario, and chains of focal pairs

Let us continue to assume that we are in the context of (6.4) and (6.5), and that we have a pair of paths which begin at some common starting point in Π_2, and which reach A^1 and A^2 in the contraction, respectively. For the sake of definiteness, we assume for the moment that the paths converge to the same point once the contraction is performed. At this stage, the two paths continue along a common trajectory into Π_1.

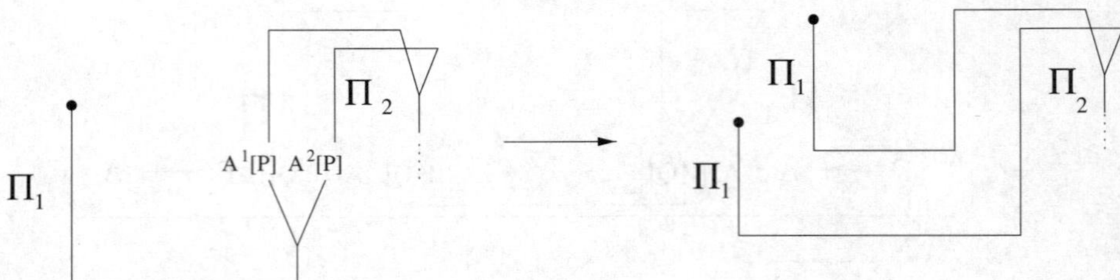

FIG. 6.3. Convergence of paths at the contraction, and the splitting induced by the duplication of subproofs

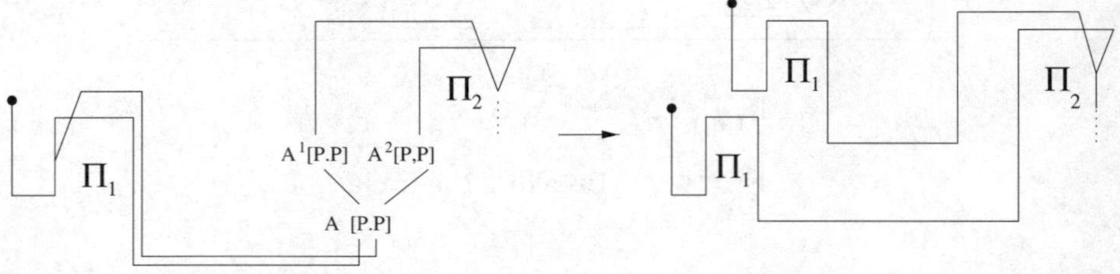

FIG. 6.4. Convergence in Π_1, with splitting of paths again after the duplication of subproofs

In the previous section, we observed that this path could end in a weak occurrence of Π_1, so that the two paths become completely split apart after the duplication of subproofs. Instead of doing that, it could continue on to a formula in Γ_1 or Δ_1 in the endsequent of Π_1, and then be reunited in the contractions that occur below, as illustrated in Fig. 6.5. In this case, the duplication of subproofs would not break apart the original focal pair in (6.4), but would simply postpone the convergence of the paths until the contractions below the two copies of Π_1 in (6.5).

This kind of process would disrupt a *chain* of focal pairs, however. Suppose

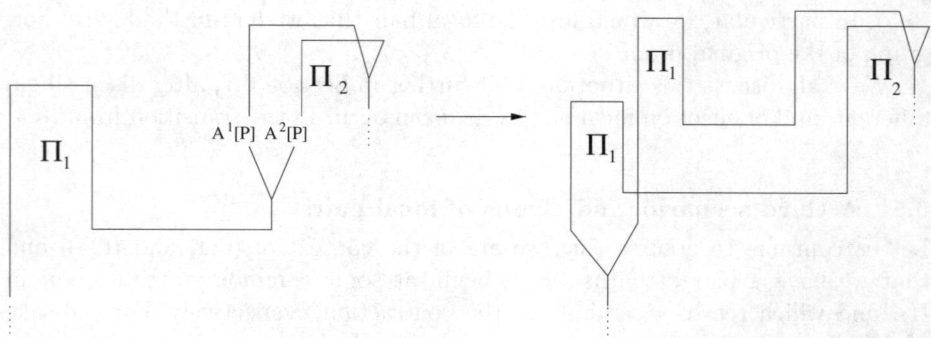

FIG. 6.5. Paths reaching the endsequent of Π_1

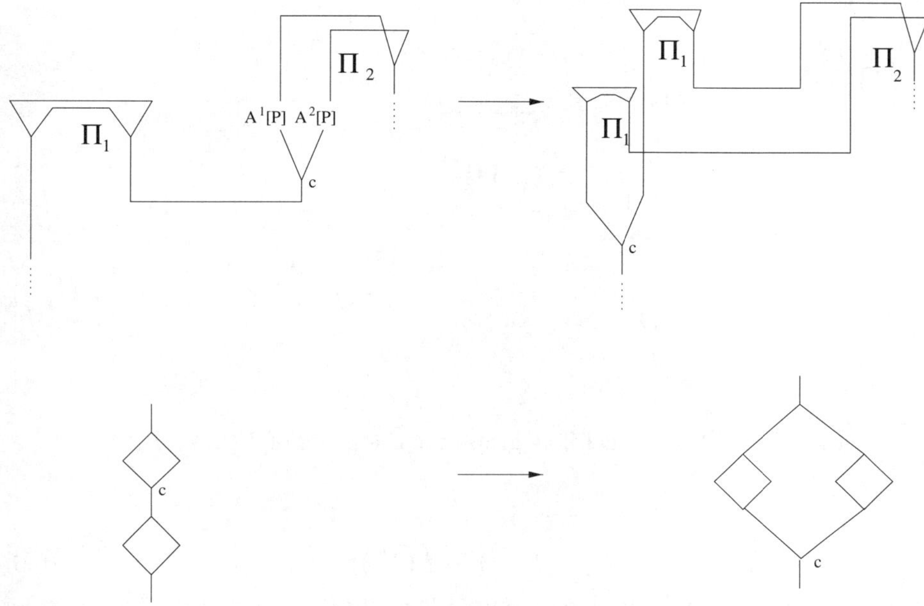

FIG. 6.6. A second focal pair in Π_1, and the effect on it of the duplication of subproofs

that our paths converge at the contraction and continue on into Π_1, where they run into a second focal pair contained in Π_1, before ending in Γ_1 or Δ_1 in the endsequent of Π_1. This possibility is depicted in the first part of Fig. 6.6, and it would give a chain of focal pairs of length 2 in the original proof, before the duplication of subproofs.

In the duplication of subproofs, we eliminate the contraction at which the first convergence takes place. At best we can only postpone the convergence from the original contraction to the ones below the two copies of Π_1 in (6.5), as in Fig. 6.6, but this would not be good enough for maintaining the chain of focal pairs of length 2 in (6.5). Instead of having two focal pairs, with one following the other, we have a kind of nesting of focal pairs, which is very different. This is illustrated in the second part of Fig. 6.6, in which we see also how the focal pair inside Π_1 is duplicated in (6.5).

Instead of having our path from the contraction of A^1 and A^2 continue into a single focal pair in Π_1, it might just as well continue into a chain of focal pairs of length n in Π_1. This would give rise to a chain of length $n + 1$ in (6.4). After the duplication of subproofs , we would again lose the chain of length $n + 1$ in the larger proof, and we would have two copies of the chain of length n from Π_1.

This type of phenomenon can occur easily in concrete examples of formal proofs. For instance, let us consider proofs of the feasibility of large numbers using cuts and contractions as in Section 4.8. As before, our basic building block

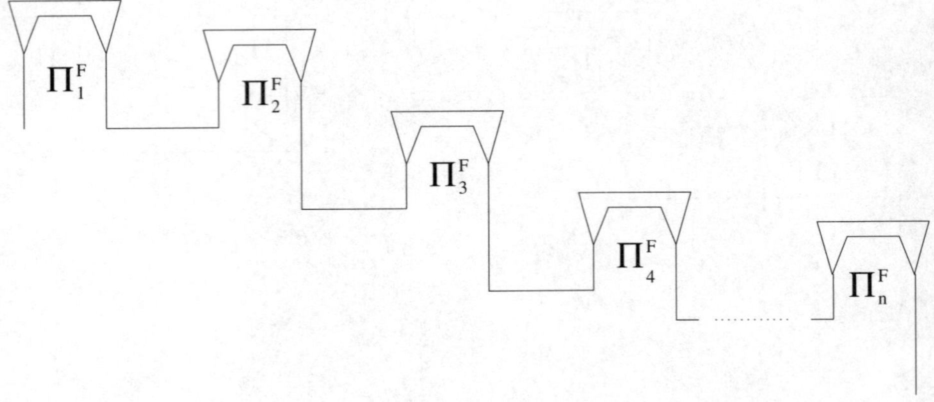

FIG. 6.7. The logical flow graph for the proof of $F(2) \to F(2^{2^n})$

is given by

$$F(2^{2^{j-1}}) \to F(2^{2^j}), \tag{6.6}$$

which can be proved for each j in only a few steps (as in Section 4.8). We can then combine a sequence of these proofs using cuts to get a proof of

$$F(2) \to F(2^{2^n}) \tag{6.7}$$

in $O(n)$ steps.

The logical flow graph for the proof of (6.7) is pictured in Fig. 6.7. The notation Π^F_j, $1 \leq j \leq n$, in Fig. 6.7 refers to the proofs of (6.6) for these values of j. The logical flow graph of each Π^F_j contains two branches, one for the contraction of two occurrences of $F(2^{2^{j-1}})$ on the left, and the other for the use of the $F : times$ rule on the right (which says that the feasibility of two terms s and t implies the feasibility of $s \cdot t$). (See [Car00] for further discussion of logical flow graphs in the context of feasibility.)

If we push a cut in this proof above the corresponding contraction by duplicating subproofs as before, then we shall see exactly the kind of phenomena described above. In the end, the logical flow graph will be transformed into a graph roughly like the one called "H" in Section 4.3. (See Fig. 4.4.)

There is more than one way to push the cuts up above the contractions in this case. Normally one would start at the "bottom" of the proof, which means starting on the far right-hand side of the picture above, but in this case one could also start at the beginning of the proof, or in the middle. If one starts at the beginning (which means the far left-hand side of the graph), then the systematic duplication of subproofs leads to an evolution of logical flow graphs like the one shown in Fig. 6.8. If one begins at the other end of the proof, then the evolution of logical flow graphs will look like the one in Fig. 6.9.

The final result is the same, independently of whether one chooses to start from the beginning or the end of the original proof, or from anywhere in between.

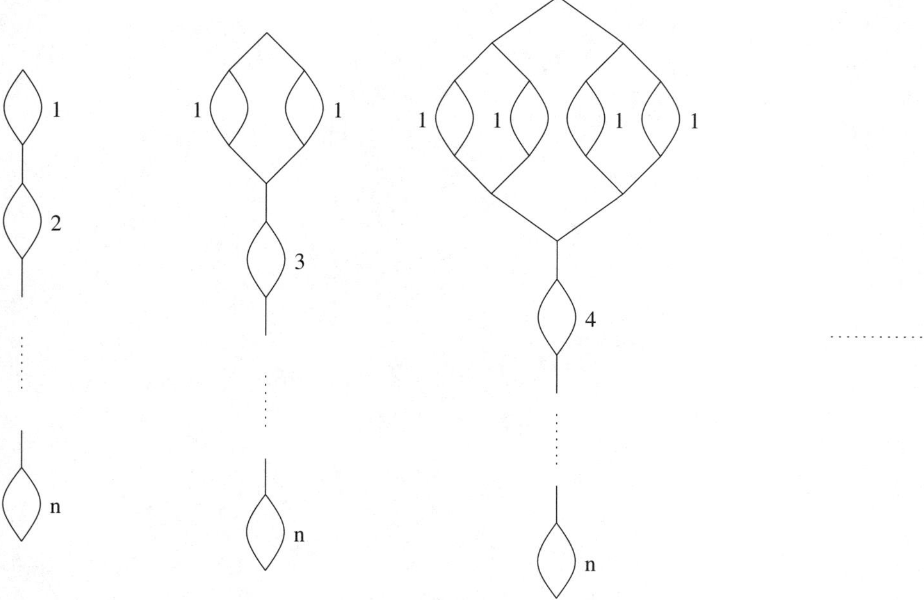

FIG. 6.8. An evolution of graphs under the duplication of subproofs, starting at the beginning of the proof

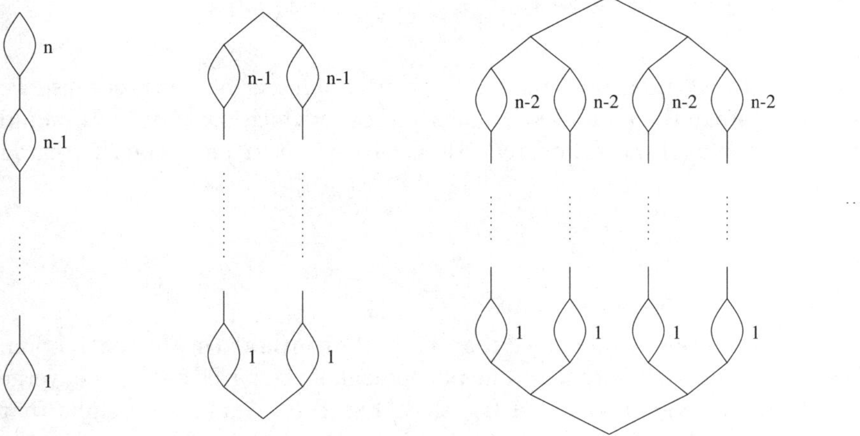

FIG. 6.9. An evolution of graphs under the duplication of subproofs, starting at the end of the proof

In the end, one obtains a graph of exponential size, which looks like the graph H in Section 4.3 (Fig. 4.4). Note that if one starts at the beginning of the proof, then the whole job is done in $n-1$ steps, i.e., with $n-1$ applications of the operation of duplicating the subproof, as in (6.4) and (6.5). If we start from

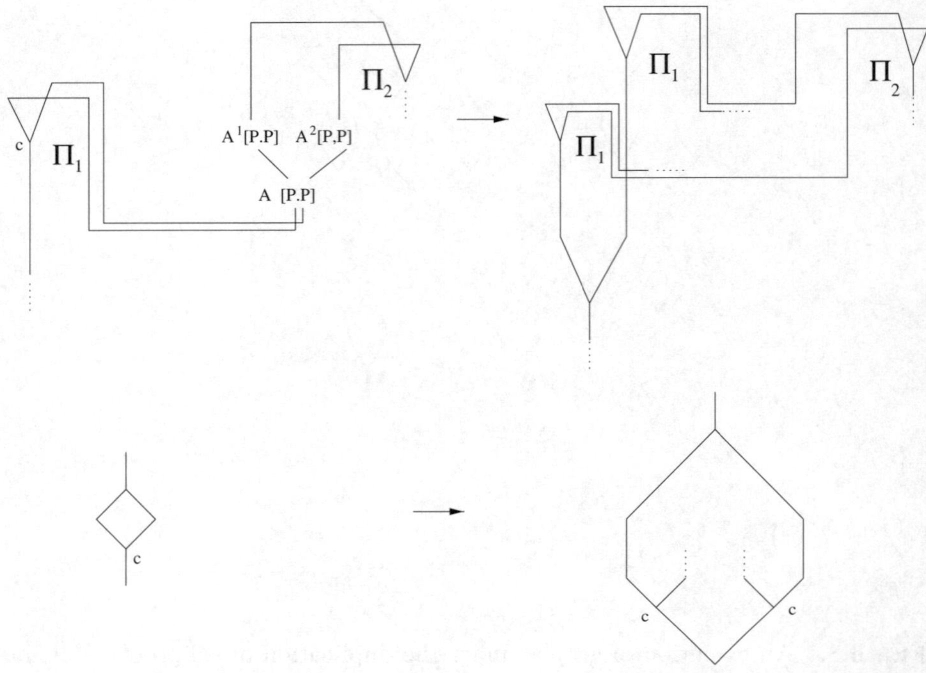

FIG. 6.10. Paths from Π_2 converging in Π_1 (after the contraction) and continuing on to the endsequent of Π_1

the end of the proof, then we need an exponential number of steps, because we double at each stage the number of smaller pieces to which the procedure would be performed next. (However, one can think of these next pieces being treated "in parallel".)

6.6 The third scenario, continued

In the preceding section, we assumed that our paths converged at the contraction, but we would have practically the same phenomena if the paths did not converge there, but did converge later on in Π_1, as in Fig. 6.10. Again we assume that after the paths converge they continue on into a formula in Γ_1 or Δ_1 in the endsequent of Π_1. In this case, the focal pair that we have in the original proof (6.4) persists in (6.5), but with the convergence postponed as before.

After passing through the point of convergence in Π_1, we might pass through a chain of focal pairs in Π_1, so that in the proof as a whole we have a chain of length $n + 1$. As before, the first focal pair in this chain would be disrupted by the elimination of the contraction at A^1, A^2, so that the chain of length $n + 1$ would not persist in (6.5). In the end, we would have two copies of the chain of length n from Π_1, just as in the first step of the evolution shown in Fig. 6.9.

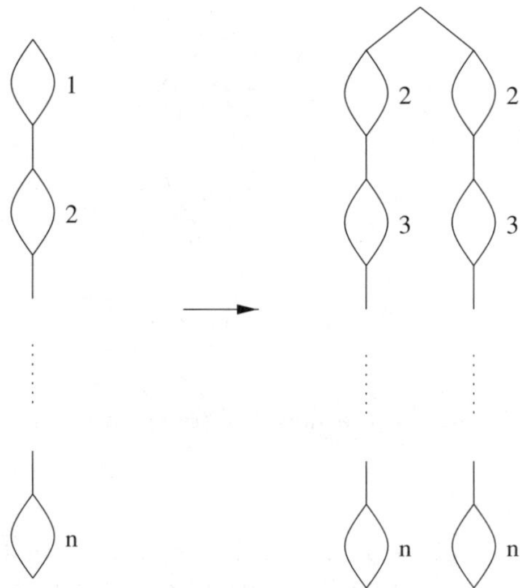

FIG. 6.11. The splitting of a chain of focal pairs of length n into two chains of length $n-1$

6.7 Chains of focal pairs in the second scenario

The configurations that we have considered in the second and third scenarios are very similar to each other. The only difference between the two lies in the possibility that our paths which begin in Π_2 might end in Π_1, or might continue on into formulae in the endsequent of Π_1.

We did not mention it before, but we could just as well consider the possibility of chains of focal pairs in the second scenario as in the third one. That is, after our pair of paths converges, either at the contraction of A^1, A^2 or later in Π_1, they could easily pass through a chain of focal pairs of length $n-1$ in Π_1 before ending in a weak occurrence in an axiom. In this case, we would have a chain of focal pairs of length n in the original proof as a whole, because of the focal pair which begins in Π_2. This chain would not persist after the duplication of the subproofs, but instead we would have two copies of the chain of length $n-1$ from Π_1 after the duplication of subproofs. This time, these two smaller chains would not come back together again, as in Section 6.5, but would diverge from each other, as in Section 6.4. This is illustrated in Figures 6.11 and 6.12.

We can see this type of phenomenon concretely in the context of feasible numbers, and the proof described in Section 4.8 again. Consider the sequent

$$\to F(2^{2^n}). \tag{6.8}$$

This can be proved in $O(n)$ steps in nearly the same way as (6.7) was. In fact, one can prove (6.8) by combining (6.7) with

$$\to F(2) \tag{6.9}$$

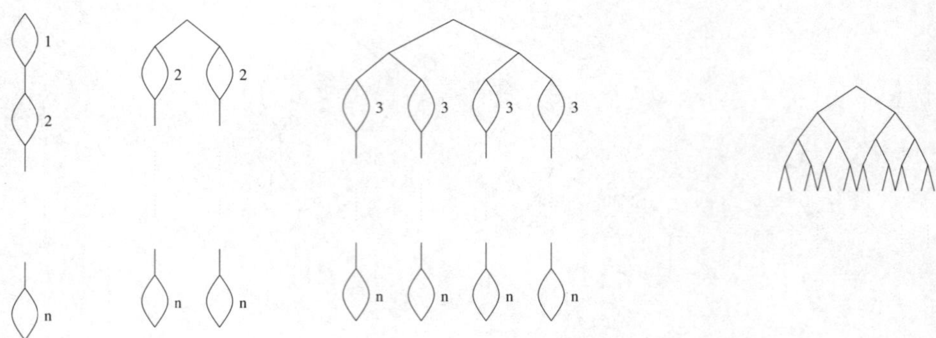

FIG. 6.12. The evolution from a chain of focal pairs to a tree (of exponential size)

using a cut. The proof of $\to F(2)$ consists of the special axiom $\to F(0)$ for feasible numbers followed by two applications of the successor rule.

The occurrence of $F(0)$ in the special axiom $\to F(0)$ behaves somewhat like a weak occurrence, in the sense that paths in the logical flow graph can end there and have nowhere else to go. This is not the case for axioms in ordinary sequent calculus, in which there are two distinguished occurrences on opposite sides of the sequent arrow which are always linked in the logical flow graph. This occurrence of $F(0)$ in $\to F(0)$ is not weak, however, and does not allow the cancellation of subproofs described in Section 6.2.

In this setting a natural instance of (6.4) is given by

$$\cfrac{\begin{matrix}\Pi_1 \\ \to F(2^{2^{j-1}})\end{matrix} \qquad \cfrac{\begin{matrix}\Pi_2 \\ F(2^{2^{j-1}}), F(2^{2^{j-1}}) \to F(2^{2^j})\end{matrix}}{F(2^{2^{j-1}}) \to F(2^{2^j})}}{\to F(2^{2^j})} \tag{6.10}$$

A key point now is that we do not have any "side" formulae in the endsequent of Π_1. In other words, the collections Γ_1, Δ_1 in Π_1 are empty here, and the scenario of Section 6.5 is simply not possible. Indeed, it is exactly the scenario of Section 6.4 which occurs here, together with the additional nuance of having a long chain of focal pairs in the proof Π_1, as discussed above.

If one takes the proof of (6.8) with cuts mentioned above, and simplifies all of the cuts over the contractions in the usual way (i.e., following (6.4) and (6.5)), then one gets in the end a binary tree of exponential size in n. That is, one has uniform binary splitting of the branches until almost the very end, where one picks up short linear graphs associated to the proof of $\to F(2)$ mentioned before.

6.8 Recapitulation

Let us pause a moment to summarize some of what we have seen so far. In the duplication of subproofs, we can easily have substantial increase in the size of the proof as a whole, and also of the underlying logical flow graphs. The geometric complexity of the logical flow graphs can also increase substantially, in the sense that there can be numerous additional contractions, and much duplication of chains of focal pairs. However, if we measure the complexity of the logical flow graph in terms of the lengths of the longest chains of focal pairs, then we have seen clearly how this is often reduced by the simplification of cuts over contractions.

This is very nice, because we know that the method of cut elimination reduces the intricacy of a proof in many ways, even if it also increases the *size* of the proof. In this analysis of the long chains of focal pairs, we see a similar effect in a very concrete and geometric way.

In the second and third scenarios, we have neglected a third possibility, which is that our paths from Π_2 could go into Π_1 through the cut and then come back out of the cut into Π_2 again (as opposed to ending in weak occurrences in Π_1, or going down to the endsequent of Π_1). This case could be analyzed in much the same manner as before. Once the paths go back into Π_2, they may or may not converge again, or encounter additional chains of focal pairs. They might eventually end in weak occurrences in Π_2, or go back through the cut into Π_1, or they might go down into the endsequent of Π_2 and continue on into the rest of the proof below.

What happens if our paths do go down into the endsequents of Π_1 or Π_2? Normally Π_1 and Π_2 are only pieces of a larger proof Π, and there could be a lot more activity in Π below Π_1 and Π_2. If there are cuts in Π below Π_1 and Π_2, then an oriented path which begins in Π_1, Π_2 could go below Π_1, Π_2 inside Π, and then come back up again, or go up in a different part of the proof. At this point the whole story could start over again.

Fortunately this does not happen in many situations of interest. In the standard procedure of cut elimination, one makes a point of simplifying the cuts from the bottom, and this would imply that there are no cuts in Π below Π_1 and Π_2. In this case, the paths could not come back up once they have gone below Π_1 and Π_2. (We shall discuss related observations further in Section 6.14.) Even if there are cuts, one may be prevented from going below Π_1 and Π_2 and then up into them again. (It may be possible to go up into other subproofs, though.) This will be true under suitable "monotonicity" assumptions on the cut formulae (e.g., when the cut formulae simply do not contain any negations), and in particular this is true for the examples related to feasible numbers mentioned in Sections 6.5 and 6.7.

This is almost the complete story about the possible behavior of paths in the logical flow graph, as it is related to the duplication of subproofs. Any oriented path in the logical flow graph of a proof can be continued until it reaches either (1) the endsequent of the proof, or (2) a weak occurrence in an axiom (or something like an occurrence of $F(0)$ in the special axiom $\to F(0)$, when non-logical axioms are permitted), or (3) until it reaches some oriented cycles around which it can wrap forever. That is, it might simply go around a particular cycle over and over again in a periodic manner, but it could also reach a union of cycles and switch back and forth between them in an unpredictable manner. This last possibility (in oriented graphs more generally) was discussed already in Chapter 5, especially Sections 5.2 and 5.4.

6.9 Proofs without focal pairs

In Sections 6.5 and 6.7, we have seen how the duplication of subproofs in the standard method of cut elimination can lead to the simplification of long chains of focal pairs, in the sense that the length of the longest chain is systematically reduced. On the basis of these examples, one might be tempted to conjecture that the presence of long chains of focal pairs is somehow necessary for exponential expansion in the passage to a cut-free proof to be unavoidable.

This is not the case. Simple counterexamples are provided by the family of sequents (3.28) in Section 3.3. To see this, let us mention the following simple observation.

Lemma 6.1 *If a formal proof* Π *contains contractions only over* negative *atomic occurrences (or only over* positive *atomic occurrences), then the logical flow graph for* Π *contains no focal pairs.*

To be more precise, this lemma applies to proofs Π in ordinary logic, without special rules of inference. It is not correct in the context of feasible numbers, for instance, as one can see from the examples discussed in Sections 6.5 and 6.7.

Proof The main point is that contractions over *negative* atomic occurrences correspond exactly to *defocusing* branch points in the logical flow graph, while contractions over *positive* atomic occurrences correspond to *focusing* branch points. To have a focal pair, one must have at least one focusing branch point and one defocusing branch point, from which the lemma follows easily. (This argument breaks down in the context of feasible numbers, because focusing branch points can also arise from the special rules of inference concerning the feasibility of sums and products, and not just from contractions.) □

In the case of the family of sequents (3.28), the proofs with cuts described in Section 3.3 contained contractions only over formulae whose atomic subformulae were all *negative*. Specifically, the contractions occurred in the derivation of (3.32) and in the passage from (3.34) to (3.29), and nowhere else. The formulae being contracted were always occurrences of F_i (defined in (3.26)) on the left side of the sequent. It follows from the lemma that the logical flow graph of these proofs contain no focal pairs. One can also check easily that they contain no nontrivial oriented cycles, and in fact they are *forests* (disjoint unions of trees).

Thus the logical flow graphs are very simple in this case, despite the fact that all cut-free proofs of (3.28) are necessarily of exponential size [Sta78, Bus88, Ore82, Ore93]. For this example, there is a natural way in which the exponential expansion that comes from the simplification of cuts over contractions is related to the presence of a long chain of focal pairs, but in a different graph associated to the proof with cuts. We shall discuss this further in Section 6.15. (Roughly speaking, the point is that there are long chains of focal pairs which reflect logical relationships in the proofs with cuts that are not reflected in the logical flow graph.)

Another interesting feature of this example is that the standard method of cut elimination leads to approximately the same kind of geometric structure in the logical flow graph as in the previous examples in which there were long chains of focal pairs. That is, one obtains graphs roughly like the one called H in Section 4.3, with many focal pairs. In fact, we shall show that this is unavoidable in a certain sense, which will be made precise in Proposition 6.11 in Section 6.12. (Note that logical flow graphs of proofs without cuts can never contain chains of focal pairs of length at least 2, as in Lemmas 6.4 and 6.5 in Section 6.12.)

One conclusion of these observations is that the standard method of cut elimination sometimes has to be able to create focal pairs in situations where there were none at the beginning. This is what happens in the case of (3.28), and we shall see how it can happen more concretely in the next section.

6.10 A fourth scenario, and the creation of focal pairs

The diagram in Fig. 6.13 represents a first step by which the duplication of subproofs in the process of cut elimination can transform a proof with very simple structure into one which is more complex. Again we think of putting ourselves back in the situation of (6.4) and (6.5), in which we are duplicating a subproof Π_1 in order to split a contraction in another subproof Π_2. Instead of looking at paths that move between Π_1 and Π_2, as we did before, we simply consider an oriented path in Π_1 which begins and ends in the endsequent. In this case, the duplication of subproofs leads to a pair of oriented paths in the new proof (6.5) which have the same endpoints (coming from the contractions below the two copies of Π_1 in the new proof).

In this way, a focal pair can be created in the logical flow graph, where none existed before. Through many repetitions of this process, one can create many focal pairs, and a very simple graph in the beginning can be converted eventually into one of exponential size which looks roughly like the graph H in Section 4.3 (Fig. 4.4).

6.11 Extensions of chains of focal pairs

The process described in the preceding section can lead not only to the creation of focal pairs, but also to the extension of existing *chains* of focal pairs. To see this, imagine that we have our proofs Π_1 and Π_2 which are being combined with a cut to make a larger proof Π^*, and that we are duplicating Π_1 in order to simplify the cut over a contraction contained in Π_2, as in (6.4) and (6.5). Imagine also that Π^* lives inside of a larger proof Π. If we have a path p inside Π_1 which begins and ends in the endsequent of Π_1, then we get a focal pair after the duplication of subproofs, as we saw before. However, we can also continue this path p below the cut in Π which connects Π_1 and Π_2, and in this continuation

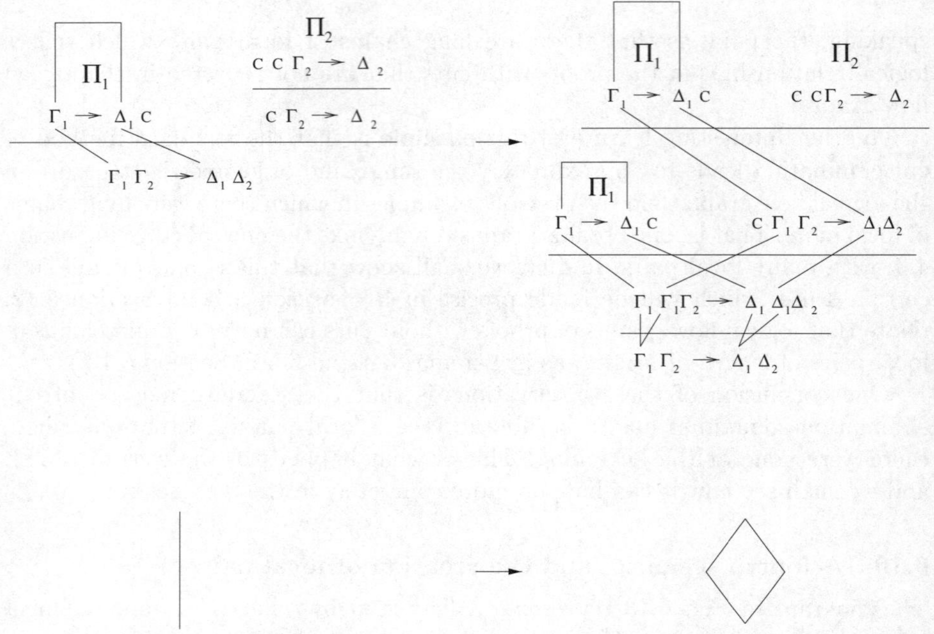

FIG. 6.13. The creation of a focal pair

p might meet additional focal pairs. In this way the creation of a new focal pair as in Section 6.10 can lead to the increase in the length of a chain of focal pairs.

For this construction it is important to have some interesting structure in the part of the proof Π below the use of the cut rule by which Π_1 and Π_2 are connected. In particular, there should be cuts below the one under consideration. If one follows the normal practice of simplifying the cuts starting from the bottom of the proof and working up, then there would be no more cuts in Π below the one that connects Π_1 to Π_2, and in this case the continuation of the path below the cut would not lead to any more focal pairs. Focal pairs might be created, but chains of focal pairs would not be extended. We shall discuss this further in Section 6.14.

6.12 Steady graphs and cut-free proofs

Logical flow graphs of proofs without cuts have very simple structure. The following definition gives a general notion of "simplicity" for an optical graph which includes the case of logical flow graphs of cut-free proofs in particular (as in Lemma 6.4 below.)

Definition 6.2 *An optical graph G is said to be* steady *if there is no oriented path which goes from a focusing branch point to a defocusing branch point.*

The other way around is allowed, i.e., oriented paths from defocusing to focusing branch points.

Examples 6.3 (a) Visibility graphs are automatically steady, because they have no focusing branch points at all.

(b) The graphs H and H^* described in Section 4.3 are steady.

Lemma 6.4 *If a proof* Π *has no cuts, then the underlying logical flow graph* G *is steady.*

Logical flow graphs of cut-free proofs do not have oriented cycles, as in [Car97] (see also Lemma 6.19 in Section 6.18), but neither are they trees necessarily, as visibility graphs are. Instead they can be like the graphs H, H^* from Section 4.3. We have seen examples of this before, and we shall discuss this in more depth later in this section.

Proof Branch points in G always come from contractions in Π. (In the context of feasible numbers one also gets branch points from the special rules of inference concerning the feasibility of sums and products of terms, but this causes no trouble for the arguments that follow.) If v is a vertex in G which is a *focusing* branch point, then any oriented path in G which begins at v must go *down* in the proof, towards the endsequent. As long as it goes straight down in the proof, the path can only pass through other focusing branch points, and no defocusing ones. In a proof with cuts an oriented path could go downwards initially but then go upwards again, towards the axioms, after turning around across a cut. In our case this cannot happen, the path cannot do anything besides going straight down towards the endsequent, and we conclude that the graph is steady.

Here is another way to make the argument. Focusing branch points in G can only occur at *positive* vertices in G, while defocusing branch points can occur only at *negative* vertices. (See Section A.3 for the notion of positive and negative occurrences of formulae in a proof and its logical flow graph.) Any oriented path in the logical flow graph that goes from a positive occurrence to a negative one must cross a cut. This is true for any formal proof, and it implies that the logical flow graph of a proof without cuts must be steady. □

The next lemma provides a way to make precise the idea that steady graphs have simple structure.

Lemma 6.5 *If* G *is an optical graph which is steady, then* G *cannot contain any chains of focal pairs of length* 2.

Proof This is an immediate consequence of the definitions, i.e., Definition 6.2 above and Definition 4.17 in Section 4.14. □

A steady graph can easily contain many focal pairs. They might interlace each other or run in parallel, as in the case of the graphs H and H^* from Section 4.3, but the lemma says that they cannot run "in sequence".

Corollary 6.6 *Let* G *be an optical graph which is steady and which has no oriented cycles. Then the visibility* $\mathcal{V}_+(v, G)$ *of* G *starting from a vertex* v *in* G *is at most quadratic in size as compared to the size of* G.

The graph H^* from Section 4.3 shows that this quadratic growth in size can occur.

Proof This will follow from Theorem 8.9 in Section 8.4. One could also give a direct argument, in the same spirit as the analysis of the graphs H and H^* in Section 4.3. In the end the two approaches are practically the same, and so we shall not pursue the more direct approach further here. □

Corollary 6.7 *If G is the logical flow graph of a proof without cuts, then the visibility of G (starting from any vertex v in G) is at most quadratic in size compared to the size of G.*

Proof We have already seen in Lemma 6.4 that such a graph is steady, and it also contains no nontrivial oriented cycles [Car97]. (See also Lemma 6.19 in Section 6.18.) Thus the assertion follows from Corollary 6.6. □

In "interesting" situations, the logical flow graph G of a proof Π without cuts will look a lot like the graphs H and H^* described in Section 4.3. More precisely, G will be a union of graphs which are disconnected from each other, coming from different atomic formulae in the proof, and it is the components of G (or at least some of them) which would resemble the graphs H and H^*. This is not true for arbitrary cut-free proofs Π, but it is true under some mild nondegeneracy conditions. Let us explain why this is so.

For this discussion we shall impose two nondegeneracy conditions on our proof Π. The first is that the number of steps in Π be large compared to the size of its endsequent. This is quite reasonable, because "short" proofs without cuts should not contain much structure. Under this assumption, there must be a large number of contractions in Π, as in the following lemma.

Lemma 6.8 *Let Π be a cut-free proof with k steps, and let t be the total number of logical symbols ($\wedge$, $\vee$, etc.) that occur in the endsequent S of Π. Suppose that m is chosen so that each formula A in S has at most m logical symbols. (Thus we could take $m = t$, for instance.) If c is the total number of contractions used in Π, then*

$$c \geq \frac{k-t}{m+1}. \tag{6.11}$$

Proof Let us begin by rewriting (6.11) as

$$(k-c) - c \cdot m \leq t. \tag{6.12}$$

Remember that there are two types of rules of inference (in sequent calculus), the *logical* rules and the *structural* rules. (See Section A.1 in Appendix A.) The *structural* rules are the cut and contraction rules, of which there are c in this case, since Π contains no cuts by assumption. Thus $k - c$ is the total number of times that a logical rule of inference is used in Π.

Let us also re-evaluate the meaning of the parameter m. We claim that if B is any formula which appears in Π, then B has at most m logical symbols. This is part of our hypotheses when B lies in the endsequent, and in general it follows from the *subformula* property for cut-free proofs, which ensures that any such B also occurs as a subformula of a formula in the endsequent. (Otherwise it would have to disappear in a cut, which is impossible here.)

To prove (6.12), we make an accounting of the way that logical symbols are added and subtracted in Π. Whenever a logical rule is used, we keep all of the logical symbols that we had before, and we add one more. When a contraction is applied, we reduce two occurrences of a formula B into one, and the total number of logical symbols is reduced by the number in B, which is at most m, as in the preceding paragraph.

There are $k - c$ logical rules in Π and c contractions. Thus in the course of the proof we add a total of $k - c$ logical symbols (to the ones that appeared already in the axioms), and we removed at most $c \cdot m$. This implies that there are at least $(k - c) - c \cdot m$ logical symbols in the endsequent, which is the same as saying that (6.12) is true, since t is defined to be the total number of logical symbols in the endsequent. This proves the lemma. □

The second requirement that we want to impose on Π concerns the role of "weak occurrences", which can be defined as follows. In an axiom

$$\Gamma, A \to A, \Delta \tag{6.13}$$

the formulae in Γ and Δ are considered to be weak occurrences. There might be more than a single choice for A here, but such a choice (for each axiom) should be fixed once and for all. Given a formula occurrence B in Π which is not in an axiom, we say that B is weak if every "direct path" in the logical flow graph that starts at an atomic occurrence in B and goes "straight up" to an axiom actually lands in a weak occurrence in that axiom.

Alternatively, one can define the weak occurrences recursively, through the following regime. If a formula occurrence B in a proof Π does not lie already in an axiom, then it was obtained from one or two previous occurrences through a rule of inference. If each of the previous formulae are weak, then we consider B to be weak also. Otherwise B is not weak.

Weak occurrences in the endsequent of a proof are not terribly interesting, because they could just as well be removed, or replaced by their negations, as far as the validity of the endsequent is concerned. However, weak occurrences are sometimes needed at intermediate stages of a proof, in order to use cuts and contractions efficiently. That is, one might combine weak and non-weak formulae using the unary logical rules $\wedge : left$ and $\vee : right$, and the resulting larger formulae might be better for applying cuts and contractions afterwards.

We do not want to allow our proof Π to become large simply by adding many weak occurrences to the axioms and then contracting them together, for instance. In order to avoid this type of degeneracy, we can restrict the way that rules of inference can be applied to weak occurrences in Π. Specifically, we shall ask that our proof Π be *reduced*, in the sense that it satisfies the following two properties.

The first is that no contraction rule or binary logical rule should ever be applied to a weak occurrence in Π. The second is that each application of a unary logical rule should involve at least one non-weak occurrence. In other words, if we use the (unary) $\vee : right$ rule in the proof to combine two formula occurrences A, B on the right side of a sequent into $A \vee B$, then we ask that at least one of A and B should be non-weak. If instead we use the (binary) $\wedge : right$ rule to combine occurrences C and D from different sequents into an occurrence of $C \wedge D$ on the right side of a larger sequent, then the first property above demands that *both* C and D be non-weak.

The next assertion implies that the restriction to reduced proofs does not entail a significant loss in generality.

Proposition 6.9 *Given a proof* Π *with* k *lines, we can transform it into a reduced proof* Π' *which has the same endsequent and at most* k *lines. If* Π *contains no cuts, then the same is true of* Π'.

This follows from some results in [Car97, Car00]. The transformation involved in the proof of Proposition 6.9 is a very natural one, and relies only on cancellation of subproofs and the addition and removal of weak occurrences. More subtle results of this nature are also given in [Car97, Car00], concerning the elimination of weak subformulae in cut-formulae.

One aspect of the nondegeneracy of reduced proofs is given in the next result.

Lemma 6.10 *Let* Π *be a reduced proof, and let* c *be the number of contractions in* Π. *Let* a *be the number of atomic occurrences contained in the distinguished (non-weak) occurrences in the axioms of* Π. *Then* $a \geq c/2$.

This is established in [Car00]. There one assumes that the distinguished occurrences in the axioms are always atomic, and one takes a to simply be the number of axioms, but the present formulation is slightly more convenient for our purposes, and it is proved in exactly the same way.

Let us assume from now on that the cut-free proof Π under consideration is reduced, and let G denote the logical flow graph of Π. From Lemma 6.10 we conclude that there are at least $c/2$ edges in G which are associated to axioms, where c denotes the total number of contractions in Π. This leads to a lower bound on the number of these "axiom edges" in G in terms of the number of lines in Π and the complexity of the endsequent, because of Lemma 6.8.

The bottom line is that there must be a large number of these axiom edges in G if the number of lines in Π is large compared to the complexity of the endsequent.

Each axiom edge E in G determines a unique "bridge" in the logical flow graph of Π, i.e., an oriented path which begins and ends in the endsequent of Π and passes through E. This is easy to check, using the assumption that Π be free of cuts. One simply starts at the two ends of E and goes straight down in the proof until one reaches the endsequent to get the two sides of the bridge.

Thus we obtain a lower bound on the number of bridges in G in terms of the number of lines in Π and the complexity of the endsequent of Π. If the number of axiom edges in Π is sufficiently large, then we may conclude that there is a large number of bridges between a particular pair of atomic occurrences in the endsequent of Π. Here is a precise statement.

Lemma 6.11 *Suppose that Π is a cut-free proof of a sequent S (which is also reduced), and let G denote its logical flow graph. Set*

$$\begin{aligned} A &= \textit{the number of axiom edges in } G \\ p &= \textit{the number of positive occurrences} \\ &\quad\ \textit{of atomic formulae in } S \\ n &= \textit{the number of negative occurrences} \\ &\quad\ \textit{of atomic formulae in } S. \end{aligned}$$

(Thus A is equal to one-half the number of atomic occurrences in the distinguished (non-weak) formulae in the axioms. Remember that the notions of positive and negative occurrences are defined in Section A.3.) Then there exist atomic occurrences P, Q in S with P positive and Q negative such that there are at least

$$\frac{A}{pn}$$

distinct bridges in G (as defined above) which go from Q to P.

Proof Indeed, the assumption that Π be free of cuts implies that the total number of distinct bridges in G must be equal to A, as mentioned before the statement of the proposition. Each of these bridges will go from a negative occurrence in the endsequent to a positive one, as one can easily check (from the definitions of "positive" and "negative"). The existence of P and Q with at least the required number of bridges then follows immediately from a simple counting argument. (That is, if P and Q did not exist, then the total number of bridges in G would have to be strictly less than A, a contradiction.) □

Let us return now to the earlier matter, in which we have a proof Π which is cut-free and reduced, and which contains a large number of lines compared to the complexity of the endsequent. From Lemmas 6.8 and 6.10 we get a lower bound on the number of axiom edges in the logical flow graph G of Π, and then Lemma 6.11 yields the existence of a pair of atomic occurrences in the endsequent S of Π which are connected by a large number of bridges. These bridges are systematically merged together in G through branch points that come from contractions in the proof.

Let K be the subgraph of G which consists of the union of the bridges between P and Q. The structure of K is similar in nature to the graphs H and H^* discussed in Section 4.3. In other words, one has roughly the same kind of systematic expansion through branching followed by systematic contraction for K as we had for H and H^*. The branching in K does not have to be as "regular" as it is for H and H^*, however, since there need not be a simple pattern to the

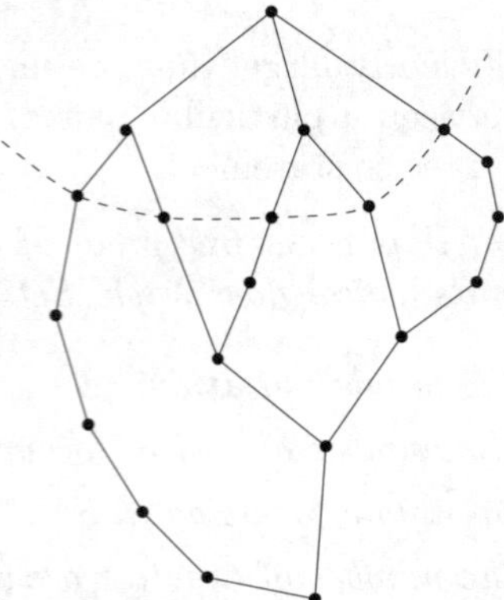

FIG. 6.14. A logical flow graph like the one called H in Section 4.3, but less balanced.

use of contractions in the proof Π. There could easily be strings of non-branching vertices in K between the branch points, for instance, coming from the use of other rules of inference (or from contractions applied to other formulae). Thus K could be "lop-sided", with much more branching on one side than another. (See Fig. 6.14 for an illustration.)

Of course, the rest of G has much the same structure as K does, in the way that bridges are merged together using contractions. The point about K is that we have a lower bound on the number of bridges between a fixed pair of atomic occurrences in terms of the relative complexity of Π and S. In other words, we know that K should contain a lot of branch points, just as for H and H^*. In other portions of G there might be less branching, but there might also be a number of other subgraphs like K, corresponding to other choices of P and Q.

To summarize, the approximate structure that we see in the graphs H and H^* from Section 4.3 is unavoidable for logical flow graphs of cut-free proofs, at least if we impose some mild restrictions on the nondegeneracy properties of the proof (and ask that the proof be large compared to the size of the endsequent). This fits very well with the examples and geometric phenomena that we have discussed in the previous sections, concerning the different ways in which a graph like H can arise from the duplication of subproofs in the standard method for simplifying cuts over contractions.

6.13 Steady graphs with oriented cycles

Lemma 6.5 and Corollary 6.6 provide simple bounds for the complexity of steady graphs, but they do not say anything about the possibility of oriented cycles. Indeed one can have nontrivial oriented cycles in a steady graph, as in Fig. 6.15.

Let G be an optical graph which is steady. It is easy to see that an oriented cycle in G cannot pass through both focusing and defocusing branch points, using the definition of a steady graph (Definition 6.2 in Section 6.12). In other words, an oriented cycle in G cannot have both a way "in" and a way "out".

FIG. 6.15. Oriented cycles in a steady graph

The cycles in Fig. 6.15, for instance, each have a way in or a way out, but not both.

In general, we have the following restrictions on the oriented cycles which can occur in steady graphs.

Lemma 6.12 *If G is an optical graph which is steady, then G satisfies (5.5) from Section 5.3. That is, if we fix a vertex v in G and consider two distinct loops L_1 and L_2 in G which can both be reached from v by oriented paths, then L_1 and L_2 will have no vertices or edges in common.*

This is not difficult to verify, and indeed it is very similar to the discussion just after the statement of Proposition 5.1 in Section 5.2. We omit the details.

Lemma 6.13 *Let G be an optical graph which is steady, and let v be a vertex in G. Let $\mathcal{L}$ be the collection of loops in G which are accessible by oriented paths starting from v (as in Section 5.3, just after Corollary 5.4). Then $\mathcal{L}$ cannot have depth more than 2. (The depth of $\mathcal{L}$ is as defined just before Proposition 5.10.)*

This is analogous to Lemma 6.5, but for steady graphs with oriented cycles. The proof is straightforward and we omit it, except to say that the main point is again the fact that an oriented cycle in G cannot have both a way in and a way out. In fact, the depth of $\mathcal{L}$ cannot be more than 1 unless v itself lies in a loop, because any loop in $\mathcal{L}$ which does not contain v has to have a way in (so as to be accessible from v) and hence no way out.

Corollary 6.14 *If G is an optical graph which is steady and v is any vertex in G, then the rate of growth of the visibility $\mathcal{V}_+(v, G)$ is at most quadratic (in the sense that (5.11) holds with $d = 2$).*

This follows immediately from Proposition 5.10 and Lemma 6.13.

6.14 Steady horizons

Imagine that we have a formal proof Π, and that we know that there are no cuts used in Π below a certain point. The part of the logical flow graph of Π which comes after that point should be relatively simple and should not interfere too much with the graph as a whole. To make this more precise let us introduce the following notion of the "horizon" of an oriented graph.

Definition 6.15 *Let G be an oriented graph, and let E_+ be a subgraph of G. We call E_+ a* positive horizon *of G if it has the property that every oriented path γ in G which begins in E_+ is wholly contained within E_+. (In saying that a path "begins in E_+" we mean that its initial vertex lies in E_+.) Similarly, we say that a subgraph E_- of G is a* negative horizon *of G if every oriented path in G which ends in E_- is entirely contained in E_-.*

Let Σ be a portion of the proof Π which contains no axioms and no cuts. Imagine for instance that Σ corresponds to the part of Π which occurs below a certain step in the proof, with no cuts below that step, and without the supporting subproofs that come before.

To be more precise, for the moment we permit Σ to be any part of Π which satisfies the following properties. We ask first of all that Σ consist of entire sequents from Π (rather than pieces of sequents). Strictly speaking, Π itself should be viewed as a tree of sequents, and we ask that Σ represent a *subtree* of this tree. We also ask that Σ contain the endsequent of Π. These two conditions amount to the statement that as soon as Σ contains a given occurrence of a sequent in Π, it contains all the ones which appear "later" in the proof Π. As before, we also require that Σ contain neither cuts nor axioms.

Let G be the logical flow graph of Π, and let K denote the part of G that corresponds to Σ. Remember that we can speak of vertices in the logical flow graph as being *positive* or *negative*, according to the sign of the corresponding atomic occurrences within Π. (See Section A.3 in Appendix A.) It is not hard to see that every edge in K either joins a pair of positive vertices or a pair of negative vertices, i.e., the signs are never mixed. This is because of the presumed absence of cuts and axioms in Σ.

Let E_+ be the subgraph of K which consists of all of the *positive* vertices in K and all of the edges between them, and let E_- be the subgraph which consists of the negative vertices and the edges between them. Thus K is the disjoint union of E_+ and E_-.

It is easy to check that E_+ and E_- are positive and negative horizons of G, respectively. This uses the assumptions that Σ contain no cuts, and that there are no gaps between the sequents in Σ and the endsequent of Π.

The idea now is that E_+ and E_- do not effect the structure of G in a very strong way. For instance we have the following general assertion.

Lemma 6.16 *Let G be an optical graph, and suppose that E_+ and E_- are subgraphs of G which are positive and negative horizons of G, respectively. Assume also that E_+ does not contain any defocusing branch points, and that E_- does not contain any focusing branch points. Let M be the subgraph of G obtained by removing E_+ and E_- from G, as well as the edges in G with an endpoint contained in E_+ or E_-. If G contains a chain of focal pairs (Definition 4.17) of length $n > 2$, then there is a chain of focal pairs of length $n - 2$ contained in M.*

In other words, the "complexity" of G as measured by long chains of focal pairs is almost the same for G as it would be if one removes E_+ and E_- from

G. Note that this lemma cannot be improved, in the sense that there may not be a chain of focal pairs in G of length $n-1$ which does not intersect E_+ or E_-. It is easy to make examples, using the graph pictured in Fig. 4.2 in Section 4.3, for instance.

The restrictions on the branch points in E_+, E_- required in the lemma hold automatically in the setting of logical flow graphs described above. This is because we chose E_+ and E_- in that case so that they contain only positive and negative vertices in the logical flow graph, respectively.

Proof Let $\{(u_i, w_i)\}_{i=1}^n$ be a chain of focal pairs of length n in G. This means in particular that $u_{i+1} = w_i$ for $i = 1, 2, \ldots, n-1$. To prove the lemma, it is enough to show that $\{(u_i, w_i)\}_{i=2}^{n-1}$ defines a chain of focal pairs in the subgraph M. To do this, we argue as follows. By assumption we know that (u_1, w_1) is a focal pair in G, and this implies that w_1 is a focusing branch point of G, by definition of a focal pair (Definition 4.15). This implies that $u_2 = w_1$ does not lie in E_-.

Similarly, $w_{n-1} = u_n$ cannot lie in E_+. Indeed, there must be an oriented path γ_n in G which begins at u_n and reaches a defocusing branch point in G, since (u_n, w_n) is a focal pair, and γ_n would have to be contained in E_+ if u_n were, because of our assumption that E_+ be an positive horizon for G. This is not possible, since we are assuming that E_+ contains no defocusing branch point.

Thus u_2 does not lie in E_-, nor can w_{n-1} lie in E_+. The horizon assumptions imply that no oriented path in G which begins at u_2 can ever meet E_-, and that no oriented path ending at w_{n-1} can intersect E_+.

If $2 \leq i \leq n-1$, then there are oriented paths in G which go from u_2 to u_i and w_i. This follows easily from the fact that $\{(u_i, w_i)\}_{i=1}^n$ defines a chain of focal pairs in G. Similarly, there are oriented paths in G which begin at u_i and w_i and end at w_{n-1} when $2 \leq i \leq n-1$. This implies that u_i and w_i lie in M when $2 \leq i \leq n-1$, since they cannot lie in E_+ or E_-, by the preceding observations.

Thus the sequence of pairs $\{(u_i, w_i)\}_{i=2}^{n-1}$ is contained in M. We know that each (u_i, w_i), $2 \leq i \leq n-1$, is a focal pair in G, but it is also a focal pair in M. In other words, if α_i and β_i are oriented paths in G that begin at u_i and arrive at w_i along different edges, as in the definition of a focal pair, then α_i and β_i are actually contained in M; otherwise one of them would meet E_+ or E_-, and this would lead to a contradiction, since E_+ and E_- are horizons which do not contain u_i or w_i.

This proves that $\{(u_i, w_i)\}_{i=2}^{n-1}$ is actually a chain of focal pairs in M, and Lemma 6.16 follows. □

The situation for logical flow graphs described above arises naturally when one applies the standard method for eliminating cuts, at least if one is careful to simplify the cuts from the bottom-up. At each stage, there will be no cuts below the point in the proof at which one is working, and one can apply Lemma 6.16 to say that the lower portion of the proof does not have much effect on the length of the longest chain of focal pairs in the logical flow graph. This permits one to concentrate on the upper portion of the proof, which can be analyzed by cases in much the same manner as we did before in this chapter.

6.15 A simplified model

In this section, we shall explain a way to track the duplication of subproofs which occurs in the simplification of cuts over contractions in terms of chains of focal pairs in a graph which is somewhat different from the logical flow graph. Instead of working with proofs we shall temporarily allow *partial proofs*, by which we mean a tree of sequents which is exactly like a proof, except that the "initial sequents" need not be axioms. That is, every sequent in the partial proof is either an initial sequent (for which no "justification" is given), or is derived from one or two sequents through the same rules as for proofs. In practice, one can imagine that the initial sequents are provable, even if the proofs are not provided.

For the present purposes, we shall restrict ourselves to partial proofs which use only the cut and contraction rules. Instead of trying to eliminate cuts we shall merely seek to transform a given partial proof into one with the same endsequent but no contractions *above* cuts. One can think of this as representing a portion of the process of cut elimination applied to a larger proof.

In order to push the cuts above the contractions, one can use the same method of duplicating subproofs as described in Section 6.2 (in the passage from (6.4) to (6.5)). This is not hard to show, but there is a small technical point which we should mention. Suppose that we have an occurrence of a formula A in a partial proof Σ which is involved in a cut at a certain stage, and that there were some contractions involving this occurrence of A earlier in Σ. It is easy to exchange the order of the rules if necessary in order to arrange for these contractions to all occur just before the cut. That is, we merely have to delay the application of the contractions if they were performed earlier, and this will not affect the rest of the partial proof. Once we have made this change, it is easy to simplify the cut over the contractions above it, using the transformation from (6.4) to (6.5).

Let us fix now a partial proof Σ which involves only cuts and contractions. We want to associate an oriented graph $\mathcal{G}$ to Σ which represents the "macroscopic" features of the partial proof, as opposed to the "microscopic" features which are represented by the logical flow graph. We shall call $\mathcal{G}$ the *macroscopic flow graph* of the partial proof Σ.

Each formula in Σ will be used to represent a vertex in $\mathcal{G}$. We use only the "whole" formula now, and not atomic subformulae, as for logical flow graphs. We also add additional vertices, one for each initial sequent in Σ. These additional vertices provide symbolic representations for the justifications of the initial sequents.

We attach edges between these vertices in the following manner. There are no edges going between any two of the "additional vertices" which represent initial sequents. However, if $*$ is an "additional" vertex associated to some initial sequent $\Gamma \to \Delta$, then we attach an edge from $*$ to every one of the formulae in

Γ and Δ. Otherwise we attach edges between vertices that represent formulae in practically the same manner as for the logical flow graph. Thus in a contraction rule

$$\frac{\Gamma \to \Delta, A, A}{\Gamma \to \Delta, A} \qquad \text{or} \qquad \frac{A, A, \Gamma \to \Delta}{A, \Gamma \to \Delta}$$

we attach an edge from each of the A's above to the one below, and we attach an edge from each formula in Γ or Δ above to its counterpart below. For the cut rule

$$\frac{\Gamma_1 \to \Delta_1, A \quad A, \Gamma_2 \to \Delta_2}{\Gamma_{1,2} \to \Delta_{1,2}}$$

we attach a ("horizontal") edge between the two occurrences of the cut formula A, and also an edge from any side formula in $\Gamma_1, \Delta_1, \Gamma_2, \Delta_2$ above the line to its counterpart below the line.

These edges come with natural orientations, in practically the same manner as for the logical flow graph. The precise definitions are as follows. Recall that a formula A appearing in a sequent $\Gamma \to \Delta$ is considered to be *positive* if it lies in Δ and *negative* if it is an element of Γ. In the cut rule above, the edge that goes between the two occurrences of the cut formula A is oriented so that it goes *from* the positive occurrence *to* the negative occurrence. Otherwise an edge between two occurrences of a formula B is oriented so that it goes from the *lower* sequent to the *upper* sequent when B occurs negatively, and so that it goes from the *upper* sequent to the *lower* sequent when B occurs positively. (Note that two occurrences of a formula B always have the same sign when they are connected by an edge which is not a horizontal edge coming from an application of the cut rule.) If a formula B lies in an initial sequent, so that there is an edge between B and the additional vertex $*$ associated to this initial sequent, then we orient the edge so that it goes *from* B *to* $*$ when B occurs negatively in the sequent, and otherwise we take the edge to be oriented *from* $*$ *to* B when B occurs positively.

This completes the definition of the macroscopic flow graph $\mathcal{G}$ associated to the partial proof Σ. Let us now consider some examples.

We begin with a partial proof which lies inside the proof of $F(2) \to F(2^{2^n})$ discussed in Section 6.5. In this partial proof the initial sequents are all of the form

$$F(2^{2^{j-1}}), F(2^{2^{j-1}}) \to F(2^{2^j}). \tag{6.14}$$

For an actual proof one would obtain this sequent from two copies of the axiom $F(2^{2^{j-1}}) \to F(2^{2^{j-1}})$ using the rule for feasibility of products, but we shall forget about this for the moment.

From here we proceed with the same partial proof as before. We apply a contraction to (6.14) to get

$$F(2^{2^{j-1}}) \to F(2^{2^j}), \tag{6.15}$$

and then we combine a sequence of these using cuts to get $F(2) \to F(2^{2^n})$, as desired. This defines a partial proof Σ which contains all of the cuts and contractions used in the original proof of $F(2) \to F(2^{2^n})$, and nothing else.

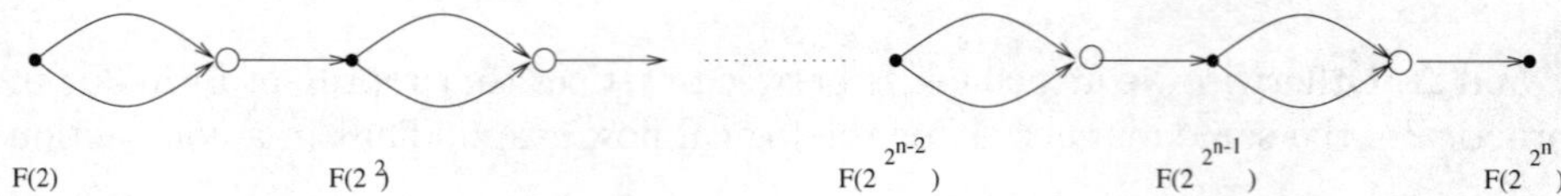

FIG. 6.16. The macroscopic flow graph $\mathcal{G}$ associated to the partial proof for $F(2) \to F(2^{2^n})$

The macroscopic flow graph $\mathcal{G}$ for this partial proof is pictured in Fig. 6.16. The nodes with larger circles represent "additional vertices" associated to the initial sequents (6.14), each of which has two incoming edges and one outgoing edge. The graph that we obtain in this manner is practically the same as the logical flow graph of the proof itself; indeed the two graphs are the same *topologically*, and differ only in the addition or removal of some vertices (at which no branching takes place). In particular, the graph $\mathcal{G}$ has a long chain of focal pairs in this case, and the effect of simplifying the cuts over the contractions on $\mathcal{G}$ is the same in essence as we described before, in terms of the logical flow graph.

Now let us consider the proof of (3.28) with cuts discussed in Section 3.3. Again we shall only consider the partial proof which reflects the roles of cuts and contractions in the proof.

In this case, the initial sequents are of the form

$$F_i,\, F_i,\, F_i,\, A_{i+1} \vee B_{i+1} \to F_{i+1}. \tag{6.16}$$

For the moment the precise definitions of the A_j's, B_j's, and F_j's do not matter, but they can be found in Section 3.3, along with a way to prove of this sequent which uses neither cuts nor contractions.

By applying contractions to the F_i's on the left side of (6.16) we can obtain the sequents

$$F_i,\, A_{i+1} \vee B_{i+1} \to F_{i+1}. \tag{6.17}$$

By combining these sequents with cuts we get a partial proof which ends with the sequent

$$F_1,\, A_2 \vee B_2, \ldots,\, A_n \vee B_n \to F_n. \tag{6.18}$$

This is practically the same as (3.28), and in any case we could make it be exactly the same by using (3.32) at the last step instead of (6.17).

This defines a partial proof Σ which contains all of the cuts and contractions used in the original proof and nothing else. (Actually, Σ reflects a minor rearrangement of the original proof, in that the contraction in (3.32) is postponed until after (3.34), but this is not a significant change.) One can simplify the cuts over the contractions using the usual method, and in fact we did this in effect already in Section 3.3, beginning in (3.42). We simply did not describe it in these terms.

This simplification of the cuts over the contractions behaves in nearly the same manner as for the previous example. There are two minor differences, which are as follows. In the present situation we have 3 copies of F_i supporting each

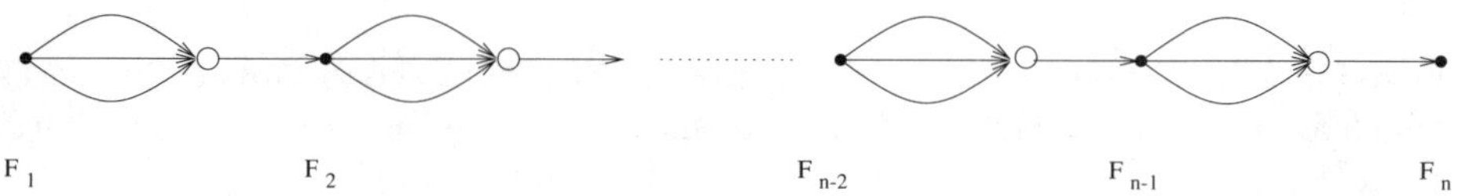

FIG. 6.17. A slightly "reduced" illustration of the macroscopic flow graph $\mathcal{G}$ for the partial proof from Section 3.3

F_{i+1} in the initial sequent (6.16), instead of two copies as before. This affects the *rate* of the exponential expansion which takes place when we simplify the cuts over the contractions, but otherwise the general behavior is the same as before. We have also the formulae $A_{i+1} \vee B_{i+1}$ in our building blocks (6.17), which have no counterpart in the previous example concerning feasible numbers. However, these formulae do not participate much in the simplification of cuts over contractions. They are merely duplicated and then contracted over and over again.

The macroscopic flow graph $\mathcal{G}$ for this example is also very similar to the one for feasible numbers. The basic structure of $\mathcal{G}$ is illustrated in Fig. 6.17. This picture is not completely faithful to $\mathcal{G}$, in that we have omitted the parts related to the formulae $A_{i+1} \vee B_{i+1}$, and we have represented the contractions over the three copies of F_i in the initial sequents (6.16) as single vertices with three outgoing edges, rather than by pairs of successive vertices with two outgoing edges each, as we technically ought to do. With these modest inaccuracies we, get a diagram which depicts in a more immediate way the main features of this proof structure, for which the simplification of cuts over contractions corresponds in essence to the splitting of defocusing branch points in Fig. 6.17.

As in the previous example with feasible numbers, one is free to choose the order in which the cuts are simplified. By starting from the "top" of the proof one can simplify all of the cuts over contractions in $O(n)$ steps, while if one starts from the "bottom" of the proof, an exponential number of steps is involved. (In the latter case, there is a kind of "parallelism" in the way that the exponentially-many steps can be effected, however.)

One should keep in mind that the logical flow graphs of these two examples are very different from each other. In the proofs concerning feasible numbers, the logical flow graph was roughly like the corresponding macroscopic flow graph, but in the example from Section 3.3 there are no focal pairs in the logical flow graph, let alone a long chain of them, as mentioned in Section 6.9.

6.16 Comparisons

The logical and macroscopic flow graphs lie at opposite extremes from each other, in terms of the way that they attempt to reflect the logical connections within a formal proof. The macroscopic flow graph treats all formulae in an initial sequent as though they were connected to each other in a significant way, without regard to their internal structure, while the logical flow graph is much more restrictive, and deals only with different occurrences of the same basic formula which are linked in a very explicit way inside the proof. (Note that the formula occurrences which are linked by the logical flow graph may involve terms with different values, as in the case of proofs of the feasibility of large numbers.)

In other words, with the macroscopic flow graph one has the danger of overestimating the logical links between different formulae, while the logical flow graph can often underestimate them. It is not at all clear how to avoid this problem through a general recipe or automatic procedure; it is too easy to disguise connections between different formulae in a proof, or to create fake circumstances which might appear to potentially contain such a disguised connection.

A related point is that it seems possible in principle that a single basic class of difficulties could be encoded over and over again in increasingly intricate ways, to systematically avoid correct analysis by an increasingly sensitive family of procedures. For instance, one might say that the proofs for feasible numbers and the proofs from Section 3.3 reflect roughly the same "difficulty" in the exponential duplication processes involved, and that this is the same kind of difficulty as in Section 3.2, but represented in a more in a more intricate way (i.e., in the reliance on cuts). A priori there is no reason why the same basic phenomenon could not be encoded in more subtle ways, which might not be accommodated by the cut and contraction rules, but instead would require more elaborate proof systems.

Note that the idea of long chains of focal pairs is relevant in each of the three situations mentioned in the preceding paragraph (i.e., feasible numbers and the proofs from Sections 3.2 and 3.3), even if they appear in significantly different ways. For the proofs of feasible numbers, the chains of focal pairs appear in both the logical and macroscopic flow graphs, while in the proofs from Section 3.3 they are not present in the logical flow graph but do play a role in the macroscopic flow graph. Neither the logical flow graph nor the macroscopic flow graph reflects the exponential activity underlying the example in Section 3.2, even though a very similar picture is relevant, as in the interpretations through paths and transitions discussed in Section 3.4. (For this example, the macroscopic flow graph does not even make much sense, because of the absence of cuts in the proof. In effect the $\supset$: $left$ rule plays a similar role, though, in representing the relevant transitions.)

6.17 A brief digression

In the macroscopic flow graph $\mathcal{G}$ defined in Section 6.15, we chose to treat all formula occurrences in an "initial sequent" as though they were connected to each other. This is too crude to be useful in general, but there is a kind of partial justification for it that one can give, and which we shall describe in this section.

Definition 6.17 *Let* Π *be a formal proof, and let* A *and* B *be two formula occurrences in* Π*. We say that* A *and* B *are* joined *if there is a finite sequence* $D_1, \ldots, D_m$ *of formula occurrences in* Π *such that* $D_1 = A$*,* $D_m = B$*, and if for each* $j < m$ *there are atomic occurrences* P *and* Q *in* Π *which lie in the same*

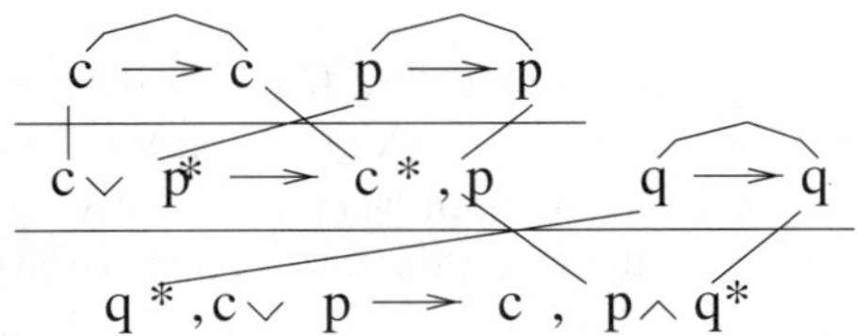

FIG. 6.18. The formula c^* in the proof is *joined* to the occurrence of q^* in the left side of the endsequent, through the sequence of formulae c^*, $c \vee p^*$, $p \wedge q^*$, q^*.

connected component of the logical flow graph of Π *and which are contained in* D_j *and* D_{j+1}*, respectively.*

In other words, the atomic occurrences P and Q should be connected by a path in the logical flow graph. One can think of this definition as saying that one looks first at the formulae which are connected to each other by the logical flow graph, and then passing to the transitive closure of this relation in order to get the notion of formulae being "joined", which is an equivalence relation. See Fig. 6.18 for an example.

One might say that this represents the most general way in which two formula occurrences in a proof can be *logically* connected, while the logical flow graph corresponds to the most restrictive type of connection.

Proposition 6.18 *Let* Π *be a proof (possibly with cuts) of a sequent* S *with* k *lines. Then there is a proof* Π' *of* S *with at most* k *lines such that all of the non-weak formulae in* S *are joined to each other in the sense of Definition 6.17.*

Recall that the notion of weak occurrences was defined in Section 6.12, in the paragraph containing (6.13). It is not hard to check that a weak occurrence in the endsequent cannot be joined to any of the other formula occurrences there.

Proof The first step is to transform Π into a proof Π' in which weak formulae are never used as auxiliary formulae for the cut rule, or for the binary logical rules ($\wedge : right$, $\vee : left$, or $\supset : left$), and to do so without increasing the number of lines in the proof. This is a special case of the conversion into a "reduced proof" given in Proposition 6.9 in Section 6.12. (The main point is that if one ever applies a binary rule of inference to a weak formula, then one could have achieved the same affect by cancelling one of the supporting subproofs and modifying and adding to the weak formulae on the other side.)

Thus we may as well assume from the start that we have a proof Π in which binary rules of inference are never applied to weak formulae. To finish the proof, we argue by induction. For proofs which are merely axioms, there is nothing to do. Otherwise, our proof Π was obtained from one or two subproofs by a rule of inference, and our induction hypothesis states that these subproof(s) satisfy the conclusions of the proposition themselves.

The rest of the argument proceeds by cases. If Π was derived from a subproof Π^* by a unary rule (either a contraction or a logical rule), then it is easy to check that the conclusions of the proposition are preserved; we only increase the number of ways that two formulae are joined in this case. Thus we assume that Π was obtained from two subproofs Π_1, Π_2 through a binary rule of inference. All of the non-weak formulae in the endsequent of Π come from the non-weak formulae in the endsequents of Π_1 and Π_2, and we already know that the non-weak formulae in the endsequent of each Π_i are joined, by the induction hypothesis. Thus we have only to find a connection between the two different collections of non-weak formulae in the endsequent of Π.

If Π was obtained from Π_1 and Π_2 by the cut rule, then the two cut formulae are non-weak, because of our initial reduction. Each is therefore joined to all of the other non-weak formulae in the endsequent of the corresponding Π_i, by induction hypothesis. They are also joined to each other, since they lie on opposite sides of the same cut. This implies that all of the non-weak occurrences in the endsequent of Π are joined to each other.

Now suppose that Π was obtained from Π_1 and Π_2 by a binary logical rule. In this case, there is a formula D in the endsequent which was obtained by combining formulae E_1 and E_2 from the endsequents of Π_1 and Π_2, respectively. Each E_i is non-weak, by our initial reduction, and hence is joined to all other non-weak formulae in the endsequent of the corresponding Π_i, because of the induction hypothesis. Of course D is joined to both of E_1 and E_2, and therefore to all non-weak formulae in the endsequent of Π. This implies that every non-weak formula in the endsequent of Π is joined to every other one, since one can make the connection through D if necessary. The proof of Proposition 6.18 is now complete. □

Let us now return to the setting of Section 6.15, in which we were looking at partial proofs in which the "initial sequents" need not be axioms. Normally we might expect our partial proofs to be extracted from complete proofs, so that the initial sequents would have supporting proofs even if they are not given explicitly.

To each such partial proof Σ we associated a macroscopic flow graph $\mathcal{G}$, in which all of the formulae occurring in an initial sequent are automatically linked to a common vertex. In general this may not reflect the structure of the supporting proof in a good way, but Proposition 6.18 tells us that we can always replace the supporting proof by one which is no larger and for which the connections between the formulae are simple, at least for the property of being joined. This would imply that the macroscopic flow graph would not create links between non-weak formulae which were not already joined.

If one knows the supporting proofs of the initial sequents in a partial proof Σ, then one might redefine $\mathcal{G}$ so that weak occurrences in an initial sequent are left unconnected to the common vertex associated to the initial sequent. Then the connections in the partial proof would reflect exactly the notion of formula occurrences being "joined" in the sense of Definition 6.17.

6.18 Proofs with simple cuts

In general, the logical flow graph of a proof with cuts can be quite complicated, with many cycles which can be nested together, for instance. See [Car00] for some examples, and see [Cara] for some techniques for interpreting the geometric structures in logical flow graphs. For proofs without cuts only very special graphs can arise. What happens if we restrict the kind of cuts that can be used? What kind of "intermediate" structures can appear?

These are broad and complicated questions, and we shall not attempt to treat them thoroughly here. In order to give an indication of some of the issues involved, however, we shall consider in this section the special case of proofs with cuts only over *atomic* formulae.

Lemma 6.19 *Let* Π *be a formal proof which has cuts only over atomic formulae. Then the logical flow graph* G *of* Π *cannot contain nontrivial oriented cycles.*

This lemma comes from [Car00]. One can analyze further the conditions under which oriented cycles may exist, but we shall not pursue this here.

Note that there can be *unoriented* cycles in the logical flow graph of a formal proof Π even when Π is free of cuts. Indeed, for cut-free proofs in which the number of steps is large compared to the complexity of the endsequent, the discussion from Section 6.12 (especially Lemma 6.11 and the remarks that follow it) shows that there must be many unoriented cycles when the proof is reduced, because there will be many pairs of bridges which begin and end at the same pair of points in the endsequent.

Proof Suppose to the contrary that there is a nontrivial oriented cycle in G. It is not difficult to show that this cycle cannot pass through the endsequent of Π; oriented paths in G can start at *negative* vertices in the endsequent, and they can end at *positive* vertices, but they have no way to go from a positive to a negative vertex in the endsequent. To go from a positive to a negative occurrence, one has to traverse a cut. (Remember that the *sign* of a vertex in the logical flow graph can be defined as in Section A.3 in Appendix A.)

On the other hand, we may assume without loss of generality that our cycle does pass through the last sequent or pair of sequents in Π that come just before the endsequent. Indeed, if this were not the case, then we could simply keep replacing Π by proper subproofs until it became true.

Under this condition, the last rule in the proof must be a cut rule. More precisely, the cut rule is the only rule in which an oriented cycle can pass through the sequents to which the rule is applied without reaching the sequent that results from the rule. This can be checked using the same argument as for showing that the cycle cannot reach the endsequent, i.e., there is no way to make a transition from positive to negative vertices in the last step of the proof unless there is a cut there. For the same reason, we know that our cycle must cross an edge associated to this final cut.

In fact, there is exactly one edge in G associated to this final cut, since we are assuming that our proof contains only atomic cuts. This edge becomes like a "one-way" road; our cycle can cross it to get from one of the supporting subproofs of the cut to the other supporting subproof, but there is no way for the cycle to return to the first one. Our cycle *has* to be able to make the return trip in order to be a cycle, from which we conclude that there could not be a nontrivial cycle to begin with. This proves the lemma. □

While the restriction to atomic cuts does prevent the formation of nontrivial oriented cycles, as above, one can still have long chains of focal pairs. This occurs in the examples of proofs of the feasibility of large numbers (without quantifiers) discussed in Sections 4.8 and 6.5, for instance.

The case of feasible numbers is slightly special, however, in the way that branch points can arise. Normally, it is only the contraction rule which can produce branch points in the logical flow graph, but in the context of feasibility, they can also arise through the special rules of inference concerning the feasibility of sums and products of terms. This makes it much easier to have long chains of focal pairs.

One can also make proofs with long chains of focal pairs and cuts only over atomic formulae in ordinary predicate logic, i.e., without using additional rules of inference like the ones for feasible numbers. These proofs are quite degenerate, however, in the sense that they rely on the application of rules of inference to weak occurrences in ways which are logically unnecessary. Without such degeneracies, one cannot have long chains of focal pairs in proofs with only atomic cuts, as in Proposition 6.20 below.

Unlike Proposition 6.20, Lemma 6.19 works just as well for proofs of feasibility. The additional rules of inference do not affect the argument given above. This is also true of our earlier observations about proofs without cuts, e.g., Lemma 6.4.

Proposition 6.20 *Let* Π *be a formal proof (in pure logic, i.e., LK), in which cuts are applied only to atomic formulae. If* Π *is reduced (in the sense described in Section 6.12, just before Proposition 6.9), then the logical flow graph* G *of* Π *does not contain a chain of focal pairs of length greater than* 2.

In other words, the logical flow graph of such a proof Π cannot behave in the manner shown in Fig. 6.19.

Let us emphasize the restriction to classical logic here, without special rules of inference as in the setting of feasibility. One should not take this too seriously, in that the basic result and method are still quite flexible, and can be used more generally (even if the statement here would not work in all situations).

If Π is not reduced, then it may be transformed into a reduced proof Π' with the same endsequent such that Π' also has no cuts over non-atomic formulae and such that the number of lines in Π' is no greater than the number of lines in Π. This follows from [Car97, Car00], and it is similar to Proposition 6.9.

For the purposes of Proposition 6.20, we do not really need the property of being a reduced proof at full strength. It will be enough to know that cuts, contractions, and binary logical rules of inference are never applied to weak

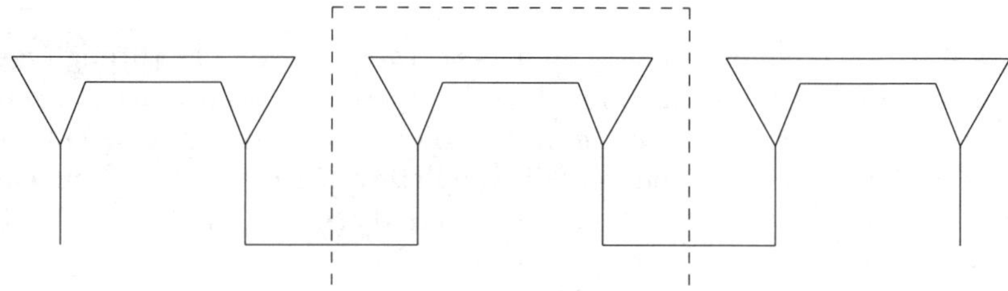

FIG. 6.19. The middle portion of this diagram cannot occur in the logical flow graph of a proof Π under the assumptions of Proposition 6.20.

occurrences, where "weak occurrences" are as defined in Section 6.17. Actually, the restriction on contractions is not essential but merely convenient, as we shall explain in Remark 6.33 below.

In the proof of Proposition 6.20 that follows, we develop a bit more machinery than is essential for the immediate purpose, in order to give a better picture of the situation in general. We begin with the following definition.

Definition 6.21 *Let Π be a proof, and let P be an occurrence of an atomic formula in Π. We call P* solitary *if it is not a proper subformula of a formula A that contains a binary connective (i.e., $\wedge$, $\vee$, or $\supset$). (The presence of negations or quantifiers is allowed.)*

Lemma 6.22 *Let Π^* be a reduced proof (in pure logic, i.e., LK), and let Z and W be two (distinct) solitary atomic occurrences in the endsequent of Π^*. Suppose that Z and W can be connected to each other in the logical flow graph of Π^* by a path that passes only through solitary occurrences in Π^*. (This path is* not *required to be oriented.) Then Z and W are of opposite sign in the endsequent (as defined in Section A.3 in Appendix A), and all other formula occurrences in the endsequent are weak.*

Before we prove this, let us make some remarks about the hypotheses. Notice first that the requirement that Π be reduced is very important here. It would be easy to make counterexamples to the lemma if we allowed Π to contain binary rules which are applied to weak formulae.

It is also important that we ask that the connection between Z and W be through *solitary* occurrences, and this is true even for reduced proofs without cuts. Indeed, consider the following proof in propositional logic.

$$\dfrac{\dfrac{\dfrac{p \to p \quad q \to q}{p,\, q \to p \wedge q} \qquad \dfrac{p \to p \quad q \to q}{p,\, q \to p \wedge q}}{p,\, p,\, q \vee q \to p \wedge q,\, p \wedge q}}{p,\, p,\, q \vee q \to p \wedge q}$$

In this proof, we start by using $\wedge : right$ rules at the top level, followed by a $\vee : left$ rule, and then a contraction at the end. The two solitary occurrences of p on the left side of the endsequent are connected to each other in the logical flow graph of this proof, but not by a path that passes only through solitary occurrences; instead the connection between these two occurrences of p has to go through the occurrence of $p \wedge q$ on the right side of the endsequent. In this case, the conclusions of the lemma fail completely, since our two occurrences of p in the endsequent have the same sign, and there are other formulae in the endsequent which are not weak.

Lemma 6.22, like Proposition 6.20, does not work in the context of feasible numbers. A simple counterexample is provided by the proof

$$\frac{F(x) \to F(x) \quad F(x) \to F(x)}{F(x),\, F(x) \to F(x^2)}$$

which uses only the $F: times$ rule. In this case, the two occurrences of $F(x)$ in the left side of the endsequent are connected to each other in the logical flow graph, and all occurrences in the proof are atomic (and hence solitary), but again the conclusions of the lemma fail to hold. Indeed this example is rather similar to the previous one, except that we can make it much simpler using the special $F: times$ rule.

Although the restriction to connections through solitary occurrences in the hypothesis of Lemma 6.22 is a rather strong one, it is also fairly natural in the context of proofs with cuts only over atomic formulae (or only over formulae without binary logical connectives). It helps to make precise the idea that the middle box in Fig. 6.19 cannot occur. The point is that the two horizontal edges in the bottom level of the picture have to correspond to cuts, and cuts over atomic formulae, by assumption.

Proof (Lemma 6.22) Notice first that if Z and W are as in the statement of the lemma, then they are necessarily non-weak. (Otherwise, they could not be connected to anything in the endsequent.)

To prove the lemma, we proceed by induction. Suppose first that Π^* consists of only an axiom. Then Z and W must be the distinguished occurrences in the axiom, and there is nothing else to do. Now suppose that

$$\Pi^* \text{ is a reduced proof,} \tag{6.19}$$

and

$$\Pi^* \text{ is obtained from one or two subproofs from a rule of inference, and Lemma 6.22 is valid for these subproofs.} \tag{6.20}$$

We also assume that we are given solitary atomic occurrences Z and W in the endsequent of Π^* which we know to be connected to each other in the logical flow graph of Π^* through solitary occurrences, as in the statement of the lemma.

Notice that every subproof of Π^* satisfies the property of being reduced, since Π^* itself does.

Suppose first that Π^* is obtained from a single subproof Π_1^* by a unary rule of inference. If the unary rule is a negation rule or a quantifier rule, then the conclusions of the lemma for Π^* follow immediately from their analogues for Π_1^*, and indeed there is practically no difference between Π^* and Π_1^* in this situation, as far as the lemma is concerned. If instead the unary rule is a logical rule which introduces a binary connective (i.e., either a $\vee$: *right*, $\wedge$: *left*, or $\supset$: *right* rule), then the story remains much the same. Our two solitary atomic occurrences Z and W in the endsequent of Π^* have immediate (solitary) counterparts Z' and W' in the endsequent of Π_1^* which are connected to each other in the logical flow graph of Π_1^* through solitary occurrences. (Note that Z and W cannot be directly involved in the unary rule in this case, because it produces a non-solitary formula by definition.) Thus the conclusions of the lemma hold for Z' and W' in the endsequent of Π_1^*, by induction hypothesis, and this implies the analogous assertions for Z and W in Π^*.

This leaves the possibility that the unary rule is a contraction rule. Let us show that this cannot happen. We should be a bit careful here, since contractions can affect the connectivity properties of the logical flow graph (unlike the unary *logical* rules of inference discussed in the previous paragraph).

Claim 6.23 Π^* *cannot be obtained from a single subproof* Π_1^* *by a contraction.*

Assume, for the sake of finding a contradiction, that Π^* is obtained from Π_1^* by a contraction rule. Let A be the formula occurrence in the endsequent of Π^* which is obtained by contracting formulae A^1, A^2 in the endsequent of Π_1^*. If we can show that at least one of A^1 and A^2 is weak, then we shall be finished, because this would be incompatible with the requirement that Π^* be reduced.

Let us distinguish between two cases, according to whether or not A contains a binary logical connective ($\vee$, $\wedge$, or $\supset$). Suppose first that A does contain such a connective. Then A cannot contain either of Z or W, since they are *solitary* occurrences by assumption. Again we let Z' and W' denote the immediate antecedents of Z and W in the endsequent of Π_1^*. The key point now is that the hypotheses of the lemma are valid for Z' and W' in the endsequent of Π_1^*, i.e., there is a connection between Z' and W' in the logical flow graph of Π_1^* which passes only through atomic occurrences. This follows from the fact that we have such a connection between Z and W in Π^*, and because the formulae A, A^1, and A^2 cannot participate in such a connection, since they contain at least one binary logical connective. (There would be a problem with this point in the argument if we did not restrict ourselves to connections through solitary occurrences, as illustrated by the first example displayed after the statement of Lemma 6.22.) Thus our induction hypothesis (6.20) implies that the conclusions of the Lemma 6.22 hold for Z' and W' in Π_1^*. This yields the weakness of both A^1 and A^2, which is more than we need.

This leaves the second case, in which A does not contain a binary logical connective. Let P denote the unique atomic occurrence contained in A, and let P^1 and P^2 denote the counterparts of P in A^1 and A^2, respectively. Suppose for the moment that P is actually the same occurrence as Z. In this case, the assumption that there be a connection between Z and W in Π^* which passes through only solitary atomic occurrences implies the existence of a similar connection in Π_1^* between W and at least one of P^1 and P^2. Again we use our induction hypothesis (6.20) to apply Lemma 6.22 to Π_1^*, to conclude that all of the formula occurrences in the endsequent of Π_1^* are weak, except for W and one of the A^i's. In particular, the other A^i is a weak formula, as desired. A similar argument can be employed when P is the same occurrence as W in the endsequent of Π^* (instead of being the same as Z).

The remaining possibility is that A does not contain a binary connective, as in the preceding paragraph, but P is distinct from each of Z and W. Let Z' and W' be the (unique) immediate predecessors of Z and W in the endsequent of Π_1^*, as before. Thus neither of Z' or W' can be the same as P^1 or P^2, since P is not the same as either of Z or W. Because Z and W can be connected to each other in Π^* through solitary atomic occurrences, we either have a connection between Z' and W' through solitary atomic occurrences in Π_1^*, or we have a connection from each of Z' and W' to at least one of P^1 and P^2. This is easy to check, since the only way that "new" connections can appear in the passage from Π_1^* to Π^* is through the contraction of A^1 into A^2. If we have a connection between Z' and W' through solitary atomic occurrences in Π_1^*, then we can use our induction hypothesis (6.20) to apply Lemma 6.22 to Z' and W' in Π_1^*, from which the weakness of A^1 and A^2 follow. Thus we suppose instead that Z' and W' are not connected to each other through solitary atomic occurrences in Π_1^*, in which case they are connected to the P^i's instead. That is, Z' is connected to P^j and W' is connected to P^k through solitary atomic occurrences in Π_1^*, where either $j = 1$ and $k = 2$ or $j = 2$ and $k = 1$. In this situation, we can use our induction hypothesis (6.20) to conclude that Z' and P^j are the only non-weak occurrences in the endsequent of Π_1^*. In particular, we should then have that W' and P^k are weak occurrences, which is impossible, since they are connected to each other in the logical flow graph of Π_1^*.

This completes the proof of Claim 6.23, to the effect that our unary rule cannot be a contraction. Since we already discussed the other unary rules of inference, we assume now that Π^* is obtained from two subproofs Π_1^* and Π_2^* by a binary rule. In this situation, each of Z and W has a unique "direct" predecessor Z', W' in the endsequent of exactly one of Π_1^* and Π_2^*, where (a priori) the choice of Π_1^*, Π_2^* may depend on Z and W. This follows from the way that the binary rules work. (That is, there is nothing like a contraction for binary rules. This is not true in the context of feasibility.) Our induction hypothesis (6.20) implies that Lemma 6.22 holds for both of the subproofs Π_1^* and Π_2^*.

Claim 6.24 Π^* *cannot be obtained from two subproofs* Π_1^* *and* Π_2^* *by a binary logical rule of inference (i.e., a* $\vee : left$, $\wedge : right$, *or* $\supset: left$ *rule).*

The remaining possibility is that Π^* is obtained by combining Π_1^* and Π_2^* using a cut rule, and we shall deal with this possibility afterwards.

To prove Claim 6.24, we assume to the contrary that Π^* is obtained from Π_1^* and Π_2^* by a binary logical rule. Notice that the rule cannot operate directly on either of Z' and W', since Z and W are solitary, while the rule itself introduces a binary connective into the formulae on which it operates.

It is not hard to see that Z' and W' have to lie in the same subproof Π_i^* of Π^*. The reason for this is that we know that Z and W can be connected to each other in the logical flow graph of Π^*, and this would be impossible if Z' and W' did not come from the same subproof (Π_1^* or Π_2^*). For this assertion, we are using strongly the fact that our rule is a logical rule, instead of a cut.

Thus it is either true that Z' and W' both lie in Π_1^*, or that they both lie in Π_2^*. We may as well assume that they both lie in Π_1^*, since the other case can be handled in exactly the same manner. The basic point now is that Z' and W' can be connected to each other in the logical flow graph of Π_1^* by a path that goes only through solitary occurrences. This follows from the corresponding statement for Z and W in Π^* (which is part of our hypothesis), and the fact that the binary logical rule adds no connections between the logical flow graphs of Π_1^* and Π_2^*. (Again, this would not be true for the cut rule.)

Once we know that Z' and W' can be connected to each other in Π_1^* through solitary atomic occurrences, we can apply Lemma 6.22 to Π_1^* (because of our induction hypothesis (6.20)) to conclude Z' and W' are the only non-weak occurrences in the endsequent of Π_1^*. This contradicts the requirement (6.19) that Π^* be reduced, since it means that the binary logical rule used to combine Π_1^* and Π_2^* was applied to weak formulae in Π_1^*. (Remember that we know that this rule is not being applied to either of Z' or W', since we know that Z and W are solitary.)

This completes the proof of Claim 6.24.

We are left with the situation in which Π^* is obtained from Π_1^* and Π_2^* by a cut rule. Let C^1 and C^2 be the formula occurrences in the endsequents of Π_1^* and Π_2^*, respectively, on which the cut rule operates. Note that neither of Z' or W' lies in C^1 or C^2, since Z' and W' were chosen to be the immediate successors of Z and W (and therefore survive to the endsequent of Π^*).

We know from our assumptions that there is a path in the logical flow graph of Π^* which connects Z and W and which passes only through solitary occurrences. We claim that this path has to cross an edge in the logical flow graph of Π^* which is associated to the cut applied to C^1 and C^2. Indeed, if this were not the case, then Z' and W' would necessarily lie in the same subproof Π_i^* of Π^* (let us say Π_1^* for the sake of definiteness), and we would have a connection between Z' and W' in Π_1^* which passes only through solitary atomic formulae. Using our induction hypothesis (6.20), we could then apply Lemma 6.22 to Π_1^* to conclude that Z' and W' are the only non-weak atomic occurrences in the endsequent of Π_1^*. In particular, the cut formula C^1 in the endsequent of Π_1^* would have to be weak, in contradiction to our hypothesis (6.19) that Π^* be reduced.

This proves that our original connection between Z and W through solitary atomic occurrences in Π^* has to pass through the cut. This implies that the cut formulae C^1, C^2 cannot contain binary logical connectives (since we have a path of *solitary* atomic occurrences which passes through them). Let P^1, P^2 denote the (unique) atomic occurrences within C^1 and C^2.

Let us assume for convenience that Z' lies in the endsequent of Π_1^*. (The other situation is handled in exactly the same way.) In this case, there must be a path in the logical flow graph of Π_1^* which connects Z' and P^1; this follows from the fact our connection between Z and W in Π^* has to pass through the cut, as mentioned above. (Specifically, one looks at the *first* moment in which the connection from Z to W crosses the cut, to get the connection from Z' to P^1.) Our induction hypothesis (6.20) permits us to apply Lemma 6.22 to Π_1^* to conclude that Z' and P^1 are the only non-weak atomic occurrences in the endsequent of Π_1^*, and that they have opposite sign.

From this we conclude that W' must lie in the endsequent of Π_2^*. For if it lay in the endsequent of Π_1^*, then it would have to be weak, and so W itself (in the endsequent of Π^*) would have to be weak also. We know that this is not possible, since W is connected to Z.

Thus W' lies in the endsequent of Π_2^*, and the same argument as for Z' and P^1 leads to the conclusion that W' and P^2 are the only non-weak atomic occurrences in the endsequent of Π_2^*, and that they have opposite signs. This tells us exactly what we wanted to know. That is, there can be no non-weak atomic formulae besides Z and W after the cut, because P^1 and P^2 are removed. Also, Z and W must have opposite signs, for the following reasons: the signs of Z and W are the same as those of Z' and W', by the definition of Z' and W'; the signs of Z' and W' are exactly opposite to those of P^1 and P^2, as obtained above; and finally the signs of P^1 and P^2 are opposites of each other, because they lie in opposite sides of the same cut (and correspond to the same atomic occurrence within the cut formula).

This completes the analysis of the situation in which Π_1^* and Π_2^* were combined using a cut rule. We saw earlier that no other binary rules besides the cut rule are possible, and we explained before that how unary rules could be treated. Thus we are finished now with the induction step of the proof of the lemma. This completes the proof of the lemma as a whole, since the "base case" (of proofs which consist only of single axioms) is immediate, as mentioned near the beginning of the proof. □

We can strengthen the conclusion of Lemma 6.22 a bit further, as follows.

Lemma 6.25 *Under the same assumptions as in Lemma 6.22, we have that there are no contractions in* Π^*, *and no binary logical rules.*

Proof This can be obtained from exactly the same kind of induction argument as before. We simply add the absence of contractions and binary logical rules to the induction hypotheses on the proper subproofs of Π^*, and the proof of Lemma 6.22 explicitly showed that no new contractions or binary logical rules could be added (as in Claims 6.23 and 6.24). □

Remark 6.26 One can also show that if Π^* is as in Lemma 6.22, then the path going between Z and W crosses all of the distinguished occurrences in all of the axioms in Π^*. This is easy to check, using the same induction argument as before.

Lemma 6.27 *Let Π be a reduced proof (in pure logic, LK) which contains cuts only over atomic formulae. Let Q and R be atomic occurrences in Π, and assume that Q determines a defocusing branch point in the logical flow graph G of Π, and that R determines a focusing branch point in G. Assume also that there is an oriented path δ in G which begins at Q and ends at R. Then Q and R cannot both be solitary.*

This puts strong restrictions on the way that a configuration like the one in the middle box in Fig. 6.19 can arise. Note that we do not assume here that Q and R occur in the same sequent in Π.

Proof Suppose to the contrary that we are given Π, Q, R, and δ with the properties described in the lemma, and that Q and R are both solitary.

Claim 6.28 *δ passes only through* solitary *occurrences in Π.*

This is easy to verify, using the requirements that Π contain cuts only over atomic formulae, and that Q and R be solitary. We omit the details.

Notice that Q and R occur in contractions in the proof Π, since they define branch points in the logical flow graph G. We also know that Q must occur *negatively* in the proof (in the sense of Section A.3), and that R must occur *positively*, since they define *defocusing* and *focusing* branch points, respectively. When the path δ leaves Q it must go "upward" in the proof, following one of the branches of the contraction, and when it arrives to R it must be headed "downward" in the proof, arriving along one of the branches of the contraction there. In particular, δ must cross a distinguished occurrence in the first axiom that it meets after leaving Q, and in the last axiom that it crosses on the way to R. It might also traverse several cuts and other axioms in between.

Let Π^* denote the subproof of Π which contains Q in its endsequent.

Claim 6.29 *δ cannot cross the endsequent of Π^* after departing from Q.*

Indeed, suppose to the contrary that δ did cross the endsequent of Π^* in this way, and let Y denote the atomic occurrence in the endsequent of Π^* that it reaches first. Of course δ arrives to Y "from above" in the proof, since δ initially departed from Q in the "upward" direction, and since we are looking at the *first* moment where δ crosses the endsequent of Π^*. Notice also that Y cannot be equal to Q; this can be derived from simple considerations of orientation, and one could also employ Lemma 6.19.

Under these conditions, we have that Π^* satisfies the requirements of Lemma 6.22, with Z and W taken to be Q and Y. From Lemma 6.25 we obtain that Π^* contains no contractions, and this contradicts the fact that Q is in fact obtained from a contraction. This proves the claim.

Now, if R were to lie in the endsequent of Π^*, or outside of Π^* altogether, then δ would have to cross the endsequent of Π^*, which is impossible, as we have just seen. Thus we obtain that R must lie within Π^*, and not in the endsequent of Π^*.

The same arguments apply if we reverse the roles of Q and R, with only modest changes to reflect the reversal of orientations. More precisely, let Π^{**} be the subproof of Π which contains R in its endsequent. The analogue of Claim 6.29 states that δ cannot cross the endsequent of Π^{**} before arriving to R, and this can be shown in exactly the same manner as before. (That is, if it did cross the endsequent of Π^{**}, then one could look at the *last* moment of crossing, etc.) Once we have this, we can conclude that Q lies within Π^{**}, and not in its endsequent, since otherwise δ would have to cross the endsequent of Π^{**}.

This yields the desired contradiction to our assumptions about Q and R, since we cannot have both that R lies strictly within the subproof Π^* that has Q in its endsequent, and that Q lies strictly within the corresponding subproof Π^{**} for R. Thus we conclude that at least one of Q and R must not be solitary, and the lemma follows. □

The next lemma reformulates some of the information in Lemma 6.27 in a way which is more convenient for the analysis of focal pairs.

Lemma 6.30 *Let* Π *be a reduced proof (in pure logic, LK) which does not contain cuts over non-atomic formulae, and let* P *and* R *be atomic occurrences in* Π *such that* (P, R) *defines a focal pair in the logical flow graph* G *of* Π *(as in Definition 4.15). Then at least one of* P *and* R *must be non-solitary. We also have that* R *must occur* positively *in the proof, and that* P *must occur* negatively *if we know that* R *is solitary.*

Note that P may not occur at a branch point in the logical flow graph of Π, which changes the setting slightly from that of Lemma 6.27.

Proof Let Π be as above, and let P and R be atomic occurrences in Π such that (P, R) defines a focal pair in G. This means that we can find oriented paths α, β in G which begin at P and which arrive at R along different edges. In particular, R must define a focusing branch point in G, and this implies that R must occur *positively* in Π.

It may be that α and β diverge immediately after departing from P, but they might also coincide for some time. Let Q be the atomic occurrence in Π which represents the first moment at which α and β diverge from each other. This means that we have an oriented path γ in G which goes from P to Q and which is an initial subpath of both α and β, and that our paths α and β then follow different edges immediately after reaching Q. Thus Q defines a defocusing branch point in G.

We can now apply Lemma 6.27 in order to conclude that at least one of Q and R is not solitary. If R is not solitary, then we are finished, and so we assume instead that Q is not solitary.

Since Q represents a *defocusing* branch point in G, it must define a *negative* occurrence in the proof Π. This means that if we start at Q and follow the path γ "backwards", towards P, then we must go "down" in the proof Π, at least initially. In fact, we have to keep going "down" in the proof until we either reach P or an occurrence in a cut formula. The latter alternative is not possible in the present situation, since Q is not solitary, and because we are assuming that Π contains cuts only over atomic formulae.

Thus γ cannot cross a cut as we go down from Q towards P, and we conclude that P lies "directly below" Q. This implies that P, like Q, must occur negatively in Π, as desired. This completes the proof of the lemma. □

Let us now use these results to prove the Proposition 6.20.

Proof (Proposition 6.20) By assumption, we have a reduced proof Π with cuts only over atomic formulae, and we want to show that the logical flow graph G does not contain a chain of focal pairs of length 3.

Let P and R be atomic occurrences in Π such that (P, R) defines a focal pair in G. Let us say that P is *terminal* in this case if it occurs negatively in Π and if there are no cuts in Π directly below P. In other words, if we start at P and go "straight down" in the proof, we should eventually land in the endsequent of Π (and not have the possibility to turn around in a cut). Similarly, we say that R is terminal if it occurs *positively* in the proof and if we also land in the endsequent of Π when we go "straight down" in the proof starting at R. In a moment, we shall explain why at least one of P and R must be terminal when (P, R) defines a focal pair.

Suppose R is terminal in the focal pair (P, R), and let ρ be the oriented path in G which begins at R and goes "straight down" in the proof and ends in the endsequent of Π. Every oriented path in G which begins at R must be a subpath of ρ; there is no possibility for deviations from cuts or contractions. This implies that R cannot be the first component of a focal pair. That is, there cannot be another atomic occurrence T in Π such that (R, T) is a focal pair in this case.

The same thing happens when P is terminal in the focal pair (P, R). In that case, there is an oriented path σ in G which begins in the endsequent of Π and goes "straight up" until it reaches P, and every oriented path in G which ends at P must be a subpath of σ. This implies that there does not exist an atomic occurrence O in Π such that (O, P) defines a focal pair in G.

The remaining point in the proof of the proposition is to establish the following.

Claim 6.31 *If P and R are atomic occurrences in Π such that (P, R) defines a focal pair, then at least one of P and R is "terminal" in the sense defined above.*

To see this, we apply Lemma 6.30, which implies in particular that either P or R is not solitary. If W is any atomic occurrence which is not solitary, then one cannot reach a cut in the proof by starting at W and going "straight down", since there are no cuts in Π over non-atomic formulae, by assumption. In other words, one can only go to the endsequent of Π if one starts at W and goes straight down. This implies that R is terminal if it is not solitary, since we already know that R must occur positively in Π (as in Lemma 6.30). If R is solitary, then Lemma 6.30 implies that P is not solitary, and that it occurs negatively in the proof, and hence is terminal for the same reason. This proves the claim.

From the claim and our earlier discussion, we conclude the following. If P and R are atomic occurrences in Π such that (P, R) defines a focal pair, then either there is no atomic occurrence T in Π such that (R, T) is a focal pair, or there is no atomic occurrence O in Π such that (O, P) is a focal pair. From this we obtain immediately that there are no chains of focal pairs of length greater than 2 in G, which is exactly what we wanted. This completes the proof of Proposition 6.20. □

Remark 6.32 Strictly speaking, the restriction to cuts over atomic formulae excludes the presence of unary connectives (quantifiers and negations) in cut-formulae, but this is irrelevant for the arguments above. In other words, we might as well allow cuts over solitary formulae (but no more). In any case, the analysis presented in this section is not intended to be definitive or exhaustive.

Remark 6.33 In Proposition 6.20, we assumed that Π was *reduced*, which means in particular that Π does not contain any contractions applied to a weak occurrence. This fact was convenient for some of the lemmas, but it was not really necessary for the proof as a whole. Specifically, this assumption was used in the proof of Lemma 6.22 to eliminate the possibility of contractions in the induction argument, but the lemma would still be true even if the proof Π^* were permitted to contain contractions of weak occurrences. One would simply have to modify the induction argument slightly to show that contractions do not disrupt the properties given in the conclusion of the lemma, even if they can be applied to weak formulae. For Lemma 6.25, one would have to modify the statement a bit, to say that every contraction used in the proof Π^* was applied to at least one weak occurrence. Similarly, in Lemma 6.27, one would have to modify the assumptions to say that the contractions associated to Q and R were not applied to weak occurrences. This would pose no difficulty for Lemma 6.30, because the contractions associated to the Q and the R which arise there cannot be applied to weak formulae. This is not hard to check, using the fact that the paths α and β split apart at Q and arrive at R along different edges. If one of the contractions were applied to a weak formula, then one of these paths would not be able to exist, as one can easily check.

In terms of the logical flow graph, the presence of contractions which can be applied to weak occurrences is not very significant geometrically. They can be used to create a lot of branch points, but where at least one of the branches is always a tree, and therefore does not contribute to oriented cycles or chains of focal pairs. The application of cut rules or binary logical rules of inference to weak occurrences is more significant, because of the way that they allow more interesting structures to be combined, without logical necessity.

7

FEASIBILITY GRAPHS

With formal proofs one has the possibility to represent objects only *implicitly*. The notion of *feasibility* provides a general mechanism for doing this, as in Sections 1.3 and 4.8. In this chapter, we present a simpler combinatorial framework in which similar effects can be achieved.

7.1 Basic concepts

Roughly speaking, a *feasibility graph* is an optical graph in which the edges and branch points are labelled, in such a way that the graph represents some kind of construction or computation.

For the sake of concreteness, let us restrict ourselves for the moment to the set Σ^* of all finite words (including the empty word ϵ) over some given alphabet Σ. In this case, we permit the edges in a feasibility graph to be labelled by an element of Σ, or by no element at all (which can be viewed as a labelling by ϵ). The idea is that as one proceeds along the graph, one will construct words through right-multiplication by the letters used as labels on the edges. This is fine when there are no *branch points*, but in general we have to decide what to do with them. We shall interpret *defocusing branch points* as being "duplicating" devices; whatever was constructed up to that point is sent along both of the edges that emerge from the given vertex. *Focusing* branch points will be used to concatenate whatever words have been constructed so far. At a focusing branch point one, must specify which incoming edge is considered to arrive on the *left*, and which arrives on the *right*, to know in what order to perform the concatenation.

To be more precise, we should be careful to distinguish between the *data* needed to define a feasibility graph and the way in which a feasibility graph is *interpreted.* To *define* a feasibility graph one specifies an optical graph G together with a labelling of each edge by at most one element of Σ, and also a designation of "left" and "right" for the incoming edges at focusing branch points. We also require that

$$\textit{feasibility graphs contain no nontrivial oriented cycles.} \tag{7.1}$$

The interpretation of a feasibility graph follows the lines mentioned above. To make this precise, let us introduce some terminology and definitions.

Let us call a vertex in a feasibility graph *extreme* if it either has no incoming edges or no outgoing edges. A vertex with no incoming edges will be called an *input vertex*, while the ones with no outgoing edges will be called *output vertices.*

A vertex u in a feasibility graph is said to be a *predecessor* of another vertex v if there is an edge in the graph that goes from u to v. Thus the input vertices

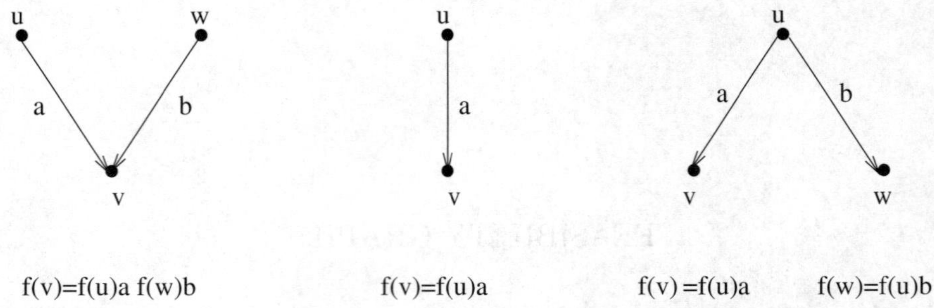

FIG. 7.1. An illustration of the notion of value functions

are precisely the ones which have no predecessors. Similarly, we call v a *successor* of u in this case, and output vertices have no successors. (We should perhaps say *immediate predecessors* and *immediate successors*, but for simplicity we omit the extra adjective in general.) Note that no vertex can be a predecessor or successor of itself when there are no nontrivial oriented cycles.

Definition 7.1 (Value functions) *Under the assumptions and notation above, a* value function *on a feasibility graph G is a function f defined on the set of vertices of G and taking values in Σ^* which enjoys the following property. Let v be an arbitrary vertex in G which is not an input vertex. If there is only* one *incoming edge arriving in v, so that v has only one predecessor u, then we ask that $f(v)$ be the same as the word obtained by multiplying the word $f(u)$ on the right by the label in Σ of the edge from u to v, if this edge is labelled. (If not, then $f(v)$ should be equal to $f(u)$.) Otherwise, v has two incoming edges, coming from vertices u_1 and u_2 (which might be the same). In this case, we again take $f(u_1)$ and $f(u_2)$ and multiply each of them on the right by the label in Σ of the corresponding edge, if there is such a label, to get two words in Σ^*. We then multiply these two words together, using the designations of the incoming edges as "left" or "right" to decide in what order to perform the multiplication, and we ask that $f(v)$ be equal to the result. (See Fig. 7.1.)*

Any such function f is a value function*, and we call it a* normalized value function *if $f(v) = \epsilon$ (the empty word) whenever v is an input vertex in G.*

Before we proceed to a general discussion of what this means, let us prove the basic existence and uniqueness result.

Lemma 7.2 *Notations and assumptions as above. Given a function f_0 from the input vertices of G to Σ^*, there is a unique value function f defined on all the vertices of G which agrees with f_0 at the input vertices.*

In particular, there is always exactly one normalized value function. It is very important here that G have no nontrivial oriented cycles, as in (7.1). The proof uses some well-known means, as in the way that Boolean circuits are shown to have well-defined values on p80 of [Pap94].

Remark 7.3 Although we have formulated Lemma 7.2 and the notion of value functions only for words over a given alphabet and certain operations over them, they both apply in great generality. One really only needs to have a set of objects and a collection of operations acting on them. The "arities" of the operations (i.e., the number of arguments) is not important either. We shall discuss this a bit further in Section 7.2, but for the moment we want to emphasize that *we shall use Lemma 7.2 freely in other contexts in this book* (no matter the underlying objects or operations on them). The reader may wish to keep this in mind when reading the proof that follows, for which the main issues concern only the geometry of G as a graph. (We should also say that the restriction to *optical* graphs plays no role either; one simply needs the underlying graph to be oriented and to contain no nontrivial oriented cycles.)

Proof To prove this, we should first organize the set of vertices in G in a convenient manner. We do this recursively, as follows. Let S_0 be the set of input vertices in G. Fix $k > 0$, and suppose that S_j has been defined for $0 \leq j < k$. Set $W_k = \bigcup_{j=0}^{k-1} S_j$. We take S_k to be the set of vertices which do not lie in W_k, but whose predecessors all lie in W_k.

We repeat this process until we reach a k for which S_k is empty. This must happen in a finite number of steps, since the S_j's are pairwise disjoint, by construction, and G is a finite graph.

Every vertex in G must lie in some S_j. Indeed, assume to the contrary that there is a vertex z which does not lie in any S_j. From the construction, we see that z has a predecessor z_{-1} which does not lie in any S_j either. We can repeat this indefinitely to obtain vertices $z_{-\ell}$ for all nonnegative integers ℓ such that $z_{-\ell}$ does not lie in any S_j, and $z_{-\ell-1}$ is a predecessor of $z_{-\ell}$. Since G is finite, we must have that $z_{-\ell} = z_{-m}$ for some ℓ, m which are distinct. This implies the existence of a nontrivial oriented cycle in G, in contradiction to our assumptions on G.

Thus the S_j's exhaust the set of vertices in G. Once we have this, it is easy to see that a value function is uniquely determined by its restriction to S_0. One can use an induction argument, and the simple fact that the restriction of a value function to any S_j, $j > 0$, is determined by its restriction to W_j. This follows from the definition of a value function, and the fact that the predecessors of any element of S_j lie in W_j.

Similarly, we can always find a value function f defined on the whole set of vertices in G given its restriction f_0 to the set S_0 of input vertices. That is, we extend f from S_0 to S_1, and then to S_2, and so on, with the extension at each step defined through the value function property. The function defined on all of the vertices at the end is easily seen to be a value function, and it agrees with f_0 on S_0 by construction. This completes the proof of Lemma 7.2. □

Lemma 7.2 indicates two natural interpretations of the notion of a feasibility graph. In the first interpretation, we simply associate to a given feasibility graph the normalized value function which it determines uniquely. In particular, we get a set of values associated to the output vertices of the feasibility graph; we think of the feasibility graph as providing an implicit construction of these values.

The "implicitness" here comes from the fact that the results of partial computations can be used more than once in the computation as a whole. This is similar to the implicitness in Boolean circuits, as opposed to Boolean expressions, or to the implicitness of descriptions of numbers through formal proofs of feasibility which do not use quantifier rules. In Chapter 16, we shall discuss stronger forms of recursion, through feasibility graphs which are used to construct other feasibility graphs. In this setting, one is permitted to duplicate whole recipes for making constructions rather than just individual words, numbers, or other basic objects, and this is quite similar to some of the basic phenomena arising through formal proofs of feasibility *with* quantifier rules allowed.

One should consider a feasibility graph to be "explicit" if each vertex has at most edge coming out of it, so that intermediate computations or constructions are not duplicated. One can always convert the implicit computations performed by feasibility graphs into explicit ones, using (negatively-oriented) visibility graphs, as we shall discuss further in Section 7.4.

In the second interpretation of a feasibility graph, we think of the graph as defining a transformation which takes in a collection of words associated to each input vertex and gives back a value function defined on the whole feasibility graph. Again, we might be particularly interested in the values at the output vertices, so that the feasibility graph represents a mapping which takes in a tuple of words at the input vertices and converts them into a tuple of words at the output vertices.

To put it another way, in the first interpretation we view the feasibility graph as representing the construction of a particular collection of words, while in the second interpretation we think of the graph as representing a mapping between collections of words (with inputs at the input vertices, and outputs at the output vertices).

Remark 7.4 One can convert the second interpretation into a special case of the first, by passing from words in Σ^* as the basic objects under construction, to functions over words. At the input vertices, one would start with copies of the identity function on Σ^*, as the initial values for the "normalized value functions" (like the empty word before). Now these initial values are themselves functions. These would be viewed as functions of different variables, even if they are all functions on the same set Σ^*. (More precisely, one would do this for the present purposes. There are plenty of variations that one could consider, and one can see this in a broader way, as in Section 7.2.) The unary operations on edges would be interpreted now as multiplying functions by single letters. The focusing branch points would be viewed as representing multiplications of functions, rather than multiplications of words. (For this, there would be the same issue of having an order for the incoming edges as before, to know which order to make the multiplication.) Defocusing branch points would give duplications of functions, as before. Although the functions at the input vertices are functions of one variable, functions of several variables can result from the multiplication of functions at the focusing branch points. Functions of the same variable can come up more than once, because of the duplications in the defocusing vertices. The values of the normalized value function on a feasibility graph, viewed in this way, would be functions with some number of variables on Σ^*, taking values in Σ^*. These functions would be the same as the ones that one gets from the second interpretation for feasibility graphs before, for constructing words given particular inputs at the input vertices, with outputs at the output vertices.

Matters like these will be considered further in Chapter 16, especially beginning in Section 16.4.

7.2 Extensions and comparisons

The basic idea of feasibility graphs is obviously very general. One only needs a set of objects X on which to work, together with some unary and binary operations defined on it. The set would play the role of Σ^* from before, the unary operations would be like adding letters to words, and the binary operations would be like multiplying words together. One would define a feasibility graph by labelling edges and focusing branch points by these operations, and then one could define the notion of value functions in exactly the same manner as before, with the same lemma of existence and uniqueness. (One should continue to assume that the graph has no oriented cycles, as in (7.1).)

For the notion of normalized value functions, one would also need a choice of an element of the set X to use as the value of normalized value functions at the input vertices of a feasibility graph, just as we used the empty word ϵ before.

One could allow operations of arity larger than 2, using graphs with correspondingly higher vertex degrees than optical graphs. For the time being, we shall concentrate on operations of arity at most 2 and optical graphs, simply because this accommodates many basic examples, and is compatible with the general terminology used so far in this book. (We shall officially dispense with the restriction to optical graphs in Section 8.6.)

There is also no problem with allowing unary operations to be assigned to vertices with only a single incoming edge (thus treating all vertices in the same fashion, whether or not they are focusing branch points). In this case, one might wish to refrain from assigning unary operations on edges at all (since they could always be simulated through the addition of extra vertices). In some contexts it will be convenient not to do this, and to permit edges to represent operations of their own, as in Section 7.1.

Boolean circuits provide another example for the general concept of feasibility graphs. We shall return to this in Section 7.11, and we shall discuss additional examples later in the chapter. As another basic setting, one can consider feasibility graphs involving *functions*. A version of this came up in Remark 7.4 in Section 7.1. We shall look more at topics concerning functions in Chapter 16.

Defocusing branch points always represent the same kind of *duplication* effect as before, i.e., with the value obtained so far used in each of the outgoing edges.

In this regard, the constructions defined by feasibility graphs are a bit like formal proofs, for which the effect of (possibly repeated) duplication can be achieved using the contraction and cut rules. One can take this further and observe that the constructions obtained through feasibility graphs can always be coded into formal proofs of "feasibility", where the general formalization of feasibility follows the special case of feasible numbers (discussed in Section 4.8). To make this more precise, one should begin by defining a (unary) feasibility predicate $F(\cdot)$, with the properties that feasibility is preserved by the basic operations in use, and that feasibility is automatic for some basic constants. In the setting of Section 7.1, in which one is dealing with words over an alphabet Σ, one would require that the empty word be feasible, and that feasibility be preserved by the unary operations of right-multiplication by a letter in Σ, and by the binary operation of concatenation of a pair of words already known to be feasible.

With this type of feasibility predicate F, it is not hard to convert a feasibility graph G into a formal proof. Actually, there are two basic ways of doing this, according to whether one prefers to think of a feasibility graph as describing the construction of particular objects, through the normalized value function, or as defining a function on arbitrary inputs. For the first interpretation, one would make a proof Π' whose endsequent expresses the feasibility of the values of the normalized value function of G at the output vertices. For the second interpretation, one would build a proof Π whose endsequent has the form

$$F(x_1), F(x_2), \ldots, F(x_k) \to F(t_1) \wedge F(t_2) \wedge \cdots F(t_l),$$

where the x_i's are variables which represent the possible values of a value function f on G at the input vertices, and the t_j's are terms which represent the values of f at the output vertices (and which involve the x_i's). In either case, the conversion from G to a proof would not entail substantial expansion in size, but for this it would be important to allow contractions and cuts, to accommodate the effect of the (possibly repeated) duplications.

The simplification of the cuts over the contractions in these proofs corresponds to pushing the defocusing branch points in the underlying feasibility graphs to the very beginning, near the input vertices, before any operations have been performed. Similar phenomena came up in Chapter 6. In particular, one does not need cuts when the defocusing branch points are all at the beginning; they are needed for duplications that come after some operations have been performed. One can see this in terms of logical flow graphs, in the way that cuts are needed to go from focusing branch points to defocusing ones (as in Lemma 6.4 in Section 6.12). For feasibility graphs, one can eliminate defocusing branch points quite easily, by passing to the visibility graph, and we shall discuss this further in Section 7.4 below. (A version of this came up before, in Section 4.12, in connection with Boolean circuits and expressions.)

Thus feasibility graphs provide a kind of model for some of the effects of formal proofs, but in a simpler combinatorial situation. Note that only propo-

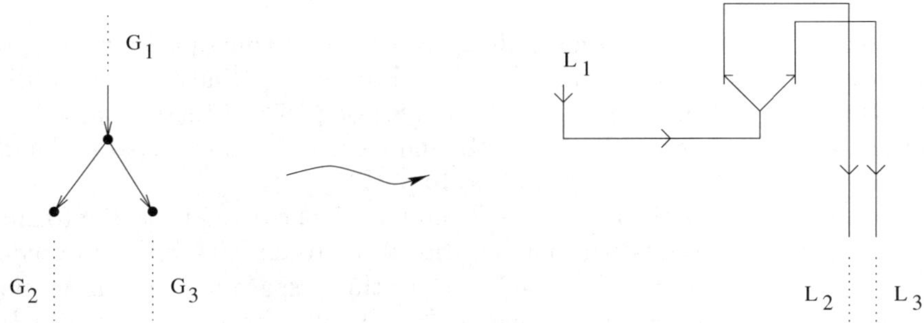

FIG. 7.2. Converting feasibility graphs into formal proofs

sitional rules of inference are needed in this general discussion (as opposed to quantifier rules), even if we do want to use feasibility *predicates*.

Remember that we already know how to extract graphs from formal proofs, using the logical flow graph. Now we are going in the opposite direction, showing how constructions through graphs can be converted back into proofs.

The conversion from feasibility graphs to formal proofs is depicted symbolically in Fig. 7.2. More precisely, the right side of Fig. 7.2 refers to portions of the logical flow graph of the resulting proof, rather than the proof itself.

One of the nice features of feasibility graphs is that they provide a setting in which it is easier to make comparisons between different constructions. Basically one can make *geometric* comparisons, through mappings between feasibility graphs. (See Section 11.5.)

7.3 Some remarks about computability

How easy is it to compute a value function associated to a feasibility graph?

For simplicity, let us ignore the complexity of the basic (unary, binary, or n-ary) operations associated to edges and vertices. These can depend arbitrarily on the context, and for the moment we would like to focus on the combinatorial aspects of the feasibility graphs. Thus we shall treat the basic operations as being single steps. We shall also ignore the sizes of the objects being constructed.

If we do this, then the computation of the value function from specific choices of initial data at the input vertices can be accomplished in a polynomial number of steps. This is easy to see from the proof of Lemma 7.2. One would produce the S_j's, as in the proof, and determine the value functions on the S_j's using the local rules and their values on the previous S_i's.

There is a sense in which one cannot do better than polynomial complexity in general, because of the "P-completeness" of the "circuit value" problem. In the circuit value problem, one seeks to compute the value of a given Boolean circuit with all inputs specified. (See p81 of [Pap94].) This is practically the simplest of all problems associated to feasibility graphs, since the "local" operations associated to edges and vertices (conjunction, disjunction, and negation) are so simple, and the possible values at any moment are just 0 and 1. "P-completeness" means that there is a *logarithmically space-bounded* reduction from any given polynomial-time problem to this one. (See [Pap94], especially Theorem 8.1 on p168, Definition 8.2 on p165, and Definition 8.1 on p160.) This type of reduction is sensitive to subclasses of P like L, NL, and some complexity classes related to parallel computation. (See p166, 377 of [Pap94].)

The circuit value problem remains P-complete if one restricts oneself to monotone circuits (i.e., circuits without negations), or to circuits which are planar. However, the problem can be solved in logarithmic space if there are no negations and the underlying graph is planar. (See [Pap94] for more information and references, especially the corollary at the top of p171 and Problem 8.4.7 on p178.)

7.4 Feasibility and visibility graphs

Let G be a feasibility graph, as in Sections 7.1 and 7.2. For the sake of concreteness, it is useful to think in terms of the construction of words, as in Section 7.1, but the matter is completely general, as indicated before.

Fix a vertex v in G, and consider the *negative* visibility graph $\mathcal{V}_-(v, G)$. This is defined in essentially the same manner that $\mathcal{V}_+(v, G)$ was (in Section 4.2), except that we use negatively-oriented paths that begin at v instead of positively-oriented paths. (One can also think in terms of positively-oriented paths which end at v, rather than begin at v.) In effect, we want to look at the *past* of a vertex instead of its *future*. This is very natural in the context of feasibility graphs, for which we might be interested in knowing from where information came.

For the present purposes, we shall treat $\mathcal{V}_-(v, G)$ as an oriented graph, in which the orientation is compatible with that of G, and a bit backwards from the one to which we are accustomed. Thus the basepoint in $\mathcal{V}_-(v, G)$ has *incoming* edges and paths, but no *outgoing* ones. More precisely, we use the orientation on $\mathcal{V}_-(v, G)$ such that the usual projection $\pi : \mathcal{V}_-(v, G) \to G$ as in Section 4.2 *preserves* orientations (instead of perhaps reversing them).

Remember that we are restricting ourselves to graphs without oriented cycles here (see (7.1)), and so the visibility $\mathcal{V}_-(v, G)$ must be finite, as in Lemma 4.8. (Of course, all of our previous results hold equally well for the negative visibility as for the positive visibility. In this regard, one can reduce from negative to positive visibilities by reversing orientations on the original graph G.)

We already know that the visibility $\mathcal{V}_-(v, G)$ is an optical graph (Section 4.5), but in fact we can make it into a feasibility graph. To do this, we use the canonical projection $\pi : \mathcal{V}_-(v, G) \to G$ to pull back the labellings of edges in G (by letters in the alphabet Σ, or by whatever unary operations are in use) to labellings of edges in the visibility $\mathcal{V}_-(v, G)$. Similarly, *focusing* branch points in $\mathcal{V}_-(v, G)$ correspond to focusing branch points in G in a simple way, and we can use the labellings on G to get labellings in $\mathcal{V}_-(v, G)$. (For the discussion about words in Σ^*, we simply need to decide which of the two incoming edges is on the "left", and which is on the "right". In other contexts, we might have to decide which operation (from some collection) is being associated to a given focusing branch point in $\mathcal{V}_-(v, G)$. All of these choices can be inherited from their counterparts in G, using π.)

Given any value function f defined on the vertices of G, we can lift it back to a value function $\pi^*(f) := f \circ \pi$ on the vertices of the visibility $\mathcal{V}_-(v, G)$. For that matter, we can define the function $f \circ \pi$ on the vertices of $\mathcal{V}_-(v, G)$ given any function f defined on the vertices of G; the important point here is that we get a value function in the sense of Definition 7.1 on $\mathcal{V}_-(v, G)$ when we apply this lifting to a value function f on G. This is not hard to check, directly from the definitions of the visibility and of value functions (Definition 7.1). (This kind of lifting will be put into a broader context in Section 11.5.)

If f is the *normalized* value function for G, then $f \circ \pi$ will be the normalized value function for the visibility $\mathcal{V}_-(v, G)$. This is because input vertices in $\mathcal{V}_-(v, G)$ must project down to input vertices in G by π, which is easy to check. We are also using the uniqueness part of Lemma 7.2.

In short, the (negative) visibility makes the "same" computation as did the original graph. The point is that it does so *explicitly*, gram by gram, because of the elimination of defocusing vertices which lead to effects of duplication. It is exactly these effects of duplication which can lead to exponential expansion in the implicit constructions given by feasibility graphs. We shall discuss this more precisely in Section 7.6 below. (We saw versions of this for Boolean circuits and Boolean expressions in Section 4.12.)

In the context of feasibility graphs, the negative visibility plays a role similar to that of cut elimination. One can make more detailed comparisons, by looking at formal proofs of feasibility which reflect the same computations or constructions as in a feasibility graph, as in Section 7.2. This is closely related to some of what we saw in Chapter 6, including Sections 6.5 and 6.7, and the proofs of feasibility discussed there. More precisely, Section 6.5 corresponds to feasibility graphs as providing constructions given certain inputs, as with value functions, while Section 6.7 corresponds to feasibility graphs as providing constructions with normalized input values, as with normalized value functions. For the latter, the negative visibility is approximately the same as cut elimination. For the former, there are some extra branchings (and stretchings) at the other side of the visibility graph that would be involved with cut elimination, which bring together some of the ends of the visibility graph.

By itself, the visibility graph makes duplications of the input vertices of the original feasibility graph. The combining of ends at the "far" side of the visibility graph brings these duplications back to individual vertices, one for each of the input vertices in the original feasibility graph (which are accessed by the visibility graph). This combining of ends comes up in logical flow graphs of formal proofs through the contractions employed in the duplication of subproofs, as in (6.5) in Section 6.2. After the combining of ends, the visibility graph becomes roughly like the graph H in Fig. 4.4 in Section 4.3, or variants of it. This is related to some of the remarks in Section 6.12. With normalized value functions, one can work with formal proofs as in Section 6.7, and these extra branchings do not arise (from cut elimination). In effect, one does not mind the duplication of the input vertices in the original feasibility graph (for which there are normalized values). They are treated like axioms, which are duplicated, rather than formulae in the bottom sequent, for which one uses contractions to recombine duplications.

In any case, aside from the comparison with cut elimination, it is reasonable to recombine ends on the other side of the visibility graph, so that input vertices from the original feasibility graph are preserved (and not duplicated), even if there are a lot of duplications afterwards. In other words, the visibility graph makes explicit all of the duplications in the feasibility graph, repeating vertices and edges for each one, while here one would only do this until one gets to the input vertices. One would push defocusing branch points all the way back to the input vertices, rather than eliminating them entirely. The defocusing branch points would be at one end, and the focusing branch points on the other, instead of eliminating the defocusing branch points. The graphs would be *steady*, in the sense of Definition 6.2 in Section 6.12, like the one called H in Fig. 4.4 in Section 4.3, or the one in Fig. 6.14 in Section 6.12. The total effect would still be much the same as that of the visibility graph.

Similar matters come up in Sections 11.5 and 16.15.

7.5 Upper bounds

Let G be a feasibility graph, as in Section 7.1. Again we shall restrict ourselves to the case of building words over an alphabet Σ, for the sake of concreteness, even though the considerations of this section apply much more broadly.

Let f be the normalized value function for G, and let w be a word in Σ^* which arises as the value of f at some output vertex o of G. What can we say about the size of w?

In general, the duplications allowed in feasibility graphs lead to the possibility of exponential expansion in the size of w as compared to the size of G. This will be illustrated concretely in Section 7.6. To get more precise bounds which take into account the structure of G, one can use the following.

Lemma 7.5 *Notations and assumptions as above. Suppose that G has no defocusing branch points. Then the number of letters used to make w is less than or equal to the number of edges in G, and is less than the number of vertices in G.*

Proof The bound in terms of the number of edges is essentially immediate, because the edges are the only places where letters are *added* (as opposed to letters that are already there being combined into new words, as at the focusing branch points). For this it is very important that there are no defocusing branch points. In fact, if every edge is associated to adding a letter (which is not required by the definitions in Section 7.1), and if there is only one output vertex in the graph, then the length of w will be equal to the number of edges in G. (One can give more formal arguments for these assertions, using induction, for instance.)

Because there are no defocusing branch points, every vertex has at most one edge flowing out of it. This implies that the number of edges in G is less than or equal to the number of vertices, since every edge flows out of some vertex. In fact, the number of edges is equal to the number of vertices minus the number of output vertices, since the output vertices are the only ones with no edges flowing out of them. In particular, the number of edges in G is less than the number of vertices, since there is an output vertex. □

If G does have defocusing branch points, then we can reduce to the situation of the lemma by lifting f to the negative visibility $\mathcal{V}_-(o, G)$, as explained in Section 7.4. Thus bounds for the size of $\mathcal{V}_-(o, G)$ lead to bounds for the size of w. (One can also derive this more directly, by making comparisons between the two at each step.) For the visibility graph, the matter is purely geometric, and it will be treated in some detail in Chapter 8. In any case, we have the very simple exponential bound given in Lemma 4.9 in Section 4.7.

Remark 7.6 We can turn the matter of bounds around, and say that lower bounds for the words being constructed provide lower bounds for the size of the visibility. If these lower bounds are much larger than the feasibility graph itself, then Theorem 8.9 in Section 8.4 implies that there must be a chain of focal pairs in the feasibility graph, with a lower bound on the length of this chain.

As usual, these considerations apply much more generally than for just the construction of words over a given alphabet. One may have to be a bit careful about the way that "sizes" of objects are measured, but the basic structure of the computations remains the same.

7.6 Concrete examples

To understand better what can happen with implicit representations through feasibility graphs, it is helpful to consider some concrete situations explicitly. We shall confine ourselves for the moment to the construction of words over alphabets, as in Section 7.1. We shall restrict our attention to *normalized value functions*, which represent implicit constructions without auxiliary inputs.

Powers of a single letter

Let us begin with exactly the situation of Section 7.1, but with an alphabet Σ that consists only of a single letter a. Thus a *word* is now just a representation of a nonnegative integer in unary notation, and our operations correspond to addition by 1 (the *successor* function in arithmetic) and addition.

Lemma 7.7 *Given any integers k and n, with $k \geq 1$ and $0 \leq n < 2^k$, we can represent a^n by a feasibility graph of size $O(k)$.*

More precisely, we can find a feasibility graph G of size $O(k)$ such that G has exactly one input vertex and one output vertex, and so that the value of the normalized value function associated to G at the output vertex is a^n. (Of course, we mean a "feasibility graph" with respect to the alphabet and structure mentioned above.)

Proof Let k and n be given. The basic graph that we shall use will be similar to the one in Fig. 4.2 in Section 4.3, and will depend on k, but not on n. The associated "labellings" will depend on n, however.

More precisely, we shall use the oriented graph whose vertices are the integers $0, 1, 2, \ldots, 2k-1$, with exactly one edge going from j to $j+1$ when j is even, and two edges going from j to $j+1$ when j is odd. To make this into a feasibility graph, we have to decide which edges are labelled by a and which are left unlabelled. All of the focusing branch points correspond to concatenation of words, since that is the only binary operation that we have here.

We leave unlabelled all edges which begin at j when j is *odd*. For the remaining edges, we use the following coding. Let $\{b_i\}_{i=0}^{k-1}$ be an arbitrary binary sequence, to be specified later. We label the edge from j to $j+1$ by a when $j = 2i$ and $b_i = 1$, and we leave it unlabelled otherwise.

This defines our feasibility graph G. It has a unique normalized value function f, as in Lemma 7.2, which we can write as $f(j) = a^{\phi(j)}$, where $\phi(j)$ is defined for $j = 0, 1, 2, \ldots, 2k$ and takes values in nonnegative integers. We have that $\phi(0) = 0$, by definitions, while in general ϕ satisfies the recurrence relations

$$\phi(2i+1) = \phi(2i) + b_i, \quad \phi(2i+2) = 2 \cdot \phi(2i+1). \tag{7.2}$$

Now, given $j \leq k$, $j \geq 1$, and a nonnegative integer $n_j < 2^j$, we can choose b_i for $i = 0, \ldots, j-1$ so that $\phi(2j-1) = n_j$. Indeed, for $j = 1$ we have that $\phi(2j-1) = \phi(1) = b_0$, and this does the job since n_1 must be 0 or 1. For $j > 1$, we have that

$$\phi(2j-1) = \phi(2j-2) + b_{j-1} = 2 \cdot \phi(2j-3) + b_{j-1}. \tag{7.3}$$

By induction, we can be free to choose $\phi(2j-3)$ as any nonnegative integer $< 2^{j-1}$, and then we can get $\phi(2j-1)$ to be any prescribed integer $< 2^j$ by choosing b_{j-1} correctly. (This argument amounts to choosing the b_i's according to the binary expansion of n_j, except that we turn things a bit backwards, so that b_i corresponds to 2^{j-i+1} instead of 2^i.)

By taking $j = k$, we see that we can get any nonnegative integer strictly less than 2^k for the value of $\phi(2k-1)$. In other words, we can reach a^n for any $n \geq 0$, $n < 2^k$, using a feasibility graph of the type described above. □

General alphabets

Now suppose that we are working in the context of Section 7.1, but with an alphabet Σ that contains at least two letters. Just as before, it is easy to make examples of feasibility graphs of size $O(k)$ which result in words of size 2^k.

The converse to this is no longer true though. Before we could get *all* words of length $\leq 2^k$ in this manner, but this is far from being the case now. Because there are at least two letters in Σ, there will be at least 2^{2^k} words over Σ of length 2^k. The number of possible feasibility graphs of size about k is far fewer.

Indeed, let n be a positive integer, and let us estimate the number of feasibility graphs with at most n vertices (up to isomorphic equivalence). Fix a set of vertices, and imagine attaching edges to it. Each fixed vertex has at most 2 edges coming out of it (under the restriction to optical graphs), and there are no more than n^2 ways of attaching at most 2 edges to any fixed vertex. Allowing independent choices at all the vertices, we get at most $(n^2)^n = n^{2n}$ different ways of attaching edges to a set of at most n vertices. (This estimate is crude, but it is enough for the moment.)

To make a feasibility graph out of such an optical graph, we have to choose labellings for the edges. That is, for each edge, we are allowed to choose either an element of Σ, or no label at all. For an optical graph, there are at most $2n$ edges (since there are at most two with any fixed initial vertex), and so we get a bound of $(S+1)^{2n}$ for the number of possible labellings for the edges of a given graph, where S denotes the number of elements in Σ.

At vertices which are focusing branch points, we should also specify an ordering between the two incoming edges. There are two choices of such orderings, and no more than n focusing branch points (since there are no more than n vertices), for a total of no more than 2^n different families of orderings for the whole graph.

Thus there are at most n^{2n} optical graphs with no more than n vertices, each of which has at most $2^n(S+1)^{2n}$ ways of being properly labelled to make a feasibility graph. This gives a bound of $2^n(S+1)^{2n}n^{2n}$ for the number of different feasibility graphs with no more than n vertices, up to isomorphic equivalence.

Think of this as being $2^{2n(\log n+c)}$, for some constant $c > 0$. For the number of words of length less than or equal to 2^k, we have at least S^{2^k} possibilities. Therefore, while feasibility graphs on n vertices can describe constructions of words of exponentially-large length compared to n, these particular words are a very small minority compared to all words of the same size.

To put it another way, the words of size approximately 2^k which admit representations by feasibility graphs of polynomial size in k are very special. It is not at all clear exactly what kind of internal symmetry or structure that they have to have, though.

The restriction to optical graphs here is not very important, and one could make analogous computations more generally.

Finitely-generated groups

Let us now decide to think of our words as representing elements of a finitely-generated group. (The earlier discussions correspond to *free* groups and semigroups.) There can be additional effects coming from the relations in the group.

For this discussion, we shall think of Σ as being a set of *semigroup* generators for the given group, e.g., a set of group generators together with their inverses.

As a simple example, consider the *Baumslag–Solitar group*, which has two generators y and x and the one relation

$$yx = x^2y. \tag{7.4}$$

(This group will arise again in Section 18.1.) For the purpose of feasibility graphs, we take Σ to be the set consisting of y and x and also their inverses. Thus distinct words can correspond to the same group element, and we want to consider the possible effects of this ambiguity.

We know from before that we can represent x^{2^k} by a feasibility graph of linear size in k, but in this group that is not very exciting, because

$$x^{2^k} = y^k x y^{-k}. \tag{7.5}$$

That is, we can actually represent the group element x^{2^k} by another *word* of linear size in k. The implicitness of the feasibility graph is not really needed.

On the other hand, we can also represent y^{2^k} through feasibility graphs of linear size in k, and we cannot achieve this simply through a tricky representation by a word of linear size. Indeed, we can define a homomorphism from our group to the infinite cyclic group $\mathbf{Z}$ by taking an arbitrary word over y and x and throwing out all the x's to simply get a power of y. It is easy to see that this actually defines a group homomorphism, because of the specific nature of the group relation. (This would not work with the roles of y and x exchanged.) Using this homomorphism, we get that any word which represents y^{2^k} in the group has length at least 2^k, because this is true in the cyclic group.

Using representations for y^{2^k} by feasibility graphs of size $O(k)$ we get the following for powers of x.

Lemma 7.8 *Notation and assumptions as above. Given a positive integer k, there is a feasibility graph L for words over Σ such that L has exactly one input vertex and one output vertex, the size of L is $O(k)$, and the value of its normalized value function at the output vertex is a word which represents the same group element as $x^{2^{2^k}}$.*

Proof Because of the identity (7.5) (with k replaced with 2^k), it is enough to choose L so that the value of its normalized valued function at the output vertex is

$$y^{2^k} x y^{-2^k}. \tag{7.6}$$

This is easy to do, using the fact that y^{2^k} and y^{-2^k} can be realized by feasibility graphs of size $O(k)$, as in Lemma 7.7. (For y^{-2^k}, one should remember that y^{-1} is included in our generating set Σ.) □

Thus the effects of implicit representation through feasibility graphs can be quite different for elements of finitely-generated groups than if we simply deal with words over an alphabet Σ as objects in their own right. We shall pursue this further in a more general way in the next section.

See [Gro93] for more information about "distortion" in finitely-presented groups, and in particular for more examples.

7.7 Measurements of complexity in groups

Let H be a finitely-generated group. Fix a set Σ of generators, which we assume contains the inverses of all of its elements. With this data, we can define a function λ on H which takes a given group element t and assigns to it the length of the shortest word over Σ that represents t.

This is a very standard measurement of complexity in a finitely-generated group. It leads to a natural geometry on H through the *word metric*, in which one defines the distance between two elements s and t to be $\lambda(s^{-1}t)$. This defines a left-invariant metric on H, i.e., the distance between s and t is not changed if we multiply them both on the left by an arbitrary element u of H.

A fundamental observation is that the function λ does not depend too strongly on the choice of generating set Σ, in the sense that a different choice Σ' of generating set would lead to a function λ' which is bounded from above and below by constant multiples of λ. This is well known and not hard to prove. The point is simply that every element of Σ can be represented by a word over Σ', and hence every word over Σ can be simulated by a word over Σ' with only linear expansion in size. This implies that λ is bounded by a constant multiple of λ', and one also has that λ' is bounded by a constant multiple of λ for the same reason. Similarly, the word metrics associated to Σ and Σ' are bounded by constant multiples of each other. (For general references on the geometry of finitely-generated groups, see [Gro84, Gro93].)

We can use feasibility graphs to define another measurement μ of complexity of words in H, as follows.

Definition 7.9 *Let H and Σ be as above, and let t be an element of H. We define $\mu(t)$ to be the size of the smallest feasibility graph M (over Σ) which represents t through the* normalized value function.

That is, M should be a feasibility graph for words over Σ, and its normalized value function at some output vertex should be a word which represents t in H.

Lemma 7.10 *If Σ' is a different set of generators for H (which contains the inverses of its elements), then the corresponding function $\mu'(t)$ is bounded from above and below by constant multiples of $\mu(t)$.*

Proof This is easy to check, and we shall omit the details. As is typical for this type of assertion, the main point is that each element of Σ can be represented as a word over Σ', and vice-versa. This permits one to convert feasibility graphs over Σ and Σ' into one another without changing the group elements represented by the values of their normalized value functions at the output vertices, and with at most linear expansion in the size of the graphs. □

This function μ is presently quite mysterious, even in free groups. This should be compared with the general ideas of Kolmogorov complexity and algorithmic information theory [Kol68, Cha87, Cha92, LV90], in which one measures the information content in a given word through the size of the shortest "computer program" which represents it (roughly speaking). This amounts to allowing arbitrary levels of implicitness in the representation of a given word, while the method of feasibility graphs restricts the implicitness severely and in a natural geometric way.

For algorithmic information theory, the measurement of information content is not computable algorithmically, while in the context of feasibility graphs, the question is more one of efficient computation. (We shall return to this theme periodically in the book, e.g., in Section 9.5 and Chapter 16.)

In groups which are not free, one has additional subtleties which can arise from the relations in the group, as we saw in the previous section. To make $\mu(t)$ small, it is not at all clear in general to what extent one should use the relations in the group to get possibly tricky reductions to small words, or to what extent one should use feasibility graphs to represent large words efficiently when they have simple patterns inside.

The matter becomes more complicated when one permits stronger forms of recursion. One avenue for doing this is discussed in [CS], where one seeks to measure the complexity of words through the minimal size of a formal proof of the "feasibility" of the given word. One can control the level of implicitness by restricting the logical nature of the formulae. For instance, one can forbid the use of quantifiers, and use only propositional rules, for the logical rules of inference. This is closely connected to the idea of feasibility graphs, and indeed one can easily code the implicit constructions described by feasibility graphs into formal proofs like this in a simple way, as mentioned in Section 7.2. One might instead allow quantifiers to be used, but only in a single layer perhaps, without alternations. In practice, this allows for another level of exponentiation in the efficiency of representations. (Compare with [Car00, CS].)

Similar effects of stronger recursion can be achieved through the use of feasibility graphs which describe the construction of other feasibility graphs, as in Chapter 16.

7.8 Trivial words in groups

Let H be a finitely-*presented* group now. Thus, in addition to a finite set Σ of generators, we also have a finite set $\mathcal{R}$ of *relations*, i.e., words which represent the identity element in H, and from which all trivial words can be obtained. Let us assume that $\mathcal{R}$ contains the empty word and the inverses of all of its elements, so that the set of trivial words consists exactly of products of conjugates of relations, and of words that can be derived from these through the cancellation of subwords of the form $u^{-1}u$ and uu^{-1}, $u \in \Sigma$.

Triviality of words in finitely-presented groups can be very tricky computationally. In deriving a given word w from products of conjugates of relations, it may be necessary to make much larger words (than w) before simplifying to w through cancellations. In this regard, it is well known that there are finitely-presented groups for which the "word problem" (of recognizing when a given word is trivial) is algorithmically unsolvable. (See [Man77].) This implies that non-recursive growth can be required to establish the triviality of some words. In other words, if there is a recursive bound for the size of the smallest derivation of the triviality of any trivial word (with respect to a given finite presentation of a group), then the word problem would be solvable for that group. This is not hard to show, since the derivations themselves are recursively enumerable (i.e., can be generated by a computer program).

There are many groups in which this does not occur, including *hyperbolic* groups [Gro87, Gd90], and, more generally, *automatic* groups [ECH$^+$92]. In these groups, one has *linear* and *quadratic* isoperimetric functions, respectively, which are functions concerning the lengths of trivial words and their representations as products of conjugates of relations. More precisely, one defines the *area* of a trivial word w in terms of the minimal integer n such that w can be written (modulo cancellations) as a product of conjugates of n relations, and the *isoperimetric function* $\phi(i)$ assigns to each positive integer i the maximal area of a trivial word of length at most i. See [Gro87, Gd90, ECH$^+$92] for more information.

Trivial words in finitely-presented groups are analogous to provable formulae in first-order predicate logic, in a number of ways. In particular, triviality of a word or provability of a formula are given in terms of the existence of certain kinds of derivations, and the derivations might be much larger than the words or formulae. The set of provable formulae is recursively enumerable, but not algorithmically decidable (at least when the underlying language is sufficiently nondegenerate, e.g., when it contains at least one unary predicate and binary function symbol). This is a well-known theorem. The recursive enumerability is like that of trivial words; it is enough to enumerate the proofs, and see which formulae arise from them. As in the case of trivial words, the algorithmic undecidability of the set of provable formulae implies that there is no recursive bound for the size of the smallest proof of an arbitrary formula (in terms of the size of the formula).

(Let us mention also the notion of *recursive groups*, in which there may be countably-infinite many generators, and a countable family of relations which is recursively enumerable. See [Man77] for some results related to these. Similar matters come up in formal logic.)

Just as one might restrict oneself to special types of groups, one can also look at questions about provable formulae in more limited situations logically. A basic case is that of propositional logic, for which one knows that the set of provable formulae is algorithmically decidable (using truth tables, for instance), but for which the existence of an algorithm which works in polynomial time is equivalent to the P=NP problem. (This reduces to the famous Cook-Levin theorem on the NP-completeness of the "satisfiability" problem for Boolean expressions [Pap94, HU79].) The existence of a propositional proof system in which provable formulae always admit proofs of polynomial size is equivalent to the NP = co-NP problem [CR79].

It is natural to think about similar possibilities for finitely-presented groups, perhaps special ones. In addition to asking about the word problem and its solvability (and the complexity of it), one can ask about the complexity of justifications of the triviality of words. This is analogous to questions about the sizes of proofs in formal logic. One can measure this directly, in terms of the realization of a given word as a contraction of a product of conjugates of relations.

This kind of direct measurement is somewhat similar to restricting oneself in the setting of provable formulae to proofs which do not use cuts, and it is well-known that propositional tautologies do not always admit cut-free proofs of polynomial size (as discussed in Sections 3.3 and 3.5). Instead, one can use more "implicit" measurements of triviality of words, based on formal proofs (as in [CS]), or on feasibility graphs.

To make the matter more concrete, let us come back to the Baumslag–Solitar group, discussed in Section 7.6 (in the last part). In [ECH+92], it is shown that this group has an *exponentially*-large isoperimetric function. (See Section 7.4 of [ECH+92], beginning on p154. Note that we are restricting ourselves to the case $p = 1$, $q = 2$, in the notation of [ECH+92].) In fact, the exponential growth of the isoperimetric function is shown to occur already for the words w_n given by $w_n = u_n v_n^{-1}$, $u_n = y^n x y^{-n}$, $v_n = x u_n x^{-1}$. (We follow here the discussion on p158-9 of [ECH+92], using the fact that $p = 1$ for extra simplification as mentioned on the bottom of p158 of [ECH+92].)

By contrast, if we allow a modest amount of implicitness, as through formal proofs with cuts, then the triviality of the words w_n can be justified much more efficiently, in a *linear* number of steps. To see this, it is helpful to think of the triviality of the words w_n in terms of the identity

$$(y^n x y^{-n})x = x(y^n x y^{-n}). \tag{7.7}$$

One can make a formal proof of this identity along the following lines. When $n = 0$, this is trivial. In general, if we know that

$$(y^k x y^{-k})x = x(y^k x y^{-k}), \tag{7.8}$$

then we can get the analogous equality for $k + 1$ as follows. Remember that we have the relation (7.4), which we can rewrite as $yxy^{-1} = x^2$. Using this and standard manipulations, we can obtain

$$\begin{aligned}(y^{k+1} x y^{-(k+1)})x &= (y^k (yxy^{-1}) y^{-k})x \\ &= (y^k x^2 y^{-k})x = (y^k x y^{-k})(y^k x y^{-k})x\end{aligned} \tag{7.9}$$

Then we apply our assumption (7.8) *twice* to get

$$(y^k x y^{-k})(y^k x y^{-k})x = x(y^k x y^{-k})(y^k x y^{-k}). \tag{7.10}$$

Using the relation as before, we can get

$$\begin{aligned}x(y^k x y^{-k})(y^k x y^{-k}) &= x(y^k x^2 y^{-k}) \\ &= x(y^k (yxy^{-1}) y^{-k}) = x(y^{k+1} x y^{-(k+1)}).\end{aligned} \tag{7.11}$$

Combining these identities we conclude that

$$(7.8) \text{ for } k \quad \text{implies} \quad (7.8) \text{ for } k+1. \tag{7.12}$$

This permits one to derive (7.7) by combining a chain of these implications from $k = 0$ to $k = n - 1$.

If one were to convert this sketch into a formal proof, one would see that *cuts* and *contractions* are used in an important way. That is, cuts are needed to combine the various implications (7.12), and contractions are involved in using the assumption (7.8) *twice* to derive its analogue for $k+1$. To simplify the cuts over the contractions one would make many duplications (as we have seen before, e.g., in Section 3.3 and Chapter 6), and this would lead to exponential growth in the number of steps in the proof. In the end (after the duplications) the "implicitness" would be gone, and the resulting proof would be little more than a repackaging of a direct derivation of w_n as a product of conjugates of relations together with some cancellations.

Notice that in the derivation sketched above, not only were there relatively few steps required, but also the words in x and y that occurred in the intermediate steps were never too large compared to the ones at the end, in (7.7). This was possible because of the use of cancellations in the intermediate steps, as opposed to forestalling the cancellations until the very end, after all of the substitutions.

Thus we see how formal proofs can be used to make measurements of complexity for the triviality of words for finitely-presented groups which are very different from more direct measurements, such as combinatorial area. One can pursue this further, in much the same manner as in [CS]. We should emphasize the relevance of specifying what kind of proofs are allowed to be used, e.g., whether quantifiers may be employed, and if so, to what extent. With quantifiers one can expect to do better than winning a single exponential, as we did above (without using quantifiers), at least sometimes. If mathematical induction over the natural numbers were also incorporated, then the triviality of w_n could be established with only a *bounded* number of steps. (Note, however, that the natural numbers are not really directly involved here, in terms of formal logic, even if it is convenient for us to use them notationally.)

At the level of polynomial versus (single) exponential complexity, it is natural to restrict oneself to formal proofs in which quantifiers are not allowed. The following is a basic example of a question that one might consider.

Problem 7.11 *Let H be a finitely-presented group defined by a set Σ of generators and a set $\mathcal{R}$ of relations. Suppose that every trivial word t of length n in Σ^* can be represented (with cancellations) as a product of conjugates of elements of $\mathcal{R}$ of total length no greater than C^n for some fixed constant C. (We assume here that Σ contains the inverses of all of its elements.) Is it true that one can make a formal proof of the triviality of such a word t without using quantifiers, and for which the total number of steps is bounded by a polynomial in n?*

To make this precise, one has to be more specific about the formalization of proofs, and the formalization of group-theoretic rules in particular, but this can be done in roughly the same manner as for feasible numbers, for instance.

One can think about this type of question in geometric terms, as follows. A trivial word for a finitely-presented group can be represented geometrically by a closed loop (in a certain 2-dimensional complex), and the combinatorial

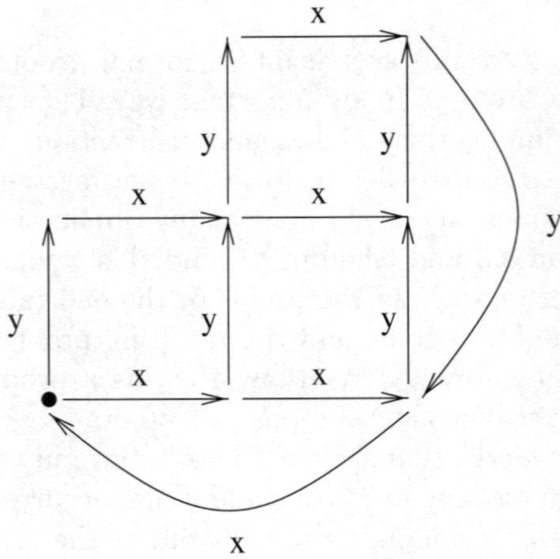

FIG. 7.3. A diagram for the group G with generators x, y and relations $yxy^{-1}x^{-1} = y^3 = x^3 = e$. (Thus G is isomorphic to $\mathbf{Z}_3 \times \mathbf{Z}_3$.) The trivial word $yxyxyx$ traces the loop that follows the outer perimeter of the diagram. This loop can be filled with five cells, namely, the three squares (which correspond to the relation $yxy^{-1}x^{-1} = e$) and the two curved-triangular cells (associated to $y^3 = e$ and $x^3 = e$).

area corresponds to the number of 2-dimensional cells in the complex which are needed to fill in the loop by a topological disk. (An example of this is shown in Fig. 7.3. See [Gro87, ECH+92] for more details.) If the number of these cells is very large compared to the size of the given word (which is the length of the loop), then the minimal disk spanning the loop might enjoy some regularity which would permit a more concise implicit description than through a simple enumeration of the cells.

Although we have emphasized formal proofs so far in this discussion, one could just as well work with feasibility graphs, or other types of derivations. As above (beginning with (7.7)), it is somewhat more pleasant to think in terms of constructing identities between words instead of trivial words, and there is nothing wrong with doing that. This is a nice point in its own right, and it applies more generally: one can think of equations as mathematical objects in their own right, and of rules for manipulating them as "admissible operations" on sets of equations. This permits one to use feasibility graphs to represent derivations of equations through the same sort of conventions as before. (For the initial inputs of normalized value functions, one could take equations of the form $x = x$, for instance.)

Notice that feasibility graphs allow exactly the kinds of duplications needed for the proof of the triviality of the words w_n for the Baumslag–Solitar group sketched above (i.e., in the utilization of (7.8) twice in the proof of (7.12)).

7.9 Examples about numbers

Let W be the set of nonnegative integers (or "whole" numbers). We shall think of W as being equipped with the operations of addition, multiplication, and successor ($n \mapsto n+1$). If we did not include multiplication, then this case would be isomorphically equivalent to the earlier example of words over an alphabet with only one element (discussed in Section 7.6). However, with multiplication, the kinds of constructions which are possible changes substantially.

In this context, we think of a feasibility graph in almost the same way as before. Each edge is either considered to be labelled, in which the intended effect is the application of the successor function, or to be unlabelled, in which case it has no effect. For the focusing branch points, one must now specify which are intended to represent additions and which are intended to represent multiplications, and this is slightly different from before. (The ordering of the incoming edges does not matter now, since addition and multiplication are commutative.)

These interpretations can be implemented through the notion of value functions in the same manner as before. In particular, one has the analogue of Lemma 7.2 for this class of feasibility graphs, and *normalized value functions* always exist and are unique. For these we assign the value 0 to all of the input vertices, instead of the empty word, as we did before.

What sort of numbers admit representations through feasibility graphs with n vertices? More precisely, this means representations through *normalized* value functions. The first main point is that numbers of *double*-exponential size can be reached by these feasibility graphs, through the repeated use of duplication and multiplication.

One can also not have more than double-exponential expansion in this case. This is not hard to show, following the general discussion of upper bounds in Section 7.5. One begins by passing to the visibility to get rid of the duplicating effects of the defocusing branch points. This leads to an "explicit" construction in which the total number of additions and multiplications is bounded by the size of the visibility. We know that the size of the visibility admits an exponential bound in the size of the underlying graph, from which one can derive a double-exponential bound on the numbers constructed by the original graph.

Not all numbers of roughly double-exponential size can arise in this manner, however. There are simply too many of these numbers, as one can establish through a counting argument like the one in Section 7.6. Specifically, the number of optical graphs on at most n vertices is bounded by n^{2n}, for the same reasons as before. We can count the number of different ways of making these graphs into feasibility graphs in the following manner. For each edge, we have the choice of labelling it or not, for a total of at most 2^{2n} such choices, since there are at most $2n$ edges. For each focusing branch point, we have to decide whether it should represent an addition or a multiplication, and this leads to a total of at most 2^n choices. Altogether, we have at most 2^{3n} different ways to make an optical graph on $\leq n$ vertices a feasibility graph, for a grand total of at most $2^{3n}n^{2n} = 2^{2n \log n + 3n}$ of these feasibility graphs. Thus, while a feasibility graph of size n can represent a number of double-exponential size in n, very few such numbers can be represented in this way, and this remains true even if we shrink considerably the range of numbers that we are trying to realize.

If we did not allow multiplications, then we would, in effect, be in the same situation as with words over an alphabet with just one element, and there would not be a gap of this nature, as we saw in Section 7.6.

7.10 Trees

We can also use feasibility graphs to make implicit constructions of trees. More precisely, let us work with *rooted trees*, which are trees in which a basepoint has been specified. We shall often write this as (T, b), with T representing the tree and b the basepoint. There is a natural unary operation on rooted trees, in which one takes a given rooted tree (T, b) and adds one new vertex and one new edge to T, with the edge going from the new vertex to the basepoint of the given tree. One adds no other edges, and the new vertex is taken to be the basepoint of the tree that results. (Actually, it is better to say that we are working with *isomorphism classes* of rooted trees.)

There is a natural binary operation, defined as follows. Let (T, b) and (T', b') be given rooted trees. We define a new tree S by taking the disjoint union of T and T' (passing to isomorphic copies, if necessary), and then identifying b and b'. The vertex that results from b and b' is taken to be the root of the new tree.

We can define feasibility graphs for building trees using these operations in the usual way. That is, each edge can either be labelled or not, according to whether or not we want to think of applying the unary operator at that moment. The focusing branch points would be interpreted using the binary operation just defined, while the defocusing branch points would be interpreted as performing a duplication in the usual way.

With these conventions, one can define the notion of a *value function* associated to a feasibility graph in the same way as in Definition 7.1. This would be a function defined on the set of vertices of the feasibility graph and taking values in sets of (isomorphism classes of) rooted trees. For the *normalized value functions*, one would use "trivial rooted trees" for the values at the input vertices, where *trivial rooted trees* have only one vertex and no edges.

Suppose that we have such a feasibility graph G, in which all of the edges are labelled by the unary operation described above. If f is the normalized value function for G, and v is a vertex in G, then $f(v)$ is given by the (isomorphism class of the) negative visibility $\mathcal{V}_-(v, G)$. This can be proved by a kind of induction argument: it is trivially true when v is an input vertex of G, and one can check that it remains true with each step of the construction described by the feasibility graph. One has to be a bit careful in making this precise, but fortunately Lemma 7.2 provides a convenient way to do this. Namely, one can argue that the negative visibility $\mathcal{V}_-(v, G)$ *defines* a normalized value function on G, and then use the analogue of Lemma 7.2 in this context to say that this is necessarily the *unique* normalized value function (up to isomorphic equivalence of rooted trees).

This example has nice consequences conceptually. We already know from Sections 7.4 and 7.5 that the construction represented by a feasibility graph G is always controlled (in any context) by the visibility of G. In the present situation, the feasibility graphs represent the visibility graphs exactly. Thus we can think of visibility graphs as both a special case of constructions made by feasibility graphs, and as something useful for working with feasibility graphs in general.

7.11 Boolean circuits

Another basic example for the notion of feasibility graphs is provided by *Boolean circuits*. For this the basic objects under consideration are simply 0 and 1, and there are two binary operations $\wedge$ and $\vee$ which to use at focusing branch points. One can think of $\neg$ as a unary operation which can be assigned to edges, but, for the sake of consistency with the standard treatment of Boolean circuits (as in Section 4.12), one can employ vertices with only one incoming edge to represent negations. In any event, this is not a serious issue.

In this situation, it is not so interesting to think of normalized value functions as describing constructions of elements of $\{0, 1\}$, but one can use the existence and uniqueness of value functions (as in Lemma 7.2) to represent computations from specified inputs. This is essentially the same as the usual way of representing a Boolean function by a Boolean circuit.

One can also reformulate this in terms of having a Boolean circuit describe the construction of a Boolean function, in the same manner as in Remark 7.4 in Section 7.1. As another option, closely related to this, one can reformulate the data in the graph as defining a feasibility graph for constructing Boolean expressions. This is analogous to feasibility graphs for constructions of words over alphabets, as in Section 7.1, with conjunctions and disjunctions playing the same role as concatenations before, and with negations instead of unary operations of adding a letter. One could use different variables at the input vertices, to get a Boolean expression from the feasibility graph, and this Boolean expression would define the same function as the original Boolean circuit. This is easy to check from the definitions. Note that the Boolean expression might be exponentially larger than the graph, because of duplications at defocusing vertices. This is similar to what happens with words over an alphabet.

As in Section 4.12, it is well known that most Boolean functions of n variables can only be computed by Boolean circuits of at least exponential size in n (on the order of $2^n/n$). (See 4.1.14 on p86 of [Pap94].) The counting arguments that we have mentioned in the previous sections are similar to this. However, as on p83 of [Pap94], it seems to be difficult to make explicit examples in which one does not have linear bounds, even though exponential size is known to be needed most of the time. It is also not known if for each Boolean circuit there is a Boolean expression which defines the same function, and for which the size of the expression is bounded by a fixed polynomial of the size of the circuit. See p386 of [Pap94] (and also Section 4.12).

7.12 Homomorphisms and comparisons

How can one make comparisons between the computations performed by different feasibility graphs? There are natural ways to do this using mappings between graphs, and we shall discuss these in some detail in Section 11.5. For the moment, we would like to mention a simpler type of comparison, in which one does not change the underlying graph, but one does the way in which it is interpreted.

In the notion of a feasibility graph, we always start with some collection of objects X, and a set of operations $\mathcal{C}$ defined on it. (We shall sometimes refer to the pair X, $\mathcal{C}$ as a *structural system*.) If we have another set of objects X' with another collection of operations $\mathcal{C}'$ defined on it, then there is an obvious notion of a homomorphism between the two. Namely, a homomorphism should consist of a mapping from X to X' and a mapping from $\mathcal{C}$ to $\mathcal{C}'$ which satisfy the following compatibility conditions. First, the mapping from $\mathcal{C}$ to $\mathcal{C}'$ should preserve arities, so that an element of $\mathcal{C}$ is necessarily associated to an element of $\mathcal{C}'$ which takes the same number of arguments. Second, if we apply an element of $\mathcal{C}$ to some tuple of elements of X, and use our homomorphism to send the result into X', then we should get the same answer as if we first applied the homomorphism to get some elements of X', and the applied the operation in $\mathcal{C}'$ which corresponds to the original one in $\mathcal{C}$ under our homomorphism.

For example, X might consist of the set of all words over some alphabet, together with the binary operation of concatenation, and unary operations corresponding to the addition of single letters. We could take X' to be the set of nonnegative integers, with the operations of sum and successor (addition by 1). The mapping from X to X' that takes a word and associates it to the integer which represents its length defines a homomorphism, with the obvious correspondence between the operations.

On the other hand, we might take for X the class of rooted trees, with the same operations as in Section 7.10. We can then define a mapping from X into the set of positive integers by taking a tree and assigning to it the number of its edges. This also gives a homomorphism, if we allow the operations of successor and addition on the positive integers. One could use the number of vertices instead, but then a different binary operation on integers would be needed, namely, $(m, n) \mapsto m + n - 1$.

This notion of homomorphism leads to a simple way of transforming one kind of feasibility graph into another. That is, if we start with a feasibility graph G that uses a set X of objects and a collection $\mathcal{C}$ of operations, and if we have a homomorphism from X, $\mathcal{C}$ to a different pair X', $\mathcal{C}'$, then we can get a new feasibility graph G' that uses X' and $\mathcal{C}'$, simply by replacing the operations in $\mathcal{C}$ that are employed by G with their counterparts in $\mathcal{C}'$ at each location.

When we do this, the computations described by G correspond to the ones described by G' in the right way. Specifically, value functions for G are transformed into value functions for G' by the homomorphism. This follows immediately from the definitions. Similarly, normalized value functions are transformed into normalized value functions, at least if our homomorphism respects the appropriate notions of "zero elements" in X and X'. In the first example mentioned above, these were the empty word ϵ and the number 0, and the homomorphism does take ϵ to 0.

This type of "reinterpretation" of a feasibility graph for one structure as a feasibility graph for another structure will be useful in Chapter 16, where we discuss stronger forms of recursion. In practice, we shall often use this idea in a slightly different way, starting from G' and going back to G, instead of the other way around. That is, the homomorphism between structural systems will still go from X, $\mathcal{C}$ to X', $\mathcal{C}'$, but the conversion of feasibility graphs will go in the other direction, and will depend on a "lifting" of operations in $\mathcal{C}'$ to operations in $\mathcal{C}$. Normally this lifting will simply be an inverse to the mapping from $\mathcal{C}$ to $\mathcal{C}'$ which is part of the homomorphism from X, $\mathcal{C}$ to X', $\mathcal{C}'$. This will ensure that the relationship between G and G' is exactly the same as before, even if we start with G' now rather than G. (Note that the correspondence between $\mathcal{C}$ and $\mathcal{C}'$ might easily be invertible, even though the mapping from X to X' is not, as in the examples above.)

8

BOUNDS FOR FINITE VISIBILITIES

Now we take up the matter of analyzing the size of the visibility when it is finite. This can be compared to similar questions about the size of regular languages when they are finite.

To control the size of the visibility when it is finite, we shall use a *stopping-time argument.* We begin with some general facts about breaking up the visibility into nice pieces, most of which are visibilities in their own right. In the finite case, we do not have the type of periodicities in the visibility that come with cycles (as in the infinite case), but there are some simple patterns nonetheless.

A much faster version of the basic method presented here was pointed out to us by M. Gromov, and it will be discussed in Section 8.7. His approach aims directly at the question of counting, and has the advantage of avoiding more easily certain inefficiencies in the estimates. The description of Gromov's argument in Section 8.7 can be read independently of the initial sections of this chapter.

8.1 The propagator rule

Let G be an optical graph, and let a vertex v of G be given. We want to explain a general procedure for breaking up the visibility of G at v into pieces. Let us first record the main statement before explaining the definitions more thoroughly.

Lemma 8.1 (The propagator rule)
Assumptions and notations:

Let $\mathcal{W}$ be a subgraph of $\mathcal{V}_+(v, G)$ which contains the basepoint, and which has the property that for every vertex in $\mathcal{W}$, there is an oriented path that goes from the basepoint in $\mathcal{V}_+(v, G)$ to the given element of $\mathcal{W}$, and that lies entirely within $\mathcal{W}$.

Let $\mathcal{B}$ denote the set of "boundary" vertices s in $\mathcal{V}_+(v, G)$ which do not lie in $\mathcal{W}$, but for which there is an edge in $\mathcal{V}_+(v, G)$ which goes from a vertex in $\mathcal{W}$ to s. Let $\mathcal{E}$ denote the set of these edges which go from a vertex in $\mathcal{W}$ to a vertex in $\mathcal{B}$.

Given $s \in \mathcal{V}_+(v, G)$, let $F(s)$ be the subtree of $\mathcal{V}_+(v, G)$ rooted at s which consists of everything in $\mathcal{V}_+(v, G)$ that "follows" s. More precisely, we define $F(s)$ to be the subgraph of $\mathcal{V}_+(v, G)$ (vertices and edges) which can be reached by oriented paths in $\mathcal{V}_+(v, G)$ starting at s.

Conclusions:

The set of vertices of $\mathcal{V}_+(v, G)$ is the disjoint union of the sets of vertices in the subgraphs $\mathcal{W}$ and $F(s)$, $s \in \mathcal{B}$, and the set of edges in $\mathcal{V}_+(v, G)$ is the disjoint union of the sets of edges in $\mathcal{W}$, in $F(s)$ for $s \in \mathcal{B}$, and in $\mathcal{E}$.

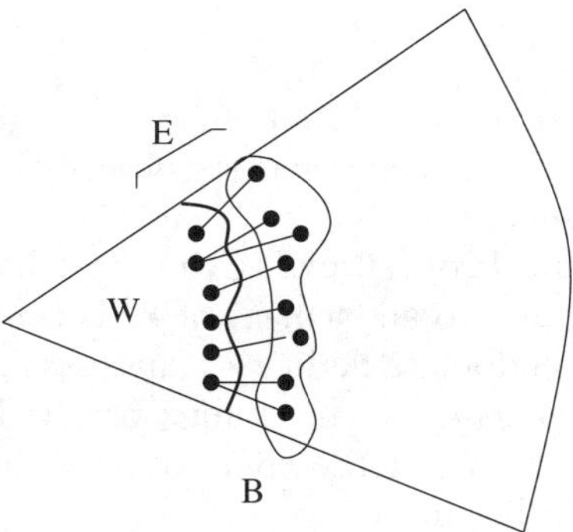

FIG. 8.1. An illustration of $\mathcal{W}$, $\mathcal{B}$, and $\mathcal{E}$ in the visibility

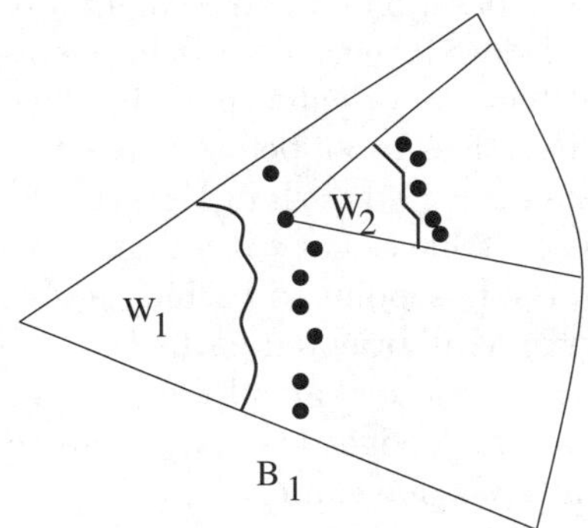

FIG. 8.2. A more refined decomposition of the visibility (see Section 8.4)

Some of the definitions above are illustrated in Fig. 8.1. (Note, however, that the definitions allow $\mathcal{W}$ to have vertices for which some outgoing edges go to vertices in $\mathcal{W}$, and some do not.)

The subtrees $F(s)$ for any s in $\mathcal{V}_+(v, G)$ amount to the same thing as the visibility graphs $\mathcal{V}_+(\pi(s), G)$. We shall discuss this in Section 8.2.

In effect, Lemma 8.1 tells us how we can decompose the visibility $\mathcal{V}_+(v, G)$ into a "central zone" $\mathcal{W}$ and a collection of subtrees $F(s)$ which are each visibilities in their own right. (We also have the edges in $\mathcal{E}$ connecting the two.) In Section 8.4, we shall use this symmetry of the description to apply apply Lemma 8.1 repeatedly, first to $\mathcal{V}_+(v, G)$, then to the subtrees $F(s)$, then to analogous subtrees within them, etc., to get a decomposition of the visibility as depicted in Fig. 8.2.

Lemma 8.1 does not rely on the special structure of visibility graphs, it is really a fact about rooted trees in general. We have stated it in the context of visibilities simply because that is where we shall want to employ it. In the applications, we shall use the fact that we are working with visibility graphs to choose $\mathcal{W}$ in a good way, and also for the extra symmetry in the $F(s)$'s mentioned above.

The proof is quite standard, but let us go through it carefully. Before we begin in earnest, let us collect some general observations. We assume that $\mathcal{W}$, $\mathcal{B}$, etc., are as in the lemma.

The uniqueness result Proposition 4.5 implies that *any* oriented path in $\mathcal{V}_+(v, G)$ from the basepoint to an element of $\mathcal{W}$ is contained in $\mathcal{W}$. Also, every edge in $\mathcal{V}_+(v, G)$ whose endpoints lie in $\mathcal{W}$ is necessarily an edge in $\mathcal{W}$. Indeed, such an edge must be crossed by the unique oriented path in $\mathcal{V}_+(v, G)$ from the basepoint to the endpoint of the edge into which the edge flows, and by assumption this endpoint lies in $\mathcal{W}$.

We know from Lemma 4.3 that there is at most *one* edge going into any vertex in $\mathcal{V}_+(v, G)$. In particular, this is true of the elements of $\mathcal{B}$. This means that any oriented path from the basepoint to an element s of $\mathcal{B}$ must pass through a vertex in $\mathcal{W}$ immediately before it reaches s. As in the preceding paragraph, from uniqueness we conclude that any oriented path in $\mathcal{V}_+(v, G)$ from the basepoint to s passes only through vertices in $\mathcal{W}$ before it reaches s.

Since there is always an oriented path in $\mathcal{V}_+(v, G)$ from the basepoint to any given vertex s in $\mathcal{V}_+(v, G)$, we have that t is a vertex in $F(s)$ exactly when there is an oriented path from the basepoint to t which passes through s. Proposition 4.5 implies that this is true of all oriented paths from the basepoint to t as soon as it is true for one of them. Note that all edges in $\mathcal{V}_+(v, G)$ which connect a pair of vertices in $F(s)$ are also edges in $F(s)$; these edges are all traversed by oriented paths starting at s, as one can check.

Proof (Lemma 8.1) Let us show first that the set of vertices in $\mathcal{V}_+(v, G)$ is the *disjoint union* of the sets of vertices in $\mathcal{W}$ and in the $F(s)$'s, $s \in \mathcal{B}$.

If $s \in \mathcal{B}$, then none of the vertices in $F(s)$ also lie in $\mathcal{W}$. Indeed, suppose that there were a vertex t in $F(s)$ which also lay in $\mathcal{W}$. Then, as above, there would be an oriented path in $\mathcal{V}_+(v, G)$ from the basepoint to t which goes through s, and the assumptions on $\mathcal{W}$ would imply that this path is contained in $\mathcal{W}$, contrary to the requirement that s lie outside $\mathcal{W}$.

Given distinct vertices $s_1, s_2 \in \mathcal{B}$, we have that $F(s_1)$ and $F(s_2)$ have no vertices in common. Indeed, assume to the contrary that there is a vertex t in common. As above, there are oriented paths from the basepoint to t which pass through s_1 and s_2, and Proposition 4.5 ensures that they must be the same path. Assume without loss of generality that this common path passes through s_1 before s_2. Then s_1 must lie in $\mathcal{W}$, since every oriented path from the basepoint to s_2 can pass only through vertices in $\mathcal{W}$ until it reaches s_2. This contradicts the assumption that $s_1 \in \mathcal{B}$.

Of course, $F(s_1)$ and $F(s_2)$ have no edges in common when they have no vertices in common.

If t is any vertex in $\mathcal{V}_+(v, G)$, then t either lies in $\mathcal{W}$ or in $F(s)$ for some $s \in \mathcal{B}$. Indeed, suppose that t does not lie in $\mathcal{W}$. Consider the oriented path in $\mathcal{V}_+(v, G)$ which goes from the basepoint to t. Remember that the basepoint of $\mathcal{V}_+(v, G)$ lies in $\mathcal{W}$. This path cannot be wholly contained in $\mathcal{W}$, since t does not lie in $\mathcal{W}$, and therefore there is a first vertex s which lies on the path, but not in $\mathcal{W}$. We have that $s \in \mathcal{B}$ by definitions, and that $t \in F(s)$.

Thus the set of vertices of $\mathcal{V}_+(v, G)$ is the disjoint union of the vertices in $\mathcal{W}$ and the vertices in the subgraphs $F(s)$, $s \in \mathcal{B}$, as claimed.

Let us proceed to the assertion about the edges. Notice first that the edges in $\mathcal{E}$ cannot lie in $\mathcal{W}$ or in any $F(s)$, $s \in \mathcal{B}$, because they cannot have *both* vertices in any one of these sets.

We want to show that the set of edges in $\mathcal{V}_+(v, G)$ is the disjoint union of the sets of edges in $\mathcal{W}$, in the subgraphs $F(s)$, $s \in \mathcal{B}$, and in $\mathcal{E}$. The disjointness of these sets of edges follows from the preceding assertion and from the disjointness of the sets of vertices in $\mathcal{W}$ and in the subgraphs $F(s)$, $s \in \mathcal{B}$. Now let e be any edge in $\mathcal{V}_+(v, G)$, and let us show that e occurs in one of these places.

Let t be the vertex in $\mathcal{V}_+(v, G)$ into which e flows (using the natural orientation for $\mathcal{V}_+(v, G)$). Lemma 4.3 implies that e is the only edge which flows into t. As usual, there is a unique oriented path in $\mathcal{V}_+(v, G)$ from the basepoint to t, and the edge e is the last edge traversed by the path. If t lies in $\mathcal{W}$, then the path and hence e lies in $\mathcal{W}$ too. If t lies in $\mathcal{B}$, then e lies in $\mathcal{E}$. The last possibility is that t lies in some $F(s)$, but is not equal to s. By assumption, there is then a nontrivial oriented path from s to t, and it must cross e on the way into t. Thus e is contained in $F(s)$ in this case. This proves the lemma. □

Let us also record one other simple fact.

Lemma 8.2 *If $\mathcal{W}$, $\mathcal{B}$, and $\mathcal{E}$ are as above, then the number of elements in each of $\mathcal{B}$ and $\mathcal{E}$ is at most twice the number of vertices in $\mathcal{W}$.*

Proof Each vertex in $\mathcal{B}$ is attached to a vertex in $\mathcal{W}$ by an edge in $\mathcal{E}$ which is oriented *from* the vertex in $\mathcal{W}$ *to* the vertex in $\mathcal{B}$. There are never more than *two* edges coming out of any given vertex in $\mathcal{V}_+(v, G)$, because of Lemma 4.4, and the requirement that G be an optical graph. In particular, this is true for the vertices in $\mathcal{W}$, and the lemma follows easily. □

8.2 Visibilities within visibilities

In the preceding section, we saw how to decompose $\mathcal{V}_+(v, G)$ into pieces in a certain way, with subgraphs $F(s)$ of $\mathcal{V}_+(v, G)$ arising among the pieces. In this section, we show that these subgraphs are visibilities in their own right.

Let G, etc., be as above, and let s be any vertex of $\mathcal{V}_+(v, G)$. Let us first try to understand $F(s)$ in a more concrete way. Recall that the vertices in $F(s)$ are the vertices in $\mathcal{V}_+(v, G)$ for which the unique oriented path from the basepoint in $\mathcal{V}_+(v, G)$ to the vertex passes through s.

Lemma 8.3 *A vertex t in $\mathcal{V}_+(v, G)$ lies in $F(s)$ if and only if the path in G represented by t (as in the definition of the visibility graph in Section 4.2) contains the path represented by s as an initial subpath.*

To say that a path β is an *initial subpath* of a path α means that β is obtained from α by starting at the initial vertex of α and following α along to some point and then stopping. This need not be quite the same as a subpath of α with the same initial vertex when there are nontrivial cycles present.

Proof If the path in G represented by t contains the path in G represented by s as an initial subpath, then the lifting $\lambda(t)$ of the path represented by t to an oriented path in $\mathcal{V}_+(v, G)$ from the basepoint to t is easily seen to contain s as a vertex. (Recall that this "lifting" is defined in Section 4.6.) This implies that $t \in F(s)$.

Conversely, if $t \in F(s)$, then it means that there is an oriented path α in $\mathcal{V}_+(v, G)$ from the basepoint to t which passes through s. Corollary 4.6 tells us that if we map this path down to G using the canonical projection $\pi : \mathcal{V}_+(v, G) \to G$ (defined in Section 4.5), then the projected path gives back the path in G represented by t. By assumption, there is an initial subpath of α from the basepoint to s, and Corollary 4.6 says that the projection of this subpath is the path in G which represents s. Thus the path represented by s is contained in the path represented by t as an initial subpath, as desired. This proves the lemma. □

The canonical projection $\pi : \mathcal{V}_+(v, G) \to G$ permits us to associate s to a vertex $\pi(s)$ in G. We want to show that the subgraph $F(s)$ of $\mathcal{V}_+(v, G)$ is isomorphic to the visibility $\mathcal{V}_+(\pi(s), G)$ in a natural way.

We first define a mapping $\phi_s : \mathcal{V}_+(\pi(s), G) \to F(s)$ as follows. Each vertex of $\mathcal{V}_+(\pi(s), G)$ represents an oriented path in G which begins at $\pi(s)$ (as in the definition of the visibility). Of course, s itself represents an oriented path in G which begins at v, and $\pi(s)$ is simply its endpoint. (See Section 4.5.) Given a vertex $u \in \mathcal{V}_+(\pi(s), G)$, we take the path in G corresponding to u (which starts at $\pi(s)$) and add it to the end of the path represented by s. This gives an oriented path in G which starts at v, follows the path corresponding to s, and then follows the path corresponding to u. The new path corresponds to a vertex in $\mathcal{V}_+(v, G)$, and we define $\phi_s(u)$ to be this vertex.

It is not hard to see that a vertex t in $\mathcal{V}_+(v, G)$ arises as $\phi_s(u)$ for some vertex u in $\mathcal{V}_+(\pi(s), G)$ if and only if the path in G that corresponds to t contains the path in G corresponding to s as an initial subpath. Indeed, any $\phi_s(u)$ has this property by construction, and to say that a given $t \in \mathcal{V}_+(v, G)$ has this property means exactly that we can break up the path in G represented by t into the path in G represented by s and a path in G which begins at the endpoint of s. The latter path determines the vertex u in $\mathcal{V}_+(\pi(s), G)$ that we want, the one for which $t = \phi_s(u)$.

The lemma above implies now that ϕ_s maps $\mathcal{V}_+(\pi(s), G)$ onto $F(s)$. It is easy to see that it is also one-to-one, by construction. One can also check easily that edges and their orientations in these graphs correspond properly. In summary, we get the following.

Lemma 8.4 *For each vertex s in $\mathcal{V}_+(v, G)$, there is a "natural" graph isomorphism ϕ_s between $\mathcal{V}_+(\pi(s), G)$ and $F(s)$ (where $F(s)$ is defined in Lemma 8.1). This isomorphism takes the basepoint of $\mathcal{V}_+(\pi(s), G)$ to the vertex s in $F(s)$.*

Thus we have "visibilities within visibilities", and the graphs $F(s)$ in the decomposition described in Lemma 8.1 are essentially visibilities in their own right.

Each visibility $\mathcal{V}_+(\pi(s), G)$ has a projection $\pi_s : \mathcal{V}_+(\pi(s), G) \to G$ of its own. As before, vertices in $\mathcal{V}_+(\pi(s), G)$ represent oriented paths in G, and π_s takes such a path and associates to it its endpoint. For the isomorphism $\phi_s : \mathcal{V}_+(\pi(s), G) \to F(s)$ defined above we have the compatibility equation

$$\pi(\phi_s(u)) = \pi_s(u) \tag{8.1}$$

for each vertex u in $\mathcal{V}_+(\pi(s), G)$. This is an immediate consequence of the definitions; the vertex $\phi_s(u)$ in $\mathcal{V}_+(v, G)$ represents the path in G which begins with the path in G represented by s and then continues with the path in G represented by u, and thus has the same endpoint as the path represented by u.

Although a key purpose of this chapter is to look at bounds for the visibility when it is finite, one should not forget about the infinite case, to which the arguments so far also apply. When there are nontrivial oriented cycles present in G, we can have an infinite family of $F(s)$'s nested inside one another, which all look exactly the same. The nature of these repetitions becomes more complicated when we have loops which intersect, as in Section 5.2.

When the visibility is infinite, Lemma 8.4 provides a version of the *Markov* or *finite type* property for trees, as on p238 of [Gro87]. The main point is that there would be infinitely many subtrees $F(s)$ in this case, but only finitely many models for them, since $F(s)$ and $F(s')$ are isomorphic as soon as s and s' project down to the same point in the underlying graph G. Conversely, one can show that an infinite rooted tree with the Markov property from [Gro87] is actually the visibility of some finite oriented graph (although perhaps not an optical graph). This follows from the same construction as in Section 9.2.

When the visibility is finite, there are only finitely many isomorphism classes of the $F(s)$'s (viewed as rooted trees, with s as root) a posteriori, but the number of different isomorphism classes compared to the size of the visibility provides an interesting measurement of the *symmetry* of the visibility graph in question. Lemma 8.4 tells us that the number of these isomorphism classes is automatically bounded by the number of vertices in the original graph G.

We shall discuss the problem of finding the most efficient representation of a given rooted tree as a visibility graph later on, beginning in Section 9.2.

8.3 The Calderón–Zygmund decomposition

Now we want to describe the basic stopping-time argument that we shall use. It is a geometric version of the Calderón–Zygmund decomposition from harmonic analysis. (See [CZ52, CS97, Gar81, Jou83, Sem99b, Ste70, Ste93].) We follow the same notations and definitions as before.

Proposition 8.5 *Let G be an optical graph, and let v be any vertex of G. Assume that (v, v) is not a focal pair. (Note that (v, v) cannot be a focal pair unless*

G contains a nontrivial oriented cycle passing through v.) Then we can find a subgraph $\mathcal{W}$ of the visibility $\mathcal{V}_+(v, G)$ with the following properties:

(a) *$\mathcal{W}$ satisfies the same conditions as in Lemma 8.1. In particular we get corresponding sets $\mathcal{B}$ and $\mathcal{E}$ of vertices and edges along "the boundary" of $\mathcal{W}$ as defined in Lemma 8.1.*

(b) *The number of elements in each of $\mathcal{B}$ and $\mathcal{E}$ is no greater than twice the number of vertices in $\mathcal{W}$.*

(c) *For each $s \in \mathcal{B}$, we have that $(v, \pi(s))$ is a focal pair.*

(d) *If t is a vertex in $\mathcal{W}$, then $(v, \pi(t))$ is* not *a focal pair.*

If we assume also that there are no oriented cycles in G (or at least that there are no oriented cycles which can be reached by an oriented path from v), then we have the following conclusions:

(i) *The restriction of $\pi : \mathcal{V}_+(v, G) \to G$ to the vertices of $\mathcal{W}$ is injective.*

(ii) *The restriction of $\pi : \mathcal{V}_+(v, G) \to G$ to $\mathcal{B}$ is at most two-to-one, which means that each vertex in G can have at most two preimages in $\mathcal{B}$.*

In this case, we have that the number of vertices in $\mathcal{W}$ is no greater than the number of vertices in G. The number of elements in $\mathcal{B}$ is at most twice that number, by (b) above. However, we can use (ii) instead to say that the number of elements in $\mathcal{B}$ is at most twice the number of vertices u in G such that (v, u) is a focal pair.

See Remark 8.7 for the case where (v, v) is a focal pair.

If we think of focal pairs as being "bad events" (in the sense that they can lead to large expansion in the visibility graph), then we are thinking to choose $\mathcal{W}$ so that it does not contain any bad events, but goes right up to places where bad events occur for the first time. This is a basic point about stopping-time arguments, to go right up to the places where the bad events occur, in order to be able to count them. By not going beyond the bad events, we maintain good control, however (and this is another basic point about stopping-time arguments, like the original Calderón–Zygmund decomposition in harmonic analysis).

The proposition does not itself provide a decomposition of the visibility $\mathcal{V}_+(v, G)$, but we can combine it with Lemma 8.1 to get the decomposition of $\mathcal{V}_+(v, G)$ that we want. That is, Lemma 8.1 will provide us with a decomposition of $\mathcal{V}_+(v, G)$ which also involves "bad" pieces of the form $F(s)$, $s \in \mathcal{B}$. The pieces are "bad" in the sense that we have no control over them. However, Lemma 8.4 tells us that these pieces are visibilities in their own right, and so we shall be able to make further decompositions inside them (in the next section).

We should point out that certain types of degenerate situations can occur. For instance, $\mathcal{W}$ might consist of v alone, with no other vertices or any edges at all. Another possibility is that v lies on a nontrivial oriented cycle in G, and that $\mathcal{W}$ is actually an *infinite* subgraph of the visibility. In particular, the restriction of the canonical projection $\pi : \mathcal{V}_+(v, G) \to G$ to the vertices of $\mathcal{W}$ would *not* be injective in this case. One can analyze this possibility further, to see more precisely how $\mathcal{W}$ and its projection back into G could behave, but we shall not pursue this here.

Proof (Proposition 8.5) Let G and v be given. Let $\mathcal{I}$ denote the set of "focal" vertices in the visibility $\mathcal{V}_+(v, G)$, which means the vertices t such that $(v, \pi(t))$ is a focal pair. Let $\mathcal{W}$ denote the part of the visibility which can be reached without touching $\mathcal{I}$. That is, the vertices in $\mathcal{W}$ are the vertices u in $\mathcal{V}_+(v, G)$ for which the oriented path from the basepoint of $\mathcal{V}_+(v, G)$ to u does not cross $\mathcal{I}$. The edges in $\mathcal{W}$ are simply the ones crossed on the way by such paths, or, equivalently, the edges in $\mathcal{V}_+(v, G)$ whose endpoints lie in $\mathcal{W}$.

With this definition, $\mathcal{W}$ satisfies automatically the requirements in the assumptions of Lemma 8.1, i.e., $\mathcal{W}$ contains the basepoint of $\mathcal{V}_+(v, G)$ (since we are assuming that (v, v) is not a focal pair), and for each vertex in $\mathcal{W}$, there is an oriented path from the basepoint to the vertex which lies entirely in $\mathcal{W}$. Thus we have (a) above, and (d) is also automatic from the definition.

As in Lemma 8.1, $\mathcal{B}$ consists of the vertices s in $\mathcal{V}_+(v, G)$ but not in $\mathcal{W}$ for which there is an edge from an element of $\mathcal{W}$ to s, and $\mathcal{E}$ is the set of edges which do this. The bounds in (b) follow from Lemma 8.2.

Part (c) says that $\mathcal{B} \subseteq \mathcal{I}$. This comes from the construction. If $s \in \mathcal{B}$, then there is a vertex t in $\mathcal{W}$ and an edge from t to s. To say that t lies in $\mathcal{W}$ means that there is an oriented path from the basepoint to t which never crosses $\mathcal{I}$. If s did not lie in $\mathcal{I}$, then there would be an oriented path from the basepoint to s which does not cross $\mathcal{I}$, and we would have that s lies in $\mathcal{W}$ too, since we chose $\mathcal{W}$ to be maximal with this property. Thus $s \in \mathcal{I}$, as desired.

Thus properties (a)-(d) are satisfied. Now we assume also that there are no oriented cycles in G which can be reached by oriented paths from v, and we prove the last assertions in the proposition. We begin with the injectivity of π on $\mathcal{W}$.

Suppose, to the contrary, that we have a pair of distinct vertices t, τ in $\mathcal{W}$ such that $\pi(t) = \pi(\tau)$. As usual, t and τ represent oriented paths in G which begin at v, and $\pi(t) = \pi(\tau)$ says exactly that these two paths have the same endpoint.

Neither of the paths represented by t and τ can be a subpath of the other, because that would imply the existence of an oriented cycle which is accessible by v.

Thus both paths start at v, but they have to diverge somewhere along the way. After diverging, they have to meet again at some vertex in G. We take w to be the *first* vertex in G at which they meet after diverging. We conclude that (v, w) is a focal pair; there are subpaths of the paths represented by t and τ which go from v to w and arrive at w through different edges. The edges are *different* because of our choice of w as the *first* vertex at which the paths meet after diverging from each other.

We want to derive a contradiction from this. By assumption, there are oriented paths in $\mathcal{W}$ from the basepoint to each of t and τ, and these paths never cross $\mathcal{I}$. However, each of these paths crosses vertices which project down to

w; this follows from Corollary 4.6, for instance. (That is, the paths in $\mathcal{V}_+(v, G)$ and the paths in G necessarily correspond to each other under the canonical projection $\pi : \mathcal{V}_+(v, G) \to G$.) Any point which projects to w must lie in $\mathcal{I}$, and therefore t and τ cross $\mathcal{I}$, a contradiction. This proves that π is injective on $\mathcal{W}$.

Remark 8.6 We actually get a little more here: if t is any vertex in $\mathcal{W}$ and τ is any vertex in $\mathcal{V}_+(v, G)$ at all with $\tau \neq t$, then we have $\pi(t) \neq \pi(\tau)$. This follows from the same argument.

We are left with the task of proving that π is at most two-to-one on $\mathcal{B}$. Let $s \in \mathcal{B}$ be given, and set $p = \pi(s)$, so that p is a vertex in G, and s represents an oriented path in G from v to p.

Since $s \in \mathcal{B}$, there is a vertex t in $\mathcal{W}$ such that there is an edge in $\mathcal{V}_+(v, G)$ which goes from t to s. By the definition of $\mathcal{V}_+(v, G)$, this means that the oriented path in G that corresponds to s can be obtained from the oriented path in G associated to t simply by adding an edge at the end. We know from above that t is uniquely determined by $\pi(t)$, since t lies in $\mathcal{W}$ and π is injective on $\mathcal{W}$.

The conclusion of this is that s is uniquely determined once we know $\pi(s)$ and the last edge e which is traversed by the oriented path in G which is represented by s. In other words, these data determine $\pi(t)$ (as the vertex in G from which e flows), and hence t itself, and then the path in G represented by s is the same as the path in G represented by t with e added at the end. There are at most two edges G which flow into $\pi(s)$, and so we conclude that there are at most two possibilities for s once $\pi(s)$ is given, as desired. This completes the proof of the proposition. □

Remark 8.7 If (v, v) is a focal pair, then Proposition 8.5 can be modified to work in almost the same way as before. As it stands, Proposition 8.5 would not be correct, because (a) requires $\mathcal{W}$ to contain the basepoint of $\mathcal{V}_+(v, G)$, and this is incompatible with (d) when (v, v) is a focal pair. To fix (d) one should change it to

(d′) If t is a vertex in $\mathcal{W}$, then $(v, \pi(t))$ is *not* a focal pair unless $\pi(t) = v$, in which case t must be the basepoint of $\mathcal{V}_+(v, G)$.

In the proof of Proposition 8.5, one should replace the set $\mathcal{I}$ by the set $\mathcal{I}'$ which consists of the same vertices in $\mathcal{V}_+(v, G)$ as $\mathcal{I}$ does, except that the basepoint in $\mathcal{V}_+(v, G)$ should be removed from $\mathcal{I}$. With these changes, the proof of Proposition 8.5 works exactly as before, to give the conclusions (a), (b), (c), and (d′). (Note that the later parts of Proposition 8.5 do not apply here, since G necessary contains an oriented cycle passing through v if (v, v) is a focal pair.)

8.4 The Corona decomposition

We can repeat the Calderón–Zygmund decomposition over and over again to get a more complicated decomposition of the visibility, which is somewhat like Carleson's Corona construction [Car62, Gar81]. (Compare also with [DS93].) It is similar as well to constructions that occur in the proof of the John–Nirenberg theorem from real analysis, although there are some differences with this too. Concerning the latter, see [CS97, Gar81, JN61, Jou83, Sem99b, Ste93].

Let G be an optical graph, and fix a vertex v in G. Apply Proposition 8.5 (or Remark 8.7, if need be) to get $\mathcal{W}_1$, $\mathcal{B}_1$, and $\mathcal{E}_1$, as above. We have added the subscript 1 to reflect the fact that this is the first stage of the construction.

Using Lemma 8.1, we can decompose the whole visibility graph $\mathcal{V}_+(v, G)$ into $\mathcal{W}_1$, the edges in $\mathcal{E}_1$, and the subgraphs $F(s)$, $s \in \mathcal{B}_1$. From Lemma 8.4, we have that each $F(s)$ is naturally isomorphic to the visibility graph $\mathcal{V}_+(\pi(s), G)$.

We can apply Proposition 8.5 (or Remark 8.7) and Lemma 8.1 to each $F(s)$, using the isomorphism with $\mathcal{V}_+(\pi(s), G)$, to decompose each of them into a $\mathcal{W}$ part, an $\mathcal{E}$-set of edges, and new subgraphs of the form $F(t)$, where t ranges through a $\mathcal{B}$-set of vertices for each choice of $s \in \mathcal{B}_1$.

Again we can apply Proposition 8.5 (or Remark 8.7) and Lemma 8.1 to each new F-part, and repeat the process indefinitely, or until we run out of $\mathcal{B}$-points. This is the Corona decomposition of the visibility. We do exhaust all of the visibility in this manner, but we may have to go through infinitely many stages to do that when the visibility is infinite. That is, every vertex in the visibility represents a *finite* path in G, and must appear within finitely many steps of the construction, no matter the presence of cycles.

For our purposes, the following is a key point.

Lemma 8.8 *Notation and assumptions as above. If there is a nonempty $\mathcal{B}$-set after n stages of the construction, then there is a chain of focal pairs starting from v which has length n.*

Proof This is an easy consequence of part (c) of Proposition 8.5, applied to each stage of the decomposition mentioned above. □

From here we can get a bound on the size of the visibility.

Theorem 8.9 *Let G be an optical graph, and let v be a vertex in G. Assume that there is no oriented cycle in G which can be reached by an oriented path from v, and that there is no chain of focal pairs in G of length n which begins at v. Then the visibility $\mathcal{V}_+(v, G)$ has at most $(2N)^n$ vertices, where N is the number of vertices in G.*

Examples like the graphs H_j^* in Section 4.3 suggest that quantities like $(N/n)^n$ might be more appropriate than the N^n in Theorem 8.9. In fact, M. Gromov pointed out how one can win an extra factor of $(n-1)!$, as we shall explain in Section 8.7. For the argument that follows, this corresponds to counting the number of elements in the $\mathcal{B}$ sets with more care, using the natural ordering on them that comes from the orientation on G, and we shall discuss this further after Lemma 8.17 in Section 8.7.

In any case, we should emphasize that something like N^n has to be involved (as opposed to the 2^n in Proposition 4.18 in Section 4.14, or C^n where C does not depend on the size of G), because of the examples of the H_j^*'s.

Proof Under the assumptions in Theorem 8.9, the Corona decomposition can proceed for at most $n-1$ stages; there will be no elements in the $\mathcal{B}$-sets at the nth stage, and so no F-sets with which to make an $(n+1)$th stage. Thus all of the vertices in the visibility $\mathcal{V}_+(v, G)$ must appear in $\mathcal{W}$ sets somewhere in the first n stages of the construction.

Our hypotheses also ensure that the number of vertices in any of the $\mathcal{W}$ sets which appear in the construction is no greater than the number of vertices in G. This follows from Proposition 8.5, except for a minor technical point; we are applying Proposition 8.5 not only to the visibility $\mathcal{V}_+(v, G)$, but also to other visibility graphs which appear in the course of the construction. We should check that there are no oriented cycles in G which can be reached by oriented paths from the vertices in G from which these other visibility graphs are defined. We have assumed this for the vertex v, but we should check it for the others that might arise. The point is that these other visibility graphs are based at vertices which can be reached by oriented paths from v. This is not hard to check, by an inductive argument. (For the first step, it is simply the fact that $\pi(s)$ can be reached from v by the path in G represented by s, and in the later stages the names change but the fact remains.) The conclusion is that oriented cycles which can be reached by these vertices can also be reached by v, and therefore do not exist, by assumption. This means that we may indeed apply Proposition 8.5 to each of the visibility graphs which arise, to say that the $\mathcal{W}$ sets which appear in the construction have no more vertices than G does.

It remains to count the total number of $\mathcal{W}$ sets. This number can increase geometrically. In the first stage, there is one. In the second stage of the construction, the number of $\mathcal{W}$ sets is bounded by the number of elements of $\mathcal{B}_1$, i.e., there will be one in each $F(s)$, $s \in \mathcal{B}_1$. The number of $\mathcal{W}$ sets in the third stage of the construction is bounded by the total number of elements in $\mathcal{B}$-sets in the second stage, and so on. That is, there is one $\mathcal{W}$ set in each $F(t)$, and the t's run through the union of the $\mathcal{B}$-sets at the given stage.

The number of elements in the first $\mathcal{B}$-set $\mathcal{B}_1$ is bounded by twice the number of vertices in $\mathcal{W}$, and hence by twice the number of vertices in G. This is true for each of the $\mathcal{B}$-sets created in any stage of the construction, i.e., created inside some $F(t)$. Keep in mind that every $\mathcal{B}$-set which appears in the construction is obtained through Proposition 8.5.

The conclusion is that there is 1 $\mathcal{W}$-set at the first stage of the construction, at most $2\,N$ at the second, and at most $(2\,N)^{j-1}$ $\mathcal{W}$-sets at the jth stage of the construction. In other words, each $\mathcal{W}$-set at level k leads to at most $2\,N$ $\mathcal{W}$-sets in the next level, as in Proposition 8.5.

The total number of $\mathcal{W}$-sets is at most

$$\sum_{j=1}^{n} (2\,N)^{j-1} \leq 2 \cdot (2\,N)^{n-1}. \tag{8.2}$$

Each $\mathcal{W}$-set has at most N elements, and so the total number of vertices in the visibility $\mathcal{V}_+(v, G)$ is $N \cdot 2 \cdot (2\,N)^{n-1}$. This proves the theorem. □

Remark 8.10 A basic flaw (or inefficiency) in the preceding argument is that we bound the number of elements in a fixed $\mathcal{B}$-set simply by twice the number of vertices in the corresponding $\mathcal{W}$-set, and we bound the latter by the total number of vertices in G. This is not so good in general, because the $\mathcal{B}$-sets are in fact controlled by just the vertices at the forward boundary of the $\mathcal{W}$-set (and not the ones in the middle or in the back), and because the $\mathcal{W}$-sets automatically enjoy some disjointness properties, i.e., the projection of a given $\mathcal{W}$-set into G is always disjoint from the projection of the rest of the visibility graph in question. This last follows from the observation in Remark 8.6 in the proof of Proposition 8.5, and it implies in particular that the projections of the $\mathcal{W}$-sets obtained in later generations are disjoint from the projection of the given one. The projections of the $\mathcal{W}$-sets are systematically pushed further and further to the ends of G, and thus are generally quite a bit smaller than G itself. This implies in turn that the $\mathcal{B}$-sets should grow at a substantially slower rate than is indicated in the theorem.

These issues are addressed by Gromov's method in Section 8.7, by taking the natural *ordering* of vertices in G into account. See Lemma 8.17 and the comments thereafter.

8.5 The derived graph

In order to analyze the behavior of the $\mathcal{B}$ sets more precisely, in terms of the geometry of G, one can consider the *derived* graph G' associated to a given optical graph G. This is defined by taking the same set of vertices as for G, but now with an edge going from a vertex u to a vertex w in G' exactly when (u, w) is a focal pair in G. (See Fig. 8.3.)

Thus, an oriented path of length n in the derived graph corresponds to a chain of focal pairs of length n in G. Similarly, the presence of distinct oriented loops in G which have a vertex in common can be detected through the existence of a vertex u in G such that (u, u) is a focal pair, which then means a vertex u which has an edge going from u to u in the derived graph. This was discussed just after the statement of Proposition 5.1, in connection with the exponential expansion of infinite visibilities.

The Corona decomposition can be related to the derived graph in the following manner. As in Section 8.4, the Corona construction produces a tree of $\mathcal{W}$-sets whose vertices eventually exhaust the vertices in the visibility graph. This tree maps into the visibility graph of G' based at v in a natural way. This is not hard to see; it comes down to part (c) of Proposition 8.5, which implies that each "step" in the Corona decomposition corresponds to a step (an edge) in G'. One could even arrange for this mapping to be injective, by adjusting the definition of G' so that there are *two* edges from u to w when (u, w) is a focal pair. We shall not pursue this now, but in Chapter 20 we shall follow similar reasoning for a more precise version of the derived graph, called the *reduced graph*.

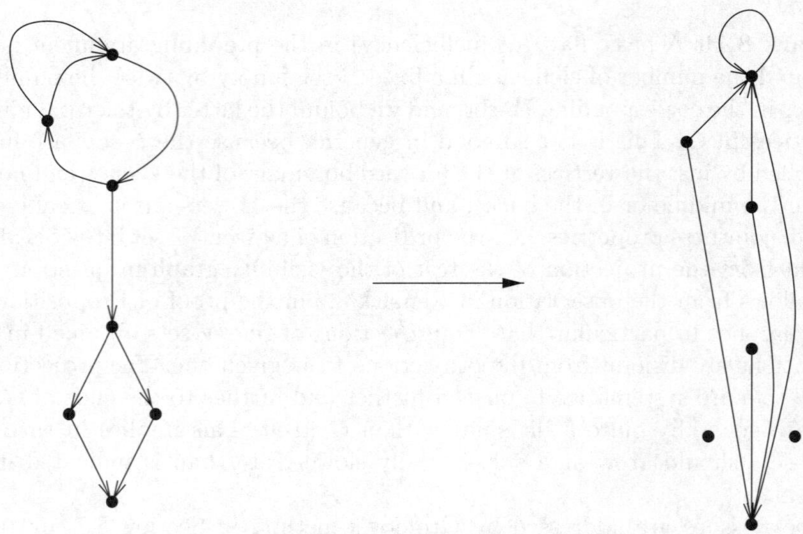

FIG. 8.3. An example of a graph G and its derived graph G'

8.6 Extensions

For various reasons of simplicity and concreteness, we have restricted our attention so far to optical graphs. In a sense, they are very natural for providing the simplest context in which this kind of exponential expansion occurs, and it is also the structure that one has for logical flow graphs for formal proofs (as in Section A.3 in Appendix A). One does not really need this restriction for our definitions and results, however.

Instead, let us think about oriented graphs with at most $k \geq 2$ inward-pointing and at most k outward-pointing edges at each vertex. The precise choice of k does not really matter, except that it will appear in some of the bounds. In practice, one often needs only a bound on the number of incoming edges at each vertex, or on the outgoing edges, but not both. The more convenient choice is not always the same at each moment, unfortunately, and so we presently ask for both, to simplify the discussion.

We call a vertex *focusing* if it has more than one incoming edge, and *defocusing* if it has more than one outgoing edge. Thus a single vertex can be both focusing and defocusing. This can easily be prevented by splitting vertices into two, with one new vertex taking all the incoming edges from the original and the other taking all the outgoing edges, and with a new edge that goes from the first of the new vertices to the second.

Similarly, one can split vertices in order to simulate arbitrary oriented graphs by optical graphs. One first applies the preceding procedure to reduce to the situation in which each vertex has at most one incoming edge or at most one outgoing edge, and then one further splits vertices with more than two incoming or outgoing edges into a series of vertices with at most two. This is not hard to

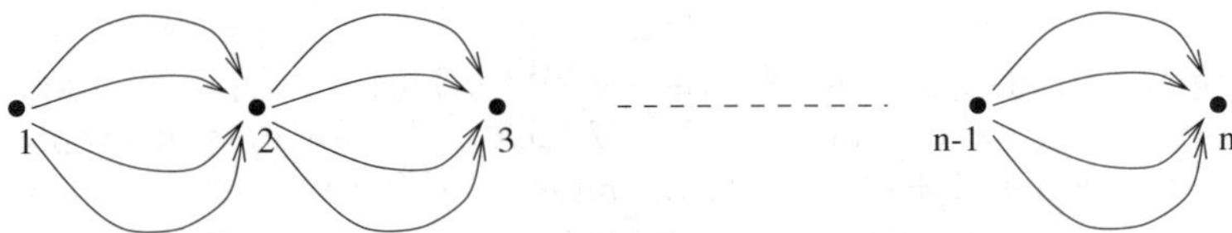

FIG. 8.4. A basic example (with no more than $k = 4$ incoming or outgoing edges at each vertex)

do.

Nonetheless, let us review how the material of this book can be applied directly to oriented graphs with at most k outgoing edges at each vertex.

Given any oriented graph G, one can define the visibility graphs of G in exactly the same way as before. They have the property that at each point there is at most one edge going in and at most k edges going out. (In fact, every point but the basepoint has exactly one edge going in; the basepoint has none.) This corresponds to Lemmas 4.3 and 4.4. One also defines the canonical projection π from the visibility to the original graph G in exactly the same manner as before. (See Section 4.5.)

One can define the canonical liftings of paths in exactly the same manner as in Section 4.6, and the uniqueness and other results from Section 4.6 carry over to this situation. For this, the number k plays no role, and the matter is purely topological and qualitative.

The material in Section 4.7 also carries over to the present situation. The qualitative questions of whether the visibility is finite or not remain unchanged, while most occurrences of the constant 2 in the various estimates should now be changed to a k. In some cases, there is a factor of 2 which comes from estimating a partial sum of a geometric series by its last term, and this need not be converted to a k.

Note well that instead of examples like that of Fig. 4.2 in Section 4.3 one can now have examples as in Fig. 8.4, with more edges between successive vertices. This makes it possible for the expansion of the visibility to be larger than it was before. We shall consider the matter of universal (exponential) upper bounds for the size of the visibility in Section 8.8, with slightly different formulations than we used before.

Fig. 8.4 should also be compared with the graphs H, H^*, H_j, and H_j^* from Section 4.3. (See Figures 4.4 and 4.5 especially.) If k is allowed to be large, then we can achieve similar effects as in those examples with graphs which simply have a large number of edges between successive pairs of vertices. For the H-graphs, we essentially went in the opposite direction, splitting apart vertices in order to reduce the number of edges attached to any vertex, while still having many different paths between certain pairs of vertices.

The considerations in Section 4.14 can be extended to the present discussion without change. With the possibility of more incoming edges at fixed vertices, one can consider extensions of the notion of focal pairs, and more intricate questions about upper and lower bounds for the visibility. This corresponds roughly to the problem of nested or parallel families of focal pairs for optical graphs like the H-graphs in Section 4.3. We shall not pursue this here.

The results of Chapter 5 apply to the more general context of non-optical (oriented) graphs with little change. The constants in some of the estimates could be affected by allowing for more incoming or outgoing edges at each point, but the basic dichotomy between exponential and polynomial growth remains the same, as well as its analysis in terms of whether loops intersect or not. Similarly, the *degree* of the polynomial rate of growth is still determined by the *depth* of the family $\mathcal{L}$ of loops whichs which are accessible from the basepoint (when these loops do not intersect), as in Section 5.3 (especially Propositions 5.10 and 5.13). Note that the 2's in the bases of the exponentials in the upper bounds in (5.3) and (5.18) should be replaced with k's.

The restriction to optical graphs in Chapter 7 did not play a very important role either. In that context, the possible numbers of incoming edges at each vertex is prescribed by the number of arguments of the operations involved, and one could just as well work with operations of higher arity as binary operations. Even in the binary case, one could allow more *outgoing* edges from the vertices. Some of the estimates should be adjusted, as in the counting arguments mentioned in Sections 7.6 and 7.9, but this is easy to do. Note that Lemma 7.5 in Section 7.5 carries over to this setting unchanged.

The construction in Section 8.1 extends to the present context without change or difficulty. Again, it is primarily qualitative and topological, with the only point being that in Lemma 8.2 one should say that the number of elements in $\mathcal{B}$ and $\mathcal{E}$ is at most k times the number of elements in $\mathcal{W}$.

The material in Section 8.2 extends without modification.

In Section 8.3, the main point is to replace the phrase "twice the number" in Proposition 8.5 (in part (b), and in two occasions in the last paragraph of the statement of the proposition) with "k times the number", and to replace "two-to-one" in (ii) with "k-to-one". With these changes, the proof goes through as before. Most of it did not rely on the notion of optical graphs anyway. There are two exceptions to this. The first is in the proof of (b), which in the general case follows from Lemma 8.2, as extended in the manner just mentioned. The second exception comes from (ii), whose proof is easily adapted to this context, using the assumption that the number of edges going into any vertex in G be at most k.

We are left with Section 8.4. Again much of it is qualitative and topological and needs no modification. The only exception is in the presence of the 2's in the estimates, which should now be mostly replaced with k's, as in the statement of Proposition 8.5. The multiplicative factor of 2 which appears in the right side of (8.2) need *not* be changed, though, because it reflects the summation of a geometric series and not the structure of the graph G. In the end the bound that results for the number of vertices of the visibility is

$$2 \cdot k^{n-1} \cdot N^n. \tag{8.3}$$

8.7 A more direct counting argument

The material in this section is based on some observations of M. Gromov, and it provides a more direct way to estimate the number of paths in a graph. As we mentioned before, it also provides a good way in which to deal with some of the inefficiencies of the earlier analysis.

Let G be an oriented graph which is free of nontrivial oriented cycles, and assume that G has at most $k \geq 2$ incoming edges at any vertex. (As in Section 8.6, we no longer restrict ourselves to optical graphs. The case where $k = 1$ is not interesting here, because it implies that there is at most one oriented path between any given pair of vertices.) Fix a pair of vertices x, y in G.

Theorem 8.11 *Notation and assumptions as above. If ℓ is the length of the longest chain of focal pairs in G which begins at x, then the number of oriented paths in G which go from x to y is at most*

$$2 \cdot k^{\ell} \cdot \frac{(N-1)^{\ell}}{\ell!}, \tag{8.4}$$

where N is the total number of vertices in G.

If $\ell = 0$ and $N = 1$, then $(N-1)^{\ell}$ should be interpreted as being 1. Also, the extra factor of 2 is easily seen to disappear in the proof when $\ell = 0$.

Proof Let γ be any oriented path in G which goes from x to y. We would like to associate to γ some combinatorial data which determines it uniquely and which provides a way for us to estimate the total number of such paths.

We first choose a sequence of vertices $x_1, x_2, \ldots, x_{j+1}$ along γ in such a way that

$$(x_i, x_{i+1}) \text{ is a focal pair for each } i \leq j; \tag{8.5}$$

$$(x_i, z) \text{ is } \textit{not} \text{ a focal pair for any vertex } z \text{ which lies (strictly) between } x_i \text{ and } x_{i+1} \text{ on } \gamma; \tag{8.6}$$

$$(x_{j+1}, z) \text{ is } \textit{not} \text{ a focal pair for any vertex } z \text{ which comes after } x_{j+1} \text{ on } \gamma. \tag{8.7}$$

We do this recursively as follows. Set $x_1 = x$. Choose x_2 to be the first vertex after x_1 on γ such that (x_1, x_2) is a focal pair, if such a vertex exists; if not, take $j = 0$ and stop. If such a vertex x_2 does exist, then we repeat the process as often as possible. That is, if x_i has been chosen, then we choose x_{i+1} to be the first vertex on γ after x_i such that (x_i, x_{i+1}) is a focal pair, if such a point exists, and otherwise we stop and set $j = i - 1$. It is easy to see from the construction that the points $x_1, x_2, \ldots, x_{j+1}$ have the properties (8.5), (8.6), and (8.7).

We want to use the sequence $x_1, x_2, \ldots, x_{j+1}$ to characterize γ as well as we can. For this we shall use the following.

Claim 8.12 *Let p and q be vertices in G, and let e be an edge in G which flows into q.* (a) *There is at most* 1 *oriented path α in G which goes from p to q, which arrives to q along e, and which has the property that (p, z) is not a focal pair for any vertex z which lies strictly between p and q on α.* (b) *There is at most* 1 *oriented path β in G which goes from p to q and which has the property that (p, z) is not a focal pair for any vertex z in β.*

The difference between (a) and (b) is that we allow (p, q) to be a focal pair in (a) but not in (b), and that we specify the final edge which is traversed by α in (a).

To prove the claim, let us consider (a) first. Suppose that α is as above, and that there is an oriented path α' in G which is different from α and which goes from p to q and arrives to q along e. Since we are assuming that G is free of oriented cycles, we cannot have that either of α or α' is a subpath of the other. Hence they must diverge from each other at some point, only to come together again later. Let z be a vertex in G such that α and α' both pass through z, but arriving along different edges. This means that (p, z) is a focal pair. On the other hand, α' is supposed to arrive to q along the same edge e as α, which means that z should lie strictly between p and q on α, in contradiction to our assumptions about α. This proves (a), and the same kind of argument works for (b) as well.

Let us return now to our path γ and the points $x_1, x_2, \ldots, x_{j+1}$.

Claim 8.13 *The oriented path γ is uniquely determined by the following data: its initial and final vertices x, y; the sequence $x_1, x_2, \ldots, x_{j+1}$ of vertices associated to γ as above; and the edges e_i, $1 < i \leq j+1$, which flow into x_i and which are traversed by γ just before reaching x_i.*

This is an easy consequence of Claim 8.12. More precisely, Claim 8.12 implies that the subpaths of γ going from x_i to x_{i+1}, $1 \leq i \leq j$, and from x_{j+1} to y are all uniquely determined by the data described in Claim 8.13, and then γ itself is obtained by concatenating these paths.

Claim 8.14 *If x and y are fixed, then there are at most*

$$\binom{N-1}{j} \tag{8.8}$$

sequences $x_1, x_2, \ldots, x_{j+1}$ of vertices in G which can be associated to an oriented path γ from x to y in the manner described above.

If $x_1, x_2, \ldots, x_{j+1}$ arises from a path γ as above, then $x_1 = x$ by construction. Also, there is a nondegenerate path going from x_i to x_{i+1} for all $i \leq j$, since (x_i, x_{i+1}) is a focal pair (as in (8.5)). Since we are assuming that G contains no oriented cycles, we may conclude that the x_i's are all distinct from each other, and that their ordering is uniquely determined from the knowledge of

$$\{x_i : 2 \leq i \leq j+1\} \tag{8.9}$$

as an *unordered* set. This proves Claim 8.14, since the x_i's for $i \geq 2$ are chosen among the $N-1$ vertices in G which are different from x_1.

We can summarize our conclusions so far as follows.

Claim 8.15 *The total number of distinct oriented paths γ in G which go from x to y is at most*

$$\sum_{j=0}^{\ell} k^j \cdot \binom{N-1}{j}. \tag{8.10}$$

Indeed, the number of paths γ associated to a particular sequence $x_1, x_2, \ldots, x_{j+1}$ is at most k^j, because of Claim 8.13 and our assumption that each vertex in G have at most k incoming edges. For fixed j, the total number of these sequences $x_1, x_2, \ldots, x_{j+1}$ is at most $\binom{N-1}{j}$, by Claim 8.14. On the other hand, j itself must satisfy $0 \leq j \leq \ell$, since the sequence $x_1, x_2, \ldots, x_{j+1}$ determines a chain of focal pairs of length j (which begins at x, by construction), and we are assuming that G does not contain any such chain of length greater than ℓ. This gives Claim 8.15.

To finish the proof of the theorem, we need to show the following.

Claim 8.16 $\sum_{j=0}^{\ell} k^j \cdot \binom{N-1}{j} \leq 2 \cdot k^\ell \cdot \frac{(N-1)^\ell}{\ell!}$.

Notice first that

$$\ell \leq N - 1. \tag{8.11}$$

In other words, if G contains a chain of focal pairs of length ℓ, and if G is free of nontrivial oriented cycles, then G must contain at least $\ell + 1$ vertices, namely the vertices involved in the chain of focal pairs (which must be distinct when there are no cycles). Using (8.11), it is not hard to check that

$$\binom{N-1}{j} \leq \frac{(N-1)^\ell}{\ell!} \tag{8.12}$$

for each $j = 0, 1, 2, \ldots, l$. To see this, one can write $\binom{N-1}{j}$ as the quotient of $(N-1)(N-2)\cdots(N-j)$ and $j!$. This is less than or equal to $(N-1)^j/j!$. One can add factors to the numerator and denominator of this last expression to get the right side of (8.12), and (8.11) ensures that the factors that one adds to the numerator are always greater than or equal to the ones in the denominator. Thus we get (8.12). On the other hand, we also have that

$$\sum_{j=0}^{\ell} k^j = \frac{k^{\ell+1} - 1}{k - 1} \leq \frac{k^{\ell+1}}{k - 1} = \frac{k}{k - 1} \cdot k^\ell \leq 2 \cdot k^\ell, \tag{8.13}$$

since $k \geq 2$, by assumption. This yields Claim 8.16. Theorem 8.11 follows from this, as mentioned above. □

Each step in the preceding argument has a counterpart in the earlier discussion of the Corona decomposition, except for one, which we can isolate as a lemma.

Lemma 8.17 *Let G be an oriented graph which is free of nontrivial oriented cycles, and fix a vertex x in G and a nonnegative integer j. Let B_j be the number of sequences $x_1, x_2, \ldots, x_{j+1}$ of vertices in G such that $x_1 = x$ and (x_i, x_{i+1}) is a focal pair for each $i \leq j$. Then*

$$B_j \leq \binom{N-1}{j}, \tag{8.14}$$

where N is the total number of vertices in G.

Proof This corresponds to Claim 8.14 above. All that we really need to know is that there is a nondegenerate oriented path in G which goes from x_i to x_{i+1} for each $i \leq j$. This implies that the x_i's are distinct, and that the ordering of the sequence $x_1, x_2, \ldots, x_{j+1}$ is determined by the graph. Thus one can reduce to counting the number of *unordered* sets of vertices $\{x_i : 2 \leq i \leq j+1\}$ inside the set of $N-1$ vertices which are different from x, and this leads to (8.14). □

In the context of the Corona decomposition, this lemma gives a better estimate for the number of elements in the $\mathcal{B}$ sets. More precisely, the total number of elements in all of the $\mathcal{B}$ sets is at most

$$\sum_{j=1}^{\ell} k^j \cdot B_j, \tag{8.15}$$

where B_j is the same as in Lemma 8.17, k bounds the number of incoming edges at every vertex in G, and ℓ is the length of the longest chain of focal pairs that begins at x. (In the setting of Section 8.4, one should take $x = v$.) This fact is not hard to verify from the definitions. Indeed, each element of a $\mathcal{B}$ set in the jth level of the construction corresponds to the endpoint of a chain of focal pairs of length j, and hence to one of the sequences counted by B_j. At each stage, we increase by a factor of k the number of elements in the $\mathcal{B}$ sets which can correspond to a single vertex in G, and this is the reason for the factor of k^j in (8.15). These assertions correspond to parts (c) and (ii) in Proposition 8.5. The 2 in part (ii) of Proposition 8.5 comes from the earlier restriction to optical graphs, and should be replaced by k in the present context, as in Section 8.6.

If one uses this better estimate for the number of elements in the $\mathcal{B}$ sets, then the bounds for the number of vertices in the visibility that follow from the Corona decomposition are the same as the ones that we get here. For this comparison one should note that the parameter n from Theorem 8.9 corresponds to $\ell + 1$ in Theorem 8.11, i.e., n was chosen to be strictly larger than the length of any chain of focal pairs, while ℓ was taken here to be the maximal length of a chain of focal pairs. Also, in the present discussion we have fixed the final endpoint of our paths, which we did not do before, and this is the source of an extra factor of N in Theorem 8.9. (The difference between the $N-1$'s in this section and the N's before does not represent a genuine mathematical difference in the arguments, but merely an increase in fastidiousness.) For the role of the power of k, see also (8.3) in Section 8.6.

Let us now give a modest refinement of Theorem 8.11.

Theorem 8.18 *Let G be an oriented graph without nontrivial oriented cycles, and let x, y be a pair of vertices in G. Let M denote the total number of edges in G, and let ℓ be the length of the longest chain of focal pairs in G which begins at x. Then the number of oriented paths in G which go from x to y is at most*

$$2 \cdot \frac{M^\ell}{\ell!}. \tag{8.16}$$

Note that $M \leq k \cdot N$ if N is the total number of vertices in G and k is the maximal number of edges which can go into any fixed vertex. Thus the present bound contains the one in Theorem 8.11. As before, we interpret M^ℓ as being equal to 1 when $M = 0$ and $\ell = 0$, and the factor of 2 in (8.16) disappears in the proof when $\ell = 0$.

Proof Let γ be any oriented path in G which goes from x to y, and choose $x_1, x_2, \ldots, x_{j+1}$ as in the proof of Theorem 8.11. Let $e_2, \ldots, e_{j+1}$ denote the edges in G which flow into $x_2, \ldots, x_{j+1}$ and which are traversed by γ just before γ arrives at the corresponding vertex x_i.

From Claim 8.13, we know that γ is determined uniquely by the edges $e_2, \ldots,$ e_{j+1} (and the fact that it goes from x to y); i.e., the knowledge of the sequence $x_2, \ldots, x_{j+1}$ can be recovered from the sequence of e_i's. On the other hand, there are at most

$$\binom{M}{j} \tag{8.17}$$

possible choices for the sequence $e_2, \ldots, e_{j+1}$. This is proved in the same manner as for Claim 8.14; the main points are that the e_i's are necessarily distinct, and that the ordering of them is automatically determined by G, since there is an oriented path going from the final vertex of e_i to the starting vertex of e_{i+1} for each $i \leq j$.

To summarize, we now have that the total number of oriented paths going from x to y is bounded by

$$\sum_{j=0}^{\ell} \binom{M}{j} \tag{8.18}$$

(since ℓ is the length of the longest chain of focal pairs in G, starting at x). We can estimate this sum in the following manner. Notice first that

$$2 \cdot l \leq M, \tag{8.19}$$

i.e., G must contain at least $2l$ edges in order to have a chain of focal pairs of length ℓ. This is easy to see from the definition, simply by considering the incoming edges in the vertices involved in the chain. Using this it is easy to check that

$$\binom{M}{j} \leq \frac{M^j}{j!} \leq \frac{M^\ell}{\ell!} \cdot 2^{j-l}. \tag{8.20}$$

Thus

$$\sum_{j=0}^{\ell} \binom{M}{j} \leq \frac{M^\ell}{\ell!} \sum_{j=0}^{\ell} 2^{j-l} \leq 2 \cdot \frac{M^\ell}{\ell!}, \tag{8.21}$$

which proves the theorem. □

Again, this argument has a counterpart in terms of the Corona decomposition, through a bound on the number of elements of the $\mathcal{B}$ sets.

8.8 Exponential bounds for general graphs

Let G be an oriented graph without oriented cycles, and let v be a vertex in G. If G has at most M edges, then what can we say about the total number of oriented paths in G which begin at v and have length at most a given number l?

If M is allowed to have oriented cycles, then one can simply choose G so that it contains only the one vertex v and M edges attached as loops, so that the number of paths of length at most l is equal to

$$\sum_{j=0}^{l} M^j. \tag{8.22}$$

It is easy to see that this is the worst-case situation for arbitrary oriented graphs. If we restrict ourselves to graphs without nontrivial oriented cycles, then we can do better, because no path can pass through the same edge twice. In fact, we have the following.

Lemma 8.19 *Let G be an oriented graph which is free of oriented cycles, and fix a vertex v in G. If M is the total number of edges in G, then the total number of oriented paths in G which begin at v and have length equal to j is at most*

$$\binom{M}{j}. \tag{8.23}$$

Proof This is simply a repackaging of an observation that was used several times in Section 8.7. Namely, if α is an oriented path in G which passes through the edges $e_1, e_2, \ldots, e_j$, then the absence of oriented cycles in G implies that the e_i's are distinct, and that the ordering of the e_i's is determined by G itself. In other words, α is uniquely determined by

$$\{e_i : 1 \leq i \leq j\} \tag{8.24}$$

as an *unordered* set of edges. This implies that the total number of paths α is bounded by (8.23), as desired. □

Corollary 8.20 *Under the same assumptions and notation as in Lemma 8.19, the total number of oriented paths in G which begin at v is at most 2^M.*

Proof No oriented path in G can pass through more any edge more than once, and hence no such path can have length greater than M. Thus the total number of paths can be bounded by the sum

$$\sum_{j=0}^{M} \binom{M}{j}, \tag{8.25}$$

and this is exactly 2^M, by the binomial theorem. □

Given an oriented graph G and a vertex v in G, let $T_j(v, G)$ denote the number of oriented paths in G which begin at v and have length exactly equal to j. Let $T_{j,M}$ denote the maximum of $T_j(v, G)$ over all choices of G and v, where G is restricted to graphs with at most M edges and no nontrivial oriented cycles. Lemma 8.19 says that $T_{j,M}$ is bounded by (8.23), but in fact there is a simple formula for $T_{j,M}$, which we give in Lemma 8.25. To do this, we first give a recurrence relation for the $T_{j,M}$'s in Lemma 8.21.

Note that both j and M are permitted to be zero here. When $j = 0$, we have that

$$T_{0,M} = 1 \tag{8.26}$$

for all $M \geq 0$, while

$$T_{j,M} = 0 \qquad \text{when } j > M, \tag{8.27}$$

since the length of every oriented path in G is at most M when G is an oriented graph with at most M edges and no nontrivial oriented cycles.

Lemma 8.21 $T_{j,M} = \max\{i \cdot T_{j-1,M-i} : 0 < i \leq M\}$ *when* $0 < j \leq M$. *(Of course, we restrict ourselves to integral values of i and j here.)*

Proof Let $j, M \geq 0$ be given, and let us first check that

$$T_{j,M} \leq \max\{i \cdot T_{j-1,M-i} : 0 < i \leq M\}. \tag{8.28}$$

Let G be an arbitrary oriented graph with at most M edges, let v be an arbitrary vertex in G, and assume that G contains no nontrivial oriented cycles. We would like to show that

$$T_j(v, G) \leq \max\{i \cdot T_{j-1,M-i} : 0 < i \leq M\}. \tag{8.29}$$

If we can do this, then (8.28) will follow, since G and v are arbitrary (i.e., one can choose G and v so that $T_j(v, G)$ is as large as possible).

Let i be the number of edges in G which flow out of v. Thus $0 \leq i \leq M$ automatically, since G contains at most M edges. If $i = 0$, then $T_j(v, G) = 0$, and (8.29) holds automatically. Thus we suppose that $i > 0$.

Let $e_1, \ldots, e_i$ be an enumeration of the edges in G which flow out of v, and let v_a denote the vertex in G into which e_a flows for each $a \leq i$. Note that the v_a's could all be the same vertex, or we might have $v_a = v_b$ for some pairs a, b and not others. However, we cannot have $v_a = v$ for any a, since we are assuming that G is free of nontrivial oriented cycles.

Given a with $1 \leq a \leq i$, let G_a denote the subgraph of G which consists of all vertices and edges in G which is accessible from v_a by an oriented path.

Claim 8.22 $T_j(v, G) = \sum_{a=1}^{i} T_{j-1}(v_a, G_a)$.

To prove this, it suffices to show that $T_{j-1}(v_a, G_a)$ is equal to the number of oriented paths in G which begin at v and cross e_a immediately after leaving v, since every nondegenerate oriented path in G which begins at v has to cross exactly one of the e_a's just after leaving v. This interpretation of $T_{j-1}(v_a, G_a)$ is easy to verify, because $T_{j-1}(v_a, G_a)$ is defined to be the number of paths in G_a of length $j-1$ which begin at v_a, and because any oriented path in G which begins at v and crosses e_a will remain in G_a afterwards (by definition of G_a). This proves Claim 8.22.

Claim 8.23 $T_{j-1}(v_a, G_a) \leq T_{j-1,M-i}$ *for each* $1 \leq a \leq i$.

To show this, we only need to check that the pair G_a, v_a satisfies the requirements for the competitors of $T_{j-1,M-1}$, i.e., that each G_a contains at most $M-i$ edges and no nontrivial oriented cycles. The absence of nontrivial oriented cycles in G_a follows from the same condition for G. As for the bound on the number of edges in G_a, we simply use the facts that G has at most M edges, and that G_a cannot contain any of the edges e_b, $1 \leq b \leq i$ (since G contains no nontrivial oriented cycles). This proves Claim 8.23.

Now that we have established these two claims, we can combine them to get the upper bound (8.29) for $T(v, G)$ that we want. This proves (8.28), since the pair G, v is an arbitrary competitor for $T(v, G)$.

It remains to establish the opposite inequality, namely

$$T_{j,M} \geq \max\{i \cdot T_{j-1,M-i} : 0 < i \leq M\}. \tag{8.30}$$

To do this, we shall (in essence) reverse the preceding argument. Fix $M > 0$ and i with $0 < i \leq M$, and let H, w be any competitor for $T_{j-1,M-i}$. In other words, H should be an oriented graph with no nontrivial oriented cycles, and at most $M-i$ edges, and which contains w as a vertex. Define a new oriented graph G by taking H and adding to it a new vertex v, and i edges going from v to w. Thus G contains at most M edges, and no nontrivial oriented cycles, because of the corresponding features of H. This implies that G, v is an admissible competitor for $T_{j,M}$. Hence

$$T_{j,M} \geq T_j(v, G). \tag{8.31}$$

On the other hand, one can argue as in the proof of Claim 8.22 to obtain that

$$T_j(v, G) = i \cdot T_{j-1}(w, H). \tag{8.32}$$

This yields

$$T_{j,M} \geq i \cdot T_{j-1}(w, H). \tag{8.33}$$

Since the pair H, w is an arbitrary competitor for $T_{j-1,M-i}$ we obtain that

$$T_{j,M} \geq i \cdot T_{j-1,M-i}. \tag{8.34}$$

This implies (8.30), since i was also chosen arbitrarily from the range $0 < i \leq M$.
The proof of Lemma 8.21 is now complete. □

Using the recursion relation in Lemma 8.21, we can compute the $T_{j,M}$'s more precisely, as follows.

Lemma 8.24 *Given nonnegative integers j and M, define $U_{j,M}$ to be the maximum value of the product*

$$\prod_{s=1}^{j} a_s, \tag{8.35}$$

where $a_1, a_2, \ldots, a_j$ is any collection of j nonnegative integers such that

$$\sum_{s=1}^{j} a_s = M. \tag{8.36}$$

(When $j = 0$ we take $U_{j,M} = 1$.) Then $U_{j,M} = T_{j,M}$ for all j and M.

Proof If $j = 0$, then both $T_{j,M}$ and $U_{j,M}$ take the value 1. (See (8.26).) They also both vanish when $j > M$, as in (8.27). To prove the equality in general, it suffices to show that $U_{j,M}$ satisfies the same recursion relation as $T_{j,M}$, i.e.,

$$U_{j,M} = \max\{i \cdot U_{j-1,M-i} : 0 < i \leq M\} \tag{8.37}$$

when $0 < j \leq M$.

This identity (8.37) really says only that maximizing the product (8.35) subject to the constraint (8.36) is the same as maximizing (8.35) first with respect to the constraint

$$\sum_{s=2}^{j} a_s = M - a_1, \tag{8.38}$$

and then maximizing over a_1's with $0 < a_1 \leq M$. (The case where $a_1 = 0$ plays no role, since the product (8.35) is then 0.) This property is easy to check, just from the definitions. This proves (8.37), and the lemma follows. □

Lemma 8.25 *Let j and M be nonnegative integers, with $0 < j \leq M$. Write $M = t \cdot j + r$, where t and r are nonnegative integers, with $0 \leq r < j$. Then*

$$T_{j,M} = (t+1)^r t^{j-r}. \tag{8.39}$$

In particular,

$$T_{j,M} = \left(\frac{M}{j}\right)^j \tag{8.40}$$

when M is an integer multiple of j.

Proof This is an easy consequence of Lemma 8.24. The main point is the following.

Claim 8.26 *If a and b are arbitrary nonnegative integers, and if c and d are nonnegative integers such that $a + b = c + d$ and $|c - d| \leq 1$, then*

$$ab \leq cd. \tag{8.41}$$

In other words, the product is maximal when a and b are as close to each other as they can be. (Note that $|c - d| \leq 1$ implies $c = d$ when $c + d$ is even, and $|c - d| = 1$ when $c + d$ is odd.) This is well-known and easy to check.

Now let j, M, t, r be as in the statement of the lemma, and let us prove that

$$U_{j,M} = (t+1)^r t^{j-r}. \tag{8.42}$$

We certainly have that $U_{j,M}$ is less than or equal to the right-hand side, because we can choose the a_s's in Lemma 8.24 so that $a_s = t + 1$ when $1 \leq s \leq r$ and $a_s = t$ when $r \leq s \leq j$. To get the reverse inequality, one can start with any sequence $a_1, a_2, \ldots, a_j$ of nonnegative integers whose sum is M and systematically modify them using Claim 8.26 in such a way that the product increases while the variation between the a_j's becomes smaller. In the end, one obtains that the optimal configuration is the one just mentioned, and the lemma follows. □

Corollary 8.27 *$T_{j,M} \leq 3^{M/3}$ for all j and M.*

Note that we have equality when $M = 3j$, as in (8.40).

Proof Let j and M be given. We may as well assume that $0 < j \leq M$, since the other cases are automatic. Let r and t be as in Lemma 8.25. We can rewrite (8.39) as

$$\frac{\log T_{j,M}}{M} = \frac{r(t+1)}{M} \cdot \frac{\log(t+1)}{t+1} + \frac{(j-r)t}{M} \cdot \frac{\log t}{t}. \tag{8.43}$$

On the other hand,

$$\frac{r(t+1)}{M} + \frac{(j-r)t}{M} = 1, \tag{8.44}$$

since $M = t \cdot j + r$, by construction. Thus we obtain that

$$\frac{\log T_{j,M}}{M} \leq \max\{t^{-1} \log t : t \in \mathbf{Z}_+\}. \tag{8.45}$$

The function $f(t) = t^{-1} \log t$ has exactly one critical point on $[1, \infty)$, at $t = e$, where it assumes its maximum among *real* numbers $t \geq 1$. Similar considerations of the derivative of f imply that $f(t)$ is increasing for t in $[1, e]$ and decreasing for t in $[e, \infty)$, and hence that the maximum of f on positive *integers* must occur at either $t = 2$ or $t = 3$. In fact it occurs at $t = 3$, because

$$e^{6 \cdot f(2)} = e^{3 \log 2} = 2^3 = 8 \quad \text{and} \quad e^{6 \cdot f(3)} = e^{2 \log 3} = 3^2 = 9 > 8. \tag{8.46}$$

Thus $\max\{t^{-1} \log t : t \in \mathbf{Z}_+\} = (1/3) \log 3$. Combining this with (8.45), we get that

$$\log T_{j,M} \leq \frac{M}{3} \cdot \log 3, \tag{8.47}$$

which proves Corollary 8.27. □

One can sharpen this estimate a bit as follows.

Lemma 8.28 *There is a universal constant $C > 0$ so that*

$$\sum_{j=0}^{M} T_{j,M} \leq C \cdot 3^{M/3}. \tag{8.48}$$

This should be compared with Corollary 8.20, whose proof shows that the sum in (8.48) is bounded by 2^M. (Note that the cube root of 3 is approximately 1.44.)

Proof The statement of Corollary 8.27 implies that the sum in (8.48) is bounded by $M+1$ times $3^{M/3}$, but the point is to get a constant C which does not depend on M. To do this, we shall simply show that for most values of j, the estimate for $T_{j,M}$ can be improved substantially.

Notice first that

$$\frac{\log t}{t} \leq \frac{\log 2}{2} \qquad \text{when } t \in \mathbf{Z}_+,\ t \neq 3. \tag{8.49}$$

Indeed, we know from the proof of Corollary 8.27 that $f(t) = t^{-1} \log t$ is increasing on $[1, e]$ and decreasing on $[e, \infty)$, and so the maximum of $f(t)$ for positive integers t which are different from 3 is attained at either $t = 2$ or $t = 4$. In fact $f(4) = f(2)$, as one can easily check, and this gives (8.49).

Let us write A for $3^{1/3}$ and B for $2^{1/2}$. Thus

$$B < A, \tag{8.50}$$

since $A^6 = 3^2 = 9$ and $B^6 = 2^3 = 8$.

Fix j and M with $0 < j \leq M$, and let t and r be chosen as in Lemma 8.25. If $t \neq 3$ and $t + 1 \neq 3$, then the same argument as in the proof of Corollary 8.27 yields

$$T_{j,M} \leq B^M, \tag{8.51}$$

because of (8.49). The total contribution of these j's to the sum in (8.48) is therefore bounded by $(M+1)B^M$, and this is bounded by a universal constant times A^M, since $A > B$.

Thus we may restrict our attention to the j's for which t is either 2 or 3. If $t = 2$, then we can rewrite (8.45) as

$$\begin{aligned}\frac{\log T_{j,M}}{M} &= \frac{r(t+1)}{M} \cdot \log A + \frac{(j-r)t}{M} \cdot \log B \\ &= \log A - \frac{(j-r)t}{M} \cdot (\log A - \log B),\end{aligned} \tag{8.52}$$

using also the identity (8.44). This implies that

$$T_{j,M} = A^M \cdot \left(\frac{A}{B}\right)^{-(j-r)t}. \tag{8.53}$$

Note that $j - r$ is positive, so that one can expect the right-hand side of (8.53) to be much smaller than A^M most of the time. Specifically, we would like to say that the sum of these $T_{j,M}$'s is bounded by a constant multiple of A^M, and for this we use the following.

Claim 8.29 *j is uniquely determined by the knowledge of M, t, and $j - r$ when $t > 1$.*

Indeed, $M = tj + r$, by construction. We can rewrite this formula as $M = (t+1)j - (j-r)$, to recover j from the other data.

The conclusion of this is that the sum of $T_{j,M}$ over the values of j corresponding to $t = 2$ is bounded by the sum

$$\sum_{l=1}^{\infty} A^M \cdot \left(\frac{A}{B}\right)^{-l\,t}, \tag{8.54}$$

because of the estimate (8.53), and the fact that the $j - r$'s are distinct positive integers for distinct choices of j. Thus the contribution of these j's to the sum in (8.48) is properly controlled, since

$$\sum_{l=1}^{\infty} \left(\frac{A}{B}\right)^{-l\,t}, \tag{8.55}$$

is a convergent geometric series.

When $t = 3$, we argue in practically the same way. We begin by rewriting (8.45) as

$$\begin{aligned}\frac{\log T_{j,M}}{M} &= \frac{r(t+1)}{M} \cdot \log B + \frac{(j-r)t}{M} \cdot \log A \\ &= \log A - \frac{r(t+1)}{M} \cdot (\log A - \log B).\end{aligned} \tag{8.56}$$

This uses also the identity $(t+1)^{-1} \log(t+1) = \log B$ (since $t = 3$) and (8.44). From here we get that

$$T_{j,M} = A^M \cdot \left(\frac{A}{B}\right)^{-rt}. \tag{8.57}$$

Of course, r is nonnegative, and it is easy to see that distinct values of j lead to distinct values of r (given M and $t = 3$), since $M = tj + r$. Thus we conclude that the sum of $T_{j,M}$ for these values of j is bounded by the sum

$$\sum_{l=0}^{\infty} A^M \cdot \left(\frac{A}{B}\right)^{-l\,t}. \tag{8.58}$$

This implies that the contribution of these values of j to the sum in (8.48) is bounded by a constant multiple of A^M, since

$$\sum_{l=0}^{\infty} \left(\frac{A}{B}\right)^{-l\,t} \tag{8.59}$$

is a convergent geometric series. This completes the proof of Lemma 8.28. □

Corollary 8.30 *Let G be an oriented graph which contains at most M edges, and no nontrivial oriented cycles. If v is any vertex in G, then the number of oriented paths in G which begin at v (and hence the number of vertices in the visibility graph $\mathcal{V}_+(v, G)$) is at most a constant multiple of $3^{M/3}$.*

This follows immediately from Lemma 8.28 and the definition of $T_{j,M}$. It is easy to build examples in which the number of paths is at least $3^{[M/3]}$, where $[M/3]$ denotes the integer part of $M/3$, simply by taking a sequence of $j+1$ vertices $0, 1, 2, \ldots, j$, where $j = [M/3]$, and attaching exactly 3 edges from the ith vertex to the $(i+1)$th vertex for each $i \leq j$.

8.9 The restrained visibility

Let G be an oriented graph, and let v be a vertex in G. One can consider variations on notion of the visibility, in which additional conditions are imposed on the paths. In this section, we shall briefly discuss the *restrained visibility graph* $\mathcal{V}_+^r(v, G)$, which is defined in the same way as before, except that now we restrict ourselves to paths which are *simple*, i.e., which do not go through any vertex more than once. One can also think of $\mathcal{V}_+^r(v, G)$ as a particular subgraph of the ordinary visibility graph $\mathcal{V}_+(v, G)$, namely the subgraph for which the vertices represent simple paths in G, and for which all edges between these vertices are included.

If G does not contain nontrivial oriented cycles which are accessible from v by an oriented path, then all the oriented paths in G which begin at v are simple anyway, and the restrained visibility coincides with the ordinary visibility. The situation becomes more complicated when G contains nontrivial oriented cycles which are accessible to v.

The restrained visibility is always finite (when G is), no matter the presence of cycles. Indeed, if G has N vertices, then simple paths in G must traverse fewer than N edges. More precise (exponential) upper bounds can be obtained through the considerations of Section 8.8. In fact, one can define "restrained" versions of the quantities $T_j(v, G)$ and $T_{j,M}$ from Section 8.8, and in the end one gets exactly the same estimates as before, because the restrained version of $T_{j,M}$ is equal to the original version. This can be proved by showing that the restrained version of $T_{j,M}$ satisfies exactly the same kind of recurrence relation as in Lemma 8.21, using nearly the same argument as before.

One can give a general lower bound on the size of the restrained visibility in the same spirit as Proposition 4.18 in Section 4.14. For this we need to make some technical adjustments in the definitions to accommodate the requirement that the paths be simple. Let us call an ordered pair of vertices (u, w) in G a *special focal pair* if there exist *simple* (nondegenerate) oriented paths α and β which go from u to w and which arrive at w along *different* edges. We call a sequence $\{(u_1, w_1)\}_{i=1}^{n}$ of special focal pairs of vertices a *special chain* if $u_{i+1} = w_i$ when $1 \leq i \leq n-1$, and if we can choose the corresponding collections of simple paths α_i, β_i, $1 \leq i \leq n$, to have the property that no α_i or β_i intersects an α_j or β_j when $i < j$ except when $j = i+1$ and the point of intersection is the ending point of the ith path and the initial point of the $(i+1)$th path. (We do not mind if α_i and β_i intersect.) With this definition, we can get exactly the same lower bounds as in Proposition 4.18, but with special chains replacing long chains of focal pairs, and the proof is almost identical to the previous one.

What about upper bounds? We can use our earlier analysis to get upper bounds which are very similar to the ones given before. Let us state this formally as follows.

Theorem 8.31 *Let G be an oriented graph with at most k inward-pointing edges at each vertex. Fix vertices x and y in G, and suppose that ℓ is the length of the longest chain of focal pairs $\{(u_i, w_i)\}_{i=1}^{\ell}$ such that $u_1 = x$ and there is a simple oriented path in G which passes through the u_i's and w_i's, in their proper order (i.e., first u_1, then $w_1 = u_2$, etc.) If G has N vertices, then there are at most*

$$2 \cdot k^{\ell} \cdot \frac{(N-1)^{\ell}}{\ell!} \tag{8.60}$$

simple oriented paths in G which go from x to y.

Note that we are continuing to use *focal pairs* in this assertion, and not *special focal pairs*, as above. We shall return to this in a moment.

Theorem 8.31 can be proved in exactly the same manner as Theorem 8.11 in Section 8.7. One has simply to restrict oneself to paths γ, α, and β which are simple, and then the possibility of oriented cycles does not matter. (In particular, the uniqueness assertion in Claim 8.12 works when α and β are simple paths, even if G contains nontrivial oriented cycles.)

The "Corona decomposition" of the ordinary visibility $\mathcal{V}_+(v, G)$ (Section 8.4) can also be employed in the analysis of the restrained visibility. This decomposition might be infinite if there are nontrivial oriented cycles, but the point is that the vertices in the *restrained* visibility can only lie in a finite part of it. More precisely, a vertex in the restrained visibility must appear within ℓ stages of the Corona decomposition, under the assumptions of Theorem 8.31 (and with $v = x$). This is not hard to check, and it is analogous to Lemma 8.8 in the original version.

The use of chains of focal pairs along simple curves for the upper bounds in Theorem 8.31, instead of special chains, as in the lower bounds, leaves an unfortunate gap that we did not have before, for the ordinary visibility $\mathcal{V}_+(v, G)$. Special chains give rise to chains of focal pairs which lie on a simple curve, but the converse is not always true. This gap is not surprising, given the more "global" nature of the requirement that a path be simple, which often leads to difficulties of this sort.

A related point is that Lemma 8.4 in Section 8.2 does not work in the same way as before. Given a vertex s in the restrained visibility $\mathcal{V}_+^r(v, G)$, let $F(s)$ be as defined in Lemma 8.1, and let us write $F^r(s)$ for the "restrained" version of $F(s)$, i.e., the part of $\mathcal{V}_+^r(v, G)$ which comes after s (including s itself). Equivalently, one can think of $F^r(s)$ as being the intersection of the ordinary $F(s)$ with the restrained visibility $\mathcal{V}_+^r(v, G)$, viewed as a subgraph of $\mathcal{V}_+(v, G)$.

For the ordinary visibility, we know from Lemma 8.4 that $F(s)$ is isomorphic to the ordinary visibility $\mathcal{V}_+(\pi(s), G)$, where $\pi : \mathcal{V}_+(v, G) \to G$ is the usual projection. We might hope that $F^r(s)$ would be isomorphic to the restrained visibility $\mathcal{V}_+^r(\pi(s), G)$, but this is not always true. The vertices in $\mathcal{V}_+^r(\pi(s), G)$ will represent simple paths in G, but these paths may intersect the path represented by s before its endpoint, and that would disrupt the earlier construction. (There is a natural mapping from $F^r(s)$ into $\mathcal{V}_+^r(\pi(s), G)$, as before, but not the other way around.)

We can get a result like Lemma 8.4 in the following manner. Let s be a vertex in $\mathcal{V}_+^r(v, G)$, so that it represents a simple oriented path in G. Let G_s denote the graph obtained from G by deleting all of the vertices through which the path in G represented by s passes, except the final vertex, and by removing all edges which meet one of these deleted vertices. Then $F^r(s)$ is isomorphic to the restrained visibility $\mathcal{V}_+^r(\pi(s), G_s)$, in the same way as in Lemma 8.4.

This version of Lemma 8.4 has weaker consequences for the "symmetry" of the restrained visibility than we had before. For the ordinary visibility, we know that if two vertices s, s' in $\mathcal{V}_+(v, G)$ satisfy $\pi(s) = \pi(s')$ — i.e., if s and s' represent paths in G with the same endpoint — then $F(s)$ and $F(s')$ are isomorphic to each other, because they are each isomorphic to $\mathcal{V}_+(\pi(s), G)$. For the restrained visibility we do not have this anymore. The subgraph $F^r(s)$ depends on the "past" of the path in G represented by s, and not just its endpoint.

The existence of isomorphisms between the $F(s)$'s provides a substantial degree of symmetry for the ordinary visibility, since we know that if the visibility $\mathcal{V}_+(v, G)$ is very large compared to the underlying graph G, then there have to be many vertices s in $\mathcal{V}_+(v, G)$ which are projected by π to the same vertex in G. Thus $\mathcal{V}_+(v, G)$ will necessarily contain many disjoint subtrees inside which are isomorphic to each other, because of Lemma 8.4. We are losing this for the restrained visibility.

We still have a kind of symmetry in the language of the story as a whole, however, in the sense that $F^r(s)$ is always isomorphic to the restrained visibility of *some* graph G_s, even if this graph will normally depend on s in a significant way. This allows for a kind of symmetry in "reasoning", in making proofs about restrained visibility graphs. If we know how to make a construction for restrained visibility graphs, then we can automatically make a similar construction for the subgraphs of the form $F^r(s)$. (Compare with the definition of the Corona decomposition in Section 8.4.)

The comparative difficulty involved in working with simple paths instead of arbitrary paths is a standard theme in computational complexity, and it will come up again in Chapter 9.

8.10 Graphs with cycles

Let G be an oriented graph, and let v be a vertex in G. Imagine that the visibility $\mathcal{V}_+(v, G)$ is infinite, so that G contains nontrivial oriented cycles that can be reached by oriented paths starting from v.

In this section, we would like to describe a different way of looking at the visibility in this case, and a different way to make use of the Corona decomposition. We begin by defining an equivalence relation on the set of vertices of G, by saying that two vertices z, w are equivalent when there are oriented paths in G which go from z to w and vice-versa. This is the same as saying that z and w are equivalent when they both lie on a single oriented cycle. It is easy to see that this does indeed define an equivalence relation on the vertices if G.

Of course, one can often expect that many vertices in G are not equivalent to any other vertex.

Let us now use this equivalence relation to define a oriented graph G_c which is a kind of "quotient" of G. For the vertices of G_c, we simply take the equivalence classes of vertices from G, using the equivalence relation that we just defined. For the edges in G_c, we use some of the edges that we had before, but not all of them. Specifically, we throw away all edges in G whose endpoints lie in the same equivalence class. If an edge in G connects vertices in different equivalence classes, then we keep the edge for G_c, and we simply view it as an edge between the two vertices in G_c which represent the equivalence classes containing the original vertices in G. This defines G_c as an oriented graph, using the same orientations that we had in G.

Lemma 8.32 *If G_c is as defined above, then G_c does not contain any nontrivial oriented cycles.*

To be more precise, we have the following.

Lemma 8.33 *Let G and G_c be as above. Suppose that we are given vertices x and y in G, and let ξ and η denote the equivalence classes which contain x and y,* *respectively, so that ξ and η are themselves vertices in G_c. If there is an oriented path in G_c that goes from ξ to η, then there is an oriented path in G that goes from x to y.*

The proof of Lemma 8.33 is straightforward, and we omit it. It is easy to derive Lemma 8.32 from Lemma 8.33, using the definition of G_c.

In the language of Section 10.1, there is an obvious "weak mapping" from G onto G_c, in which we collapse the equivalence classes and throw away some of the edges, in the manner described above. The main point is that we never throw away an edge in G in the mapping to G_c unless its endpoints in G correspond to the same vertex in G_c. For this reason, every oriented path in G can be mapped down to an oriented path in G_c (perhaps with shorter length). Lemma 8.33 provides a kind of converse to this, to the effect that paths down in G_c can always be lifted to G.

If we take an oriented path γ in G and project it down to a path γ' in G_c, then we can think of γ' as an extraction of the non-cycling parts of γ from the cycling parts. This is similar to what we did before, in Chapter 5, but less precise. An advantage of the projection into G_c is that it is a single recipe which applies at once to *all* paths in G, and in particular which can be used in counting the number of non-cycling parts of paths in G. This is related to the comments at the end of Section 5.4, and to Remark 5.11.

One can think of G as a collection of "completely cyclic" subgraphs, which are glued together in a way that is described by the projection from G onto G_c. By a "completely cyclic" graph we mean an oriented graph in which each pair of vertices in the graph is contained in an oriented cycle. If ξ is one of the equivalence classes of vertices in G (with respect to the equivalence relation that we defined before), then the subgraph of G consisting of the vertices in ξ and all edges with endpoints in ξ is completely cyclic, and it is maximal among completely cyclic subgraphs of G.

Completely cyclic graphs fall into two categories, the ones which consist of only a single oriented loop, and the ones which contain at least two nontrivial distinct oriented loops. In the first case the visibility is always just an infinite ray (when the loop is nontrivial), while in the second case one gets an exponentially-growing infinite tree with strong periodicity properties. Our original graph G is a kind of amalgamation of the completely-cyclic subgraphs mentioned in the previous paragraph, and this leads to mixture of patterns at the level of the visibility. The projection from G to G_c induces a "weak mapping" between the corresponding visibility graphs, as in Section 10.1, and this induced mapping reflects the way that the visibilities of the completely cyclic subgraphs of G are connected to each other inside the visibility of G.

9

SOME RELATED COMPUTATIONAL QUESTIONS

In this chapter, we look at some computational questions related to graphs and their visibilities. Additional questions come up in the next few chapters.

As in Section 8.6, our previous treatment of optical graphs can be extended to oriented graphs in general, with only modest changes. From now on we work in the general setting.

9.1 The size of the visibility

Let G be an oriented graph, and let $\mathcal{V}_+(v, G)$ be the corresponding visibility at some fixed vertex v. Assume for the moment that $\mathcal{V}_+(v, G)$ is finite, which is the same as saying that there is no oriented cycle in G which is accessible from v by an oriented path, as in Lemma 4.8 in Section 4.7.

Let f denote the function on vertices of G which counts the number of their preimages in $\mathcal{V}_+(v, G)$ under the canonical projection $\pi : \mathcal{V}_+(v, G) \to G$ (defined in Section 4.2). This is the same as saying that $f(u)$ counts the number of distinct oriented paths in G from v to u. We call this the *counting function on G for the visibility* $\mathcal{V}_+(v, G)$. Of course, f depends on v, even if we suppress this from the notation.

This function enjoys the following property. Given vertices p and q in G, we say that p *precedes* q if there is an edge in G which goes from p to q (with respect to the orientation), and we say that q is a *successor* of p in this case. Then

$$f(u) = \sum \{f(w) : w \text{ is a vertex in } G \text{ which precedes } u\} \tag{9.1}$$

for all vertices u in G, except for v, for which we have $f(v) = 1$. This is not hard to see. In this formula, one should be careful to take multiplicities into account; if there are k edges in G that go from w to u, then $f(w)$ should be counted k times in the sum.

There is a simple way to compute this function, but before we get to that, let us mention some preliminary matters. For this we allow G to have nontrivial oriented cycles which are accessible from v.

Define a sequence of subsets $\{Z_j\}_{j=0}^{\infty}$ of the set of vertices in G as follows. Set $Z_0 = \{v\}$, and, for $j \geq 1$, define Z_j recursively as the set of vertices in G which have at least one predecessor in Z_{j-1}. This is equivalent to taking Z_j to be the set of vertices in G at the end of an oriented path in G which starts at v and has length j, as one can verify.

Set $Z_+ = \bigcup_{j\geq 0} Z_j$. This is the same as the set of vertices in G which can be reached by an oriented path in G which begins at v.

Lemma 9.1 *If G has at most n vertices, then $Z_+ = \bigcup_{j=0}^{n-1} Z_j$. In this case, $Z_j \neq \emptyset$ for some $j \geq n$ if and only if there is a nontrivial oriented cycle in G which is accessible from v, in which case $Z_j \neq \emptyset$ for all j.*

Proof If u is a vertex in G which can be reached by an oriented path that starts at v, then u can be reached by an oriented path from v that does not cross itself. This can be obtained by removing cycles, as needed (without affecting the initial and final endpoints of the path). A path which does not go through the same vertex twice has length equal to 1 less than the number of vertices that it does cross, and length less than n in particular. Thus every vertex in G which can be reached by an oriented path starting at v can be reached by such a path with length less than n, so that $Z_+ = \bigcup_{j=0}^{n-1} Z_j$.

If there is an oriented path in G which begins at v and has length greater than or equal to n, then that path must cross itself, and hence go around a nontrivial oriented cycle. Conversely, if there is an oriented path in G which begins at v and goes around a nontrivial oriented cycle, then one can get paths of any length, by going around the cycle repeatedly to get longer paths (and taking subpaths to get shorter paths). This gives the second part of the lemma. □

Let Z_- be the set of vertices in G which do not lie in Z_+. To decide whether a vertex u in G lies in Z_+ or Z_- is the same as the "reachability problem", starting from v; i.e., is there an oriented path in G which goes from v to u. There are simple algorithms for determining this in polynomial time (as in [Pap94]), and the equality of Z_+ with the union of the Z_j's, $0 \leq j < n$, gives one way to do this, since it is easy to generate the Z_j's.

Similarly, one can decide whether there is a nontrivial oriented cycle in G which is accessible from v in polynomial time, by checking if $Z_n \neq \emptyset$. (One can also determine whether there are nontrivial oriented cycles in G at all, or whether there is one passing through a given vertex.)

Now let us return to the question of computing f. We assume that there are no nontrivial oriented cycles in G which are accessible from v, so that f is finite everywhere.

By definition, $f(v) = 1$, and $f(u) = 0$ when u lies in Z_-. To compute the values of f at the remaining vertices in G, one can use the same kind of argument as in the proof of Lemma 7.2 in Section 7.1. To reduce to that setting in a more precise way, one can remove from G the vertices in Z_-, and the edges which have at least one of their endpoints in Z_-. This will not affect the computation of f at the vertices in Z_+, and the resulting graph will not have any nontrivial oriented cycles at all. The vertex v will be the only input vertex (with no incoming edges), and the other vertices can be accessed by oriented paths starting at v.

In computing f, binary representations of positive integers should be used (or some other base larger than 1), rather than unary representations. The values of f could be exponentially large in general, but no more than that, as we have seen, and this gives a suitable bound for the sizes of the binary representations of the values. The complexity of the sums involved in the computation are thus suitably limited as well. (Compare with Section 7.3.)

In this manner, f can be computed in polynomial time. This is not hard to see. One could also compute f through the adjacency matrix for the graph G, as in Section 12.2. For this the complexity would again be polynomial.

What about focal pairs and chains of focal pairs?

Suppose u and w are vertices in G. Then (u, w) is a focal pair in G if and only if there are vertices w_1, w_2 and edges e_1, e_2 in G such that e_i goes from w_i to w, $i = 1, 2$, $e_1 \neq e_2$, and there are oriented paths from u to w_1 and w_2. (See Definition 4.17 in Section 4.14. Note that w_1 and w_2 may be equal, in which case one can use the same path from u to $w_1 = w_2$.) The problem of deciding whether there are oriented paths in G from u to w_1 and w_2 is the same as two instances of the reachability problem. In particular, it can be solved in polynomial time, and hence one can decide whether (u, w) is a focal pair in polynomial time.

As in Section 8.5, we can define the "derived" graph G' by taking the vertices of G' to be the same as those for G, and joining two such vertices u and w by an (oriented) edge from u to w exactly when (u, w) is a focal pair. In view of the preceding observation, G' can be computed from G in polynomial time.

A chain of focal pairs in G which begins at v is essentially the same as an oriented path in the derived graph G' which begins at v. If one wants to know about the possible lengths of chains of focal pairs in G, starting at v, then one can look at lengths of oriented paths in G', starting at v. This can be treated using sets Z'_j associated to G' which are defined in the same ways as the sets Z_j were before (associated to G). That is, there is an oriented path in G' which begins at v and has length j exactly when $Z'_j \neq \emptyset$.

More precisely, it may be that there are chains of focal pairs of arbitrary length, starting at v, because of the presence of an oriented cycle in G' which is accessible from v. This can be determined in polynomial time, and, if it does not happen, then one can determine the length of the longest chain of focal pairs in G starting at v, which is always smaller than the number of vertices in $G' = G$. These assertions use the analogue of Lemma 9.1 for G' instead of G.

From our earlier work in Section 4.14 and Chapter 8, we know that the length of the longest chain of focal pairs in G, starting at v, can be related to the approximate size of the visibility graph. On the other hand, the actual size of the visibility can be computed in polynomial time, as well as the whole function f on the vertices of G.

However, questions about focal pairs and chains of focal pairs have some additional features. A basic point in this direction is that the reachability problem can be solved in nondeterministic logarithmic space. See p49 of [Pap94]. In other words, the reachability problem lies in the class NL. In fact, it is complete for NL, as on p398 of [Pap94]. By Savitch's theorem, the reachability problem can be solved deterministically in $(\log n)^2$ space. See p149 of [Pap94]. The question of whether a given ordered pair (u, w) defines a focal pair can be solved in a simple way in terms of the reachability problem, as indicated above, and this leads to a solution for it in nondeterministic logarithmic space as well. Specifically, to check if a given ordered pair (u, w) is a focal pair, one can "guess" a pair of vertices w_1, w_2 (in the customary sense of nondeterministic algorithms), check if they have edges going from them to w (which are distinct if $w_1 = w_2$), and then use the algorithm for reachability to check if there are oriented paths from u to each of w_1 and w_2. The latter involves successive "guessing" of adjacent vertices, to go from one vertex in G to another. (See p49 of [Pap94].)

One can look at chains of focal pairs in a similar manner. In order to decide if there is a chain of focal pairs of a given length m starting at a given vertex v, one can "guess" a first vertex v_1 such that (v, v_1) might be a focal pair, and check this using the method based on reachability. If this is a focal pair, then one can "guess" a second vertex v_3, and check if (v_1, v_2) is a focal pair, etc. Each time that one gets a focal pair, one should increase a counter by 1, and check if the length m has been reached. The various vertices being guessed can be presented in logarithmic space, through a standard coding of the original graph G. For this question about chains of focal pairs, one need not be concerned with values of m which are greater than the number of vertices in the graph; if there is a chain with length equal to the number of vertices in the graph, then a vertex is repeated in the chain, and there are chains of arbitrary length.

Analogous considerations apply to questions about cycles. The existence of a nontrivial oriented cycle in G which can be accessed from a vertex v is equivalent to the existence of a vertex u in G such that there is an oriented path from v to u, and an oriented path from u to a predecessor of u. This lends itself to nondeterminism again, and the use of the nondeterministic method for the reachability problem. Similarly, the existence of a pair of distinct oriented loops in G which are accessible from v, as in Section 5.2, can be reformulated in terms of the existence of a vertex w in G such that there is an oriented path in G from v to w, and (w, w) is a focal pair. (This was indicated shortly after the statement of Proposition 5.1 in Section 5.2.) This can also be treated through nondeterminism and the reachability problem.

In a different direction, note that many basic questions about paths and cycles become more difficult if one requires that the paths and cycles be simple. This is not an issue for the reachability problem, because the existence of an oriented path from one vertex to another is equivalent to the existence of a simple oriented path between the same vertices, i.e., cycles can be removed from a path without affecting its endpoints. For many other questions, one has NP-completeness. See [GJ79, Pap94].

If an oriented graph G has no nontrivial oriented cycles, then oriented paths are automatically simple, and the distinction does not really arise. Also, if one makes assumptions about loops not meeting, as in Section 5.3, then numerous questions about them become easier.

9.2 The visibility recognition problem

Let (T, b) be a finite rooted tree. Recall that a *tree* is a connected graph in which there are no nontrivial unoriented paths which begin and end at the same vertex, but otherwise do not go through any vertex more than once. A *rooted* tree is a tree with a distinguished vertex (in this case b).

The *visibility recognition problem* asks the following question: Given a finite rooted tree (T, b) and a positive integer k, can we realize T as the visibility of an oriented graph G with size $\leq k$? In other words, when is there an oriented graph G of size $\leq k$ and a vertex v in G so that T is isomorphic to the visibility $\mathcal{V}_+(v, G)$, by a mapping which takes the root b of T to the basepoint of $\mathcal{V}_+(v, G)$?

If k is at least as big as the size of T, then the answer to this question is automatically "yes", because one can take $G = T$ (with the orientation described in Remark 9.4 below) and $v = b$. In general, one might be able to choose G to be much smaller, depending on the possible symmetry in T.

Proposition 9.2 *The visibility recognition problem can be solved in polynomial time (as a function of the size of T). In fact, there is a polynomial-time algorithm for finding an oriented graph M and a vertex b_0 in M such that T is isomorphic to $\mathcal{V}_+(b_0, M)$ by an isomorphism which takes b to b_0, and so that M is as small as possible.*

The minimality property of M, b_0 can be strengthened as follows. If G is any other oriented graph which contains a vertex v such that T is isomorphic to the visibility $\mathcal{V}_+(v, G)$ through an isomorphism which takes the root b to the basepoint of $\mathcal{V}_+(v, G)$, then there is a mapping from a subgraph of G that contains v onto M, and which sends v to b_0. If G has the same size as M, then this mapping is an isomorphism from G onto M.

We shall not discuss mappings between graphs in a systematic way until Chapter 10, and indeed we shall postpone parts of the proof of this proposition to Sections 9.4 and 10.7. In this section, we shall explain the basic construction for M, and why it can be done in polynomial time.

There is an "implicit" version of the visibility recognition problem which is more delicate. See Section 9.3.

Remark 9.3 The basic constructions and definitions given in this section do not rely on the *finiteness* of T, and they will also be used in the context of infinite trees in Chapter 10. For instance, we need T to be finite in order to speak about polynomial-time complexity, but not for the actual definition of M. If T is allowed to be infinite, then M could be infinite as well. Typically we shall restrict ourselves to situations where M is finite, although this is not necessary for much of what we do either.

Remark 9.4 (Orientations for rooted trees) If (T, b) is a rooted tree, then there is a special orientation that we can give to T, in which everything is oriented *away* from the root b. We can define this orientation precisely as follows.

A path α in T is simple if it does not pass through any vertex more than once. For each vertex v in T, there is a simple path that starts at b and ends at v. One can start with any path from b to v (whose existence is guaranteed by the assumption that T be connected), and remove cycles repeatedly, as needed, until there are none left. This gives a simple path. Because T is a tree, a simple path from b to v (or between any two vertices) is also *unique*. This is not hard to check.

There is a unique (and consistent) way to orient the edges of T so that the simple paths in T which begin at b always follow a positive orientation. This is not hard to show, using the uniqueness of simple paths from b to any given vertex. With this orientation, b has no incoming edges, and every other vertex has exactly one incoming edge, coming from the unique simple path in T from b to the given vertex. In particular, b is determined by the orientation.

A visibility graph $\mathcal{V}_+(v, G)$ automatically comes with a natural root, namely, the basepoint of $\mathcal{V}_+(v, G)$ (i.e., the vertex in $\mathcal{V}_+(v, G)$ which represents the degenerate path in G that contains v, and does not pass through any edges). The standard orientation for the visibility (defined in Section 4.2) is the same as the one determined by this choice of root through the construction above. This is not hard to verify, using the fact that the "canonical liftings" of oriented paths in G which begin at v to paths in the visibility which begin at the basepoint are always oriented and simple. (See Section 4.6.)

From now on, when we are speaking about a rooted tree (T, b), we shall automatically treat it as being oriented in the manner discussed above. Sometimes it will be more convenient to think in terms of rooted trees, and sometimes in terms of oriented trees. In practice, it will not really matter which we choose, because the orientation and the root will each determine the other. In particular, when we speak of two trees as being "isomorphic", we shall normally intend that the isomorphism preserve roots or orientations, and it will not matter which, because each will imply the other.

Let us now describe the main construction behind Proposition 9.2. Let (T, b) be an arbitrary rooted tree, which we allow to be infinite for the general construction.

The idea is to try to "fold" T as much as possible. To do this, we begin with the following. If s is a vertex in T, then we take $F(s)$ to be as in Lemma 8.1, i.e., the subtree of T which comes *after* s, which means the part of T (vertices and edges) which can be reached by an oriented path which begins at s. Thus $F(s)$ includes s, and can be viewed as a rooted tree in its own right, with root s.

Notice that if s and t are vertices in T and t lies in $F(s)$, then $F(t)$ is contained within $F(s)$.

Let us define an equivalence relation on the set of vertices of T, by saying that two vertices s and s' are equivalent if $F(s)$ and $F(s')$ are isomorphic as rooted trees. This is clearly an equivalence relation, and so we may speak about the set of equivalence classes of vertices of T.

In fact, we use this set of equivalence classes for the set of vertices in M. Given two of these equivalence classes u and w, we need to decide whether there should be an oriented edge in M which goes from u to w, and if so, how many. For this we simply look at what happened back in T. Let s be a vertex in T which lies in the equivalence class of vertices determined by u. Let $A(s)$ denote the set of vertices z in T for which there is an edge going from s to z. Thus $A(s)$ lies within $F(s)$, and consists of the "first" vertices in $F(s)$ which come *after* s. We attach no edges from u to w in M if there are no elements of $A(s)$ which lie in the equivalence class determined by w. If there are exactly k elements of $A(s)$ which lie in the equivalence class determined by w, then we put exactly k edges in M that go from u to w.

Claim 9.5 *This number k does not depend on the representative s of the equivalence class determined by u.*

Indeed, if s' is any other such representative, then $F(s)$ is isomorphic to $F(s')$ (as a rooted tree), and this induces a one-to-one correspondence between the sets $A(s)$ and $A(s')$. This correspondence also respects the equivalence relation that we have defined on vertices of T, since $F(z)$ is contained in $F(s)$ when z lies in $A(s)$ (or when z lies in $F(s)$, for that matter); for if z lies in $A(s)$ and z' is its counterpart in $A(s')$ under the isomorphism between $F(s)$ and $F(s')$, then we also get an isomorphism between $F(z)$ and $F(z')$, simply by restricting the one between $F(s)$ and $F(s')$ to $F(z)$. This implies that $A(s)$ and $A(s')$ must contain the same number of elements which lie in the equivalence class determined by w, which is what we wanted.

This finishes the description of M as an oriented graph (which may be infinite when T is). For the special vertex b_0 in M, we take the equivalence class of vertices in T which contains b. Note that b cannot be equivalent to any of the other vertices in T when T is finite, because $F(b)$ will be strictly larger in size than $F(s)$ for any other vertex s in T. When T is infinite, we can easily have that $F(b)$ is isomorphic to $F(s)$ for other vertices s in T.

Definition 9.6 (Minimal representation) *The pair M, b_0 (constructed as above) is called the* minimal representation *for the rooted tree (T, b).*

Remark 9.7 If (T', b') is another rooted tree which is isomorphic to (T, b), and if M', b'_0 is the minimal representation for (T', b'), then M', b'_0 is necessarily isomorphic to M, b_0. This is easy to see from the construction, which did not use anything special about the representation of (T, b). We do not quite get a canonical isomorphism between M and M', however. The isomorphism between T and T' does lead to a canonical correspondence between the vertices of M and M', but for the edges there can be some freedom. Indeed, while the *number* of edges going between a particular pair of vertices in M is determined by the structure of (T, b) in a natural way, in general there is no particular labelling of these edges, or other method for deciding how these edges should be matched with the ones going between the corresponding vertices in M'.

Lemma 9.8 *If (T, b) is a rooted tree and M, b_0 is its minimal representation, then T is isomorphic to the visibility $\mathcal{V}_+(b_0, M)$ by a mapping which sends the basepoint b of T to the usual basepoint of the visibility $\mathcal{V}_+(b_0, M)$.*

This is not hard to check directly, but the task will be more pleasant after we have developed some machinery about mappings between graphs. See Remark 10.33 in Section 10.7 for more details.

The fact that M is minimal in terms of *size* will be established in Corollary 9.25 in Section 9.4. The stronger minimality properties of M described in the second part of Proposition 9.2 will be established in Lemma 10.34 and Remark 10.36 in Section 10.7. The remaining point in the proof of Proposition 9.2 is the following.

Lemma 9.9 *If (T, b) is a finite rooted tree, then the minimal representation M, b_0 can be obtained from (T, b) in polynomial time.*

Proof We simply have to go back and examine the various steps of the construction and the way in which they can be made effective.

To each vertex v in T, let us associate a number j, which is the length of the shortest path in T which goes from b to v. This number can be computed recursively with no trouble: at b the number is 0, at the vertices adjacent to b the number is 1, at the vertices adjacent to those but not considered yet the number is 2, etc.

This same recursive process can be used to derive the orientation for T defined in Remark 9.4. That is, as we proceed from a vertex at level j to an adjacent one at level $j+1$, we orient the corresponding edge so that it goes *from* the vertex at level j *to* the one at level $j+1$.

A similar process can be used to construct the subtrees $F(s)$ defined above. That is, once we reach a vertex s, we simply keep track of all the vertices and edges which follow s in T. Again it is easy to do this recursively.

The remaining point is to be able to decide when two vertices s, t in T are equivalent to each other according to the equivalence relation defined before. In other words, one should decide when $F(s)$ and $F(t)$ are isomorphic to each other as rooted trees. In fact, one can decide when two planar graphs are isomorphic in linear time ([HW74]; see also p285 of [GJ79]). Of course trees are always planar, and in order to limit oneself to isomorphisms which preserve the roots of the trees, one can add some number of vertices to the graphs, and edges between these vertices and the roots (and no other edges), so that the roots become distinguished from the other vertices. (The problem of deciding when $F(s)$ and $F(t)$ are isomorphic to each other can also be treated through the method of Proposition 9.11 in Section 9.4. This is discussed in Remark 9.16.)

Using these observations, it is easy to see that the construction of the minimal representation M, b_0 can be achieved through a polynomial-time algorithm. (For some related aspects, Section 9.3.) □

This completes the proof of Proposition 9.2, except for the points which will be treated in Sections 9.4 and 10.7.

Let us record one more simple property of minimal representations, for future reference.

Lemma 9.10 *If M, b_0 is the minimal representation of some rooted tree (T, b), then every vertex and edge in M can be reached by an oriented path which begins at b_0.*

Proof This is easy to derive from the construction, using the fact that everything in T can be reached by an oriented path beginning at b. (This also involves the way that we defined our orientation for (T, b), as in Remark 9.4.) □

9.3 An implicit version

Consider the following "implicit" version of the visibility recognition problem. Suppose that we are given an oriented graph H, a vertex w in H, and a positive integer k. When is it true that there is an oriented graph G of size at most k and a vertex v in G so that the visibilities $\mathcal{V}_+(w, H)$ and $\mathcal{V}_+(v, G)$ are isomorphic? (By "isomorphic" we mean as rooted trees, which is equivalent to being isomorphic as oriented graphs in this case, as indicated at the end of Remark 9.4.)

This question contains the earlier "explicit" version as a special case, by taking H, w to be T, b. In this case, T and H would have the same size, but in general T could be exponentially-larger in size than H, or even infinite (when H contains oriented cycles).

It is not clear whether this implicit version of the visibility recognition problem can also be solved in polynomial time, however. This does turn out to be true if we restrict ourselves to graphs H which are free of nontrivial oriented cycles. For such graphs, one can produce the minimal representation M, b_0 for the visibility graph in polynomial time, as in Proposition 9.17 in Section 9.4.

In general, when H is allowed to contain nontrivial oriented cycles, the implicit version of the visibility recognition problem lies in the class NP. This will be discussed in Section 10.7, especially Corollary 10.39.

For the original "explicit" version of the problem (as stated in Section 9.2), the existence of a polynomial-time solution is clearly not the final word. One can expect to be able to do better than that.

A basic point is that while a rooted tree (T, b) might be exponentially-larger in size than its minimal representation M, b_0, the *depth* of T is controlled by the depth of M, and hence by the size of M (unless T is infinite and M contains nontrivial oriented cycles). (Compare with Section 4.4.) This suggests the use of *parallelism* in the analysis of the explicit version of the visibility recognition problem, as in Section 15.2 of [Pap94] (beginning on p369).

One can also look at the matter in terms of nondeterminism and bounds on space. The effective witness that we have for the visibility recognition problem (in Lemma 10.37) involves the existence of a mapping from the tree T into a graph G of size at most k, a mapping which satisfies a certain property (of being a "local $+$-isomorphism") that is completely local. If k is much smaller than the size of T (which is likely to be the most interesting situation), then the nondeterminism required for this problem would be quite modest in terms of *space*. One could "guess" G right from the start, but to build a suitable mapping f from T to G, one could make individual local "guesses", which would be nearly independent of each other. That is, one would first "guess" where in G to map the root of T, and then one would systematically proceed down the branches of the tree, "guessing" at each stage how to map the next edges and vertices in T into G. At each moment, the number of new "guesses" needed would be controlled by the local geometry of G, and by k in particular, but one would not care about what took place in earlier stages of the construction, nor what happens along other branches of T. One would only need to know where one is in T and G, but *not* how one got there.

Neither of these considerations of nondeterminism or parallelism would work so well in the case where (T, b) is given only implicitly, as the visibility of some graph H, even if H is free of nontrivial oriented cycles. For the implicit version, there is not so much reason to emphasize the situation where k is much smaller than the size of H; one could easily be interested in situations where H and k are of roughly the same size, and where the visibility of H is much larger, perhaps exponentially larger. If k is much smaller than H, then it is another matter, and one might find some benefit in parallelism, or space bounds.

For the "guessing" of effective witnesses, there is another problem, which is that choices for values of the mapping at one part of the graph no longer enjoy the same kind of independence from choices made at other parts of the graph as in the case of trees, even when there are no nontrivial oriented cycles.

These issues of parallelism and controlled space for the visibility recognition problem have natural analogues in the more basic context of evaluating Boolean circuits. The general form of the latter problem is P-complete, as in Theorem 8.1 on p168 of [Pap94]. This means that one should probably not expect to be able to do better than polynomial-time computability for this problem, in terms of parallelism or controlled space, for instance. For Boolean expressions and circuits which are trees (which are practically the same thing), one can do better, as discussed on p386 and p396 of [Pap94].

9.4 The visibility isomorphism problem

Let G and H be two oriented graphs, let v be a vertex in G, and let w be a vertex in H. The *visibility isomorphism problem* is to decide if the visibilities

$$\mathcal{V}_+(v, G) \quad \text{and} \quad \mathcal{V}_+(w, H) \tag{9.2}$$

are isomorphic to each other as rooted trees (using the basepoints as roots). This is the same as asking whether they are isomorphic as oriented trees, as in Remark 9.4.

This problem turns out to lie in the class NP, as we shall see in Section 10.9. For the moment, we want to prove the following.

Proposition 9.11 *The visibility isomorphism problem can be decided in polynomial time if we restrict ourselves to oriented graphs G and H which do not contain nontrivial oriented cycles.*

Proof Without loss of generality, we may assume that $G = H$. (Otherwise, we can simply replace G and H both with their disjoint union.) The vertices v and w can still be different, though.

To prove the proposition, we shall to give a recursive test for deciding when the visibilities of G at two different vertices are isomorphic. We begin by setting some notation.

If u is any vertex in G, let $O(u)$ denote the collection of edges in G which are attached to u and which are oriented *away* from u. If $e \in O(u)$, let us write $\nu(e)$ for the vertex in G into which e flows. Note that $\nu(e)$ cannot be u, since G contains no nontrivial oriented cycles, by assumption.

Fix a vertex u for the moment, and let us define a graph $V(u)$ associated to u as follows. Let $\mathcal{U}$ denote the disjoint union of the visibilities $\mathcal{V}_+(\nu(e), G)$ over $e \in O(u)$. Note that a single vertex v may arise as $\nu(e)$ for several edges $e \in O(u)$, in which case $\mathcal{U}$ will contain as many copies of the visibility $\mathcal{V}_+(v, G)$ as there are edges $e \in O(u)$ with $\nu(e) = v$. To get $V(u)$, we take $\mathcal{U}$ and add to it a new vertex b_u and several edges flowing out of b_u, exactly one edge from b_u to the basepoint of each visibility graph $\mathcal{V}_+(\nu(e), G)$ in $\mathcal{U}$. It is easy to see that $V(u)$ is a *tree*, since the individual visibilities in $\mathcal{U}$ are, and we can view it as a rooted tree, with root b_u.

If $O(u)$ happens to be empty, then so is $\mathcal{U}$, and $V(u)$ consists only of the one vertex b_u (and no edges).

Claim 9.12 *$V(u)$, b_u is isomorphic (as a rooted tree) to $\mathcal{V}_+(u, G)$.*

This is not hard to check, just using the definition of the visibility. That is, each nondegenerate oriented path in G which begins at u must cross exactly one of the edges e in $O(u)$ after leaving u, and the continuation of the path after e corresponds to a vertex in the visibility graph $\mathcal{V}_+(\nu(e), G)$. One can also see Claim 9.12 as a special case of the "propagator rule" (Lemma 8.1 in Section 8.1), with $\mathcal{W}$ taken to consist only of the basepoint of $\mathcal{V}_+(u, G)$. With this interpretation, the various copies of visibilities contained in $\mathcal{U}$ correspond to the subtrees $F(s)$ in Section 8.1 (through the isomorphism provided by Lemma 8.4 in Section 8.2), and the edges in $V(u)$ going from b_u to the visibilities contained in $\mathcal{U}$ correspond to the set of edges $\mathcal{E}$ in Lemma 8.1.

Claim 9.13 *Let x and y be vertices in G. Then the visibilities of G at x and at y are isomorphic to each other if and only if there is a bijection ϕ from $O(x)$ onto $O(y)$ such that the visibility of G at $\nu(e)$ is isomorphic (as a rooted tree) to the visibility of G at $\nu(\phi(e))$ for every edge $e \in O(x)$.*

This is easy to check, using Claim 9.12.

The assumption that G contain no nontrivial oriented cycles permits us to use Claim 9.13 as an effective recursive test for the isomorphic equivalence of the visibilities of G, without getting trapped into endless loops. To make this precise, we shall use the following enumeration of the vertices in G.

Claim 9.14 *We can arrange the vertices in G in a finite sequence $v_1, \ldots, v_n$ in such a way that for each j the successors of v_j are among the vertices v_i with $i < j$, and this arrangement can be found by a polynomial-time algorithm.*

Remember that the *successors* of a vertex u are simply the vertices of the form $\nu(e)$, $e \in O(u)$.

Claim 9.14 is contained in Problem 1.4.4 on p14 of [Pap94], but let us include a proof for the sake of completeness. We simply choose the v_j's recursively in the manner suggested by the statement. That is, we choose v_1 to be any vertex which has no successors, and if $v_1, \ldots, v_j$ have been chosen, then we take v_{j+1} to be any vertex in G whose successors are all among the v_i's with $i \leq j$. It is clear that this procedure will work in polynomial time if it works at all; we have only to make sure that there is never a time in which some vertices in G remain unchosen, but we cannot find one that meets our requirements.

For this we use our assumption that G contain no nontrivial oriented cycles. (We have used a similar argument before, in Lemma 7.2.) Suppose that we have chosen $v_1, \ldots, v_j$ already, and that there is at least one vertex in G which is not among the v_i's. We want to find v_{j+1}. (We allow $j = 0$ here, in which case none of the v_i's has been selected yet.) If v is any vertex in G which is not yet chosen, then either all of the successors of v are among the vertices already chosen, or they are not. If so, then we take v for v_{j+1}. If not, then we replace v by any one of its omitted successors, and repeat the process. The absence of nontrivial oriented cycles ensures that no vertex in G will be examined more than once during this procedure, and so a suitable choice for v_{j+1} has to appear in a limited amount of time.

Thus our recursive procedure for selecting the v_j's does not stop before we run out of vertices in G altogether, and Claim 9.14 follows.

Using this enumeration of the vertices of G, we can analyze the visibilities of G in the following way.

Claim 9.15 *There is a polynomial-time algorithm which assigns to each v_j a code (e.g., a binary string) such that v_i and v_j have the same code if and only if the corresponding visibilities $\mathcal{V}_+(v_i, G)$ and $\mathcal{V}_+(v_j, G)$ are isomorphic.*

To prove this, we choose this coding recursively, as follows. To v_1 we associate the code 1. Suppose now that codes have been assigned to $v_1, \ldots, v_j$, in such a way that two vertices have the same code exactly when the corresponding vertices are isomorphic. To assign a code to v_{j+1}, we need to determine whether its visibility is isomorphic (as a rooted tree) to any of those that came before.

Let Γ_{j+1} denote the collection of codes assigned to the vertices $\nu(e)$, $e \in O(v_{j+1})$, counting multiplicities. (Thus, for each code that appears, we count the number of edges e in $O(v_{j+1})$ which give rise to that particular code.) All of the vertices $\nu(e)$ for $e \in O(v_{j+1})$ lie among the v_i's with $i \leq j$, by Claim 9.14, and so they have been assigned codes already.

The key point now is that the visibility of G at v_{j+1} is isomorphic to the visibility at some v_i, $i \leq j$, if and only if the collection of codes Γ_{j+1} is the same (counting multiplicities) as the corresponding collection Γ_i associated to v_i. This is easy to check, using Claim 9.13. In other words, Γ_{j+1} and Γ_i are the same if and only if there is a one-to-one correspondence ϕ between $O(v_{j+1})$ and $O(v_i)$ of the type described in Claim 9.13.

If the visibility of G at v_{j+1} is isomorphic to the visibility at some v_i with $i \leq j$, then we assign to v_{j+1} the same code as is assigned already to such a v_i. Otherwise, we assign to v_{j+1} a new code. By repeating this process, we can assign codes to all of the v_j's, and with the property that two vertices have the same code exactly when their visibilities are isomorphic as rooted trees. It is also clear that the entire procedure can be carried out in polynomial time. This proves Claim 9.15, and Proposition 9.11 now follows easily (since all of the vertices of G are among the v_j's). □

Remark 9.16 The procedure described in the proof of Proposition 9.11 can also be used in the context of Lemma 9.9 in Section 9.2, for deciding when two vertices s, t in the rooted tree (T, b) determine isomorphic subtrees $F(s)$, $F(t)$. This is because $F(s)$ is isomorphic (as a rooted tree) to the visibility $\mathcal{V}_+(s, T)$ in this case. This is not hard to show, but one can also avoid the issue and simply work directly with the $F(s)$'s instead of the visibility graphs. (Notice that Claims 9.12 and 9.13 can be applied directly to the $F(s)$'s instead of the visibility graphs.) The rest of the proof of Proposition 9.11 then goes through in exactly the same manner as before.

Let us now use Proposition 9.11 to treat a special case of the implicit version of the visibility recognition problem (from Section 9.3).

Proposition 9.17 *Let G be an oriented graph which does not contain nontrivial oriented cycles, and let v be a vertex in G. Let (T, b) be the rooted tree which is the visibility $\mathcal{V}_+(v, G)$, equipped with its natural basepoint, and let M, b_0 be the minimal representation of (T, b) (Definition 9.6 in Section 9.2). Then the size of M is no greater than the size of G, and M, b_0 can be obtained (up to isomorphic equivalence) from G, v through a polynomial-time algorithm.*

In fact, the construction also provides a certain kind of mapping from a subgraph of G (mentioned in the next lemma) onto M. This mapping can be obtained from G in polynomial time, and represents the relationship between G and M more explicitly. This will be made precise in Lemma 10.34 and Remark 10.35 in Section 10.7, and the broader discussion of mappings between graphs in Chapter 10 is also useful for seeing how minimal representations are realized and compared with other graphs.

The following lemma provides the first step in the proof of the proposition, and we record it separately for future reference.

Lemma 9.18 *Let G be an oriented graphs, and let v be a vertex in G. Let G_0 denote the subgraph of G which consists of all vertices and edges in G which can* *be reached by an oriented path beginning at v. Then G_0 can be obtained from G through a polynomial-time algorithm, and the visibility $\mathcal{V}_+(v, G_0)$ is (canonically) isomorphic to the visibility $\mathcal{V}_+(v, G)$ by a mapping which preserves the basepoints of the visibilities.*

Proof The fact that G_0 can be extracted from G in polynomial time follows from the fact that the "reachability" problem of deciding when a particular vertex can be reached by an oriented path starting at v can be solved in polynomial time. (See [Pap94], and Section 9.1 above, especially Lemma 9.1 and the remarks thereafter.) The existence of the isomorphism between the visibilities of G and G_0 is an easy consequence of the definition of the visibility. The point is that all of the oriented paths in G which begin at v are also contained in G_0, by definition, and this is all that matters for the visibility graphs. □

Note that looking at multiple instances of the "reachability" problem in a graph G lends itself to parallelism, as on p362-3 of [Pap94]. One can also consider the determination of the subgraph G_0 as above in terms of bounds on space, using the corresponding results for the reachability problem, as on p49 and p149 of [Pap94]. In this regard, the bounds would concern the space used for computation, and not writing the answer on an output tape, just as in Section 2.5 in [Pap94]. Indeed, G_0 could have roughly the same size as G, so that the size of the output would not be too small.

Proof (Proposition 9.17) We may as well assume that every vertex and edge in G is accessible by an oriented path which begins at v, since otherwise we could replace G with the subgraph G_0 mentioned in Lemma 9.18.

Let (T, b) and M, b_0 be as in the statement of Proposition 9.17. If s is any vertex in T, let $F(s)$ denote the subtree of T which consists of everything in T "after" s, including s itself, as in Section 9.2. Remember that in Section 9.2 we defined an equivalence relation on the set of vertices of T, by saying that two vertices s and t are equivalent when $F(s)$ and $F(t)$ are isomorphic as rooted trees. Let us first reinterpret this relation directly in terms of G.

Claim 9.19 *If s is any vertex in T, then $F(s)$ is isomorphic (as a rooted tree) to the visibility $\mathcal{V}_+(\pi(s), G)$, where π denotes the canonical projection from $T = \mathcal{V}_+(v, G)$ down to G (as defined in Section 4.5).*

This is equivalent to Lemma 8.4 in Section 8.2. Notice that every vertex in G arises as $\pi(s)$ for some vertex s in $\mathcal{V}_+(v, G)$ here, because we are assuming that every vertex in G can be reached by an oriented path that begins at v.

Claim 9.20 *A pair of vertices s, t in T are equivalent (in the sense that $F(s)$ and $F(t)$ are isomorphic as rooted trees) if and only if the visibilities $\mathcal{V}_+(\pi(s), G)$ and $\mathcal{V}_+(\pi(t), G)$ are themselves isomorphic as rooted trees.*

This is an immediate consequence of the previous claim.

In the original construction in Section 9.2, we took the set of vertices in M to be the set of equivalence classes of vertices in T. Now we can do this directly at the level of G. That is, we say that two vertices in G are equivalent if the corresponding visibility graphs are isomorphic (as rooted trees), and this clearly defines an equivalence relation. From Claim 9.20 we see that there is a natural one-to-one correspondence between the equivalence classes of vertices in T (with respect to the original equivalence relation) and equivalence classes of vertices in G (with respect to this new relation). This uses also the fact that every vertex in G arises as the projection of a vertex from $T = \mathcal{V}_+(v, G)$, as mentioned above. Thus, we might as well simply *define* the set of vertices in M to be the set of equivalence classes of vertices in G with respect to this new equivalence relation. We would then take b_0 to be the equivalence class which contains v. (Note that $v = \pi(b)$, since the root b of T was chosen to be the basepoint of the visibility $T = \mathcal{V}_+(v, G)$.)

We can deal with the edges in M through the same basic recipe as in Section 9.2. Specifically, let ξ and η be two equivalence classes of vertices in G, which we now think of as vertices in M. The number of oriented edges in M going from ξ to η is determined as follows. Let x be any vertex in G which lies in the equivalence class determined by ξ, and let j denote the number of edges in G which flow out of x and end in a vertex in G which lies in the equivalence class determined by η. Then we attach exactly j edges in M which go from ξ to η.

Claim 9.21 *This number j does not depend on the specific choice of vertex x in the equivalence class ξ.*

Indeed, suppose that x' is any other vertex in G which represents ξ. This means that there is an isomorphism between the visibilities of G at x and x' (as rooted trees). This isomorphism induces a one-to-one correspondence ϕ between the sets $O(x)$ and $O(x')$ of outgoing edges at x and x', as in Claim 9.13. From Claim 9.13 we also know that if e is an element of $O(x)$ and if $\nu(e)$ denotes the vertex into which e flows (and similarly for $\nu(\phi(e))$), then the visibilities of G at $\nu(e)$ and $\nu(\phi(e))$ are isomorphic as rooted trees. This implies that the number of e's in $O(x)$ for which $\nu(e)$ lies in the equivalence class determined by η is the same as the number of edges e' in $O(x')$ with the analogous property, which is exactly what we wanted. This proves Claim 9.21.

The remaining point about this number j is that it is the same as what we would have obtained from the construction in Section 9.2, using the equivalence classes of vertices in T which correspond to ξ and η. This is not hard to check, using Claim 9.20, and also the fact that if s is any vertex in T, then the canonical projection π from $T = \mathcal{V}_+(v, G)$ into G induces a one-to-one correspondence between the edges in T which flow away from s and the edges in G which flow away from $\pi(s)$. This last assertion is the same as Lemma 4.4 in Section 4.5.

The next claim summarizes what we have accomplished so far.

Claim 9.22 *The minimal representation M, b_0 for the visibility graph $\mathcal{V}_+(v, G)$ (viewed as a rooted tree, with its usual basepoint as root) can be obtained (up to isomorphic equivalence) from G and v through the procedure described above.*

Let us mention one other small fact.

Claim 9.23 *Let ξ be an equivalence class of vertices in G (and hence a vertex in the graph M constructed above), and let x be a vertex in G which lies in this equivalence class. Then the total number of edges coming out of x in G is equal to the total number of edges in M coming out of ξ.*

This follows easily from the way that we attached edges between the equivalence classes of vertices in G. Using this we get the following.

Claim 9.24 *The size of M is no greater than the size of G.*

Indeed, there cannot be more vertices in M than in G, since the vertices of M are given by equivalence classes of vertices in G. The fact that the edges in M are no more numerous than in G can be obtained using Claim 9.23.

The last remaining point in the proof of Proposition 9.17 is that the derivation of M, b_0 from G, v can be accomplished in polynomial time. This is true because of Proposition 9.11, which ensures that one can decide which pairs of vertices in G are "equivalent" in the sense of isomorphic visibilities in polynomial time. Once one has this, it is easy to see that the rest of the construction of M, b_0 above can be obtained through a polynomial-time algorithm. This completes the proof of Proposition 9.17. □

The basic construction and analysis of the preceding proof is pretty robust, and we shall refer to it several times, in slightly different contexts. In the next corollaries, we mention a couple of particular facts that one can derive from the same basic arguments.

Corollary 9.25 *Let G be an oriented graph, with or without oriented cycles, and let v be a vertex in G. Let M, b_0 be the minimal representation of the visibility $\mathcal{V}_+(v, G)$, viewed as a rooted tree. Then M is a finite graph with no more vertices or edges than G has.*

This follows from the same argument as in the proof of Proposition 9.17. In this case the tree $T = \mathcal{V}_+(v, G)$ can be infinite, but this causes no trouble, and part of the content of Corollary 9.25 is that the minimal representation is necessarily finite in this case. (Even if $T = \mathcal{V}_+(v, G)$ is infinite, it still has the virtue of being *locally finite*, i.e., the number of edges attached to any given vertex is always finite, and controlled by G.) The main difference is that we cannot apply Proposition 9.11 at the end, to say that the construction of M can be accomplished in polynomial time.

Note that Corollary 9.25 establishes the minimality of M (for a fixed rooted tree (T, b)) in terms of size, as promised in Proposition 9.2.

Corollary 9.26 *If the visibility isomorphism problem can be solved in polynomial time for oriented graphs in general (whether or not they have nontrivial oriented cycles), then the minimal representation M, b_0 of the visibility graph $\mathcal{V}_+(v, G)$ can be obtained from G, v (up to isomorphism) in polynomial time, whether or not the oriented graph G contains nontrivial oriented cycles.*

Again, this follows from the same proof as before, with the assumption of a polynomial-time solution for the visibility isomorphism problem being used in place of Proposition 9.11.

9.5 Computations with implicit descriptions

Imagine that we are given a class of objects, and some notion of "implicit descriptions" for them. We can then consider the following pair of questions.

$$\text{How can one tell when a given object admits an implicit description of at most a given size?} \tag{9.3}$$

For this one could allow the initial object itself to be given only implicitly, so that the problem becomes one of finding a more efficient implicit description, or of switching from one kind of implicit description to another. One might also use other measurements besides size (such as number of steps, in some contexts).

$$\text{How can one tell when two implicit descriptions represent the same underlying object?} \tag{9.4}$$

The visibility recognition and isomorphism problems can be seen as special cases of these questions, in which the basic objects are rooted trees, and the "implicit descriptions" are given through visibility graphs. The questions themselves make sense much more broadly, however.

Another basic situation would be to allow implicit descriptions through arbitrary "computer programs" (Turing machines), as in Kolmogorov complexity and algorithmic information theory. In this case, there are well-known incomputability results for the sizes of minimal representations (see [LV90, Man77]), and one has little hope for being able to determine when different computer programs lead to the same object more efficiently than by simply executing the programs and comparing the results when they halt (if they do halt).

Of course it is reasonable to expect that questions like the ones above become more difficult to treat as the level of implicitness increases. The use of arbitrary Turing machines is an extreme case of this.

Feasibility graphs provide a much more restrained form of implicitness in which to consider issues like these. As in Chapter 7, one can use feasibility graphs in a variety of situations, including words over a given alphabet, numbers, or Boolean functions. The representation of rooted trees through visibility graphs can also be seen as a special case of feasibility graphs, as discussed in Section 7.10.

Remember that for feasibility graphs we always restrict ourselves to oriented graphs without nontrivial oriented cycles. (See Section 7.1, especially (7.1).) In the context of rooted trees, this amounts to using only visibilities of graphs that do not contain nontrivial oriented cycles. For this case, we have polynomial-time solutions to our questions above coming from Propositions 9.2, 9.11, and 9.17, and this includes the possibility of starting with objects which are defined only implicitly for the first question.

In general, the situation for feasibility graphs is not as simple as for rooted trees. Consider Boolean circuits, for instance. With what kind of efficiency can one decide when two Boolean circuits represent the same Boolean function, or when a given Boolean circuit of size n represents a function which can actually be represented by a circuit of size $k\,(< n)$?

It is not so easy to say how a Boolean circuit should have to look in order to represent a particular given function (which might be provided only implicitly at that). There is too much flexibility in the way that Boolean functions can be built up. Similar difficulties occur in many other situations, particularly "algebraic" ones. For instance, in a finitely-presented group, it can be very difficult to tell when two words represent the same element of the group, depending on the nature of the relations. It might even be impossible to do this algorithmically, as in the unsolvability of the word problem [Man77]. (For "automatic groups", one has normal forms for words which allow the solution of the word problem in quadratic time. See [ECH$^+$92], and also Section 17.5.)

For feasibility graphs, this problem appears already in the context of free groups, or free semigroups, i.e., words over an alphabet Σ. How, for instance, should one be able to determine when a given word admits a short description through a feasibility graph? This is clearly related to some kind of internal symmetry or patterns in the word, but it is not clear how to make this especially precise. The problem becomes worse when the original word is only given implicitly, as through a feasibility graph, which might exploit some of the internal structure of the word, at the same time that it obscures other important structure.

With rooted trees, one can be much more precise about what kind of structure is needed to have concise implicit descriptions through visibility graphs, or how these different implicit descriptions are related to each other. We have seen this in Sections 9.2 and 9.4, and we shall pursue it further in Chapter 10, through the use of mappings between graphs.

Many of the geometric tools that we have for graphs and their visibilities also apply to feasibility graphs, i.e., for making comparisons between different feasibility graphs, or for figuring out how to "fold" one kind of feasibility graph into another. We shall discuss this further in Sections 11.4 and 11.5. The problem is that in most situations the geometric tools are not "complete" in the way that they can be for graphs and their visibilities, in the sense of accounting for all of the relevant comparisons and conversions.

One can also use feasibility graphs to represent constructions which are more "implicit" than the ones described in Chapter 7, by using feasibility graphs to construct other feasibility graphs. We shall discuss this further in Chapter 16. In this case, the problems of recognition and comparison for implicit descriptions become more difficult, so that even when the objects being constructed are finite rooted trees, one does not such effective geometric tools as we have here.

For another situation where intermediate forms of implicitness arise in a natural way, see the discussion of "regular expression equivalence" problems on p503-4 of [Pap94]. In this case, different levels of implicitness are obtained by adjusting the types of operations that are allowed to be used in regular expressions, and in their representation of formal languages. (See Section 1.1 for the basic paradigm, using the operations $+$, $\cdot$, and $*$.) The equivalence problem asks when two expressions represent the same formal language (as in the question (9.4)), and the complexity class of this problem can often be analyzed rather precisely in terms of the operations allowed, as discussed on p504 of [Pap94].

Although we have focused on the questions (9.3) and (9.4) here, one should keep in mind the broader issues entailed in making computations at an implicit level, or in trying to extract information of interest in convenient or efficient ways from implicit representations. (See also the discussion of "succinct problems" beginning on p492 of [Pap94].) Similar issues arise in many other subjects as well, e.g., in looking for computable invariants in geometry, topology, and dynamical systems, or in the context of data compression. Part of the problem is simply to find good ways to make comparisons. This is a fundamental aspect of a lot of mathematics, and it is a source of numerous difficulties in many areas (such as signal processing).

10

MAPPINGS AND GRAPHS

In this chapter, we introduce basic machinery for comparison and interaction between oriented graphs and their visibilities, through *mappings* between graphs. Some of the basic definitions and lemmas will follow standard ideas about covering spaces from topology, and they will enable us to provide effective witnesses for the visibility recognition and isomorphism problems, as discussed in Sections 9.2, 9.3, and 9.4. In Chapter 11 we shall look at the way that mappings between graphs can be used to compare calculations and constructions lying below *feasibility graphs* (Section 7.1).

As in Section 8.6, we no longer restrict ourselves to optical graphs (e.g., for the definition of the visibility).

10.1 Mappings and weak mappings

In this section, we record a few basic definitions and properties of mappings between graphs. For the sake of completeness, we shall be fairly precise.

Recall that we allow graphs to have edges for which both endpoints are the same vertex, and to have multiple edges between a given pair of vertices. (See the beginning of Chapter 4.)

Definition 10.1 *Let G and H be graphs (which we allow to be infinite). By a* mapping f *between G and H, written $f : G \to H$, we mean a mapping from vertices in G to vertices in H, and a mapping from edges in G to edges in H, such that the obvious compatibility conditions are satisfied, i.e., the endpoints of the image of an edge e in G are the same as the images of the endpoints of e. We let f denote also the induced map on vertices, and the induced map on edges.*

We allow this definition to be applied to *infinite graphs*, in order to accommodate visibility graphs and trees automatically. Remember from the beginning of Chapter 4 that the word "graph" is normally interpreted to mean "finite graph", unless explicit specification is made to the contrary, as we do for this definition. In the text below, we continue to follow the convention that "graph" means "finite graph".

An isomorphism between graphs is given by a mapping which is invertible. This is equivalent to requiring that the mapping define bijections between the sets of vertices and edges of the graphs. The compatibility conditions (for the vertices and edges) for the inverse mapping will hold automatically, because of the analogous conditions for the original mapping.

It will turn out to be convenient to extend the definition of mappings between graphs in the following manner. Recall that a *partially defined* mapping from a

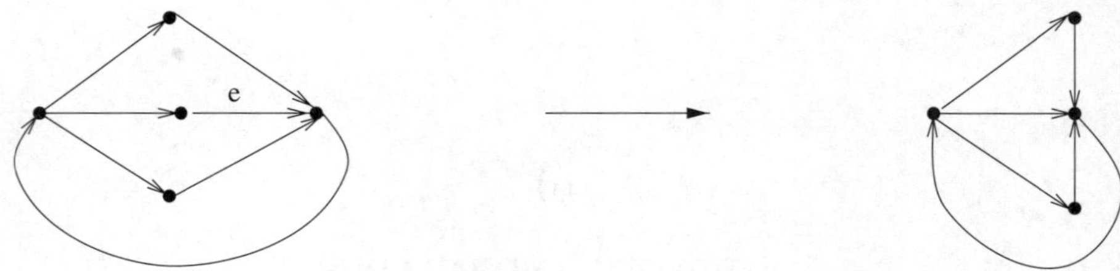

FIG. 10.1. An example of a weak mapping (which is left undefined on the edge e)

set A to a set B simply means a mapping defined on a subset of A and taking values in B.

Definition 10.2 (Weak mappings) *Let G and H be graphs. By a* weak mapping *f between G and H, also written $f : G \to H$, we mean a mapping from vertices in G to vertices in H together with a* partially defined *mapping from edges in G to edges in H which satisfies the usual compatibility condition (i.e., if e is an edge in G and $f(e)$ is defined, then f maps the endpoints of e to the endpoints of $f(e)$), and which also has the property that $f(e)$ is defined whenever e is an edge in G whose endpoints are mapped to distinct vertices in H. (Again, we allow this definition to be applied to both* finite *and* infinite *graphs.)*

In other words, f is permitted to be undefined on edges e for which both endpoints are mapped to the same vertex in H. See Fig. 10.1 for an example. A nice feature of weak mappings is that they map paths to paths in a natural way. More precisely, a path p in G is given by a succession of vertices $v_0, v_1, v_2, \ldots, v_k$ and edges $e_1, e_2, \ldots, e_k$ connecting them, and the application of a weak mapping $f : G \to H$ yields a similar succession of vertices and edges in H, but perhaps with some steps skipped, corresponding to the e_i's at which f is not defined. (Redundant vertices would would be taken out, as well as the edges at which f is not defined. This is necessary for the definition of a path anyway, i.e., one does not allow repeated vertices without an edge between them.) This does not disturb the "continuity" of the image path, because of the requirements in the definition of a weak mapping.

Note the image path under a weak mapping may be shorter — traverse fewer edges — than the original path, and that this cannot occur with the image of a path under an ordinary (non-weak) mapping between graphs.

If $f : G \to H$ and $h : H \to K$ are mappings between graphs, then one can define the *composition* $h \circ f$ as a mapping from G to K. It also makes sense to talk about the *composition of weak mappings*. Specifically, if $f : G \to H$ and $h : H \to K$ are weak mappings, then we define $h \circ f : G \to K$ as a weak mapping in the following manner. For vertices there is no problem in defining the composition, only for edges. We consider an edge e in G to be in the domain of $h \circ f$ only when e lies in the domain of f and $f(e)$ lies in the domain of h, in which case $h \circ f(e) = h(f(e))$. With this definition we have that $h \circ f$ is a weak mapping; if e is an edge in G which is not in the domain of $h \circ f$, then either it is not in the domain of f, in which case its endpoints have the same image under f and therefore under $h \circ f$, or $f(e)$ is defined but does not lie in the domain of h, in which case again the endpoints of e have the same image under $h \circ f$. This shows that $h \circ f$ is a weak mapping.

It is easy to see that if $f : G \to H$ and $h : H \to K$ are weak mappings, and p is a path in G, then the image of p under $h \circ f$ is the same as the path that we get by taking the image of p first under f, and then under h.

If G and H are *oriented* graphs, then we shall typically restrict ourselves to mappings and weak mappings $f : G \to H$ which *preserve orientations*. This simply means that if e is an edge in G on which f is defined, then the initial vertex of $f(e)$ in H should be the same as the image under f of the initial vertex of e in G, and similarly for the final vertices. Mappings and weak mappings between oriented graphs send oriented paths to oriented paths, as one can easily check.

As in Section 8.6, we can extend our original definition of visibility graphs to arbitrary oriented graphs. Thus, if G is any oriented graph, and v is a vertex in G, then the visibility $\mathcal{V}_+(v, G)$ is a graph whose vertices represent oriented paths in G which begin at v. One attaches an edge in $\mathcal{V}_+(v, G)$ from a vertex s in $\mathcal{V}_+(v, G)$ to a vertex t exactly when the oriented path in G represented by t is obtained from the oriented path in G represented by s by adding a single edge at the end. The canonical projection from $\mathcal{V}_+(v, G)$ to G (as in Section 4.5) is always an orientation-preserving mapping between these graphs.

Mappings between graphs lead to mappings between visibilities. More precisely, suppose that G and H are oriented graphs, and let a mapping $f : G \to H$ and a vertex v in G be given. If f preserves orientations, then f induces a mapping $\widehat{f} : \mathcal{V}_+(v, G) \to \mathcal{V}_+(f(v), H)$. That is, the vertices of $\mathcal{V}_+(v, G)$ represent oriented paths in G which begin at v, the images of these paths under f are oriented paths in H which start at $f(v)$, and hence represent vertices in $\mathcal{V}_+(f(v), H)$. Thus f induces a mapping from vertices in $\mathcal{V}_+(v, G)$ to vertices in $\mathcal{V}_+(f(v), H)$. It is not hard to see that one also has a natural mapping from edges to edges, and that $\widehat{f}$ preserves orientations as a mapping between the visibilities. This all comes from unwinding the definitions.

An example of a mapping between graphs and the induced mapping between visibilities is shown in Fig. 10.2. In this case, the visibility of the image is much larger than the visibility of the domain, even though the mapping f is surjective on edges and vertices. The visibility of the domain is like an infinite ray, and it is mapped into just one of the many branches of the visibility of the image. (The number of branches in the visibility of the image increases exponentially.)

Similarly, if $f : G \to H$ is a *weak* mapping which preserves orientations, then we get a weak mapping $\widehat{f} : \mathcal{V}_+(v, G) \to \mathcal{V}_+(f(v), H)$. This is also not hard to check, simply by unwinding the definitions.

There is an obvious compatibility property between the induced mapping $\widehat{f} : \mathcal{V}_+(v, G) \to \mathcal{V}_+(f(v), H)$ and the canonical projections of $\pi_G : \mathcal{V}_+(v, G) \to G$

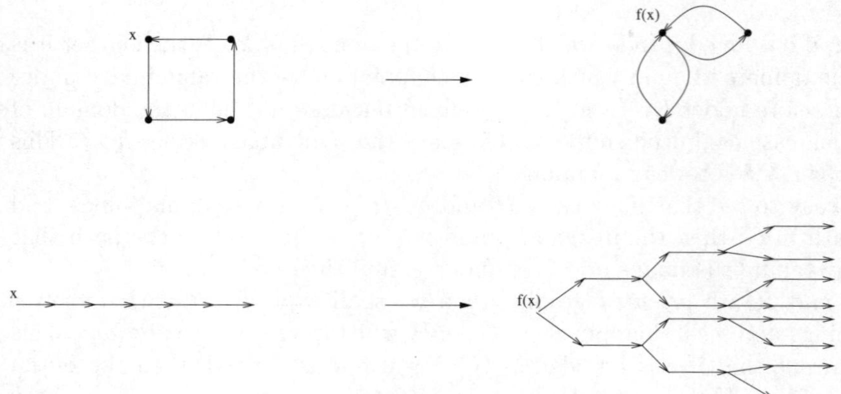

FIG. 10.2. This picture indicates a mapping between graphs, and the induced mapping between the associated visibility graphs (at the vertices x and $f(x)$). One can take f so that its image goes around all four edges in the graph on the right. The image of the mapping between visibility graphs follows only one branch in the tree on the lower part of the right side.

and $\pi_H : \mathcal{V}_+(f(v), H) \to H$ from Section 4.5. Namely,

$$\pi_H \circ \widehat{f} = f \circ \pi_G. \tag{10.1}$$

This can be verified directly from the definitions, and it applies equally well to both mappings and weak mappings.

There is also an obvious "homomorphism property" for the induced mappings between the visibilities, which is the following. Let $f : G \to H$ be an orientation-preserving mapping or weak mapping, and fix a vertex v in G. Suppose that K is another oriented graph, and that $h : H \to K$ is an orientation-preserving mapping or weak mapping. Then we get $\widehat{h} : \mathcal{V}_+(f(v), H) \to \mathcal{V}_+(h(f(v)), K)$ and $\widehat{h \circ f} : \mathcal{V}_+(v, G) \to \mathcal{V}_+(h(f(v)), K)$ in addition to $\widehat{f}$ as above, and we have that

$$\widehat{h \circ f} = \widehat{h} \circ \widehat{f}. \tag{10.2}$$

This is easy to check. At the level of vertices, it simply says that the image of a path under $h \circ f$ is the same as first taking the image under f, and then under h. For edges, it is similarly just a matter of definitions, with a small amount of extra care for the case of weak mappings. (Keep in mind that $\widehat{h \circ f}$, $\widehat{h}$, and $\widehat{f}$ are only weak mappings if $h \circ f$, h, and f are. Of course $h \circ f$ may only be a weak mapping as soon as just one of h and f is.)

Note that our basic observations about mappings between graphs and induced mappings between visibilities run into trouble immediately if we try to work with *restrained visibilities* (Section 8.9) instead of ordinary visibilities. Since the restrained visibilities are defined in terms of the global requirement that the paths be simple, one would have to either impose a global condition like injectivity on the mappings, or else allow the induced mappings to be defined only on certain parts of the restrained visibilities.

10.2 Computational questions

A number of the familiar NP-complete problems about graphs deal with the existence of mappings between graphs. This includes the travelling salesman problem and the Hamiltonian path problem, for instance, since paths can be viewed as mappings between graphs whose domains are "linear" graphs. Let us mention now a couple of other problems about mappings between graphs which are of a slightly different nature.

Proposition 10.3 *Let G be a graph, and let T denote the "triangle" graph with exactly three vertices and one edge between every pair of distinct vertices. The problem of deciding whether there exists a mapping from G into T is NP-complete.*

Proof This is really just a reformulation of the well-known "3-coloring" problem, which asks whether for a given graph G there is a way to assign to each vertex in G one of three colors in such a way that adjacent vertices have different colors. Specifically, a 3-coloring exists if and only if there is a mapping from G into T. Thus the proposition follows from the fact that the 3-coloring problem is NP-complete (Theorem 9.8 on p198 of [Pap94]). □

Of course, one can reformulate k-colorings in general in similar terms.

Proposition 10.4 *Let G be a graph, let k be a positive integer, and let C_k be a complete graph on k vertices (so that C_k contains exactly one edge between every pair of distinct vertices). The problem of deciding whether there is a mapping from C_k into G is NP-complete.*

Thus Proposition 10.3 deals with mappings *into* a graph with very simple structure, while Proposition 10.4 deals with mappings *from* graphs with simple structure.

Proof This is a small modification of the "clique" problem, in which one is given a graph G and a number k, and one is asked whether it is possible to find a set of k vertices in G such that any two distinct vertices in this set are adjacent to each other. For the clique problem, one may remove all of the edges in G with both endpoints at the same vertex at the beginning, since this does not affect the possibility of having a k-clique, and then the existence of a k-clique is equivalent to the existence of a mapping from C_k into G. (If there do exist edges with both endpoints at the same vertex, then the mapping problem automatically has the answer "yes", because one can simply collapse C_k onto a single loop in G.) Since the clique problem is NP-complete (Corollary 2 on p190 of [Pap94]), the same is true of the mapping problem described in Proposition 10.4. □

Note that Propositions 10.3 and 10.4 would not work at all if one permitted arbitrary *weak mappings*, since whole graphs could then be collapsed to single vertices. In this regard, the notion of a "mapping" between graphs is already rather nondegenerate, and hence the kind of NP-completeness results as above.

So far we have considered only graphs without orientations, but we could just as well work with oriented graphs and orientation-preserving mappings between them. There is a basic trick for passing from unoriented graphs to oriented ones, which is given by the following construction.

Definition 10.5 *Let G be an (unoriented) graph. By the* edge double *of G we mean the* oriented *graph $\widetilde{G}$ which is obtained from G in the following manner. We use the same set of vertices for $\widetilde{G}$ as for G, but we double the number of edges, by doubling the number of edges between every fixed pair of vertices. We require that these edges be oriented in such a way that for any pair of distinct vertices a, b there are as many edges going from a to b as there are going from b to a.*

Lemma 10.6 *Let G and H be graphs, and let $\widetilde{G}$ and $\widetilde{H}$ be their edge doubles. Then there exists a mapping from G to H if and only if there exists a mapping from $\widetilde{G}$ to $\widetilde{H}$ which preserves orientations. Similarly, G and H are isomorphic as unoriented graphs if and only if $\widetilde{G}$ and $\widetilde{H}$ are isomorphic as oriented graphs.*

This is easy to check.

Proposition 10.7 *Let G be an oriented graph.*

(a) Let $\widetilde{T}$ be the oriented graph with three vertices and exactly one edge from x to y for any ordered pair of distinct vertices x, y. (Thus there are now 6 *edges in all.) The problem of deciding whether there exists an orientation-preserving mapping from G into $\widetilde{T}$ is NP-complete.*

(b) Let a positive integer k be given, and let $\widetilde{C}_k$ be an oriented graph with k vertices and exactly one oriented edge from z to w for any ordered pair z, w of distinct vertices in $\widetilde{C}_k$. Then the problem of deciding whether there is an orientation-preserving mapping from $\widetilde{C}_k$ into G is NP-complete.

Proof The fact that each problem lies in the class NP is immediate, since they are each formulated in terms of the existence of suitable "effective witnesses" (or "succinct certificates"). See p181-2 of [Pap94] for more details about this way of verifying the NP property. NP-completeness can be derived from Propositions 10.3 and 10.4 by using edge-doubling to show that the problems for oriented graphs contain their analogues for unoriented graphs as a special case. □

10.3 Local +-isomorphisms

Definition 10.8 *Let G and H be oriented graphs (possibly infinite), and let $f : G \to H$ be a mapping between them which preserves orientations. We say that f is a* local +-isomorphism *if for each vertex $u \in G$ we have that f induces a one-to-one correspondence between the edges in G attached to u and oriented away from u and the edges in H attached to $f(u)$ and oriented away from $f(u)$.*

An example of a local +-isomorphism is shown in Fig. 10.3.

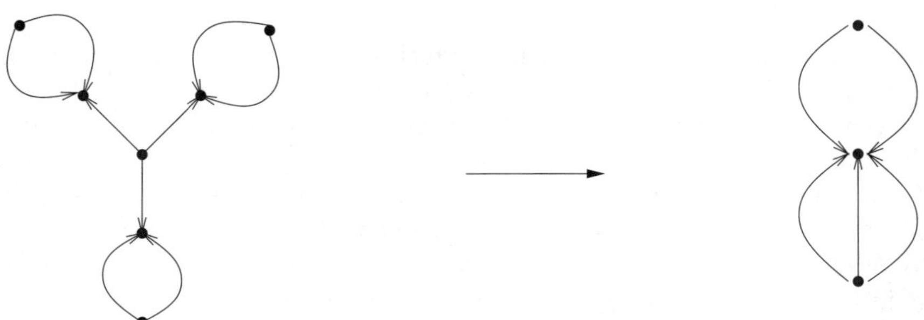

FIG. 10.3. An example of a local +-isomorphism between two graphs (with the central vertex on the left mapped to the bottom vertex on the right)

For mappings which are locally +-isomorphic, we can "lift" oriented paths in a simple way, to get the following.

Lemma 10.9 *Let G and H be oriented graphs, and let $f : G \to H$ be a mapping which preserves orientations and is a local +-isomorphism. Then the induced mapping $\widehat{f} : \mathcal{V}_+(v, G) \to \mathcal{V}_+(f(v), H)$ between the visibilities is an isomorphism, for every vertex v in G.*

This was also discussed in [CS99], but in a slightly different form.

Proof The argument is pretty straightforward, and is analogous to classical arguments for covering surfaces (as in [AS60, Mas91]).

Let us check first that $\widehat{f}$ induces a bijection on vertices. This comes down to statements about oriented paths and their images under f. To say that $\widehat{f}$ is an injection on vertices means that two distinct oriented paths p and q in G which both start at v cannot have the same images in H under f. This is pretty easy to see. By assumption, p and q start at the same vertex, but since they are distinct, they will agree along some common subpath α, and then follow different edges immediately afterwards. Their images will do the same, follow a common subpath (which is the image of α under f), and then diverge at the corresponding moment (at the end of $f(\alpha)$). For this we use the assumption that f is a local +-isomorphism to know that distinct edges in G emerging from the endpoint of α must be sent to distinct edges in H. Thus we have that the image paths diverge at some point and hence are distinct. This says exactly that $\widehat{f}$ is injective on vertices.

Now let us check surjectivity of $\widehat{f}$ on vertices. This is the same as saying that if τ is some oriented path in H which begins at $f(v)$, then there is an oriented path σ in G which begins at v and which is mapped by f to τ. One constructs σ one step at a time. One knows the starting point for σ. To get the first edge that σ traverses (assuming that τ is nondegenerate), we use the assumption that f is a local +-isomorphism to know that there is an edge in G flowing from v whose image in H is the first edge traversed by τ. This gives us the second vertex in σ, and we can get the second edge in the same way. Repeating this process as needed, we get an oriented path σ in G which starts at v and whose image in H under f is precisely τ. This proves that $\widehat{f}$ is surjective on vertices.

It remains to show that $\widehat{f}$ is a bijection on edges. Let p and q be oriented paths in G which begin at v. If p is obtained from q by adding an edge at the end, then the same is true for the image paths, and conversely. If we now think of p and q as defining vertices in the visibility $\mathcal{V}_+(v, G)$, then this says that there is an edge in $\mathcal{V}_+(v, G)$ which goes from q to p if and only if there is an edge in $\mathcal{V}_+(f(v), H)$ which goes from the image of q under $\widehat{f}$ to the image of p under $\widehat{f}$. This is equivalent to saying that $\widehat{f}$ defines a bijection between edges in $\mathcal{V}_+(v, G)$ and $\mathcal{V}_+(f(v), H)$, since we already know that it defines a bijection between vertices (and since there is at most one edge going from one vertex to another in a visibility graph, by the definition of visibility graphs). This proves the lemma. □

The canonical projection from the visibility to the original graph is always a local +-isomorphism. We state this more precisely next.

Lemma 10.10 *Let G be an oriented graph, and let v be a vertex in G. Consider the canonical projection $\pi : \mathcal{V}_+(v, G) \to G$ (defined in Section 4.5). This is a local +-isomorphism.*

Proof This is easy to see from the definitions. (Compare also with Lemma 4.4 in Section 4.5.) □

The lifting of paths which occurs in Lemma 10.9 is really just a matter of definitions in this case. Remember that the visibility graph of $\mathcal{V}_+(v, G)$, using the basepoint of $\mathcal{V}_+(v, G)$ as a starting point, is isomorphic to $\mathcal{V}_+(v, G)$ itself in a canonical way, as we pointed out in Section 4.6, shortly after Corollary 4.7.

The property of being a local +-isomorphism is preserved under compositions, as in the next result.

Lemma 10.11 *Let G, H, and K be oriented graphs (possibly infinite), and let $f : G \to H$ and $h : H \to K$ be orientation-preserving mappings which are local +-isomorphisms. Then the same is true of $h \circ f : G \to K$.*

Proof This is a straightforward consequence of the definitions, and we omit the details. □

Let us now record some modest refinements of some of the definitions and observations of this section.

Definition 10.12 *Let G and H be (possibly-infinite) oriented graphs, and let $f : G \to H$ be an orientation-preserving mapping. We say that f is* locally +-injective *if for each vertex u in G we have that every edge in G flowing away from u is mapped to a* distinct *edge in H flowing away from $f(u)$. We say that f is* locally +-surjective *if for each vertex u in G and each edge in H flowing away from $f(u)$ there is an edge in G flowing away from u which is sent by f to the given edge in H.*

Thus $f : G \to H$ is a local +-isomorphism if and only if it is both locally +-injective and locally +-surjective.

Lemma 10.13 *Suppose that G, H, and K are oriented graphs, and that $f : G \to H$ and $h : H \to K$ are orientation-preserving mappings. If f and h are both locally +-injective, then the same is true of $h \circ f$. If they are both locally +-surjective, then the same is true of $h \circ f$.*

Proof This is an easy exercise. □

Lemma 10.14 *Let G and H be oriented graphs, and let $f : G \to H$ be an orientation-preserving mapping. If f is locally +-injective, then the induced mapping $\widehat{f} : \mathcal{V}_+(v, G) \to \mathcal{V}_+(f(v), H)$ between the visibilities is (globally) injective. If f is locally +-surjective, then $\widehat{f} : \mathcal{V}_+(v, G) \to \mathcal{V}_+(f(v), H)$ is (globally) surjective.*

This was also discussed in [CS99], but in a slightly different way.

Proof This follows from the same reasoning as in the proof of Lemma 10.9, and we leave the argument as an exercise. □

Let us record a couple of other simple facts.

Lemma 10.15 *Let G and H be oriented graphs, and let $f : G \to H$ be an orientation-preserving mapping. If f is locally +-injective and injective on vertices, then f is also injective on edges. If f is locally +-surjective and surjective on vertices, then it is also surjective on edges.*

Proof This is an easy exercise from the definitions. If we are in the injective case, and f maps a pair of edges in G to the same edge in H, then injectivity on vertices implied that the two edges in G must have the same initial vertex, and then local +-injectivity implies that the edges must themselves be the same. Similarly, for the surjective case, given an edge in the image, we can take its initial vertex and find a preimage for that in G using surjectivity on vertices. Local +-surjectivity then implies that the given edge in H also has a preimage in G. □

Remark 10.16 (Clarification of terminology) Let G and H be graphs, and let $f : G \to H$ be a mapping between them. If we say that f is "injective", "surjective", or "bijective" without further qualification, then we shall mean that both the induced mappings on edges and on vertices enjoy the corresponding property. Also, let us emphasize that if G and H are oriented, and we say that f is a local +-isomorphism, or local +-injection, etc., then it should automatically be assumed that f preserves orientations.

Lemma 10.17 *Let G and H be oriented graphs, and suppose that $f : G \to H$ is a mapping which preserves orientations. If G contains nontrivial oriented cycles, then the same is true for H. Conversely, if f is also a surjection and a local +-surjection, and if H contains a nontrivial oriented cycle, then G must contain one as well.*

Proof The first part is immediate from the definitions. Notice that it would not work for *weak* mappings, which can collapse nontrivial cycles to single vertices.

For the second part, suppose that H contains a nontrivial oriented cycle and that f is both surjective and locally $+$-surjective. Fix a vertex w in H such that the visibility $\mathcal{V}_+(w, H)$ is infinite. For instance, one can simply choose w so that it lies on a nontrivial oriented cycle. Let v be a vertex in G such that $f(v) = w$. Then the visibility $\mathcal{V}_+(v, G)$ must be infinite, since f induces a mapping from $\mathcal{V}_+(v, G)$ to $\mathcal{V}_+(w, H)$ which is *surjective*, as in Lemma 10.14. From this it follows that G contains a nontrivial oriented cycle (which is accessible by an oriented path beginning at v), as in Lemma 4.8. (One can also check this more directly, by lifting an oriented path in H which goes around a cycle many times to an oriented path in G which is longer than the total number of vertices in G. This would force the lifted path to pass through at least one vertex twice, and hence to provide an oriented cycle in G.) This proves the lemma. □

Note that the second part of the lemma would not work if G were permitted to be infinite. For instance, H might consist of a single loop, and G might be an infinite ray which is wrapped around H by the mapping f infinitely many times.

In the context of finite graphs, this corresponds to the fact that the cycles in G might have to be very long compared to the ones in H. To see this concretely, let G_k denote the oriented graph consisting of k vertices arranged as in a circle. That is, there should be an edge going from the jth vertex to the $(j+1)$th vertex for each $j < k$, and an edge going from the kth vertex to the first vertex, and no other edges. If m and n are positive integers such that m divides n, then we can map G_n to G_m simply by wrapping G_n around G_m n/m times. In terms of vertices, we take the jth vertex in G_n and send it to the ith vertex in G_m, where $i \equiv j$ modulo m. This gives a mapping from G_n onto G_m which is a local $+$-isomorphism (and which behaves just as well in the negatively-oriented directions). Both G_n and G_m contain oriented cycles, as in the context of Lemma 10.17, but not every cycle in G_m can be lifted to a *cycle* in G_n. The shortest cycle in G_m has length m, while the shortest cycle in G_n has length n, which may be much larger.

One can also enlarge G_n to get graphs which still admit a local $+$-isomorphism into G_m, but so that the larger versions of G_n contain vertices that do not lie on cycles. This is easy to do, by adding some vertices and oriented edges (or paths) from them to vertices in G_n.

Instead of cycles, we can also look at focal pairs (Definition 4.15).

Lemma 10.18 *Let G and H be oriented graphs, and suppose that $f : G \to H$ is orientation-preserving and a local $+$-injection. If u and w are vertices in G such that (u, w) forms a focal pair, then there is a vertex y in H such that* (a) *$(f(u), y)$ forms a focal pair and* (b) *there is an oriented path in H going from y to $f(w)$.*

Proof To say that (u, w) is a focal pair in G means that there is a pair of oriented paths α, β in G which begin at u and which arrive at w along different incoming edges. We can use f to map these paths into H, to get two oriented paths γ, δ going from $f(u)$ to $f(w)$. These paths might not arrive to $f(w)$ along different incoming edges, however, so that $(f(u), f(w))$ may not be a focal pair in H.

If we only asked that f be orientation-preserving, then γ and δ might even be the same path in H. In the present situation, this possibility is prevented by the assumption that f be a local +-injection. This assumption ensures that if α and β agree for j steps and then split apart (as they must do at some point, since they arrive at w along different edges), then the same must be true for γ and δ.

To prove the lemma, we take y to be the vertex in H which represents the first moment at which γ and δ come together again after diverging. They must come together again, since they both end at $f(w)$, but they might do this before reaching $f(w)$. It easy easy to see that y has the required properties, using subpaths of γ and δ to obtain oriented paths in H which begin at $f(u)$ and arrive at y along different edges, and to obtain an oriented path in H going from y to $f(w)$. This completes the proof of Lemma 10.18. □

Corollary 10.19 *Suppose that G and H are oriented graphs and $f : G \to H$ is orientation-preserving and a local +-injection. If G contains a chain of focal pairs (Definition 4.17) of length n which begins at some vertex u, then H contains a chain of focal pairs which begins at $f(u)$ and has length n too.*

This is easy to check, using Lemma 10.18.

Remark 10.20 Suppose that $f : G \to H$ is a *local −-injection* instead of a local +-injection, which is defined in the same way as before except that one asks that f be injective on the set of *incoming* edges at any given vertex, instead of on the set of outgoing edges. In this case focal pairs are mapped to focal pairs, and similarly for chains of focal pairs. That is, one does not need to introduce an auxiliary vertex y in this case, as in Lemma 10.18, and this is easy to verify. In the end, the difference between local +-injections and local −-injections in this context is relatively minor, because the asymmetry between positive and negative orientations in the definition of focal pairs is relatively minor.

These observations about focal pairs are roughly analogous to the *first* part of Lemma 10.17, about the preservation of cycles under the application of a mapping. As for the second part of Lemma 10.17, it can easily happen that H has a lot of focal pairs or chains of focal pairs, but G does not, even when we have a local +-isomorphism from G onto H. For instance, G might be the visibility of H starting from a vertex w, with the mapping taken to be the canonical projection, as in Lemma 10.10. (In this case, we should ask that all vertices and edges in H be accessible from w by oriented paths, to ensure that the canonical projection be a surjection.)

In the end, the situations for oriented cycles and focal pairs (or chains of focal pairs) are more similar than they might appear to be at first. This corresponds to the remarks made after the statement and proof of Lemma 10.17.

One can think of visibility graphs and measurements of their sizes as reflecting a kind of "entropy" in oriented graphs, as we have mentioned before. (See Sections 4.4 and 4.13.) Lemmas 10.9 and 10.14 show that this kind of "entropy" behaves well under mappings between graphs.

The notion of "minimal representations" (Definition 9.6 in Section 9.2) provides another way to measure "information content" in oriented graphs. More precisely, one can use the minimal representations of visibility graphs to make invariants of oriented graphs under local +-isomorphisms, since they induce isomorphisms between the visibility graphs. In general these invariants need not behave so well under local +-injections or local +-surjections, however.

10.4 Some interpretations

The notion of a local +-isomorphism has a nice interpretation in terms of "combinatorial dynamical systems". As in Section 4.13, suppose that one has an oriented graph G together with the following data. If v is a vertex in G, then a set of points $S(v)$ should be specified, which may change with v. To each edge e in G there should be associated a mapping $\phi_e : S(u) \to S(w)$, where u and w are the vertices at the beginning and end of e (with respect to the orientation on G).

From this data one can build an oriented graph X and a mapping $\Phi : X \to G$ as follows. For the vertices of X one takes the *disjoint* union of the sets $S(v)$, where v runs through all vertices in G. (If these sets were not disjoint to begin with, then they can be made disjoint, e.g., by identifying $S(v)$ with the Cartesian product $S(v) \times \{v\}$.) If x and y are elements of X, with $x \in S(u)$ and $y \in S(w)$, say, then one attaches an edge ξ going from x to y for each edge e in G going from u to w such that $\phi_e(x) = y$ (if there are any such edges e). This defines X as an oriented graph, and the mapping $\Phi : X \to G$ comes directly from the construction. Specifically, we set $\Phi(z) = v$ when z is a vertex in X, $z \in S(v)$, and if ξ is an edge in X that goes between two vertices x and y in X, then we set $\Phi(\xi) = e$, where e is the edge in G which goes from $u = \Phi(x)$ to $w = \Phi(y)$ and which corresponds to ξ as in the definition of X.

It is not hard to see that $\Phi : X \to G$ is actually a local +-isomorphism in this situation. This follows from the construction in a straightforward manner. Conversely, if one has a local +-isomorphism $\Phi : X \to G$ from some oriented graph X into G, then one can use it to define the same kind of system over G in a simple way. Indeed, for each vertex v in G, one can take $S(v)$ to be the set of vertices in X which are mapped to v by Φ. If e is an edge in G which goes from a vertex u to a vertex w, then one can define a mapping $\phi_e : S(u) \to S(w)$ by saying that $y = \phi_e(x)$, where $x \in S(u)$ and $y \in S(w)$, exactly when there is an edge ξ in X that goes from x to y and which is mapped to e by Φ. In this case, we have that ϕ_e is single-valued and defined on all of $S(u)$, precisely because of our assumption that $\Phi : X \to G$ be a local +-isomorphism.

This correspondence provides an interesting perspective on computational questions of the following nature: Given an oriented graph X, when is there a local +-isomorphism from X onto a particular graph G? Or onto a graph of size at most k, for a given value of k? These questions are easily seen to lie in the complexity class NP, and it seems unlikely that they can be solved in polynomial time in general.

We shall discuss these and similar problems in the next sections. We shall also see how for each oriented graph X there is a canonical "minimal folding graph", which is a graph of minimal size onto which X admits a local +-isomorphism. It is closely connected to the notion of "minimal representations" from Definition 9.6 in Section 9.2, to which we shall return in Section 10.7.

It is natural to think of local +-isomorphisms as providing a way to say that two graphs are similar to each other in terms of their local structure. If the domain is much larger than the image, then the existence of a local +-isomorphism can be seen as an expression of symmetry in the domain (that is, it will have many copies of the same basic "patterns" found in the image).

For the correspondence described above, there are natural versions under other conditions, e.g., local +-injectivity.

10.5 The local +-injection problem

Definition 10.21 *Let G and H be oriented graphs. The* local +-injection problem *asks whether there is a mapping $f : G \to H$ which is a local +-injection.*

Proposition 10.22 *The local +-injection problem is NP-complete.*

That the local +-injection problem lies in NP follows from the fact that it is formulated in terms of the existence of a suitable "effective witness" (as on p181-2 of [Pap94]). As for the NP-completeness, we shall provide three different ways of seeing this, each with a slightly different character. (One could also consider other kinds of "local injections", without involving orientations in particular.)

The first method works through the 3-coloring problem, as in Section 10.2.

Lemma 10.23 *Let G be an oriented graph, and let k be a positive integer. Let $\widetilde{T}_k$ be an oriented graph with exactly 3 vertices and k edges going from a to b whenever a and b are distinct vertices in $\widetilde{T}_k$. Then the problem of deciding whether there is a local +-injection from G into $\widetilde{T}_k$ is NP-complete.*

The number k is allowed to vary here. The main point is that it should be at least as large as the maximum number of edges coming out of any given vertex in G. If this is not the case, then there cannot be any local +-injections from G into $\widetilde{T}_k$.

Proof If k is at least as large as the maximum number of edges coming out of any vertex in G, then the existence of a local +-injection from G into $\widetilde{T}_k$ is equivalent to the existence of an orientation-preserving mapping from G into $\widetilde{T}_1$. Indeed, every orientation-preserving mapping from G into $\widetilde{T}_k$ can be "projected" to an orientation-preserving mapping from G into $\widetilde{T}_1$, simply by composition with the obvious mapping from $\widetilde{T}_k$ to $\widetilde{T}_1$ (in which the vertices are held fixed and the

k edges from a to b in $\widetilde{T}_k$ are collapsed to a single edge in $\widetilde{T}_1$, for every pair of distinct vertices a, b). Conversely, orientation-preserving mappings from G into $\widetilde{T}_1$ can always be "lifted" to local +-injections into $\widetilde{T}_k$ in a simple way. (This uses the requirement that k be at least as large as the number of outgoing edges at any vertex in G.)

Thus the problem in Lemma 10.23 is equivalent to the one in part (a) of Proposition 10.7, when k is at least as large as the maximum number of edges coming out of any vertex in G. This implies NP-completeness, as desired. □

Lemma 10.23 provides one approach to the NP-completeness of the local +-injection problem, in which the target graph H is chosen to have a very special form. We can also use the NP-completeness of the clique problem to get NP-completeness for the local +-injection problem for a special class of graphs G in the domain.

Lemma 10.24 *Let H be an oriented graph, and let k be a positive integer. Let $\widetilde{C}_k$ be an oriented graph with k vertices and exactly one oriented edge from a to b for every ordered pair a, b of distinct vertices. Then the problem of deciding whether there is a local +-injection from $\widetilde{C}_k$ into H is NP-complete.*

Again the number k is permitted to vary (with H).

Proof This is practically the same as Proposition 10.4 and part (b) of Proposition 10.7. Let L be an *unoriented* graph, and suppose that we want to decide whether L contains a k-clique, i.e., a set of k vertices such that any two distinct vertices from this set are adjacent. We may as well assume that L contains no edges for which the two endpoints are the same vertex, since that does not effect the existence of a k-clique. If H denotes the *oriented* graph obtained from L by edge-doubling (Definition 10.5), then it is not difficult to check that L contains a k-clique if and only if there is a local +-injection from $\widetilde{C}_k$ into H. This uses the fact that any orientation-preserving mapping from $\widetilde{C}_k$ into H is automatically injective on vertices, because of the absence of edges in H which begin and end at the same vertex.

Thus the problem in Lemma 10.24 contains the clique problem as a special case, and this implies NP-completeness. □

The third approach to the NP-completeness of the local +-injection problem is slightly more complicated than the previous two, but it involves a construction of broader utility.

Definition 10.25 (Cones over graphs) *Let G be an oriented graph, and fix a positive integer j. We define a new oriented graph G^c, called the* j-cone over G, *as follows. We start with G itself, and add exactly one new vertex p (the "cone point"). For each vertex v in G, we add exactly j edges going from p to v, and no other edges.*

Note that G^c contains G as a subgraph, and that the cone point of G^c admits no incoming edges, while every other vertex in G^c admits at least j incoming edges. On the other hand, vertices in G have as many outgoing edges in G^c as they have in G.

Lemma 10.26 *Let G and H be oriented graphs, and choose $j \in \mathbf{Z}_+$ so that the product of j with the number of vertices in G is strictly larger than the number of edges which come out of any single vertex in H. Let G^c and H^c be the graphs associated to G and H as in Definition 10.25, with this choice of j. Then the existence of a local +-injection $f : G^c \to H^c$ is equivalent to the existence of an orientation-preserving injection $g : G \to H$.*

Remember that a mapping between graphs is called "injective" when it is injective on both edges and vertices, as in Remark 10.16.

Proof Suppose first that we have an orientation-preserving injection $g : G \to H$, and let us extend it to a similar mapping from G^c to H^c. Let p be the cone point of G^c, as in Definition 10.25, and let q be the cone point of H. We choose $f : G^c \to H^c$ so that $f = g$ on G, $f(p) = q$, and so that f induces a one-to-one correspondence between the edges in G^c that go from p to v and the edges in H^c which go from q to $g(v)$ for each vertex v in G. This is possible, because both of these sets of edges have exactly j elements. This defines $f : G^c \to H^c$ as an orientation-preserving mapping, and it is easy to see that f has to be an injection, since g is. (The main point is that f is injective on the set of all edges coming out of p because g is injective on vertices.)

Conversely, suppose that we have a local +-injection $f : G^c \to H^c$. Let us check that f has to map p to q. The number κ of edges flowing out of p in G^c is equal to the product of j and the number of vertices in G, by construction. The number of edges flowing out of $f(p)$ in H^c has to be at least κ, since f is a local +-injection. If $f(p)$ were not equal to q, then the number of outgoing edges in H^c at $f(p)$ would be the same as in H, and would be less than κ by our choice of j. This proves that $f(p) = q$, since f is assumed to be a local +-injection.

No other vertex in G^c can be mapped to q by f, because q has no incoming edges in H^c, while each vertex in G^c besides p has at least j incoming edges (coming from p). Thus the restriction of f to G defines an orientation-preserving mapping from G to H. To finish the proof of the lemma, it suffices to show that $f : G^c \to H^c$ is an injection,

Let u and v be vertices in G, and suppose that $f(u) = f(v)$. If $u \neq v$, then the $2j$ edges in G^c that go from p to either u or v have to be mapped by f into the j edges in H which go from q to $f(u) = f(v)$. This is impossible, since we are assuming that f is a local +-injection. Thus we conclude that f is actually injective on the vertices of G. From this we obtain also that f must be injective on edges, since it is a local +-injection. (See Lemma 10.15.) This completes the proof of Lemma 10.26. □

Lemma 10.26 provides a third approach to the NP-completeness of the local +-injection problem, by showing that a polynomial-time solution to the local +-injection problem would lead to a polynomial-time solution of the problem of deciding when one graph can be embedded into another. The latter is

NP-complete, as mentioned in Problem 9.5.23 on p212 of [Pap94]. The NP-completeness of the embeddability problem can be seen as a direct consequence of the NP-completeness of the clique problem, but there are other special cases of the embeddability problem which are NP-complete, and which are somewhat different from the clique problem.

Thus we have a number of ways in which to see well-known NP-complete problems as special cases of the local +-injection problem. What happens for local +-isomorphisms?

Definition 10.27 *Let G and H be oriented graphs. The* local +-isomorphism problem *asks whether there exists a local +-isomorphism from G into H.*

This problem again lies in NP, but it is not clear whether it should be NP complete. The approaches to the NP-completeness of the local +-injection problem mentioned above simply do not work in this case, because of the "exactness" that local +-isomorphisms have which is missing from local +-injections or orientation-preserving mappings in general. We do have the following, however.

Proposition 10.28 *If the local +-isomorphism problem can be solved in polynomial time, then one can decide when two oriented graphs are isomorphic in polynomial time as well.*

The problem of deciding when two graphs are isomorphic is somewhat notorious for resisting classification along the lines of P and NP, as mentioned in [Pap94]. Note that the graph-isomorphism problem for oriented graphs implicitly contains its counterpart for unoriented graphs, as in Lemma 10.6.

Proof Let G and H be arbitrary oriented graphs, and suppose that we want to decide whether G and H are isomorphic to each other or not. Let G^c and H^c be the "cone" graphs obtained from G and H as in Definition 10.25, with the parameter j again chosen large enough so that the product of j with the number of vertices in G is strictly larger than the number of edges which come out of any fixed vertex in H.

Claim 10.29 *There is an isomorphism from G onto H if and only if there is a local +-isomorphism from G^c into H^c.*

This is very similar to Lemma 10.26. If there is an isomorphism from G onto H, then it extends to an isomorphism from G^c onto H^c in a straightforward manner, just as before. Conversely, if $f : G^c \to H^c$ is a local +-isomorphism, then it is a local +-injection in particular, and exactly the same arguments as before apply. Thus we get that f takes the cone point of G^c to the cone point of H^c, and that its restriction to G (viewed as a subgraph of G^c) takes values in H. This restriction is injective, exactly as before, and one can use the same reasoning to show that the restriction of f to G defines a surjection onto H. More precisely, one can use the local +-surjectivity of f at the cone points to conclude that f maps the set of vertices in G *onto* the set of vertices in H, and the surjectivity property for edges can then be derived from local +-surjectivity, as in Lemma 10.15.

This proves Claim 10.29, and Proposition 10.28 follows easily. □

We shall return to the local +-isomorphism problem in Section 10.14.

10.6 A uniqueness result

Let us pause a moment to give the following characterization of the visibility and its associated canonical projection. The precise formulation of this characterization is chosen for practical convenience of future reference. (Otherwise one might speak more directly in terms of trees, etc.)

Lemma 10.30 *Let G and T be oriented graphs, with T allowed to be infinite (but locally finite). Fix vertices v in G and b in T, and assume that $f : T \to G$ is an orientation-preserving mapping which is a local +-isomorphism and satisfies $f(b) = v$. Assume that b has no incoming edges in T, and that every other vertex in T has at most one incoming edge, and admits an oriented path to it from b. Then there is an isomorphism ρ from T onto the visibility $\mathcal{V}_+(v, G)$ such that $f = \pi \circ \rho$, where $\pi : \mathcal{V}_+(v, G) \to G$ is the usual projection from the visibility back to G (as defined in Section 4.5).*

This is a uniqueness result in the sense that $\pi : \mathcal{V}_+(v, G) \to G$ automatically satisfies all the properties assumed of $f : T \to G$.

Lemma 10.30 is in truth just a small variation on Lemma 10.9. In fact, T must be isomorphic to its own visibility graph under the assumptions of the lemma, and the possible infiniteness of T poses no real trouble for the previous arguments. For the sake of clarity, we write down a direct proof, though.

Proof The assumptions on T actually imply that it is a *tree*. For our purposes, the main point is that if t is any vertex in T, then there is a *unique* oriented path in T from b to t. Existence is part of our assumptions, and we need only check uniqueness. We argue by induction on the lengths of the paths. Suppose that α and β are oriented paths in T which go from b to an arbitrary vertex t, and that the minimum of their lengths is some nonnegative integer n. If $n = 0$, then $t = b$, and both α and β must be the trivial path at b, because there are no incoming edges at b, by assumption. Suppose instead that $n \geq 1$, and that we have uniqueness when the minimal length is $n - 1$. Then $t \neq b$, since b has no incoming edges, and in fact α and β have to arrive at t along the same edge, since t has at most one incoming edge, by hypothesis. If α' and β' are the paths in T obtained by removing the last step from each of α and β, then α' and β' are oriented paths which begin at b and end at the same vertex, and the minimum of their lengths is $n - 1$. Our induction hypothesis implies that $\alpha' = \beta'$, from which we conclude that $\alpha = \beta$. This implies the uniqueness of oriented paths in T from the basepoint b to any given vertex t.

This enables us to define $\rho : T \to \mathcal{V}_+(v, G)$ in the following way. Given a vertex t in T, we let α be the unique oriented path in T going from b to t, and we use f to map α to an oriented path γ in G which goes from v to $f(t)$. This path γ determines a vertex in the visibility $\mathcal{V}_+(v, G)$. This defines ρ as a mapping on vertices, and one can extend it to a compatible mapping on edges (which preserves orientations) in a straightforward manner.

Let us prove that ρ is an isomorphism, starting with injectivity. Suppose that we have two distinct vertices t and t' in T which are mapped by ρ to the same vertex in $\mathcal{V}_+(v, G)$. Let α and α' be the unique oriented paths in T which begin at b and end at t and t', respectively. To say that $\rho(t) = \rho(t')$ means that α and α' are mapped to the same path in G. Since t and t' are distinct, we must have that α and α' agree until they reach some vertex s in T, at which point they diverge, i.e., follow different edges; the only other possibility is that one of α and α' is a proper subpath of the the other, but this is precluded by the fact that α and α' have the same length, since they map down to the same path in G. The images of α and α' in G under f must then diverge at $f(s)$, since f is a local +-isomorphism. Thus the images of α and α' are in fact distinct, which proves that f is injective on vertices. Injectivity on edges follows easily from there. (Compare also with Lemma 10.15 in Section 10.3.)

As for surjectivity, suppose that we are given an oriented path γ in G which begins at v and has length n. We want to find an oriented path α in T which begins at b and is mapped to γ by f. (In this event, the vertex in $\mathcal{V}_+(v, G)$ represented by γ will be the image of the endpoint t of α under ρ.) We argue by induction. If $n = 0$, then there is nothing to prove, since γ must be the trivial path at v, and we know that $f(b) = v$. If $n \geq 1$, then we let γ' denote the initial subpath of γ of length $n - 1$, which includes all but the last step. By induction hypothesis, there is an oriented path α' in T which begins at b and is mapped to γ' by f. The assumption that f be a local +-isomorphism permits us to extend α' by one step to get a path which is mapped to γ by f, as desired. This proves the surjectivity of ρ on vertices, and it is not difficult to establish the surjectivity on edges through similar reasoning (or using Lemma 10.15).

Thus ρ is an isomorphism. The formula $f = \pi \circ \rho$ follows immediately from the definition of ρ, and the lemma follows. □

Remark 10.31 In practice, one often starts with a rooted tree (T, b), and then *defines* an orientation on it so that it has the properties assumed in Lemma 10.30. One can always do this, as in Remark 9.4 in Section 9.2.

10.7 Minimal representations

Let us return now to the topic of "minimal representations" from Section 9.2, for which the language and machinery of local +-isomorphisms is quite useful.

Recall from Definition 9.6 in Section 9.2 that every rooted tree (T, b) has an associated minimal representation M, b_0. Following our usual customs (from the beginning of Chapter 4), we allow our trees to be infinite but locally finite (i.e., with only finitely-many edges attached to any particular vertex), unless the contrary is explicitly stated. However, we shall make the standing assumption that

$$\text{all rooted trees } (T, b) \text{ considered here have finite minimal representations } M,\ b_0. \tag{10.3}$$

This is largely unnecessary, but it is sometimes convenient in dealing with the "minimality" properties of the minimal representation. It also helps to keep the general discussion simple, and it is consistent with our usual convention that "graphs" be finite. Notice that (10.3) holds automatically for rooted trees which come from visibility graphs, as in Corollary 9.25.

This assumption will be in force whenever we are considering minimal representations of rooted trees (in the next few sections). We shall also be free to treat rooted trees as being oriented, using the orientation described in Remark 9.4. We shall often do this, as in the next lemma, without mentioning it explicitly.

Lemma 10.32 *Let (T, b) be a finite rooted tree, and let M, b_0 be its minimal representation. There is a mapping $p : T \to M$ which is a local $+$-isomorphism, and sends b to b_0. This mapping p is also surjective on both vertices and edges.*

Proof Let T, M, etc., be given as above. Remember that the vertices of M are equivalence classes of vertices of T, by construction. (See Section 9.2.) Thus we automatically have a canonical quotient mapping from the vertices of T onto the vertices of M. This mapping sends b to b_0, by the definition of b_0.

There is not a canonical mapping from the edges of T to the edges in M, but there almost is. Fix vertices s in T and w in M, and let u be the vertex in M which corresponds to s. Let us write $T(s, w)$ for the set of edges in T which go from s to a vertex in T which lies in the equivalence class determined by w, and let $M(u, w)$ denote the set of edges in M which go from u to w. The main point now is that $T(s, w)$ and $M(u, w)$ have exactly the same number of elements. This comes from the construction of M in Section 9.2.

Every edge e in T lies in $T(s, w)$ for exactly one choice of s and w. Indeed, given e, we simply take s to be its "starting" vertex, and we choose w so that it represents the equivalence class of vertices in T that contains the other endpoint of e. Thus, to define our mapping p on the edges in T, it suffices to define it on each $T(s, w)$ separately.

We can choose p so that it defines a one-to-one correspondence between $T(s, w)$ and $M(u, w)$ for each selection of s and w. We can do this because $T(s, w)$ and $M(u, w)$ always have the same number of elements. In general, there is no particularly "canonical" way to choose the bijection between them, though.

No matter how these one-to-one correspondences between the $T(s, w)$'s and $M(u, w)$'s are chosen, in the end we get a mapping from edges in T to edges in M which is compatible with the canonical mapping between vertices mentioned before. Thus we get a mapping $p : T \to M$. It is not hard to check that this mapping is a local $+$-isomorphism, and that it is surjective on both vertices and edges. This completes the proof of Lemma 10.32. □

Remark 10.33 In Lemma 9.8 in Section 9.2, we asserted that if T and M are as above, then T is isomorphic to the visibility $\mathcal{V}_+(b_0, M)$ by a mapping which takes the basepoint b of T to the usual basepoint for the visibility. This can be derived as a corollary to Lemmas 10.32 and 10.30, using also Remark 10.31.

Lemma 10.34 *Let (T, b) be a rooted tree, and let M, b_0 be its minimal representation. Let G be an oriented graph, let v be a vertex in G, and suppose that T is isomorphic to $\mathcal{V}_+(v, G)$ by a mapping which takes b to the basepoint of $\mathcal{V}_+(v, G)$. Assume also that every vertex in G can be reached by an oriented path from v. Then there is a local $+$-isomorphism from G onto M which takes v to b_0.*

Notice that Lemma 10.9 provides a converse to this.

Proof We may as well assume that T simply *is* the visibility graph $\mathcal{V}_+(v, G)$, and that b *is* the usual basepoint for $\mathcal{V}_+(v, G)$, since the minimal representations of (T, b) and $\mathcal{V}_+(v, G)$ are isomorphic to each other (as in Remark 9.7).

Recall that the minimal representation M, b_0 can be derived (up to isomorphic equivalence) directly from G, through the procedure described in the proof of Proposition 9.17 in Section 9.4. This procedure gives rise to a mapping from G to the minimal representation, in exactly the same way as in the proof of Lemma 10.32. More precisely, in this procedure, the vertices of M are represented as equivalence classes of vertices in G, and this leads immediately to a mapping from vertices in G to vertices in M. The corresponding mapping on edges can be chosen in the same manner as in the proof of Lemma 10.32. This leads to a mapping between graphs, and it is not hard to see that it is a surjection and a local $+$-isomorphism, which is exactly what we wanted. □

Remark 10.35 If G has no oriented cycles, then we saw in Proposition 9.17 that the minimal representation M, b_0 could be obtained from G (up to isomorphic equivalence) in polynomial time. We also mentioned in Corollary 9.26 that the same conclusion holds in general if there is a polynomial-time solution to the visibility isomorphism problem. In both cases, the local $+$-isomorphism from G onto M mentioned in Lemma 10.33 can be obtained effectively in polynomial time, and through the same basic construction. Similarly, the mapping p in Lemma 10.32 can also be obtained in polynomial time.

Remark 10.36 Using Lemma 10.34, we can finish the proof of Proposition 9.2 in Section 9.2. All that remains is to establish the second part of Proposition 9.2, concerning the stronger minimality properties of the minimal representation. Specifically, let G be any oriented graph for which there is a vertex v in G such that the visibility $\mathcal{V}_+(v, G)$ is isomorphic to (T, b) as a rooted tree. If G_0 is the subgraph of G consisting of all vertices and edges which are accessible by oriented paths beginning at v, then the visibility of G_0 at v is isomorphic to the visibility of G at v, and hence is isomorphic (as a rooted tree) to (T, b). This permits us to apply Lemma 10.34 (with G replaced by G_0) to conclude the existence of a local $+$-isomorphism from G_0 onto M which takes v onto b_0. If G has the same size as M does, then this mapping must be injective, and hence an isomorphism, and G_0 must be all of G. (Otherwise the size of M would be strictly less than that of G.) This gives the second part of Proposition 9.2, as desired.

Note that these minimality properties of the minimal representation work just as well for infinite trees as finite trees (even though Proposition 9.2 is stated explicitly only for finite trees).

Let us now use our machinery of local +-isomorphisms to give an "effective witness" for the visibility recognition problem.

Lemma 10.37 *Let (T, b) be a rooted tree, and fix a positive integer k. Then the following are equivalent:* (1) *There is an oriented graph H of size at most k and a vertex w in H such that (T, b) is isomorphic to the visibility $\mathcal{V}_+(w, H)$ by an isomorphism which takes b to the basepoint of $\mathcal{V}_+(w, H)$;* (2) *There is an oriented graph H of size at most k and a local +-isomorphism $h : T \to H$.*

Proof Indeed, (1) implies (2) simply because the canonical projection $\pi : \mathcal{V}_+(w, H) \to H$ from Section 4.5 is always a local +-isomorphism, as in Lemma 10.10. Conversely, if (2) holds, then we can get an isomorphism as in (1) (with $w = h(b)$) from Lemma 10.30. This uses also Remark 10.31. □

The next lemma provides a similar effective witness for the "implicit" version of the visibility recognition problem (as described in Section 9.3).

Lemma 10.38 *Let (T, b) be a rooted tree, and assume that T is isomorphic to the visibility $\mathcal{V}_+(v, G)$ of some oriented graph G, where the isomorphism sends b to the basepoint of the visibility. Assume also that every vertex in G can be reached by an oriented path which begins at v. Then the following are equivalent:* (i) *There is an oriented graph H of size at most k and a vertex w in H such that (T, b) is isomorphic to the visibility $\mathcal{V}_+(w, H)$ by an isomorphism which takes b to the basepoint of $\mathcal{V}_+(w, H)$;* (ii) *There is an oriented graph H of size at most k and a local +-isomorphism $\theta : G \to H$.*

This lemma is a bit more subtle than the previous one, in that the graph H in (i) cannot always be used in (ii).

Proof That (ii) implies (i) follows immediately from the definitions and Lemma 10.9.

Conversely, assume that (i) holds, and let us try to prove (ii). We cannot convert the isomorphism between T and the visibility of H directly into a mapping from G to H, and so instead we argue as follows. Let M, b_0 be the minimal representation for (T, b), so that (T, b) is isomorphic to the visibility of M at b_0 (as in Remark 10.33). We also know that the size of M is not greater than the size of H, by Corollary 9.25. Thus the size of M is also at most k.

In other words, if (i) holds for some graph H, then it also holds with H, w replaced by the minimal representation M, b_0. Now we can use Lemma 10.34 to obtain the existence of a local +-isomorphism from G onto M that sends v to b_0. Thus (ii) holds, with H taken to be M, and the lemma follows. □

Corollary 10.39 *The implicit version of the visibility recognition problem (as stated in Section 9.3) lies in the class NP.*

Proof In this problem one, is given an oriented graph G, a vertex v in G, and a positive integer k, and one asks whether there is an oriented graph H of size $\leq k$ and a vertex w in H such that the visibility $\mathcal{V}_+(v, G)$ is isomorphic (as a rooted tree) to the visibility $\mathcal{V}_+(w, H)$. We may as well restrict our attention to graphs G in which all of the vertices and edges can be reached by an oriented path starting at v, because of Lemma 9.18. In this case, we can use Lemma 10.38 to say that the existence of such a graph H is equivalent to the existence of a local +-isomorphism from G into an oriented graph of size at most k. The latter is clearly an NP problem, because it is stated directly in terms of the existence of an effective witness which is of controlled size and whose validity can be verified in polynomial time. □

Corollary 10.40 *Consider the computational problem in which one is given an oriented graph M and a vertex b_0 in M, and one is asked to decide whether M, b_0 is isomorphic to the minimal representation of its own visibility graph $\mathcal{V}_+(b_0, M)$ (as a rooted tree). This problem lies in the class co-NP.*

Proof Let k be the size of M.

Claim 10.41 *M, b_0 is isomorphic to the minimal representation of $\mathcal{V}_+(b_0, M)$ if and only if there is* not *an oriented graph G of size strictly less than k such that $\mathcal{V}_+(b_0, M)$ is isomorphic (as a rooted tree) to $\mathcal{V}_+(v, G)$ for some vertex v in G.*

This follows easily from the precise form of the minimality properties of the minimal representation mentioned in Remark 10.36. Once we have this, Corollary 10.40 follows immediately from Corollary 10.39. □

Remark 10.42 If the visibility isomorphism problem (Section 9.4) admits a polynomial-time solution, then so does the problem described in Corollary 10.40. This is because a polynomial-time solution to the visibility isomorphism problem leads to a polynomial-time solution to the implicit version of the visibility recognition problem, as in Corollary 9.26. Similarly, the problem described in Corollary 10.40 can be solved in polynomial time when M is free of nontrivial oriented cycles, because of the solution to the implicit version of the visibility recognition problem in that case which is given by Proposition 9.17.

10.8 Mappings and effective witnesses

In Lemmas 10.37 and 10.38, we saw how mappings between graphs can provide natural effective witnesses for some computational questions about graphs, and we shall see another example of this in Lemma 10.43 in Section 10.9. What about effective witnesses for the "complements" of these problems, i.e., for the nonexistence of certain types of mappings between graphs?

Of course, this is a familiar theme in traditional geometry and topology, i.e., the search for invariants which reflect the obstructions to the existence of certains of mappings. In the present context of graphs, one might try to do this using adjacency transformations, for instance. (See Chapter 12 for definitions and basic properties, including Lemma 12.4 for the relationship between local +-isomorphisms and adjacency transformations.)

The problem of deciding when two graphs are isomorphic is well-known for resisting classification in terms of the usual complexity classes (see [Pap94]), and this indicates that one should not be overly-optimistic about the possibility of finding good classes of geometric invariants for graphs. Some of the complexity questions that we are considering here — like the existence of a local +-isomorphism into a graph of at most a given size — might be more flexible in this regard. Similarly, it would be nice to have some kind of computable criterion for a graph to be isomorphic to a minimal representation (of a rooted tree), or more generally to be isomorphic to a "minimal folding graph ", in the sense of Definition 10.56 in Section 10.11.

We shall see that the isomorphism problem for minimal representations and minimal folding graphs is somewhat easier than for graphs in general, and this suggests that it might be easier to find good invariants for dealing with them than for arbitrary graphs. A related issue is to find computable invariants which reflect the behavior of minimal representations or minimal folding graphs which are given only implicitly, e.g., as the minimal representation of the visibility of some other (non-minimal) graph. This is closely connected to looking for computable quantities which are invariant under surjective local +-isomorphisms, as we shall see in Section 10.16.

A different avenue to consider would be that of finding some kind of "proof system" for constructing all graphs which fail to have a certain property, like the existence of a local +-isomorphism into some particular graph, or into a graph of a given size. This should be compared with the "Hajós calculus" for constructing graphs which do not admit 3-colorings. (See Section 2.3. Remember also that the 3-coloring problem can be reformulated in terms of the existence of mappings into particular graphs, as in Proposition 10.3 and Lemma 10.23 in Sections 10.2 and 10.5.)

We shall encounter similar themes in Section 13.3, in connection with the NP-complete "visibility surjection problem".

10.9 The visibility isomorphism problem

We can also use local +-isomorphisms to obtain effective witnesses for the visibility isomorphism problem (described in Section 9.4).

Lemma 10.43 *Let G and H be oriented graphs, and fix vertices v in G and w in H. Assume that every vertex and edge in G can be reached from v by an oriented path, and that every vertex and edge in H can be reached by an oriented path from w. Then the following are equivalent:* (1) *there is an isomorphism between the visibilities $\mathcal{V}_+(v, G)$ and $\mathcal{V}_+(w, H)$ which takes the basepoint of $\mathcal{V}_+(v, G)$ to the basepoint of $\mathcal{V}_+(w, H)$;* (2) *there is an oriented graph M and orientation-preserving local +-isomorphisms $g : G \to M$ and $h : H \to M$ such that $g(v) = h(w)$.*

Note that there may not be a local $+$-isomorphism directly between G and H in this situation. Condition (2) is the next best thing, and it has the nice feature of being symmetric in G and H, as it should be.

As usual, the restriction to graphs G and H in which all vertices and edges are accessible from the basepoints is not a serious one, because of Lemma 9.18.

Proof This is almost the same as Lemma 10.38. The fact that (2) implies (1) follows immediately from Lemma 10.9. That is, the existence of local $+$-isomorphisms from G and H into a common graph M leads to isomorphisms from the visibilities of G and H to the visibility of M based at the vertex $g(v) = h(w)$, and this leads to an isomorphism between the visibilities of G and H.

Suppose instead that (1) holds, and let M, b_0 be the minimal representation for the rooted tree represented simultaneously by the visibilities $\mathcal{V}_+(v, G)$ and $\mathcal{V}_+(w, H)$ of G and H (in the sense of Definition 9.6). We can then get local $+$-isomorphisms $g : G \to M$ and $h : H \to M$, as in Lemma 10.34, and we also have that $g(v)$ and $h(w)$ are both equal to b_0. This proves (2), as desired. □

Corollary 10.44 *Given a pair of oriented graphs G and H and vertices v in G and w in H, the problem of deciding whether the visibilities $\mathcal{V}_+(v, G)$ and $\mathcal{V}_+(w, H)$ are isomorphic (by an isomorphism which preserves basepoints) lies in the class NP.*

This follows easily from Lemmas 10.43 and 9.18.

The next lemma concerns the possibility of producing the mappings mentioned in condition (2) in Lemma 10.43 in an effective manner.

Proposition 10.45 *Let G and H be oriented graphs, and let v and w be vertices in G and H, respectively. Assume that either* (a) *G and H contain no nontrivial oriented cycles, or* (b) *that the visibility isomorphism problem can be solved in polynomial time, whether or not the graphs in question contain nontrivial oriented cycles.*

Suppose also that the visibilities $\mathcal{V}_+(v, G)$ and $\mathcal{V}_+(w, H)$ are isomorphic to each other by an isomorphism which preserves the basepoints, and that every vertex and edge in G can be reached by an oriented path that begins at v, and that every vertex and edge in H can be reached by an oriented path that begins at w.

Under these conditions, there is an oriented graph M and mappings $g : G \to M$, $h : H \to M$ such that M, g, and h can be obtained from G, v, H, and w in polynomial time, g and h are surjections and local $+$-isomorphisms, and $g(v) = h(w)$.

Remember that if G and H are free of nontrivial oriented cycles, then the visibility isomorphism problem can be solved for them (and all of their vertices) in polynomial time, as in Proposition 9.11.

Proof Let G, H, etc., be as above. The argument is the same in essence as for the proof that (2) implies (1) in Lemma 10.43, except that we have to be more careful about how the mappings and graphs are produced. The following is the first main step.

Claim 10.46 *Given G and v, we can construct in polynomial time an oriented graph M', a vertex b_0' in M', and a mapping $g' : G \to M'$ such that M', b_0' is isomorphic to the minimal representation of the visibility $\mathcal{V}_+(v, G)$, g' is a surjection and a local $+$-isomorphism, and $g'(v) = b_0'$.*

This follows by combining some of our earlier results. The fact that we can build an isomorphic copy of the minimal representation of the visibility in polynomial time comes from Proposition 9.17 and Corollary 9.26. The existence of the mapping g comes from Lemma 10.34, and the fact that it can be produced in polynomial time was discussed in Remark 10.35. This proves the claim.

In the arguments that follow, we shall not only assume that M' and g' have the properties described above, but also that they were constructed through the methods of the results mentioned in the previous paragraph. We can do the same for H instead of G, to get an oriented graph M'', a vertex b_0'' in M'', and a mapping $h' : H \to M''$ of the same nature as for G.

The only problem now is that g' and h' map into different graphs.

Claim 10.47 *M' and M'' are isomorphic to each other, by an isomorphism which takes b_0' to b_0''.*

Indeed, we are assuming that the visibilities $\mathcal{V}_+(v, G)$ and $\mathcal{V}_+(w, H)$ are isomorphic to each other (as rooted trees), and this implies that they have isomorphic minimal representations. (See Remark 9.7.) This implies the existence of an isomorphism between M', b_0' and M'', b_0'', since they are isomorphic to the minimal representations of the visibilities of G and H at v and w (respectively), by construction. The remaining issue is to make the isomorphism between M' and M'' effective.

Claim 10.48 *If ξ and η are vertices in M' such that the visibilities $\mathcal{V}_+(\xi, M')$ and $\mathcal{V}_+(\eta, M')$ are isomorphic to each other (in a way that preserves basepoints, as usual), then ξ and η must actually be the same vertex in M'. (The analogous statement holds for M'' as well.)*

This comes from the basic constructions, going back to the proof of Proposition 9.17. Let ξ and η be given, as in the statement of Claim 10.48, and let x, y be vertices in G such that $g'(x) = \xi$, $g'(y) = \eta$. Since g' is a local $+$-isomorphism, it induces isomorphisms between the visibility of G at x and the visibility of M' at ξ, and also between the visibility of G at y and the visibility of M' at η, by Lemma 10.9. Our assumption about the visibilities of M' at ξ and η now implies that the visibilities of G at x and y are isomorphic to each other too. This means that x and y lie in the same equivalence class of vertices in G (as defined in the proof of Proposition 9.17), and hence that they are mapped to the same vertex in M' by g'. This last follows from the way that g' was constructed (in the proof of Lemma 10.34). Thus we conclude that $\xi = g'(x)$ and $\eta = g'(y)$ must be the same vertex in M', which is what we wanted. This proves Claim 10.48.

Claim 10.49 *For each vertex α in M', there is exactly one vertex β in M'' such that the visibilities $\mathcal{V}_+(\alpha, M')$ and $\mathcal{V}_+(\beta, M'')$ are isomorphic (by an isomorphism which preserves basepoints). Conversely, for each vertex β in M'', there is exactly one vertex α in M' with the same property.*

To see this, notice first that every isomorphism between M' and M'' induces an isomorphism between the corresponding visibility graphs. This implies the existence of β given α, or of α given β, and the uniqueness assertions follow from Claim 10.48. This proves Claim 10.49.

Our next task is to make certain that this correspondence between vertices can be found in polynomial time.

Claim 10.50 *Under the assumptions of Proposition 10.45, given vertices α in M' and β in M'', one can decide in polynomial time whether the visibilities $\mathcal{V}_+(\alpha, M')$ and $\mathcal{V}_+(\beta, M'')$ are isomorphic to each other (by isomorphisms which preserve the basepoints).*

Indeed, if G and H are free of nontrivial oriented cycles, then the same is true of M' and M'', by Lemma 10.17. This uses also the fact that $g' : G \to M'$ and $h' : H \to M''$ are surjections and local $+$-isomorphisms. In this case, the claim follows from Proposition 9.11. (Note that the sizes of M', M'' are less than or equal to the sizes of G, H, as in Corollary 9.25, or Claim 10.46.) If either G or H is not free of oriented cycles, then the assumptions of Proposition 10.45 imply that the visibility isomorphism problem is solvable in polynomial time for all graphs, and there is nothing to do. This proves the claim.

Claim 10.51 *Under the assumptions of Proposition 10.45, there is a polynomial-time algorithm for finding an isomorphism f between M' and M'' such that $f(b_0') = b_0''$.*

From Claim 10.49, we know that there is a unique one-to-one correspondence between the vertices of M' and M'' which is compatible with the visibilities, and Claim 10.50 implies that this correspondence can be computed in polynomial time. To complete this to an isomorphism between oriented graphs, we have just to define a compatible bijection between edges in M' and M''.

Remember that there is an isomorphism between M' and M'' which takes b_0' to b_0'', as in Claim 10.47. The action of this isomorphism on vertices has to be the same as the one just defined, since isomorphisms between oriented graphs induce isomorphisms between the corresponding visibility graphs.

From this it follows that if α and γ are arbitrary vertices in M', and if β, δ are their counterparts in M'' under the correspondence defined above, then the number j of oriented edges in M' which go from α to γ must be the same as the number of oriented edges in M'' which go from β to δ. That is, the number of these edges must be the same, because theses edges have to be mapped to each other under the isomorphism between M' and M'' that we know exists.

Once we know this, it is easy to see that a one-to-one correspondence between the edges of M' and M'' can also be chosen in polynomial time, and in a way that is compatible with the correspondence between vertices already defined. This gives us the isomorphism between M' and M'' that we seek. Note that it takes b_0' to b_0'', since that is true for the isomorphism mentioned in Claim 10.47, and since the two isomorphisms induce the same mappings on vertices (if possibly not on edges). This completes the proof of Claim 10.51.

The conclusions of Proposition 10.45 now follow by combining Claim 10.46 (and its analogue for H) with Claim 10.51. □

Remark 10.52 In the preceding proof, we encountered a number of facts about minimal representations which are true in general and interesting in their own right. We shall develop similar themes in more detail in Section 10.11.

10.10 Minimal representations and DP

DP is the complexity class which consists of problems which can be described by the intersection of two languages, with one language in NP and the other in co-NP. A basic example of a problem in DP is the "exact" version of the travelling salesman problem, in which one is given a positive integer k and a graph G with (integer) distances assigned to its edges, and one asks whether the shortest tour through all of the cities has length exactly equal to k. In order for this to be true, one needs to know that (a) there is a tour of length *at most* k, and (b) that there is no tour of length strictly *less* than k. Condition (a) describes an NP language, while (b) corresponds to a co-NP language, and this shows that the exact version of the travelling salesman problem lies in DP.

See Section 17.1 of [Pap94] for more information. Note that the exact version of the travelling salesman problem is actually *complete* for the class DP, and there are numerous other examples of DP-complete problems which are variants of standard examples of NP-complete problems, including "critical" versions of the satisfiability and colorability problems, or of the existence of Hamiltonian paths.

One should not confuse DP with the intersection of NP and co-NP, which consists of languages that simultaneously lie in NP and in co-NP.

Lemma 10.53 *Consider the following computational problem: one is given oriented graphs G and M and vertices v in G and b_0 in M, and one asks whether M, b_0 is isomorphic to the minimal representation for the rooted tree $\mathcal{V}_+(v, G)$ (equipped with its usual basepoint). This problem lies in the class DP.*

Proof To see this, we have just to observe that the requirement that M, b_0 be isomorphic to the minimal representation of $\mathcal{V}_+(v, G)$ is equivalent to the combination of two simpler statements, namely (1) that $\mathcal{V}_+(v, G)$ be isomorphic to $\mathcal{V}_+(b_0, M)$ and (2) that M, b_0 be isomorphic to the minimal representation of $\mathcal{V}_+(b_0, M)$. The first is an instance of the visibility isomorphism problem, which we know to lie in NP (Corollary 10.44). The second is an instance of the co-NP problem described in Corollary 10.40. This implies that the problem described in the lemma lies in DP. □

Corollary 10.54 *The problem described in Lemma 10.53 can be resolved in polynomial time if the visibility isomorphism problem can be, or if we restrict ourselves to graphs G and M which are free of oriented cycles.*

Proof This is an easy consequence of the proof of Lemma 10.53. Remember that the visibility isomorphism problem can be solved in polynomial time when the given graphs are free of oriented cycles, as in Proposition 9.11. This leaves the problem of deciding when M, b_0 is isomorphic to the minimal representation of $\mathcal{V}_+(b_0, M)$ in polynomial time, and this can be handled as in Remark 10.42 at the end of Section 10.7. □

10.11 Minimal folding graphs

Given an oriented graph G, can one find a "minimal" oriented graph M such that there is a local $+$-isomorphism from G onto M?

This is similar to the earlier questions about minimal representations of rooted trees and visibility graphs, except that we do not want to assume that everything in G can be reached by an oriented path starting from a single vertex v. Nonetheless, we can resolve the problem through practically the same kinds of constructions as used before, especially in the proof of Proposition 9.17.

More precisely, let us fix an oriented graph G, and say that two of its vertices q and q' in G are equivalent if the corresponding visibilities

$$\mathcal{V}_+(q, G) \quad \text{and} \quad \mathcal{V}_+(q', G) \tag{10.4}$$

are isomorphic to each other, by an isomorphism which preserves the basepoints. This defines an equivalence relation on the set of vertices of G, and we use the resulting set of equivalence classes for the set of vertices in M.

To define edges in M, we proceed as follows. Let ξ and η be two equivalence classes of vertices in G, and let x be a vertex in G which represents ξ. Let j denote the number of outgoing edges in G which arrive to a vertex in the equivalence class defined by η.

Lemma 10.55 *This number j does not depend on the choice of x in the equivalence class defined by ξ.*

The same point was discussed before, in Claim 9.21 in Section 9.4. In any case, it is not hard to check from the definitions.

Thus j depends only on ξ and η, and we complete the construction of M by attaching exactly j edges going from ξ to η, and applying the same procedure to all choices of ξ and η.

Definition 10.56 *Given an oriented graph G, the oriented graph M just described is called the* minimal folding graph *of G.*

See Fig. 10.4 for some examples.

Let us first mention one of the basic minimality properties of the minimal folding graph. This theme will be developed further in this and the next sections.

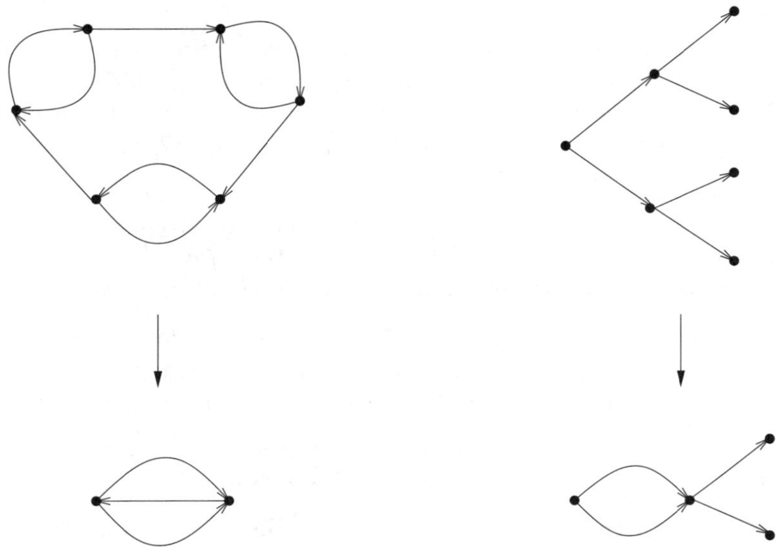

FIG. 10.4. Two examples of graphs and their minimal folding graphs

Lemma 10.57 *Let G be an oriented graph, and let M be its minimal folding graph.*

(a) The number of vertices in M is the same as the number of isomorphism classes of rooted trees which arise as the visibility of G at some vertex (with the root of the visibility always taken to be the basepoint, as usual).

(b) If H is an oriented graph, and $f : G \to H$ is a local $+$-isomorphism, then H contains at least as many vertices as M does.

Proof Part (a) follows immediately from the construction of the minimal folding graph. (Remember from Remark 9.4 that it does not matter if we prefer to think of visibility graphs as oriented trees or rooted trees (or both at the same time) for the purpose of defining "isomorphic equivalence".)

To get (b), we use Lemma 10.9, which says that f induces isomorphisms which preserve basepoints between the visibilities of G and H. Thus the isomorphism classes of rooted trees which occur as visibility graphs of G also arise from visibility graphs of H, so that (b) follows from (a). □

The next lemma makes precise the way in which the earlier notion of minimal representations is included in the present setting.

Lemma 10.58 *Let G be an oriented graph, let v be a vertex in G, and suppose that every other vertex and edge in G can be reached by an oriented path beginning at v. Let M be the minimal folding graph of G, and let M', b_0' denote the minimal representation of the visibility graph $\mathcal{V}_+(v, G)$ (viewed as a rooted tree, with its basepoint as root.) Then M and M' are isomorphic to each other as oriented graphs.*

In effect, this was shown in the proof of Proposition 9.17, since the construction of the isomorphic copy of the minimal representation given there is the same as that of the minimal folding graph. (Note that the absence of oriented cycles is not needed for this part of the proof of Proposition 9.17. See also Corollary 9.25 and the discussion thereafter.)

Lemma 10.59 *The minimal folding graph M of a given graph G can be constructed (up to isomorphism) in polynomial time if G is free of nontrivial oriented cycles, or if the visibility isomorphism problem can be solved in polynomial time.*

This is analogous to Proposition 9.17 and Corollary 9.26 in Section 9.4. It is also easy to verify directly from the construction, using Proposition 9.11 to deal with the visibility isomorphism problem when G has no nontrivial cycles.

Lemma 10.60 *Let G be an oriented graph, and let M be the minimal folding graph of G. Then there is a local $+$-isomorphism ϕ from G onto M.*

As usual, this mapping ϕ is canonical in its action on vertices, but this is not quite true for the action on edges.

Proof The proof of this is practically the same as for Lemmas 10.32 and 10.34 in Section 10.7. We have a canonical mapping from the set of vertices in G onto the set of vertices in M, in which we simply send each vertex in G to the corresponding equivalence class (using the equivalence relation defined above). For the edges, we do not quite have a canonical mapping, but we almost do. Let a vertex x in G and a vertex η in M be given, and let ξ be the vertex in M which represents the equivalence class of vertices in G that contains x. Let $G(x, \eta)$ denote the collection of all edges in G which go from x to a vertex in the equivalence class determined by η, and let $M(\xi, \eta)$ denote the collection of edges in M which go from ξ to η. Thus $G(x, \eta)$ and $M(\xi, \eta)$ have the same number of elements, by construction, and this is true for all choices of x and η. This permits us to choose ϕ so that it induces a one-to-one correspondence between $G(x, \eta)$ and $M(\xi, \eta)$ for all choices of x and η. There may be many ways to select the individual one-to-one correspondences between $G(x, \eta)$ and $M(\xi, \eta)$, but for the moment, we do not mind.

This defines ϕ as a mapping from G to M. It is easy to see that ϕ is surjective on vertices and edges, and that it is a local $+$-isomorphism. This proves the lemma. □

Remark 10.61 If we are in the situation of Lemma 10.59, so that the visibility isomorphism problem can be decided in polynomial time, then we can also construct a mapping $\phi : G \to M$ as in Lemma 10.60 in polynomial time. This follows easily from the construction.

Lemma 10.62 *Let G be an oriented graph, and let M be its minimal folding graph. If u_1 and u_2 are distinct vertices in M, then the corresponding visibilities*

$$\mathcal{V}_+(u_1, M) \quad and \quad \mathcal{V}_+(u_2, M) \tag{10.5}$$

are not *isomorphic to each other (by an isomorphism which preserves the basepoints).*

Proof Indeed, let G and M be given as above, and let u_1 and u_2 be two vertices in M for which the corresponding visibility graphs are isomorphic. We want to show that u_1 and u_2 must actually be the same vertex.

Let q_1 and q_2 be vertices in G which lie in the equivalence classes determined by u_1 and u_2, respectively. Thus the mapping ϕ in Lemma 10.60 sends q_1 to u_1 and q_2 to u_2 by construction. This implies that the visibilities $\mathcal{V}_+(q_1, G)$ and $\mathcal{V}_+(u_1, M)$ are isomorphic to each other, and similarly for q_2 and u_2, since ϕ is a local $+$-isomorphism. This uses Lemma 10.9. Employing our assumption that the visibilities of M at u_1 and u_2 be isomorphic, we conclude that $\mathcal{V}_+(q_1, G)$ and $\mathcal{V}_+(q_2, G)$ must be isomorphic to each other (by an isomorphism which preserves the basepoints, as usual).

Thus q_1 and q_2 actually lie in the same equivalence class of vertices in G. This implies that u_1 and u_2 must be equal to each other, by the definition of M, and the lemma follows. □

Corollary 10.63 *Let K be an oriented graph. The following are equivalent: (a) K is isomorphic to the minimal folding graph of some oriented graph G; (b) the visibility graphs of K at distinct vertices are not isomorphic to each other (by isomorphisms which preserve the basepoints); (c) K is isomorphic to its own minimal folding graph.*

Proof The fact that (a) implies (b) comes from Lemma 10.62, while (c) implies (a) is trivial. The passage from (b) to (c) is an easy consequence of the definition of the minimal folding graph. □

Definition 10.64 *Let M be an oriented graph. We shall say that M is* minimal *if it is isomorphic to the minimal folding graph of some oriented graph (and hence to the minimal folding graph of itself, by Corollary 10.63).*

Lemma 10.65 *Let M and N be oriented graphs, and let $f : M \to N$ be a mapping between them which is a local $+$-isomorphism. If M is minimal, then f is injective. In particular, f is an isomorphism if it is also a surjection.*

Proof Indeed, let $f : M \to N$ be as above, and suppose that two vertices u, w in M are mapped to the same vertex z in N. Then the visibilities of M at u and w are both isomorphic to the visibility of N at z, by Lemma 10.9. This implies that the visibilities of M at u and w must be isomorphic to each other, and hence that $u = w$, because of Lemma 10.62. Thus f is injective on vertices. It must also be injective on edges, since it is a local $+$-isomorphism, as in Lemma 10.15. This proves Lemma 10.65. □

Corollary 10.66 *An oriented graph M is minimal if and only if there does* not *exist a local $+$-isomorphism from M into an oriented graph with fewer vertices or fewer edges than M.*

Proof Indeed, if M is minimal, then every local $+$-isomorphism from M into another graph is an injection, by Lemma 10.65, and hence the image cannot have fewer edges or vertices. Conversely, suppose that M has the property that there is no local $+$-isomorphism from it into a graph with fewer vertices or edges. We can apply this to the local $+$-isomorphism ϕ from M onto its own minimal folding graph, given by Lemma 10.60, to conclude that the minimal folding graph of M has at least as many edges and vertices as M does. This forces ϕ to be an injection, and hence an isomorphism, since we already know that ϕ is a surjection. Thus M is isomorphic to its own minimal folding graph, and is therefore minimal. This proves Corollary 10.66. □

Corollary 10.67 *The problem of deciding whether or not a given oriented graph M is minimal lies in co-NP.*

This is an easy consequence of Corollary 10.66. Alternatively, one could use the characterization of minimal graphs in terms of distinct visibilities, as in Corollary 10.63, and the fact that the visibility isomorphism problem lies in NP (by Corollary 10.44).

Remark 10.68 One can decide in polynomial time whether a given oriented graph M is minimal or not if either M is free of nontrivial oriented cycles, or if the visibility isomorphism problem can be solved in finite time. This also follows from the characterization of minimal graphs in terms of distinct visibilities, employing Proposition 9.11 when there are no nontrivial oriented cycles. (One could also derive this using Lemma 10.59, but the preceding method is a bit more direct.)

Lemma 10.62 also implies the following rigidity property for mappings *into* minimal graphs.

Lemma 10.69 *Let H and M be oriented graphs, and let $g : H \to M$ and $h : H \to M$ be local $+$-isomorphisms. If M is minimal, then g and h define the same mappings on vertices. In particular, every local $+$-isomorphism from M into itself fixes every vertex in M.*

Proof If w is any vertex in H, then the visibility of H at w is isomorphic to the visibilities of M at $g(w)$ and $h(w)$, by Lemma 10.9. Thus the visibilities of M at $g(w)$ and $h(w)$ must be isomorphic to each other, which implies that $g(w) = h(w)$, because of Lemma 10.62. This proves Lemma 10.69. □

Lemma 10.69 makes it easier to find local $+$-isomorphisms into minimal graphs when they exist, especially in the context of trying to find an isomorphism between two minimal graphs. (See also Remark 10.90 in Section 10.13.)

10.12 Universal constructions

We shall now describe a kind of "universal minimal folding graph" $\mathcal{RT}$ with the property that the constructions of minimal representations and minimal folding graphs discussed before (in Sections 9.2 and 10.11) can be reformulated as mappings into this universal space.

For the vertices of $\mathcal{RT}$ we would like to take the "set" of isomorphism classes of all locally-finite rooted trees. (Hence the name "$\mathcal{RT}$", which stands for "rooted trees".) To make this more formally correct, we shall not work with literally *all* locally-finite rooted trees, but only ones whose vertices lie in a fixed countable set (like $\mathbf{Z}_+$). We can also ask that edges be represented by ordered pairs of vertices, since there will not be multiple edges between any two vertices (because we are working with trees). Let $\mathcal{C}$ denote this more restricted set of locally-finite rooted trees, and notice that every locally-finite rooted tree has an isomorphic copy contained in $\mathcal{C}$. Isomorphic equivalence between rooted trees defines an equivalence relation on $\mathcal{C}$, and for the set of vertices in $\mathcal{RT}$ we take the set of equivalence classes in $\mathcal{C}$. In the end, this the same in essence as the idea of isomorphism classes of all locally-finite rooted trees.

It is not hard to show that $\mathcal{RT}$ is an uncountable set with the cardinality of the continuum. If one prefers, one could restrict oneself to locally-finite rooted trees which arise as visibility graphs of (*finite*) graphs, in order to have a countable set.

To make $\mathcal{RT}$ into an (infinite) oriented graph, we attach edges to its vertices in the following manner. Let (T, b) be any locally finite rooted tree, which we think of as representing a vertex in $\mathcal{RT}$. Let $s_1, \ldots, s_k$ denote the vertices in T which are adjacent to b. There may be no such vertices (in which case T consists only of the vertex b and no edges), but our restriction to locally-finite trees ensures that there are only finitely many of the s_j's. Now imagine removing b from T, along with the k edges attached to b. What remains in T is a disjoint union of k trees $T_1, \ldots, T_k$, where T_j contains s_j for each j. Thus we get k rooted trees (T_j, s_j), $1 \leq j \leq k$, which represent vertices in $\mathcal{RT}$ in their own right. Note that some of these rooted trees may be isomorphic to each other, and therefore represent the same point in $\mathcal{RT}$. Some of the (T_j, s_j)'s could also be isomorphic to (T, b).

In any case, we attach k outgoing edges $e_1, \ldots, e_k$ to the vertex in $\mathcal{RT}$ determined by (T, b), with the edge e_j going from there to the vertex determined by (T_j, s_j). This collection of edges depends only on the isomorphism type of (T, b), and not on its particular representation.

By doing this for all isomorphism types of locally-finite rooted trees, we get all of the edges for $\mathcal{RT}$, which then becomes an oriented (infinite) graph.

Remark 10.70 The graph $\mathcal{RT}$ is *not* locally finite, and indeed there are always infinitely many edges going *into* any vertex. This is not hard to check; given any rooted tree (S, c), we can realize it as a (T_j, s_j) for infinitely many choices of (T, b). However, there are only finitely many edges coming *out* of any vertex in $\mathcal{RT}$, by construction.

If G is any (finite) oriented graph, we automatically get a mapping

$$\Phi : G \to \mathcal{RT}, \tag{10.6}$$

which is defined as follows. If v is any vertex in G, then we choose $\Phi(v)$ to be the vertex in $\mathcal{RT}$ which represents the same rooted tree as the visibility $\mathcal{V}_+(v, G)$ (with its basepoint as root). It is not hard to extend this to a mapping between edges, in such a way that Φ becomes a local +-isomorphism. (We have made this type of observation a number of times now. The main point is the following. Fix a vertex v in G and a vertex β in $\mathcal{RT}$, and set $\alpha = \Phi(v)$. Then the number of oriented edges in $\mathcal{RT}$ which go from α to β is the same as the number of outgoing edges e from v which arrive to a vertex w such that $\Phi(w) = \beta$. This can be checked directly at the level of trees and visibility graphs, and it implies that Φ can be extended to the edges in a way which leads to a local +-isomorphism from G into $\mathcal{RT}$.)

This mapping is canonically defined on the vertices of G, but not on the edges, for which there may be some local permutations (as in the context of Lemmas 10.32 and 10.60 in Sections 10.7 and 10.11). However, the image of G under Φ is canonical (and does not depend on the particular choices of local correspondences between the edges). In fact, the image of G under Φ provides another representation of the minimal folding graph of G. All that Φ really does is to repeat the construction of the minimal folding graph of G, but inside this universal graph $\mathcal{RT}$.

Similar observations apply to rooted trees and their minimal representations. To be precise, let (T, b) be any locally-finite rooted tree, and let us think of T as being oriented, using the orientation described in Remark 9.4. If s is any vertex in T, let $F(s)$ denote the subtree of T consisting of all vertices and edges which can be reached by oriented paths starting at s (as in Section 9.2). Thus for each vertex s in T, we get a rooted tree $(F(s), s)$, and this defines a mapping from vertices in T to vertices in the universal space $\mathcal{RT}$. This mapping can be extended to edges in such a way as to get a local +-isomorphism from T into $\mathcal{RT}$, in the same way as before.

This construction for rooted trees is the same in essence as the one for graphs and their visibilities, because the subtree $F(s)$ of T is isomorphic in a canonical way to the visibility of T at s. One has only to allow the notion of visibility to be applied to locally finite trees even if they are infinite, which does not really cause any trouble. For trees, however, it is a little simpler to forget about visibility graphs and simply think in terms of the subtrees $F(s)$ as above.

In this case, the image of T under our mapping into $\mathcal{RT}$ provides another formulation of the definition of minimal representations from Section 9.2. We also get the following.

Lemma 10.71 *Let (T, b) be any locally-finite rooted tree, and let α denote the vertex in $\mathcal{RT}$ which represents this rooted tree. Then the visibility of $\mathcal{RT}$ at α is isomorphic to T by an isomorphism that takes the basepoint of the visibility graph to the root b of T.*

In other words, the visibility of our universal graph $\mathcal{RT}$ always looks like the rooted tree which is represented by the vertex in $\mathcal{RT}$ from which the visibility is taken.

Strictly speaking, the definition of the visibility is not supposed to be applied to infinite graphs like $\mathcal{RT}$, but there is not really any trouble in extending the usual definition to this case, especially since $\mathcal{RT}$ is locally finite in the "positive" directions (as in Remark 10.70). In particular, this ensures that the visibility of $\mathcal{RT}$ (in *positive* directions) is locally-finite.

Perhaps the easiest way to prove the lemma is to use the fact that we have a local +-isomorphism from T into $\mathcal{RT}$ as described above, and that this mapping takes b to α. This puts us in a position to apply something like Lemma 10.9 or Lemma 10.30 in Section 10.6. Although the particular statements given before do not apply here, it is easy to see that the same arguments still work. (The main point is simply that every oriented path in $\mathcal{RT}$ beginning at α corresponds in a unique way to an oriented path in T that begins at b, and this is true because it is true for each individual step along the way, by construction.)

Corollary 10.72 *If α and β are distinct vertices in $\mathcal{RT}$, then the visibility graphs of $\mathcal{RT}$ at α and β are* not *isomorphic as rooted trees.*

Indeed, Lemma 10.71 implies that the visibilities at α and β are isomorphic as rooted trees to the rooted trees represented by α and β themselves, which cannot be isomorphic to each other since α and β represent distinct vertices in $\mathcal{RT}$. (In other words, if α and β represented isomorphic rooted trees, then they would have to be the same vertex in $\mathcal{RT}$, by construction.)

To summarize a bit, Corollary 10.72 shows that $\mathcal{RT}$ is its own minimal folding graph, as in Corollary 10.63, while our earlier observations imply that $\mathcal{RT}$ contains all other minimal folding graphs. In fact, minimal folding graphs lie in $\mathcal{RT}$ in a unique way (through local +-isomorphisms), because the positions of their vertices are pinned down by the nature of their visibility graphs. (Compare with Lemma 10.69.)

Let us mention one more property of $\mathcal{RT}$.

Lemma 10.73 *If α, β are any two vertices in $\mathcal{RT}$, then there is an (unoriented) path in $\mathcal{RT}$ which goes from α to β in two steps.*

Of course the behavior of the *oriented* paths in $\mathcal{RT}$ is much more restricted.

Proof Let α, β be given, and let (T, b), (S, c) be rooted trees in the isomorphism classes represented by α and β, respectively. Let (U, d) be the rooted tree obtained by taking the disjoint union of S and T, adding a new vertex d, and also an edge from d to b and from d to c. Let γ denote the vertex in $\mathcal{RT}$ which corresponds to the isomorphism class of (U, d). From the way that we defined $\mathcal{RT}$, it follows that there are oriented edges in $\mathcal{RT}$ which go from γ to α and from γ to β, which proves the lemma. □

One can take the idea of universal constructions a bit further, as follows. Fix some set-theoretic universe in which to work, and let **RT** denote the set of literally *all* locally-finite rooted trees. That is, we treat two trees as being different as soon as they are different objects set-theoretically, whether or not they are isomorphic to each other. It is easy to make **RT** into an oriented graph, by attaching edges in roughly the same manner as for $\mathcal{RT}$. This graph is far from being locally-finite, but it is true that every vertex has only finitely many outgoing edges, for the same reason as before.

In this case we get some kind of huge forest (disjoint union of trees). There is a natural mapping from this forest to $\mathcal{RT}$, in which one takes a given rooted tree and assigns to it the corresponding isomorphism class in $\mathcal{RT}$. This defines a mapping between vertices, and one can also choose a correspondence between edges to get a local +-isomorphism in the end. One can think of $\mathcal{RT}$ as being the minimal folding graph of **RT**, in which case this mapping corresponds to the one described in Lemma 10.60.

Although we shall not rely on the graph $\mathcal{RT}$ in a serious way in this book, it sometimes provides an interesting perspective.

10.13 The visibility spectrum

Definition 10.74 *Let G be an oriented graph. By the* visibility spectrum *of G we mean the collection $\mathcal{S}(G)$ of isomorphism classes of rooted trees which arise as visibility graphs of G (at arbitrary vertices in G). (For this we always use the basepoints of the visibility graphs as their roots.)*

To make this more precise (set-theoretically), one can think of $\mathcal{S}(G)$ as being a subset of the set of vertices in the graph $\mathcal{RT}$, as described in Section 10.12. In fact, one can simply think of $\mathcal{S}(G)$ as being the image under the mapping Φ in (10.6) of the set of vertices in G.

In practice, we shall often not need to know $\mathcal{S}(G)$ as a precise set. Instead, we shall simply want to be able to make comparisons between the visibility spectra of different graphs, e.g., to say when two graphs have the same visibility spectrum, or when the visibility spectrum of one is contained in the other.

Note that we do not count multiplicities here, i.e., the number of times that a given rooted tree arises as the visibility of a particular graph G. One certainly could do this, and it might be interesting to look at the distribution of the multiplicities as a measurement of structure in the graph (e.g., how "diffuse" it is).

One of the main points of this section will be to show how relations between visibility spectra are manifested at the level of mappings between graphs.

Lemma 10.75 *Let K and L be oriented graphs, and suppose that there is a local +-isomorphism f from K into L. Then the visibility spectrum of K is contained in that of L. If also f maps K onto L, then the visibility spectra of K and L are the same.*

Proof This is an easy consequence of the fact that f induces isomorphisms between the corresponding visibility graphs, as in Lemma 10.9. □

Corollary 10.76 *The visibility spectrum of an oriented graph is always the same as that of its minimal folding graph (Definition 10.56).*

This follows immediately from Lemmas 10.75 and 10.60.

Lemma 10.77 *The number of elements in the visibility spectrum of a given oriented graph G is never greater than the number of vertices in G, and the two numbers are equal if and only if G is minimal (Definition 10.64).*

Proof The first part of the lemma is immediate from the definition of the visibility spectrum, while the second can be derived from Corollary 10.63. □

We can also rephrase this by saying that there is a canonical mapping from the set of vertices of G onto $\mathcal{S}(G)$ (which takes a vertex v in G and assigns to it the isomorphism class of its visibility graph), and that this mapping is a one-to-one correspondence if and only if G is minimal.

Proposition 10.78 *Let G and H be oriented graphs, and let M and N be their minimal folding graphs. Then the visibility spectrum of G is contained in that of H if and only if there is a local $+$-isomorphism from M into N, and the visibility spectra of G and H are equal to each other if and only if M and N are isomorphic.*

This provides a kind of partial converse to Lemma 10.75. The more "direct" converse is not true: although local $+$-isomorphisms between graphs always lead to comparisons between the corresponding visibility spectra, in general we can go back in the opposite direction (from comparisons between visibility spectra to mappings) only at the level of minimal folding graphs. (See Proposition 10.101 in Section 10.16 for some related observations.)

Proof The "if" parts of both assertions follow easily from Lemma 10.75 and Corollary 10.76. To prove the "only if" parts, we proceed as follows.

Suppose that the visibility spectrum of G is contained in that of H. From Corollary 10.76 we get that the visibility spectrum of M is automatically contained in that of N. This leads to a one-to-one mapping f from the vertices of M into the vertices of N, where f maps a vertex u in M to a vertex w in N exactly when the visibility of M at u is isomorphic to the visibility w of N. This uses Lemma 10.62 to know that there is at most one choice of w possible for a given u, and at most one u for each w.

To extend f as a mapping between edges, we need to show the following.

Claim 10.79 *Let u_1 and u_2 be two vertices in M. Then the number of edges in M that go from u_1 to u_2 is the same as the number of edges in N which go from $f(u_1)$ to $f(u_2)$.*

This is another minor variation of a fact that we have used several times now, as in Lemma 10.55 in Section 10.11 and Claim 9.21 in Section 9.4, for instance. In fact, one can derive this version from the one in Lemma 10.55, by taking G to be the disjoint union of M and N, and noticing that the equivalence relation on vertices used in the context of Lemma 10.55 is the same as the relation between vertices in M and in N under consideration here.

Once we have Claim 10.79, we can extend f to a mapping between edges so that f induces a one-to-one correspondence between the edges that go from u_1 to u_2 in M and the edges that go from $f(u_1)$ to $f(u_2)$ in N for every pair of vertices u_1, u_2 in M. As usual, this mapping between is not uniquely determined in general, because of the possibility for permuting the edges between a given pair of vertices.

By construction, the mapping between edges is compatible with the mapping between vertices, so that $f : M \to N$ is a bona fide mapping between graphs. It is clearly a local +-injection too, and the only possible way that f could fail to be a local +-isomorphism would be if there were a vertex u in M such that $f(u)$ has an outgoing edge in N which flows into a vertex in N that does not lie in the image of f. To see that this cannot happen, it is enough to know that the total number of edges in N coming out of $f(u)$ is the same as the total number of edges in M that flow from u. This last assertion holds because the visibility of M at u is isomorphic to the visibility of N at $f(u)$, by construction.

Thus we have a local +-isomorphism from M into N, which is what we wanted. This mapping is also injective, as one can see from the construction (and which must also be true in general, by Lemma 10.65).

There is one last point, which is to show that f is an isomorphism between M and N when the visibility spectra of G and H are the same. Under this assumption, we have that the visibility spectra of M and N must also be the same, as in Corollary 10.76. This implies that every vertex in N must arise as the image of a vertex in M under f, because of the way that we defined the action of f on vertices. Thus f actually induces a bijection between the vertices of M and N, and it is easy to check that the same must be true for the edges. (See also Lemma 10.15.) Thus f gives an isomorphism between M and N, and the proposition follows. □

Corollary 10.80 *Let G and H be oriented graphs, and assume that there is a local +-isomorphism from G onto H. Then the minimal folding graphs of G and H are isomorphic to each other.*

Indeed, the visibility spectra of G and H are the same in this case, as in Lemma 10.75, and so the corollary follows from Proposition 10.78.

Corollary 10.81 *Let G and H be oriented graphs, and let M be the minimal folding graph of G. Assume that there is a local +-isomorphism from G onto H. Then the size of H is at least as large as the size of M, and H and M must be isomorphic to each other if they have equal size.*

This gives a more precise formulation of the "minimality" of the minimal folding graph.

Proof The assumption that there be a local +-isomorphism from G onto H implies that M is isomorphic to the minimal folding graph of H (Corollary 10.80). Hence there is a local +-isomorphism from H onto M (because of Lemma 10.60). This implies that the size of H must be at least as big as the size of M. In the case of equality, the mapping from H onto M must also be injective, and hence an isomorphism. □

Corollary 10.82 *If G is an oriented graph and k is a positive integer, then the minimal folding graph of G has size at most k if and only if there is a local $+$-isomorphism from G onto an oriented graph H of size at most k.*

Proof The "if" part follows from Corollary 10.81, and the "only if" part follows from the fact that there is always a local $+$-isomorphism from G onto its minimal folding graph (Lemma 10.60). □

Corollary 10.83 *The problem of deciding whether the minimal folding graph of a given oriented graph G has size at most k is in the class NP.*

This follows immediately from Corollary 10.82.

Corollary 10.84 *Let G be an oriented graph, and let k be a positive integer. The problem of deciding whether G admits a local $+$-isomorphism onto an oriented graph H of size at most k can be solved in polynomial time if G is free of nontrivial oriented cycles, or if the visibility isomorphism problem can be solved in polynomial time for arbitrary oriented graphs.*

Proof Corollary 10.82 shows that this problem is equivalent to deciding whether the minimal folding graph M of G has size at most k, and the minimal folding graph can even be constructed (up to isomorphism) in polynomial time under the conditions above, as in Lemma 10.59. □

Corollary 10.85 *If G and M are oriented graphs, then M is isomorphic to the minimal folding graph of G if and only if M is minimal and there is a local $+$-isomorphism from G onto M.*

Proof If M is isomorphic to the minimal folding graph of G, then M is automatically minimal (Definition 10.64), and there is a local $+$-isomorphism from G onto M by Lemma 10.60. Conversely, if there exists a local $+$-isomorphism from G onto M, then G and M must have isomorphic minimal folding graphs, as in Corollary 10.80. If M is also minimal, then M is isomorphic to its own minimal folding graph, by Corollary 10.63, and hence to the minimal folding graph of G. □

Corollary 10.86 *Given a pair of oriented graphs G and M, the problem of deciding whether M is isomorphic to the minimal folding graph of G lies in the class DP (discussed in Section 10.10).*

This is analogous to Lemma 10.53 and Corollary 10.54 in Section 10.10.

Proof This is an easy consequence of Corollary 10.85 and the fact that deciding minimality is co-NP (Corollary 10.67), since determining the existence of a local $+$-isomorphism from G onto M is clearly an NP problem. □

Corollary 10.87 *Let G and H be oriented graphs. The minimal folding graphs of G and H are isomorphic to each other if and only if there exist an oriented graph K and surjective local $+$-isomorphisms $g : G \to K$ and $h : H \to K$.*

This is analogous to Lemma 10.43 in Section 10.9.

Proof If the minimal folding graphs of G and H are isomorphic to each other, then we can take K to be the common minimal folding graph, and use Lemma 10.60 to get the desired mappings. Conversely, if K, g, and h exist as above, then Corollary 10.80 implies that the minimal folding graphs of G and H are each isomorphic to that of K, and hence to each other. □

Corollary 10.88 *The problem of determining whether a given pair of oriented graphs G and H have isomorphic minimal folding graphs lies in the class NP.*

This is an easy consequence of Corollary 10.87. (Compare with Corollary 10.44, for the visibility isomorphism problem.)

Corollary 10.89 *Let G and H be oriented graphs, and assume that either G and H contain no nontrivial oriented cycles, or that the visibility isomorphism problem can be solved in polynomial time (for all oriented graphs, whether or not they have nontrivial oriented cycles). Then the problem of deciding whether G and H have isomorphic minimal folding graphs can be solved in polynomial time, and the problem of deciding whether H is itself isomorphic to the minimal folding graph of G can be solved in polynomial time.*

Proof For this it is better to use Proposition 10.78 than Corollary 10.87. From Proposition 10.78 we know that isomorphic equivalence of the minimal folding graphs can be tested in terms of the equality of the visibility spectra, and this is exactly what can be solved in polynomial time under the hypotheses of the corollary. (This uses Proposition 9.11 in Section 9.4 to handle the visibility isomorphism problem in the case of graphs without nontrivial oriented cycles.) As for the second part, H is isomorphic to the minimal folding graph of G if and only if (1) H is itself minimal and (2) the minimal folding graphs of G and H are isomorphic to each other. We have already seen that (2) can be checked in polynomial time under the conditions of the lemma, while for (1) we have a similar fact from Remark 10.68. □

Remark 10.90 Let G and H be oriented graphs, and assume that either they are both free of nontrivial oriented cycles, or that the visibility isomorphism problem can be solved in polynomial time. Then the minimal folding graphs M and N of G and H can be constructed (up to isomorphism) in polynomial time, as in Lemma 10.59. If M and N are actually isomorphic, then the method of the proof of Proposition 10.78 can also be used to produce an isomorphism between M and N in polynomial time in this case. Indeed, the fact that one can make comparisons between visibility graphs in polynomial time in the present situation (using Proposition 9.11 when G and H contain no nontrivial oriented cycles) permits one to compute the (unique) correspondence f between vertices in

M and N in polynomial time. Once one has this, it is easy to fill in a compatible mapping between edges in M and N (because one knows that the number of edges going from a vertex u_1 to a vertex u_2 in M has to be the same as the number of edges from $f(u_1)$ to $f(u_2)$ in N).

These observations are analogous to Proposition 10.45 in Section 10.9. A general point is that one should expect it be easier to find isomorphisms or local +-isomorphisms between minimal graphs, because of the uniqueness statement provided by Lemma 10.69.

Remark 10.91 Given a (finite) collection of isomorphism types of rooted trees, when do they form the visibility spectrum of a finite graph? There is a simple answer to this question in terms of "successors" of rooted trees. Let us say that a rooted tree (S, c) is a "successor" of a rooted tree (T, b) if the vertex τ in the universal space $\mathcal{RT}$ from Section 10.12 which represents (T, b) has an edge going to the vertex σ in $\mathcal{RT}$ which represents (S, c). This is the same as saying that if you remove b from T, along with all of the edges attached to it, then S is isomorphic to one of the components of the remaining forest, by an isomorphism which takes c to a vertex which is adjacent to b in T. It is not hard to show that a finite collection $\mathcal{C}$ of isomorphism types of (locally-finite) rooted trees is the visibility spectrum of a finite graph if and only if $\mathcal{C}$ contains the successors of all of its elements. In fact, one can realize the minimal graph with this visibility spectrum simply by taking the subgraph of $\mathcal{RT}$ which consists of the vertices associated to the elements of $\mathcal{C}$ and all edges in $\mathcal{RT}$ that go between these vertices. One can also produce it more concretely through the same construction as for minimal folding graphs. There is an analogous point concerning the finiteness of the minimal representation of a given rooted tree (T, b) (as in Definition 9.6), which is characterized by the requirement that there be a finite set of isomorphism types of rooted trees which contains (T, b) as well as all of its successors, the successors of its successors, etc. This can be checked directly from the construction of the minimal representation in Section 9.2.

10.14 The local +-isomorphism problem

Recall from Definition 10.27 in Section 10.5 that in the local +-isomorphism problem one is given a pair of oriented graphs G and H, and one is asked if there is a mapping from G into H which is a local +-isomorphism. This is somewhat different from some other problems concerning local +-isomorphisms that we have considered so far, in which the target graph was *not* given in advance, but was part of the structure whose existence is in question. (See Lemmas 10.37, 10.38, and 10.43, Corollary 10.66, and the various corollaries in Section 10.13, for instance.)

Proposition 10.92 *If the local +-isomorphism problem can be solved in polynomial time for graphs without nontrivial oriented cycles, then the same is true for oriented graphs in general.*

This should be compared with Propositions 9.11 and 9.17, Corollaries 10.54 and 10.89, and Remarks 10.68 and 10.90, in which the absence of oriented cycles seemed to make the problems in question easier to treat.

Proof Let G and H be arbitrary oriented graphs. We want to show how to transform G and H into oriented graphs G' and H' which are free of nontrivial oriented cycles, and which have the property that a local $+$-isomorphism from G into H exists if and only if there is a local $+$-isomorphism from G' into H'.

To do this, we use a "double-decker" construction. Fix a positive integer k, which should be chosen strictly larger than the number of edges going from one vertex p to a different vertex q in H, for any choices of p and q.

For the vertices of G' we take the set of ordered pairs (v, i), where v is a vertex in G and $i \in \{1, 2\}$. For the edges we do the following. If v and z are any vertices in G, and if G contains $j = j(v, z)$ edges going from v to z, then we attach j edges in G' going from $(v, 1)$ to $(z, 2)$. In addition to these edges, we attach another k edges going from $(v, 1)$ to $(v, 2)$ in G' for every vertex v in G. We call these the "special vertical edges" from $(v, 1)$ to $(v, 2)$, and we use the phrase "induced edges" for edges in G' that correspond to edges in G. Note that there can be induced edges from from $(v, 1)$ to $(v, 2)$ in G', coming from edges in G which go from v to itself.

These are all of the edges that we attach to G'. In the end we obtain an oriented graph which is obviously free of nontrivial *oriented* cycles, since all of the edges go from the first level to the second. We construct H from H' in exactly the same manner.

Claim 10.93 *If $f : G \to H$ is a local $+$-isomorphism, then there is a local $+$-isomorphism $f' : G' \to H'$.*

Indeed, given any orientation-preserving mapping $f : G \to H$, we can define an associated mapping $f' : G' \to H'$ in the following manner. If v is any vertex in G, then we set $f'(v, i) = (f(v), i)$ for $i = 1, 2$. This defines f' as a mapping between vertices. Now let e be an edge in G, which goes from some vertex v to another vertex z, so that $f(e)$ is an edge in H which goes from $f(v)$ to $f(z)$. By construction, e has a counterpart e' in G' which goes from $(v, 1)$ to $(z, 2)$, and we define $f'(e')$ to be the edge in H' which is induced by $f(e)$ (and goes from $(f(v), 1)$ to $(f(z), 2)$).

If instead we have a special vertical edge in G' that goes from $(v, 1)$ to $(v, 2)$ for some vertex v in G, then we map it to a special vertical edge in H' which goes from $(f(v), 1)$ to $(f(v), 2)$. We do not care how the special vertical edges from $(v, 1)$ to $(v, 2)$ in G' correspond to the special vertical edges from $(f(v), 1)$ to $(f(v), 2)$ in H', so long as this correspondence be a bijection for each vertex v in G. (Note that we have the same number k of such edges in both cases.)

This defines $f' : G' \to H'$ as an orientation-preserving mapping between graphs. It is easy to see that f' is also a local $+$-isomorphism if f is. This proves Claim 10.93.

Next we want to show that local +-isomorphisms from G' into H' lead to local +-isomorphisms from G into H.

Claim 10.94 *Suppose that $g : G' \to H'$ is a local +-isomorphism. Given a vertex v in G, let (x, i_1) and (y, i_2) be the vertices in H' given by $g(v, 1)$ and $g(v, 2)$, respectively. Then $i_1 = 1$, $i_2 = 2$, and $y = x$.*

To see this, notice first that (x, i_1) must have edges coming out of it in H', since the same is true of $(v, 1)$ in G'. This implies that $i_1 = 1$. Similarly, (y, i_2) must have incoming edges, since $(v, 2)$ does, and this yields $i_2 = 2$.

Now let us show that $y = x$. Remember that there are k special vertical edges in G' going from $(v, 1)$ to $(v, 2)$, and therefore at least k edges in H' going from $g(v, 1)$ to $g(v, 2)$. This uses the assumption that g be a local +-isomorphism, and it would work just as well if g were a local +-injection. If x were not equal to y, then all of the edges in H' going from $(x, 1)$ to $(y, 2)$ would be have to be induced from edges in H that go from x to y. The number of these edges would then be strictly less than k, because of the way that we chose k. This contradicts the fact that there are at least k edges going from $g(v, 1)$ to $g(v, 2)$, and Claim 10.94 follows.

Claim 10.95 *If $g : G' \to H'$ is a local +-isomorphism, then there is a local +-isomorphism $f : G \to H$ (which is induced from g in an almost-canonical way).*

Using Claim 10.94, we can define f on the vertices of G in such a way that

$$g(u, i) = (f(u), i) \tag{10.7}$$

for every vertex u in G and for $i = 1, 2$. The next step is to define f as a mapping on edges.

Fix a vertex v in G, and let $C(v)$ denote the set of vertices z in G such that $f(z) = f(v)$. If e is an edge in G which flows out of v and ends in a vertex z which does *not* lie in $C(v)$, then we can define $f(e)$ in a canonical way as follows. Let e' be the "induced edge" from $(v, 1)$ to $(z, 2)$ in G' that corresponds to e. Then g maps e' to an edge a' in H' which goes from $(f(v), 1)$ to $(f(z), 2)$. Since z does not lie in $C(v)$, we have that $f(z) \neq f(v)$, and therefore a' is an induced edge (as opposed to a special vertical edge). We set $f(e) = a$, where a is the edge in H that goes from $f(v)$ to $f(z)$ and corresponds to the induced edge a'.

Now let e be an edge in G which goes from v to a vertex z in $C(v)$, and let e' be the corresponding induced edge in G' that goes from $(f(v), 1)$ to $(f(v), 2)$ (since $f(z) = f(v)$ in this case). (Note that z might be equal to v.) Unfortunately, the image of e' under g may not be an induced edge in this case, because the induced edges in the image can be mixed with the special vertical edges. We have to be a little more careful in choosing the value of f at e.

Let $\gamma(v)$ denote the total number of edges in G which go from v to an element of $C(v)$ (including v itself), and let $\lambda(f(v))$ denote the total number of edges in H which begin and end at $f(v)$. We would like to show that

$$\gamma(v) = \lambda(f(v)). \tag{10.8}$$

This would permit us to choose f in such a way that it induces a one-to-one correspondence between the edges in G which go from v to an element of $C(v)$ and the edges in H which go from $f(v)$ to itself.

To prove (10.8), we use the requirement that g be a local $+$-isomorphism. In addition to the induced edges in G' and H', we also have the special vertical edges, and in particular we have k special vertical edges in G' which go from $(v,1)$ to $(v,2)$, and k special vertical edges in H' which go from $(f(v),1)$ to $(f(v),2)$. All of the special vertical edges in G' which go from $(v,1)$ to $(v,2)$ are mapped to edges in H' which go from $(f(v),1)$ to $(f(v),2)$, but they may not be mapped to special vertical edges. They might instead be mixed with the induced edges in H', which is the same problem as before.

However, the total numbers are conserved. The total number of edges in H' which go from $(f(v),1)$ to $(f(v),2)$ is equal to $\lambda(f(v))+k$. This accounts for both the induced edges and the special vertical edges. Similarly, $\gamma(v)+k$ represents the total number of edges in G' which go from $(v,1)$ to $(z,2)$ for some element z of $C(v)$, including v itself, for which one has k special vertical edges. These $\gamma(v)+k$ edges in G' are precisely the ones which begin at $(v,1)$ and which are mapped by g to edges in H' that go from $(f(v),1)$ to $(f(v),2)$. Since g is a local $+$-isomorphism by assumption, it induces a one-to-one correspondence between these two collections of edges in G' and H', and we conclude that

$$\gamma(v) + k = \lambda(f(v)) + k. \tag{10.9}$$

This proves (10.8).

From (10.8) it follows that there is a one-to-one correspondence between the edges in G which go from v to an element of $C(v)$ and the edges in H which go from $f(v)$ to itself. Fix such a correspondence, and define f on the edges in G that go from v to an element of $C(v)$ so that it agrees with this correspondence. This completes the definition of f on edges in G that come out of v, since we took care of the case of edges that go from v to vertices not in $C(v)$ before.

By doing this for every vertex v in G, we can define f on all edges in G. It is easy to see from the construction that $f : G \to H$ is a legitimate mapping between graphs (i.e., that the mappings on vertices and edges are compatible with each other), and that f preserves orientations. One can also check that f is a local $+$-isomorphism, using the assumption that g be a local $+$-isomorphism.

This finishes the proof of Claim 10.95. From Claims 10.93 and 10.95 we see that the existence of a local $+$-isomorphism from G to H is equivalent to the existence of such a mapping from G' to H'. This proves Proposition 10.92, since the construction of G' and H' from G and H can easily be carried out in polynomial time. □

Remark 10.96 The graphs G' and H' constructed in the proof of Proposition 10.92 are not only free of nontrivial oriented cycles, but their visibility graphs are also very simple. This is because they do not contain any oriented paths of length greater than 1.

Let us now consider a modest refinement of the local +-isomorphism problem.

Definition 10.97 *Let G and H be oriented graphs. The* surjective local +-isomorphism problem *asks whether there is a local +-isomorphism from G onto H.*

Proposition 10.98 *The surjective local +-isomorphism problem lies in NP. If it admits a polynomial-time solution for graphs which do not contain nontrivial oriented cycles, then there is also a polynomial-time solution for the general case. The existence of a polynomial-time solution for the surjective local +-isomorphism problem implies that there is a polynomial-time solution for the problem of deciding when two oriented graphs are isomorphic to each other.*

Proof The fact that the surjective local +-isomorphism problem lies in NP is immediate from the definition. The reduction to the case of graphs which do not contain nontrivial oriented cycles can be accomplished in the same way as in the proof of Proposition 10.92.

If the given graphs G and H have the same number of vertices and edges, then the surjective local +-isomorphism problem is equivalent to the problem of deciding when G and H are isomorphic to each other. On the other hand, for the graph-isomorphism problem, one might as well reduce to the case of graphs of exactly the same size at the outset, since that condition is necessary for the existence of an isomorphism, and is verifiable in polynomial time. From this it follows easily that a polynomial-time solution for the surjective local +-isomorphism problem leads to one for the graph-isomorphism problem. (Compare with Proposition 10.28.) □

One can also view the graph-isomorphism problem as being the "symmetric" version of the surjective local +-isomorphism problem, i.e., in which one is given oriented graphs G and H and one asks whether there is both a local +-isomorphism from G onto H and from H onto G.

The method of Proposition 10.92 can also be used to show that the isomorphism problem for oriented graphs admits a polynomial-time solution in general if it admits one for graphs which are free of nontrivial oriented cycles. (For that matter, the method of Proposition 10.92 can also be used in the context of local +-injections or local +-surjections, with only modest changes.)

10.15 Comparisons with k-provability

The problem of "k-provability" in formal logic asks whether a given formula admits a proof with at most k steps. This has been shown to be algorithmically undecidable in a certain formalization of first-order predicate logic by Buss [Bus91].

This problem is somewhat analogous (at least in the context of propositional logic, where it is clearly algorithmically decidable) to that of taking an oriented graph G and asking whether it admits a local $+$-isomorphism into a graph of size at most k. Remember that formal proofs always have oriented graphs below them, namely, the *logical flow graph*, as in Section A.3. One can think of this question about graphs as a kind of geometric model for the problem of deciding when a formal proof can be compressed to one of size at most k through better use of cuts and contractions. (Size and number of steps are not the same, but let us leave this aside for the moment.)

A given proof might be larger than necessary, through duplication of subproofs, and a roughly similar issue of duplication is present in the context of graphs and local $+$-isomorphisms.

Oriented graphs provide a model setting for issues related to "comparisons" between formal proofs, as in Section 2.1. Local $+$-isomorphisms provide a basic means for making comparisons between graphs, and other mappings can be used too, but this need not be the whole story. A tricky feature of formal proofs is that there can be codings which hide important connections between different formula occurrences. For instance, the logical flow graph only makes connections between atomic occurrences which represent the same basic formula (i.e., the same propositional variable in propositional logic), but there can be crucial connections between occurrences of variables with different names, through formulae which express implications or equivalence between the different variables. This occurs in the proofs described in Section 3.3, for example. (See also Sections 6.9, 6.15, and 6.16.) Analogous matters come up in the context of graphs, as in the proof of Proposition 10.92, or the use of cones in Section 10.5, for instance.

Remember that formal proofs can be used to represent implicit constructions of mathematical objects (like numbers, or words over an alphabet, or elements of a finitely-generated group), through the notion of *feasibility*. (See [CS].) Questions about the existence of proofs of a given size, or with at most a certain number of steps, can be seen as questions about the existence of implicit representations with corresponding bounds, as in (9.3) in Section 9.5.

10.16 A partial ordering between graphs

If G and H are oriented graphs, let us write $G \succ H$ if there exists a local $+$-isomorphism from G onto H.

Lemma 10.99 *If G and H are oriented graphs such that $G \succ H$ and $H \succ G$, then G and H are isomorphic to each other.*

Proof Indeed, if $G \succ H$, then G has at least as many edges and vertices as H does. This implies that any orientation-preserving mapping from H onto G must be injective, and hence an isomorphism. Thus H and G must be isomorphic to each other, since the assumption $H \succ G$ implies the existence of such a mapping from H onto G. □

Corollary 10.100 *$\succ$ defines a partial ordering on isomorphism classes of oriented graphs.*

Proof Indeed, $\succ$ is reflexive by definition, it is transitive because of Lemma 10.13, and Lemma 10.99 implies that $G \succ H$ and $H \succ G$ only when G and H represent the same isomorphism class. □

It is natural to say that an oriented graph M is minimal for $\succ$ if it is true that $M \succ L$ implies that M is isomorphic to L for any other oriented graph L. This is equivalent to the notion of minimality given in Definition 10.64, by Lemma 10.65. Note that Corollary 10.85 says that for any oriented graph G there is at most one oriented graph M (up to isomorphism) such that M is minimal and $G \succ M$, namely, the minimal folding graph.

Let us define a new relation $\sim$ between oriented graphs, by taking the transitive symmetric closure of $\succ$. In other words, we write $G \sim H$ when there is a finite sequence of oriented graphs $L_1, \ldots, L_j$ such that $L_1 = G$, $L_j = H$, and

$$\text{either} \quad L_i \succ L_{i+1} \quad \text{or} \quad L_{i+1} \succ L_i \tag{10.10}$$

for each i, $1 \leq i \leq j-1$.

Proposition 10.101 *Let G and H be oriented graphs. Then the following are equivalent:*

(1) $G \sim H$;
(2) *The visibility spectra of G and H (Definition 10.74) are the same;*
(3) *The minimal folding graphs of G and H are isomorphic to each other;*
(4) *There is an oriented graph N such that $G \succ N$ and $H \succ N$;*
(5) *There is an oriented graph P such that $P \succ G$ and $P \succ H$.*

Proof We know from Lemma 10.75 that the visibility spectra of two graphs coincide when there exists a local +-isomorphism from one onto the other. Thus the first condition implies the second one.

The second and third conditions are equivalent to each other because of Proposition 10.78, and Corollary 10.87 yields the equivalence of the third and fourth conditions. Each of the fourth and fifth conditions implies the first, by definition of the relation $\sim$.

To finish the proof, it suffices to show that the fourth condition implies the fifth one. Thus we suppose that there is an oriented graph N and surjective local +-isomorphisms $g : G \to N$ and $h : H \to N$, and we want to find a graph P such that $P \succ G$ and $P \succ H$.

To do this, one can simply take P to be the *fiber product* associated to $g : G \to N$ and $h : H \to N$. The notion of fiber product for graphs is discussed in some detail in Section 15.6, and it produces a graph P together with a pair of "projections" from P to each of G and H. These projections must be local +-isomorphisms, because g and h are, as in Lemma 15.6. Similarly, the projections must be surjective as well, because of the surjectivity of g and h, as in Lemma 15.7. This completes the proof of Proposition 10.101, modulo the material to be covered in Section 15.6. □

10.17 Monotonicity properties

Consider the local +-isomorphism problem, in which one is given a pair of oriented graphs G and H, and one is asked whether there exists a mapping $g : G \to H$ which is a local +-isomorphism. What happens if we add edges or vertices to G, or to H? In either case there is no clear monotonicity property for the local +-isomorphism, in that the addition of edges or vertices on either end could both create or break patterns which permit a local +-isomorphism to exist.

On the other hand, we can think of local +-isomorphisms as defining their own ordering on graphs, and then the local +-isomorphism problem behaves in a simple way. Specifically, if G', H' are two other oriented graphs, and if there exists a local +-isomorphism from G' to G and from H to H', then the existence of a local +-isomorphism from G to H implies the existence of a local +-isomorphism from G' to H'. In other words, there is more likely to be an answer of "yes" for the pair (G', H') than for (G, H). This is because the composition of local +-isomorphisms is again a local +-isomorphism, as in Lemma 10.11.

Now consider the problem where one is given an oriented graph G and a number k, and one is asked whether G admits a local +-isomorphism into some oriented graph of size at most k. If G_0 is another oriented graph which admits a local +-isomorphism into G, then an answer of "yes" for G automatically implies the same answer for G_0. In this case, there is an extra subtlety: if there is actually a local +-isomorphism from G_0 *onto* G, then it is also true that an answer of "yes" for G_0 implies the same answer for G. This does not follow directly from the definitions, as in the case of the other direction, from G to G_0; instead one uses the fact that G and G_0 have the isomorphic minimal folding graphs (Corollary 10.80), and that the minimal folding graph determines the optimal value of k (Corollary 10.82). (Compare also with Lemma 10.38, where the existence of a local +-isomorphism into a graph of size at most k depends only on the visibility of G at a certain vertex v.)

This reversal of the naive monotonicity property clearly does not work in the case of the local +-isomorphism problem. That is, we might have local +-isomorphisms from G' onto G, from H onto H', and from G' onto H', but none from G into H. Indeed, we might have that G is even smaller in size than H. (One can take $G' = H$ and $H' = G$, for example.)

Of course, this type of reversal of monotonicity depends heavily on the special nature of local +-isomorphisms, and does not work in general for other kinds of mappings.

A similar phenomenon occurs for the computational problem in which one is given oriented graphs K and L, and one is asked whether there exists a third oriented graph N such that K and L admit local +-isomorphisms onto N. This is because an answer of "yes" for this question occurs if and only if $K \sim L$, where $\sim$ is the equivalence relation defined in Section 10.16 (in the paragraph containing (10.10)). This follows from Proposition 10.101.

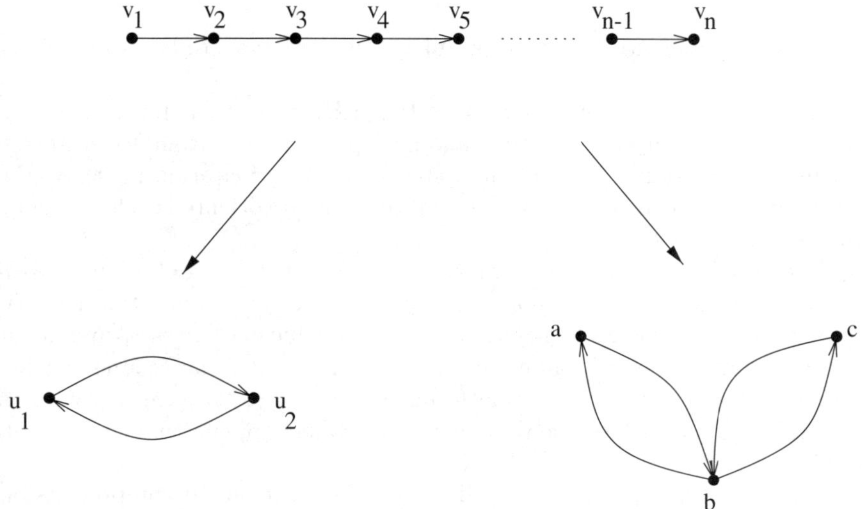

FIG. 10.5. Two mappings

10.18 Possible behavior of mappings

Let G and H be oriented graphs, and $f : G \to H$ an orientation-preserving mapping. How might f behave? What are some basic examples and constructions?

Of course one can have *embeddings*, but let us consider some other types of examples.

The first is a simple projection. Let H be given, an arbitrary oriented graph, and let G be a disjoint union of some identical copies of H, with no edges between the different copies. Then we can simply project G down to H in the obvious way, so that each point in the image has several preimages.

Here is a different kind of "covering", in which the domain is connected. Imagine that H consists of a pair of vertices u_1, u_2 with an oriented edge from u_1 to u_2 and vice-versa, and that G consists of n vertices $v_1, v_2, \ldots, v_n$, with an edge from v_j to v_{j+1} for each j, $1 \leq j \leq n-1$, and no others. One can map G to H by "wrapping" G around the loop, with v_j mapped to u_1 when j is odd, and to u_2 when j is even. (See Fig. 10.5.)

As a more complicated situation, one can take H to have three vertices a, b, c, with oriented edges from b to each of a and c, and oriented edges from each of a and c to b, but none between a and c. If we take G to be the "linear" graph as above, then there are plenty of ways to map G into H. Each time that we pass through b, we can decide anew whether to go to a or to c. (This choice of H is also depicted in Fig. 10.5.)

These two examples rely heavily on the presence of nontrivial *oriented cycles* in H, and indeed, in the presence of such cycles, it is easy to make noninjective mappings. What if we restrict ourselves to graphs H which have no nontrivial oriented cycles? In this case we automatically have injectivity of orientation-preserving mappings from "linear" graphs, and so we have to lose the injectivity in other ways.

If H is an oriented graph and u is a vertex in H, then we can form the visibility graph $\mathcal{V}_+(u, H)$, and consider the canonical projection π from $\mathcal{V}_+(u, H)$ to H (as defined in Section 4.5). This mapping is always orientation-preserving, and typically far from being one-to-one, even if there are no oriented cycles. (Compare with Fig. 4.1.)

What other kinds of configurations can occur? Let us think about this in a general way for a moment. Suppose that $f : G \to H$ is an orientation-preserving mapping between oriented graphs, and that we have vertices v and w in G which are mapped to the same vertex in H. If v and w are connected by an oriented path, then there is a nontrivial oriented cycle passing through their common image. Let us assume that this is not what happens, and consider other possibilities.

It may be that v and w simply lie in different connected components of G. If not, then the remaining possibility is that v and w are connected by a non-oriented path. This path may alternate several times between parts which are positively and negatively oriented. Let us consider two special cases.

In the first case, we imagine that there is a vertex p in G such that there are oriented paths from p to each of v and w. This is certainly possible, and occurs in the case where G is the visibility and the mapping is the canonical projection, for example.

For the second case, we imagine that there are oriented paths from v and w to some common vertex q. It is easy to make examples where this happens. One can take two linear graphs and make them merge at a certain point, for instance. Let us describe a general construction of "merging" which has this kind of effect.

Let G_1 and G_2 be two oriented graphs, and let A_1 and A_2 be subgraphs of G_1 and G_2, respectively. Suppose that A_1 and A_2 are isomorphic, and that $\theta : A_1 \to A_2$ is an isomorphism between them. We define the *merge* of G_1 and G_2 over A_1, A_2, θ to be the graph G obtained as follows. One first takes the disjoint union of copies of G_1 and G_2. One then identifies edges and vertices in A_1 with their counterparts in A_2, using θ, to obtain a new graph G. Thus G contains copies of each of G_1 and G_2, but in such a way that the subgraphs A_1 and A_2 become a single subgraph A of G, with A isomorphic to A_1 and A_2. The edges and vertices in G_1, G_2 that do not lie in A_1 or A_2 remain separate in the merged graph G.

An example of the merge of two graphs is given in Fig. 10.6. In this case, the subgraphs A_1 and A_2 correspond to the "perimeters" of the squares, while the edges in the middle are being duplicated.

Let us consider now the special case where G_1 and G_2 are two copies of a single oriented graph H, with A_1 and A_2 corresponding to the same subgraph B of H, and with θ taken simply to be the identity. We call G the *duplication of H along B*. In this situation, we have a natural orientation-preserving mapping

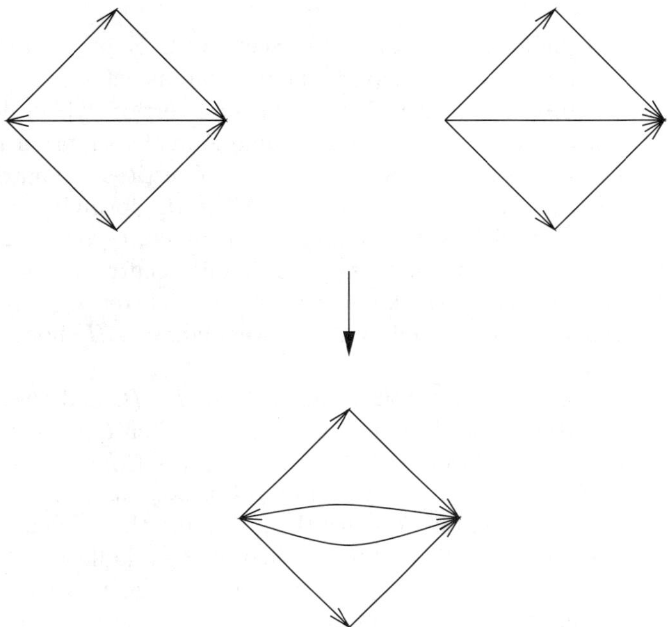

FIG. 10.6. An example of the merge of two graphs

$$f : G \to H, \tag{10.11}$$

corresponding to the identity mapping on the two copies G_1 and G_2. This mapping is one-to-one on the part of G that comes from B, and two-to-one on the rest of G.

One can use this merging operation many times, and over different subgraphs, to build up more complicated mappings.

There is a modest refinement of this duplication construction which will be useful in a moment. Imagine that we start with an oriented graph H and a subgraph B as before, but that now we permit ourselves to label edges in H by either 1, 2, or both 1 and 2. All edges contained in B should be labelled by both 1 and 2. Let G_1 be a copy of the subgraph of H consisting of all of the vertices in H and all of the edges which have 1 as a label, and take G_2 to be a copy of the analogous subgraph, but with 2 instead of 1. We now choose G to be the merge of G_1 and G_2 along the copies of B which lie inside them. (This was the reason for requiring that the edges in B be labelled by both 1 and 2, to make certain that G_1 and G_2 both contain a complete copy of B.) The earlier construction corresponds to marking all edges in H by both 1 and 2, and the new graph G is a subgraph of the old one.

This refinement is convenient for comparing these general duplication procedures for graphs with the duplication of subproofs which occurs in the standard method of cut elimination. Recall that the "duplication of subproofs" refers to transformations of the form (6.4) to (6.5) in Section 6.2. Suppose that we have a proof Π as in (6.4), and let H be its logical flow graph (as defined in Section A.3). Let B denote the subgraph of H which contains all vertices which do not come from the subproof Π_1 or the contraction formula A in the subproof Π_2 in (6.4), and all edges in H whose endpoints lie in this set of vertices. To mark the edges in H by 1, 2, we proceed as follows. Let A^1, A^2 be the formula occurrences in the subproof Π_2 of Π in (6.4) which are being contracted to give A. The edges in H which connect atomic occurrences in A^1 to atomic occurrences in A should be marked by 1, while the edges which connect atomic occurrences in A^2 to atomic occurrences to A should be marked by 2. All other edges in H should be marked with both 1 and 2.

If G denotes the oriented graph obtained from H, B, and these markings through the duplication procedure mentioned about, then G is nearly the same as the logical flow graph of the proof (6.5) that results from the duplication of subproofs in this case. More precisely, G is *topologically* the same as the logical flow graph of the new proof, but to make them literally the same as graphs, one should "stretch" some edges, by adding vertices in the middle.

Note that the marking of edges in H is important in this case, to accommodate the splitting of branch points at atomic occurrences in the contraction formula A in (6.4). This splitting of branch points is a key feature of the duplication of subproofs, as we saw in Chapter 6.

10.19 Possible behavior of mappings, continued

In the previous section, we took a general look at possible behavior of orientation-preserving mappings between graphs. Now we want to specialize somewhat, and ask about possible behavior for orientation-preserving mappings which are local $+$-isomorphisms.

The canonical projection from a visibility graph $\mathcal{V}_+(v, G)$ to G is always a local $+$-isomorphism (as in Lemma 10.10). The operation of duplication of an oriented graph along a subgraph described above does not always produce a local $+$-isomorphism, but there is a simple condition which ensures that this be true.

Definition 10.102 (+-complete subgraphs) *Let H be an oriented graph, and let B be a subgraph of H. We say that B is* $+$-complete *if the following two properties hold. First, if p is a vertex in B and q is a vertex in H, and if there is an oriented path in H from p to q, then q should also be a vertex in B. Second, any edge in H whose endpoints belong to B should also lie in B.*

This is equivalent to saying that the natural inclusion of B into H should be a local $+$-isomorphism.

It is easy to make examples of this. For instance, given any oriented graph H and a vertex v in H, we can take B to be the subgraph of vertices and edges in H which can be reached by oriented paths which start at v. This is illustrated by Fig. 10.7, where B is the subgraph corresponding to the inner square.

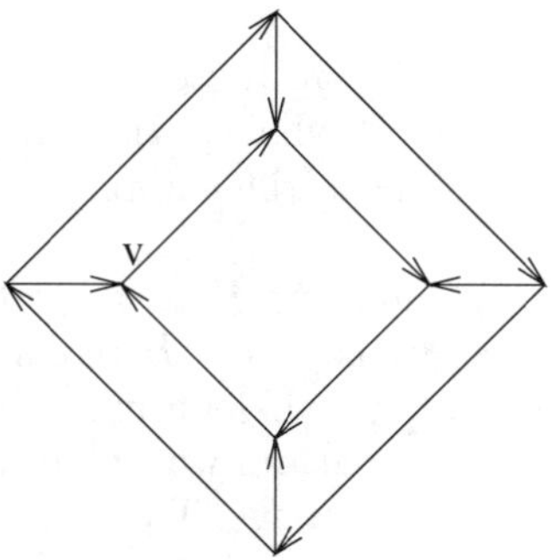

FIG. 10.7. The inner square is a +-complete subgraph of the graph as a whole.

Lemma 10.103 *Let H be an oriented graph, let B be a +-complete subgraph, and let K be the duplication of H along B, as defined in Section 10.18. Then the associated orientation-preserving mapping $f : K \to H$ (as in (10.11) in Section 10.18) is a local +-isomorphism.*

Proof This is easy to check, directly from the definitions. (Note that it does not work in general for the more refined version of the duplication of H over B which uses markings of the edges in H by 1, 2, as described in Section 10.18.) □

Let us consider another situation. Suppose that we start with an oriented graph G and two subgraphs A_1, A_2 of G, and that A_1 and A_2 are isomorphic to each other. Fix an orientation-preserving isomorphism $\alpha : A_1 \to A_2$. We ask that α be equal to the identity mapping on the intersection of A_1 and A_2, if it is nonempty. This implies that α takes vertices and edges in $A_1 \backslash A_2$ to vertices and edges in $A_2 \backslash A_1$. We define a new oriented graph K by taking the quotient by α. In other words, we identify every vertex in A_1 with its image in A_2 under α, and similarly for edges (and we leave alone the vertices and edges in G which are not in A_1 or A_2). The earlier constructions of merging and duplication from Section 10.18 can be seen as special cases of this one, in which we start with a graph which is the disjoint union of two others.

Let us call K the *quotient of G by α*. From the construction, we automatically get a canonical "quotient" mapping $g : G \to K$, and g preserves orientations.

Lemma 10.104 *Let G, A_1, A_2, $\alpha : A_1 \to A_2$, K, and g be as above. Assume also that A_1 and A_2 are +-complete subgraphs of G. Then $g : G \to K$ is a local +-isomorphism.*

Proof This is not hard to check, just using the definitions. (One might consider separately the cases where a given vertex in G does not lie in A_1 or in A_2, or it lies in one and not the other, or it lies in both.) □

In the context of duplication of an oriented graph H along a subgraph B, one could take G to be the disjoint union of two copies of H, and A_1 and A_2

to be the corresponding copies of B in these copies of H. The quotient process above would then give the graph K which is the duplication of H along B. The mapping $g : G \to K$ is not the same as the mapping considered in Lemma 10.103, which goes from K to H.

On the other hand, one could view H as a quotient of K. Specifically, K contains two subgraphs which are copies of H, coming from the copies of H in G. These two copies of H in K are not disjoint in general, and their intersection is the copy of B which lies in K in a natural way. One can employ these two copies of H in K in the role of A_1 and A_2 above. The construction gives an automatic isomorphism between these two copies of H, which is equal to the identity on their intersection. These two copies of H in K are $+$-complete subgraphs, as one can check, using the $+$-completeness of B inside of H. The quotient mapping from K to H is the same as the mapping in Lemma 10.104.

11

MAPPINGS AND COMPARISONS

Mappings between graphs are useful in part because they provide a general way to make comparisons, and to express the existence of certain kinds of patterns or symmetry. This is true in a purely geometric way, and also for constructions or computations in other contexts, including the use of *feasibility graphs*. We shall consider some basic illustrations of these principles in the present chapter.

11.1 Locally +-stable mappings

Let us introduce a new class of mappings between oriented graphs, a modest weakening of the notion of local +-isomorphisms.

Definition 11.1 *Let G and H be oriented graphs, and let $f : G \to H$ be an orientation-preserving mapping. We say that f is* locally +-stable *if for each vertex v in G we have either that there are no edges in G flowing out of v, or that f induces a one-to-one correspondence between the edges in G that flow out of v and the edges in H that flow out of $f(v)$. (We allow this definition to be applied also to* infinite *graphs.)*

Thus locally +-stable mappings are always locally +-injective, and at any given vertex they are either locally +-surjective or the graph in the domain simply stops there. (See Fig. 11.1 for an example.)

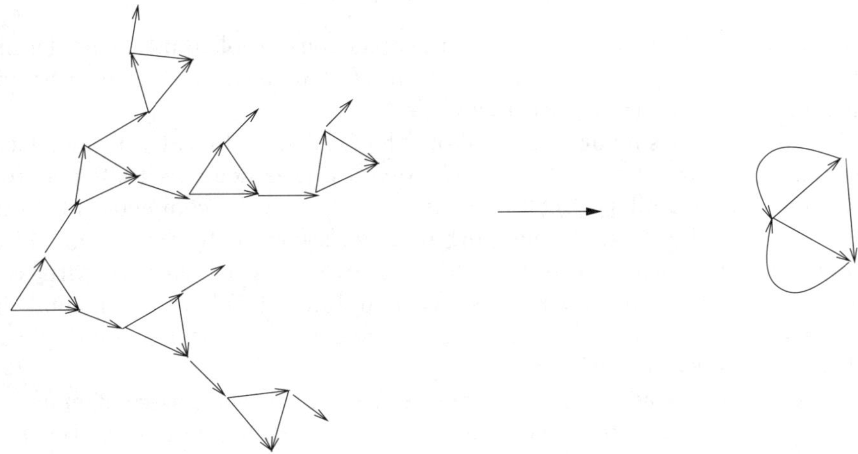

FIG. 11.1. An example of a locally +-stable mapping

If we think of local $+$-isomorphisms as being analogous to the notion of covering surfaces from topology, then locally $+$-stable mappings are like covering spaces with some kind of boundary. It permits the flexibility of wrapping a large graph G around a smaller graph H with cycles finitely many times, without having any nontrivial oriented cycles in G. (Compare with Lemma 10.17 in Section 10.3.) This can be very useful, for "temporary" or limited periodicity which might be present. A basic example is provided by the graphs depicted in Figures 4.2 and 4.3 in Section 4.3, for which there is an obvious locally $+$-stable mapping from the first to the second, which reflects the basic pattern in Fig. 4.2, and the way in which it stops suddenly.

To put this a bit differently, notice that local $+$-isomorphisms between graphs without nontrivial oriented cycles always preserve "depth", i.e., the length of the longest oriented path in a graph which begins at a given vertex. (See Definition 4.1 in Section 4.4.) This is a kind of rigidity property of local $+$-isomorphisms, which restricts them from making comparisons between different levels in a graph, even if they can be very effective at making comparisons across fixed levels. With locally $+$-stable mappings, one can also make comparisons of the patterns at different levels in a graph.

On the other hand, for locally $+$-stable mappings, there may not be much regularity in the way that the repetitions stop, and this can be a nontrivial omission in the information about patterns which they provide. One can try to ameliorate this, through extra conditions or data.

As in Chapter 10, there are natural computational questions about locally $+$-stable mappings. Here is one.

Definition 11.2 (The locally $+$-stable mapping problem) *Suppose that G is an oriented graph, and that k is a positive integer. The* locally $+$-stable mapping problem *is to decide whether there is an oriented graph H of size at most k such that G admits a locally $+$-stable mapping into H.*

Instead of size, one could consider the analogous problem using the depth of H (the length of the longest oriented path in H, starting at any vertex, assuming that there are no nontrivial oriented cycles).

This question is similar to ones about the existence of local $+$-isomorphisms into a graph of size at most k, as in Chapter 10. (See Lemmas 10.37 and 10.38 in Section 10.7, Corollary 10.82 in Section 10.13, and the comments in Section 10.15.) The locally $+$-stable mapping problem seems to be somewhat harder, though. It clearly lies in the class NP, since there is an effective witness by definition, but it is not apparent that it should be solvable in polynomial time when H is free of cycles, as was the case in the context of local $+$-isomorphisms, by Corollary 10.84 in Section 10.13.

In Sections 11.2 and 11.3, we shall discuss some other classes of mappings, which can also be used to measure symmetry and patterns in graphs. Before we get to that, let us mention a couple of other points related to locally $+$-stable mappings.

Given an oriented graph G and a subgraph G_0 of G, let us say that G_0 is *locally +-stable* if the inclusion of G_0 into G is a locally +-stable mapping. This is the same as saying that for each vertex in G_0, either every edge in G which flows out of it also lies in G_0, or none of them do.

Let $f : G \to H$ be a locally +-stable mapping between oriented graphs, and let v be a vertex in G. As in Section 10.1 we get an induced mapping

$$\widehat{f} : \mathcal{V}_+(v, G) \to \mathcal{V}_+(f(v), H) \tag{11.1}$$

between visibilities. This induced mapping is injective, by Lemma 10.14. Thus $\widehat{f}$ embeds the visibility $\mathcal{V}_+(v, G)$ into $\mathcal{V}_+(f(v), H)$ as a subgraph, and it is easy to see that this embedding is also locally +-stable, since f is.

11.2 Locally +-uniform mappings

The following is a weakening of the property of being a local +-surjection (Definition 10.12).

Definition 11.3 *Let G and H be oriented graphs, and let $f : G \to H$ be an orientation-preserving mapping. We say that f is* locally +-uniform *if it enjoys the following property. Let u and w be any pair of vertices in G, and let $O(u, G)$ and $O(w, G)$ denote the set of edges in G which flow out of u and w, respectively. If $f(u) = f(w)$, then we require that f map $O(u, G)$ and $O(w, G)$ to the same set of edges in H.*

In other words, for a local +-surjection we would have that f maps $O(u, G)$ onto the set $O(f(u), H)$ of all edges in H which flow out of $f(u)$, and we would get the same thing for $O(w, G)$ automatically. For the locally +-uniform condition, we drop the requirement that we get everything in $O(f(u), H)$, and ask instead that the image of $O(u, G)$ under f depend only on $f(u)$, and not on the specific choice of u. This is automatically true when f is (globally) injective on vertices.

Let us record a couple of simple facts about locally +-uniform mappings.

Lemma 11.4 *Let G and H be oriented graphs, and let $f : G \to H$ be an orientation-preserving mapping between them. Then f is locally +-uniform if and only if there is a subgraph H_0 of H such that f maps G into H_0, and is a local +-surjection when viewed as a mapping into H_0.*

Proof This is a straightforward consequence of the definitions. (One simply takes H_0 to be the image of f for the "only if" part.) □

Lemma 11.5 *Let G and H be oriented graphs, and let $f : G \to H$ be an orientation-preserving mapping between them. Then f is both locally +-injective and locally +-uniform if and only if there is a subgraph H_0 of H which contains the image of f such that f is a local +-isomorphism as a mapping into H_0.*

Proof This is again a straightforward consequence of the definitions. □

The class of mappings which are both locally +-injective and locally +-uniform provides another extension of the notion of a local +-isomorphism. This generalization is somewhat different from the condition of local +-stability, but in both cases one has local +-injectivity together with a condition which ensures some uniformity in the way that the local patterns in one graph are distributed in another.

11.3 Mappings and symmetry

Let G and K be graphs, and think of K as being much smaller than G. Let us take a broader look at the way that the existence of a mapping $f : G \to K$ with suitable nondegeneracy properties can reflect symmetries or patterns in G.

If we impose no restrictions on the mapping f, and if K contains an edge e whose endpoints are the same vertex w (so that e defines a loop of length 1), then there is no information to be had from the mere existence of a mapping from G into K. This is because one can always obtain such a mapping by sending all of the vertices of G to w, and all of the edges to e.

If K does not contain an edge of this type, then mappings from G into K automatically enjoy the "local injectivity" property that adjacent vertices in G are sent to distinct vertices in K. This case is already quite nontrivial, as in Proposition 10.3 in Section 10.2.

Although the mere existence of a mapping between a particular pair of graphs can reflect substantial information about the domain graph G, it may not say too much about the symmetry or patterns in G. For this purpose, one might consider mappings with stronger injectivity properties, as in the following definition.

Definition 11.6 (α-injective on vertices) *Let G and K be graphs, and let $f : G \to K$ be a mapping between them. Given a positive integer α, we say that f is* α-injective on vertices *if it is true that $f(u) \neq f(v)$ whenever u and v are distinct vertices in G at distance $\leq \alpha$ from each other (as measured by the number of edges traversed by the shortest (unoriented) path between u and v).*

This condition is illustrated in Fig. 11.2. For these examples, one can take all three graphs to be oriented, and the mappings between them to be orientation-preserving. The orientations for the graphs on the right are as marked in the picture, and in all three cases the orientations follow the rules $b \to c$, $b \to a$, and $a \to c$ along the triangles, and $a \to b$ and $c \to b$ for edges between triangles.

This type of local injectivity property implies that the local configurations in G are constrained by those in K. This may not say too much about how these configurations actually look, especially when K is rich with edges. To make up for this, one might restrict oneself to mappings which are locally surjective as well as injective, but this is a very strong (and rigid) condition.

In all of these situations, with different types of mappings, one is looking at ways in which one graph might be "folded" so that it can fit inside of another (typically smaller) one. Questions of this type — roughly like the existence of

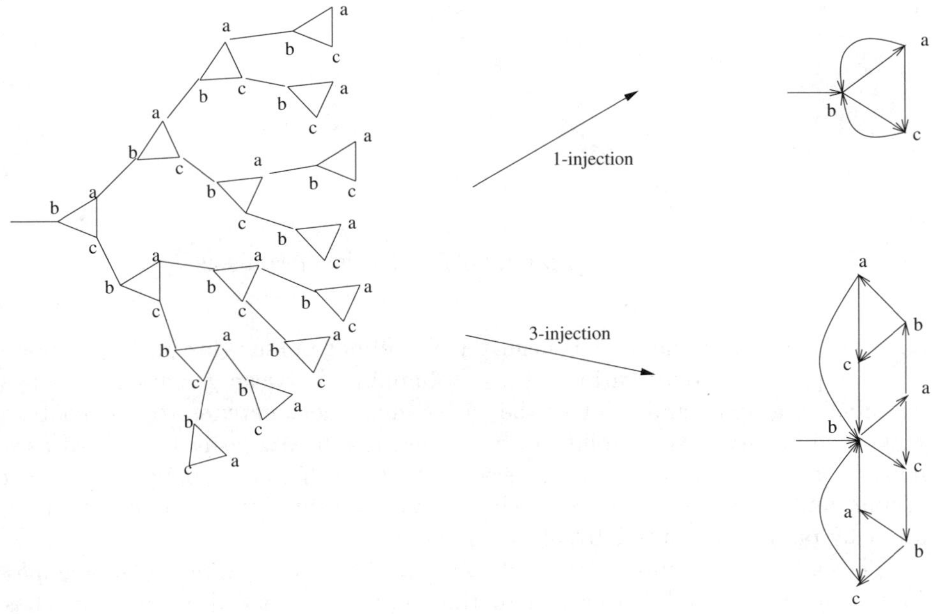

FIG. 11.2. Examples of α-injective mappings

immersions or good projections — do not seem to be as much studied in computational complexity as questions related to embeddability. There are some exceptions to this, connected to "bandwidth" problems, concerning the existence of a linear ordering of the vertices of a given graph with bounds on the distortion of adjacency. (See p200-1 of [GJ79], p215 of [Pap94].) There is also the (NP-complete) "digraph D-morphism" problem (p203 of [GJ79]), which is nearly the same as taking two oriented graphs and asking whether there is an orientation-preserving mapping between them which is a local +-surjection.

11.4 Labelled graphs

Fix some set S of "symbols". Let us call a (possibly infinite) graph a *labelled graph* if it comes equipped with mappings from the edges and/or vertices into S.

Part of the motivation for this comes from the notion of *feasibility graphs* discussed in Chapter 7. In that case, the labels can correspond to some operations on an underlying set, but for the moment this does not matter, and we are content to have labellings by abstract symbols. (We are also ready to allow the labels to be defined only at some edges, or some vertices, etc. That can also be accomplished by adding a special symbol to S to mark unlabelled objects.)

By a *mapping between labelled graphs* we mean simply a mapping in which a vertex or edge in the domain has the same label as its counterpart in the image.

One can also consider more involved notions, in which the mappings are permitted to change the labels in accordance with some set of rules, e.g., following

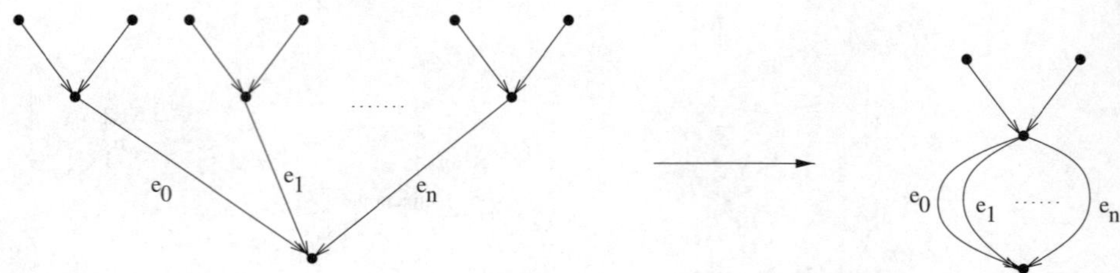

FIG. 11.3. A family of local −-isomorphisms

a suitable notion of "homomorphism". We shall not pursue this in this section.

Our questions about various kinds of mappings between graphs make sense for labelled graphs, and a lot of the earlier machinery extends to this context without difficulty. The visibility of a labelled graph can easily be viewed as a labelled graph, for instance. In general, if $f : G \to H$ is any mapping between graphs, and if we have labels for H already, then we can simply define a labelling for G by pulling back the labels from H via f.

The earlier treatments of *minimal representations* and *minimal folding graphs* (from Chapters 9 and 10) extend to the context of labelled graphs. One has just to take the extra information of the labellings into account at every step, i.e., in looking at isomorphic equivalence of rooted trees and visibility graphs, or in making correspondences between edges coming out of different vertices. As long as one is consistent in maintaining compatibility with the labels, the labels themselves do not cause any trouble, and the basic structure of the earlier arguments and constructions remains the same as before. (For instance, the additional structure of labels does not create significant difficulties for Claims 9.5, 9.12, 9.13, or 9.21, or for Lemma 10.55, and these were some of the most basic ingredients in the earlier constructions.)

11.5 Feasibility graphs

Let G be a feasibility graph with respect to some structure. (See Chapter 7 for definitions and examples.) Thus G is oriented and has no nontrivial oriented cycles, and comes equipped with some kind of labellings for edges and focusing branch points, indicating operations to be performed. Additional information, such as orderings of incoming edges at a given vertex, may also have to be prescribed.

For the purposes of feasibility graphs, it is better to work with *local −-isomorphisms* instead of local +-isomorphisms. These are defined in exactly the same manner as in Definition 10.8 in Section 10.3, except that one looks at the *incoming* edges at a given vertex instead of the *outgoing* edges. See Fig. 11.3 for a family of examples.

Suppose now that we have another oriented graph H, and an orientation-preserving mapping $\phi : H \to G$ which is a local −-isomorphism. Then we can "pull back" the feasibility structure from G to H, i.e., to determine when and how edges in H should be labelled, or which operations should be associated to branch points (if there is any choice), or how incoming edges at focusing branch points in H should be ordered. In each case, one simply uses ϕ to map into G, and one follows the choices given there. Note that the property of being a local $-$-isomorphism ensures that the focusing branch points in H have exactly the same number of incoming edges as their counterparts in G; this ensures that compatibility with the arities of the operators is maintained. This need not be true for the outgoing edges, which reflect duplications in the feasibility graph, and for which this kind of compatibility is not an issue.

If f is a *value function* on G (as in Definition 7.1, for the case of words over an alphabet), then the "pull-back" $\phi_*(f) = f \circ \phi$ of f to a function on the vertices of H is also a value function there, with respect to the feasibility structure on H derived from G. This follows from the definitions; the main point is that the concept of a value function is purely local, and that our mapping $\phi : H \to G$ respects the local structure needed for this to work.

There is a similar statement for *normalized value functions*. Recall that input vertices are vertices with no incoming edges, and that normalized value functions are value functions whose restrictions to the input vertices are required to be a fixed constant value (prescribed in advance). This constant value might normally be very simple, like the empty word in the context of words over a finite alphabet, or the number 0 in the context of numbers (as in Section 7.9), etc. For the correspondence between normalized value functions on G and H, the main point is that a vertex in H is an input vertex if and only if its image in G under ϕ is, since ϕ is a local $-$-isomorphism.

As a basic example, let H be the negative visibility of G based at some vertex in G, and let ϕ be the canonical projection from the visibility graph back into G. This case was already discussed (in slightly different terms) in Section 7.4. As a variant of this, one might take the negative visibility graph, and identify input vertices in it that are mapped by the canonical projection to the same vertex in G. If one does this, then the input vertices of the new graph would correspond to input vertices in G in an injective manner, and otherwise the graph would behave like the visibility graph. The defocusing branch points would be pushed all the way back to the input vertices, rather than being eliminated completely, as the negative visibility graph does. This is illustrated in Fig. 11.4. See also Section 7.4 in this regard.

These considerations of mappings and feasibility graphs can be extended as follows. Let us call a mapping $\phi : H \to G$ *locally $-$-stable* if it satisfies the same condition as in Definition 11.1, except that one uses the *incoming* edges at a given vertex instead of the *outgoing* edges. This is the same as saying that the mapping behaves like a local $-$-isomorphism, except possibly at input vertices of H, where there are no incoming edges. In this situation, H can inherit a feasibility structure from G in the same manner as before, but now it is no longer true that input vertices in H are necessarily mapped to input vertices of G.

If $\phi : H \to G$ is a locally $-$-stable mapping, then value functions on G can

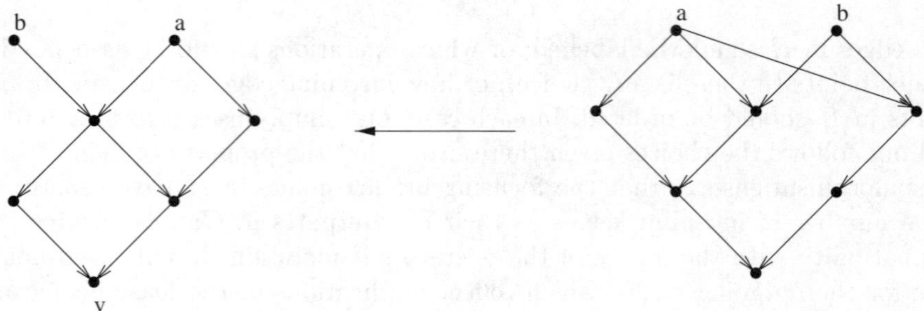

FIG. 11.4. The defocusing branch point in the graph on the left is pushed back in the graph on the right to the input vertices.

be pulled back to value functions in H, in the same way as before. The pull-back of a *normalized* value function will no longer be a normalized value function in general, however.

In connection with pulling back feasibility structures, let us relax the assumptions on G. In order to be a feasibility graph, G is supposed to be free of oriented cycles, as in (7.1) in Section 7.1. Let us drop this requirement for the moment, but keep the rest of the feasibility structure, so that G becomes a kind of labelled oriented graph. The local structure of G still describes certain types of computations or constructions, but we cannot apply Lemma 7.2 to get existence and uniqueness of value functions (globally) on G.

With locally $-$-stable mappings $\phi : H \to G$, we can have nontrivial oriented cycles in G and none in H, even if ϕ maps H onto G. This would not be the case if ϕ were a local $-$-isomorphism, or a local $-$-surjection, as in Lemma 10.17 in Section 10.3. This permits one to have "large" graphs H which are wrapped around a much smaller graph G, and with H having no nontrivial oriented cycles, so that it becomes a feasibility graph. In other words, one can pull back a feasibility structure from G, in the same manner as before, and get a feasibility graph based on H, whether or not G has nontrivial oriented cycles.

To illustrate this possibility, consider the oriented graphs H_n defined as follows. Each H_n has exactly $n+1$ vertices, written as $0, 1, 2, \ldots, n$, and two oriented edges from the jth vertex to the $(j+1)$th vertex, $0 \leq j < n$. Thus the H_n's are free of nontrivial oriented cycles. Let G be the graph with one vertex to which two edges are attached as loops. It is easy to define locally $-$-stable mappings from H_n into G for every n. (See Fig. 11.5.) One can add labels for the feasibility structure, and so that the mappings from the H_n's into G are compatible with the labellings.

One could also weaken the $-$-stability condition to allow mappings $\phi : H \to G$ which are merely *local $-$-injections.* This condition is defined in the same way as in Definition 10.12, but with the obvious change from $+$ to $-$. (See Fig. 11.6 for a family of examples. Note that they are not locally $-$-surjective.) In this case, there is a problem with giving H the structure of a feasibility graph. We

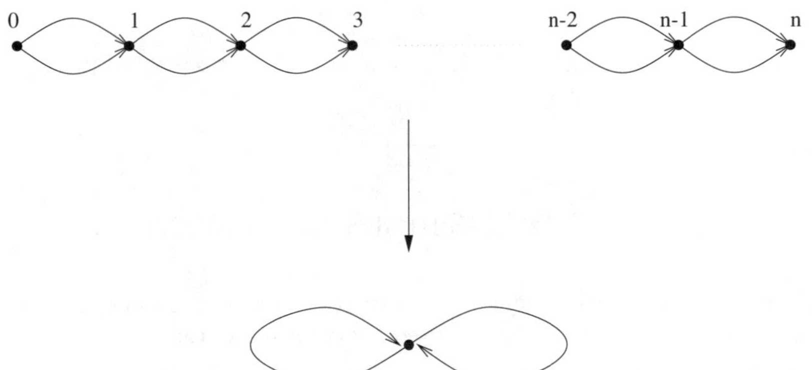

FIG. 11.5. The locally −-stable mapping from H_n to G

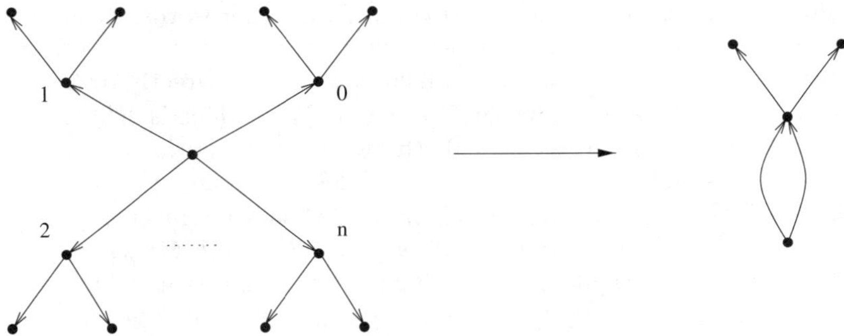

FIG. 11.6. A family of locally −-injective mappings

can still pull back all the labellings (of edges and designations of focusing branch points) as before, but now the vertices in H may not have as many incoming edges as they should for the arities of the operations that they are supposed to represent. For instance, there might be only one edge going into a vertex that was representing a binary operation in G. (This occurs in Fig. 11.6, since every vertex on the left has at most one incoming edge, while the middle vertex on the right has two incoming edges.) One could resolve this problem through suitable conventions, e.g., missing arguments will always be given normalized or "trivial" values (like the empty word, or the number 0), or will be left free, like input vertices.

Remark 11.7 In the preceding discussion, we always assumed that the feasibility structure on the graph H was the same as the one that we could "pull back" from G using the mapping $\phi : H \to G$. One could extend this by allowing the operation in G not to be literally the same as the one in H, but to be related to it by a "homomorphism" in the sense of Section 7.12. Alternatively, one could treat the transformation of feasibility graphs through homomorphisms as a separate operation, which is applied after the one of pulling-back as above.

12

ADJACENCY MATRICES AND COUNTING

In this chapter, we review *adjacency matrices* associated to graphs, and some classical ways in which they can be used for counting paths.

12.1 The adjacency matrix

Let G be a finite oriented graph. It will be convenient to be a bit more functorial than in the standard literature, and so we write $\mathcal{F}(G)$ for the vector space of real-valued functions on the set of vertices of G. In other words, we are basically looking at $\mathbf{R}^n$, where n is the number of vertices in G, but it is convenient to use a realization of this vector space which is tied more directly to G.

If u is a vertex in G, let e_u be the function in $\mathcal{F}(G)$ which is defined by taking $e_u(w) = 1$ when $w = u$ and $e_u(w) = 0$ otherwise. By doing this for all vertices u in G, we get a "standard basis" for $\mathcal{F}(G)$. We let $\langle \cdot, \cdot \rangle$ denote the standard inner product on $\mathcal{F}(G)$, for which the e_u's form an orthonormal basis.

Next we define a linear transformation $A : \mathcal{F}(G) \to \mathcal{F}(G)$ which represents the *adjacency* of vertices in G. We can define A through its matrix with respect to the standard basis on $\mathcal{F}(G)$, by requiring that $\langle A(e_u), e_w \rangle$ be the number of oriented edges from u to w. Note that this need not be a symmetric transformation, since we are taking orientations into account. Of course, the matrix entries of A are nonnegative integers.

The following is a key and well-known property of the adjacency transformation.

Lemma 12.1 *For any positive integer j and vertices u, w in G, $\langle A^j(e_u), e_w \rangle$ is equal to the number of distinct oriented paths of length j that go from u to w.*

Proof This is easy to check (using induction, for instance). □

The next lemma is also commonplace, and there are variants of it and its proof concerning similar points.

Lemma 12.2 *If G contains an oriented cycle, then $A^j \neq 0$ for all positive integers j. If G does not contain an oriented cycle, and if n is the number of vertices in G, then $A^n = 0$.*

Proof Indeed, if G contains an oriented cycle, then it contains an oriented path of any given length j, and this implies that $A^j \neq 0$. If G has no oriented cycles, then no oriented path can go through any vertex more than once, and hence must traverse $< n$ edges. This implies that $A^n = 0$. □

The notations and assumptions used above will be in force throughout this chapter.

12.2 Counting in the visibility

Fix a vertex v in G, and consider the visibility $\mathcal{V}_+(v, G)$. Given a nonnegative integer j, let N_j denote the number of vertices in $\mathcal{V}_+(v, G)$ which are at distance j from the basepoint. This is the same as the number of distinct oriented paths in G which begin at v and have length j. Thus we have $N_0 = 1$, and we can compute N_j for $j \geq 1$ as follows:

$$N_j = \sum_{w \in G} \langle A^j(e_v), e_w \rangle. \tag{12.1}$$

We are abusing our notation slightly here, writing "$w \in G$" for "w is a vertex in G".

This is a standard formula, and it reflects the possibility for exponential growth in the sizes of the N_j's in a simple way.

Consider the power series

$$\sum_{j=0}^{\infty} N_j \, t^j. \tag{12.2}$$

This converges for small values of t, because the N_j's cannot grow faster that a single exponential. Combining (12.2) with (12.1), we get that

$$\begin{aligned}\sum_{j=0}^{\infty} N_j \, t^j &= \sum_{j=0}^{\infty} \sum_{w \in G} \langle A^j(e_v), e_w \rangle \, t^j \\ &= \sum_{w \in G} \sum_{j=0}^{\infty} \langle t^j \, A^j(e_v), e_w \rangle = \sum_{w \in G} \langle (I - t\,A)^{-1}(e_v), e_w \rangle\end{aligned} \tag{12.3}$$

for small values of t. This uses the familiar Neumann expansion

$$(I - t\,A)^{-1} = \sum_{j=0}^{\infty} A^j \, t^j, \tag{12.4}$$

which holds when t is sufficiently small. (One could work with formal power series here as well.)

From this we see that the power series (12.2) actually represents a *rational* function of t, i.e., a quotient of polynomials. Indeed, Cramer's rule permits us to realize $(I - t\,A)^{-1}$ as the cofactor transpose of $I - t\,A$ divided by the determinant of $I - t\,A$. Thus we have the quotient of a polynomial of degree at most $n - 1$ by a polynomial of degree at most n, where n is the number of vertices of G. Both polynomials have integer coefficients, since the matrix entries of A are integers, and the polynomial in the denominator $(\det(I - t\,A))$ takes the value 1 at $t = 0$.

This is a classical and well-known argument, and it is quite nice, given its simplicity and conclusions (including ones from Lemma 12.3 below). In addition to situations with graphs, this type of result comes up in connection with growth functions for regular languages and for L-systems. For the latter, and further developments in this direction, see [RS80]. Concerning regular languages, a basic method for dealing with their growth functions (which count the number of words of a given length) is to represent the language through a deterministic finite automaton, as in Chapter 14, and then apply the method reviewed here to the graph associated to the automaton. The use of a *deterministic* automaton ensures that the counting of paths corresponds exactly to the counting of words, without redundancies. Growth functions and possible rationality of them also occur in the study of finitely-generated groups, as in [Can84, Far92, Gro87]. They appear in dynamical systems as well, in connection with counting of periodic orbits. See [CP93, LM95, Man71, Rue94], for instance, for more information and references.

Lemma 12.3 *Let $R(t)$ be any rational function of t which is not singular at the origin. (That is, $R(t)$ can be expressed as a quotient of polynomials in t, where the polynomial in the denominator does not vanish at $t = 0$.) Let*

$$\sum_{j=0}^{\infty} R_j \, t^j \tag{12.5}$$

denote the power series expansion of $R(t)$ about $t = 0$. Then the sequence of Taylor coefficients R_j can be written as the sum of a sequence with only finitely-many nonzero terms, and another sequence which is itself a finite linear combination of sequences of the form

$$\{j^p \alpha^j\}_{j=0}^{\infty}, \tag{12.6}$$

where p is a nonnegative integer and α is a complex number. The α's which are used here are the reciprocals of the singular points of $R(t)$ in the complex plane. If $R(t) = P(t)/Q(t)$, where $P(t)$ and $Q(t)$ are polynomials, then the exponents p used in (12.6) are less than the degree of $Q(t)$, and the sequence with only finitely-many nonzero terms has nonzero terms only for j less than or equal to the degree of $P(t)$.

Note that the α's can easily be complex even if $R(t)$ takes real values when t is real. This will be clear from the proof. Also, every sequence of the type described in the lemma does occur as the power series expansion around $t = 0$ of a rational function; this is not hard to check, by reversing the argument that follows. Concerning rational functions $R(t)$ which are singular at $t = 0$, one can always reduce to the nonsingular case by compensating with some extra powers of t.

Proof The main point is that $R(t)$ can be realized as a linear combination of functions of the form

$$\frac{t^l}{(1 - \alpha t)^m}, \tag{12.7}$$

where l and m are nonnegative integers, and α is a complex number. To see this, suppose that we are given $R(t)$ as the quotient $P(t)/Q(t)$, where $P(t)$, $Q(t)$

are polynomials, with $Q(0) \neq 0$. Since polynomials can always be factored over the complex numbers, we can write $Q(t)$ as a constant multiple of a product of functions of the form $(1 - \alpha t)$, $\alpha \in \mathbf{C}$. This also uses the fact that $Q(0) \neq 0$. To get the representation for $R(t)$ mentioned above, one employs this factorization of the denominator $Q(t)$ and the method of partial fractions. The α's which arise in this manner are reciprocals of the roots of $Q(t)$. The l's used here are less than or equal to the degree of $P(t)$, and the m's are less than or equal to the degree of $Q(t)$ (and are equal to the orders of the corresponding zeros of $Q(t)$).

Thus one is reduced to rational functions of the special form (12.7), and these are easy to treat. The main point is that

$$\frac{1}{1 - \alpha t} \quad \text{has} \quad \sum_{j=0}^{\infty} \alpha^j t^j \quad \text{for its power series expansion at } t = 0, \tag{12.8}$$

and that the power series of $1/(1 - \alpha t)^m$ can be derived from this (when $m > 1$) by differentiating $m - 1$ times. The process of differentiation has the effect of introducing polynomials of j into the Taylor coefficients, and shifting the indices backward, and these are easily accommodated by linear combinations of the sequences (12.6). Similarly, the factor of t^l in the numerator of (12.7) gives rise to a shift forward in the sequence of Taylor coefficients, and this can be accommodated by linear combinations of the sequences (12.6) and a sequence with only finitely-many terms. (The latter compensates for the vanishing of the coefficients of t^j when $j < l$, coming from the multiplication by t^l.) When $m = 0$, (12.7) reduces to t^l, and this case is covered by a sequence with only finitely-many nonzero terms. It is not hard to check that the α's, p's, and sequences with only finitely-many nonzero terms satisfy the conditions indicated at the end of the statement of the lemma. This proves Lemma 12.3. □

Returning to the story of the N_j's, we can apply the lemma to the rational function whose power series is given by (12.2), to conclude that the sequence $\{N_j\}$ can be expressed as a finite linear combination of sequences of the form (12.6), together with a sequence which is nonzero for only finitely many j's. We also know that the reciprocals of the α's in (12.6) come from the complex zeros of the polynomial $Q(t)$ in the denominator of $R(t)$, and in our case we can take $Q(t)$ to be $\det(I - tA)$. (It may be possible to make cancellations with the numerator, so that a polynomial of lower degree could be used in the denominator.) The zeros of $\det(I - tA)$ are exactly the reciprocals of the nonzero *eigenvalues* of A, at least if we permit ourselves to use *complex* vectors, and not just real ones, as in Section 12.1. Thus the α's themselves are exactly the nonzero (complex) eigenvalues of the adjacency transformation A.

Note that the polynomial factors j^p in (12.6) do not arise in this situation when A is diagonalizable. This is not hard to verify. When A is not diagonalizable, there are nontrivial factors of j^p for a particular choice of α exactly when the Jordan canonical form for A has a nontrivial nilpotent part associated to the eigenvalue α (with the exponent p given in terms of the degree of nilpotency). We shall see examples of this in Section 12.3.

These are all fundamental techniques in mathematics. More broadly, one can often apply algebraic or spectral methods to linear transformations in order to obtain information about associated problems of counting or asymptotic behavior. A common type of reasoning, somewhat different from the above, is to show that the leading behavior of a given asymptotic process is related to the eigenvalues of largest modulus of an associated linear transformation, and that these eigenvalues and their eigenvectors can be analyzed further in terms of underlying geometric or combinatorial structure, etc. See [IM76, LM95, Rue94] for some examples along these lines.

In the context of visibility graphs, the lemma above has the nice feature that it gives strong information about the behavior of the N_j's, and in a simple manner. For more aspects of this kind of representation, and related questions, see [RS80].

The story of visibility graphs also provides a nice context in which to see some deficiencies of these methods. A basic problem is that they are most useful in their treatment of *asymptotic* properties of the N_j's, while for j's of moderate size, they are much less effective.

As an extreme case, consider the situation in which the visibility is finite, so that only finitely many of the N_j's are nonzero. Let us assume for simplicity that every vertex in our graph G can be reached by an oriented path which begins at our fixed vertex v. We can always arrange for this to be true, by removing the other portions of G, and this will not affect the visibility $\mathcal{V}_+(v, G)$ or the N_j's. Under this condition, the adjacency transformation A associated to the graph G is nilpotent, as in Lemma 12.2. Specifically, $A^n = 0$ as soon as n is greater than the length of the longest oriented path in G, which begins at v in this case. In particular, one can take n to be the number of vertices in G.

If $A^n = 0$, then

$$(I - tA)^{-1} = \sum_{j=0}^{n-1} t^j A^j \tag{12.9}$$

for all complex numbers t, and the sum on the right-hand side is just a polynomial in t of degree less than n. In other words, our rational function has no singularities in this case. One can also show that the determinant of $I - tA$ is equal to 1 for all t, using the identity

$$\begin{aligned} \det(I - tA) \cdot \det\Big(\sum_{j=0}^{n-1} t^j A^j\Big) &= \det(I - tA) \cdot \det(I - tA)^{-1} \\ &= \det I = 1. \end{aligned} \tag{12.10}$$

The left side is a product of polynomials in t, and since the product is a nonzero constant, the polynomials have to be constant themselves. (They are both equal to 1, since their value at $t = 0$ is 1.)

Thus we do not obtain so much in this situation from the algebraic methods described before. The behavior of the N_j's can be quite interesting, though, at least if n is not too small. We no longer have a sharp distinction between "exponential" and "polynomial" rates of growth in the realm of finitely many n's, but one can try to analyze the difference between large and small rates of growth, as we did in Chapter 8.

Even when A is not nilpotent, there are problems of roughly the same nature with the algebraic techniques, e.g., if one wants to know about the behavior of the N_j's for modest values of j. In this case it can easily happen that the nilpotent effects are dominant. This point also came up in Chapter 5, especially in Sections 5.3 and 5.4.

To put the possibilities for "nilpotent" behavior into perspective, let us look at a couple of very simple examples. Consider the graphs pictured in Figures 4.2 and 4.3 in Section 4.3. Let us call these graphs G_1 and G_2, respectively, and let A_1, A_2 be the associated adjacency transformations. The matrix for A_2 looks like $\binom{0\,2}{1\,0}$, and one can check that this matrix has distinct eigenvalues and is therefore diagonalizable. The algebraic/spectral methods work well for the analysis of A_2 and its powers. This is not the case for the transformation A_1, which is nilpotent. On the other hand, it is clear that the powers of A_1 follow the powers of A_2 in their behavior, at least for some time (with the amount of time depending on the starting point in the vector space). One might say that the algebraic and spectral methods are still lurking around in this case, through the comparison with A_2, even if it is not clear how to apply them directly to A_1. This also makes sense geometrically, since the visibilities of the two graphs evolve in exactly the same manner for some time. (This can be made precise using the notion of locally +-stable mappings, from Section 11.1.)

Of course, G_1 is very special, and one could easily make examples which are much less regular in their behavior. In general, one cannot expect such simple comparisons as the one between G_1 and G_2.

12.3 Some concrete examples

Let us look at the spectral properties of the adjacency transformations of some specific oriented graphs. We shall only look at graphs with cycles, since otherwise the adjacency transformations are nilpotent and have no nonzero eigenvalues, as in Lemma 12.2.

For the first example, we take a "circular" graph G with k vertices $u_1, u_2, \ldots$, u_k, and exactly one edge going from u_i to u_{i+1} when $1 \leq i < k$, and one edge going from u_k to u_1. Thus G consists of a single oriented loop, and nothing else.

If $f(u_i)$ is a function on the set of vertices of our graph G, then the adjacency transformation $A = A_G$ is described by the equations

$$\begin{aligned} Af(u_i) &= f(u_{i-1}) \qquad \text{when } 1 < i \leq k, \\ Af(u_1) &= f(u_k). \end{aligned} \tag{12.11}$$

What are the possible eigenvalues of this transformation? In other words, for which complex numbers λ can we find an f which is not identically 0 and for which $Af = \lambda f$? It is easy to check that there are no nontrivial f's for $\lambda = 0$, and so we restrict ourselves to $\lambda \neq 0$. We can rewrite $Af = \lambda f$ as

$$f(u_i) = \lambda^{-1} f(u_{i-1}) \qquad \text{when } 1 < i \leq k, \tag{12.12}$$

and $f(u_1) = \lambda^{-1} f(u_k)$. It is easy to see that this can happen (with f nontrivial) if and only if λ is a kth root of unity, i.e., $\lambda^k = 1$. (Throughout this section, we shall permit our functions f to take complex values, as is customary in trying to determine the complete set eigenvalues of a matrix.) Of course, there are k of these (complex) roots of unity (including 1 itself), and this leads to k distinct eigenvalues for A, and a basis of k eigenvectors in our vector space (of complex-valued functions on the vertices of G). Thus the adjacency transformation can be diagonalized in this case, and the diagonalization is a discrete version of the Fourier transform.

This diagonalization is compatible with the analysis of the preceding section, in the following sense. No matter what basepoint v in G one chooses, the total number N_j of oriented paths of length j in G which begin at v is equal to 1 for all j. In particular, the N_j's remain bounded (and are nonzero), which is consistent with the fact that the eigenvalues are all roots of unity. There are no nontrivial polynomial factors like j^p in (12.6) here, and this corresponds to the fact that our matrix is diagonalizable, as opposed to having a Jordan canonical decomposition with nontrivial nilpotent parts.

Now consider a more complicated graph H, which consists of two loops connected by a single strand. To be more precise, suppose that H has vertices x_i, $1 \leq i \leq m$, and an edge going from x_i to x_{i+1} for each $i < m$. We also suppose that we have two additional edges, going from x_a to x_1 and from x_m to x_b, where $1 \leq a < b \leq m$. The adjacency transformation $A = A_H$ can be written explicitly as

$$\begin{aligned} Af(x_i) &= f(x_{i-1}) \qquad\qquad \text{when } i \neq 1, b, \\ Af(x_1) &= f(x_a), \\ Af(x_b) &= f(x_{b-1}) + f(x_m). \end{aligned} \tag{12.13}$$

What can we say about the spectral theory of this transformation?

We can obtain some eigenvectors for A in the following way. Let L denote the subgraph of H which consists of the vertices $x_b, x_{b+1}, \ldots, x_m$ and the edges between them, so that L is an oriented loop with $m - b + 1$ vertices. Let λ be a root of unity of this degree, i.e.,

$$\lambda^{m-b+1} = 1. \tag{12.14}$$

Thus λ is an eigenvalue for the adjacency transformation associated to the loop L (as in the previous example), and it turns out to be an eigenvalue for $A = A_H$ as well. Indeed, consider the function $f(x_i)$ defined on the vertices of H by

$$\begin{aligned} f(x_i) &= 0 && \text{when } i < b \\ &= \lambda^{-(i-b)} && \text{when } b \le i \le m. \end{aligned} \tag{12.15}$$

It is not hard to see that f satisfies $Af(x_i) = \lambda f(x_i)$ for all i, and so defines a (nonzero) eigenvector for A with eigenvalue λ. Note that we obtain $m - b + 1$ such eigenvalues and eigenvectors for A in this way.

The strand between the two loops in H leads to some nilpotent vectors for the adjacency transformation. To see this, let us fix an integer j such that $a < j < b$ (if there are any), and define a function g on the vertices of H in the following way. We set $g(x_j) = 1$, and we take g to be zero at all other vertices except for one, which we have to choose a bit carefully. Let j' be the (unique) integer such that $b \le j' \le m$ and such that

$$b - j \equiv m + 1 - j' \quad \text{modulo } m + 1 - b. \tag{12.16}$$

We set $g(x_{j'}) = -1$ and $g(x_i) = 0$ when $i \neq j, j'$. With this special choice of j', we have that $A^{b-j} g = 0$. This is not hard to check. As one applies A over and over again, the vertex where $A^i g$ takes the value 1 moves steadily "upward" in H, towards x_b, while the place where $A^i g$ takes the value -1 goes around the loop L, perhaps many times (if $b - j$ is large compared to $m - b + 1$). The special choice of j' ensures that the values of 1 and -1 will arrive at x_b at exactly the same moment, where they cancel each other out, leaving only the zero vector.

This recipe gives $b - a - 1$ nilpotent vectors for A, and in fact they are all generated from the one for $j = a + 1$ under the powers of A. Thus we are left with looking at the "spectral effects" from the bottom loop K, consisting of the vertices $x_1, x_2, \ldots, x_a$, together with the edges between them. Let us try to build eigenvectors for H from those of K.

Let ω be a complex number such that

$$\omega^a = 1. \tag{12.17}$$

There are a of these roots of unity, and they each define an eigenvector for the loop K, as in the earlier story for the graph G. We can try to define an eigenvector h on all of H and with eigenvalue ω in the following way. First we set

$$h(x_i) = \omega^{1-i} \qquad \text{when } 1 \le i < b. \tag{12.18}$$

This is the only choice possible for the values of $h(x_i)$ when $i < b$, subject to the normalization $h(x_1) = 1$. Leaving aside the choice of $h(x_b)$ for the moment, we should also take

$$h(x_i) = \omega^{b-i} h(x_b) \qquad \text{when } b < i \le m. \tag{12.19}$$

In order for h to define an eigenvector for A, we need to be able to choose $h(x_b)$ so that

$$h(x_b) = \omega^{-1}\, Ah(x_b) = \omega^{-1}(h(x_{b-1}) + h(x_m)) = \omega^{-b+1} + \omega^{b-m-1} h(x_b). \tag{12.20}$$

We can do this so long as

$$\omega^{m-b+1} \neq 1, \tag{12.21}$$

in which case we produce an eigenvector h for A with eigenvalue ω.

The condition (12.21) is the same as saying that our prospective eigenvalue ω is not among the set of eigenvalues that we obtained before, from the loop L at the "top" of H (as in (12.14)). If $a = m - b + 1$, then *all* of the ω's will be among the earlier set of λ's. There will always be some overlap between the two sets of numbers, since 1 is admissible by both.

Let us proceed further, assuming now that ω satisfies

$$\omega^{m-b+1} = 1, \tag{12.22}$$

so that it is among the earlier collection of eigenvalues. In this case, we modify the choice of h somewhat, as follows. We keep the same choices for $h(x_i)$ when $i < b$, as in (12.18), but for $i \geq b$ we change to

$$h(x_i) = T \cdot \omega^{-i} \cdot (m - i + 1), \tag{12.23}$$

where T is a parameter that will be determined later. Using these choices, let us compute $(A - \omega I)h$, where I denotes the identity transformation. Notice first that

$$(A - \omega I)h(x_i) = 0 \qquad \text{when } i < b, \tag{12.24}$$

as one can easily check. This simply amounts to saying that h does behave like an eigenvector when $i < b$. For $i > b$, we have that

$$\begin{aligned}(A - \omega I)h(x_i) &= h(x_{i-1}) - \omega h(x_i) \\ &= T\omega^{-i+1}(m - (i-1) + 1) - T\omega^{-i+1}(m - i + 1) \\ &= T\omega^{-i+1}.\end{aligned} \tag{12.25}$$

For $i = b$, we have that

$$\begin{aligned}(A - \omega I)h(x_b) &= h(x_{b-1}) + h(x_m) - \omega h(x_b) \\ &= \omega^{-b+2} + T\omega^{-m} - T\omega^{-b+1}(m - b + 1).\end{aligned} \tag{12.26}$$

(Remember that $h(x_{b-1})$ is given by (12.18) rather than (12.23).) We would like to have that

$$(A - \omega I)h(x_b) = T\omega^{-b+1}, \tag{12.27}$$

to match with the formula for $i > b$ given in (12.25).

Notice that $\omega^{-m} = \omega^{-b+1}$, because of our assumption (12.22). Thus we can rewrite (12.26) as

$$(A - \omega I)h(x_b) = (\omega + T - T(m - b + 1))\omega^{-b+1}. \tag{12.28}$$

To get (12.27), we can simply choose T so that

$$\omega - T(m - b + 1) = 0, \tag{12.29}$$

i.e., $T = \omega(m - b + 1)^{-1}$.

To summarize, we may choose T so that $T \neq 0$ and

$$\begin{aligned}(A - \omega I)h(x_i) &= 0 && \text{when } i < b \\ &= T\omega^{-i+1} && \text{when } i \geq b.\end{aligned} \tag{12.30}$$

In this way, we get a vector h which is not an eigenvector for A, but which has the property that $(A-\omega I)h$ is an eigenvector with eigenvalue ω. Specifically, it is a multiple of the eigenvector f in (12.15), with $\lambda = \omega$. From this we obtain that $(A-\omega I)^2 h = 0$, so that h is a nilpotent vector for $A-\omega I$. This is exactly what the Jordan decomposition tells us to look for, in addition to ordinary eigenvectors.

We have now accounted for the entire Jordan canonical form of the adjacency transformation associated to H. Indeed, we saw before how the $m - b + 1$ eigenvectors of the upper loop L in H can be extended to give eigenvectors for all of H, and how the $b-a-1$ vertices in the "strand" between the two loops in H lead to the same number of nilpotent vectors for the adjacency transformation. We have also accounted for all the eigenvectors in the lower loop K, showing that they can either be extended to eigenvectors for all of H (when the corresponding eigenvalue ω for K is *not* an eigenvalue for L), or that they can be extended to a function h on H which has the nilpotency property just established (when ω is already an eigenvalue for L). This accounts for everything, because the total number of vectors that we have produced (eigenvectors and vectors with suitable nilpotency properties) is the same as the dimension of our vector space (i.e., the total number of vertices in H, which is m).

The ω's which are eigenvalues for both K and L are the same as the ω's which satisfy both (12.17) and (12.22), and their total number depends on a, b, and m. There is always at least one of these ω's, namely $\omega = 1$. If a and $m - b + 1$ are relatively prime, then $\omega = 1$ is the only common solution to (12.17) and (12.22). If $m-b+1$ is a multiple of a, then every solution of (12.17) also satisfies (12.22).

The fact that there is always at least one ω which is an eigenvalue for both K and L — and hence at least one nontrivial nilpotent vector for $A - \omega I$, with $\omega \neq 0$ — fits perfectly with the fact that the visibility of H starting from x_1 grows *quadratically*, as in Propositions 5.10 and 5.13. Keep in mind that Propositions 5.10 and 5.13 dealt with the total number of paths of length *at most* equal to j, which is the same as

$$\sum_{i=0}^{j} N_i, \tag{12.31}$$

where N_i denotes the number of oriented paths in H which begin at x_1 and have length exactly i. In the present circumstances, the N_i's are of linear growth, and this leads to the quadratic growth of the sum. One can see the linear growth of the N_i's geometrically, in the freedom that oriented paths beginning at x_1 have, in being able to traverse the bottom loop K some number of times, and then go up to the loop L, where they can wrap around some more, before ending with total length exactly equal to i. Algebraically, the linear growth of the N_i's corresponds to having $|\alpha| = 1$ and $p = 1$ in (12.6), and not $|\alpha| > 1$ or $p \geq 2$. That is exactly what we have in this case. The α's in (12.6) come from the nonzero complex eigenvalues of A (as explained just after the proof of Lemma 12.3 in Section 12.2), and in this case the eigenvalues are all roots of unity. The presence of factors j^p in (12.6) with $p = 1$ reflects the fact that we have vectors h which satisfy

$$(A - \omega I)^2 h = 0 \quad \text{but} \quad (A - \omega I)h \neq 0 \tag{12.32}$$

for certain nonzero choices of ω (namely, the common solutions of (12.17) and (12.22)). We do not have vectors h with higher-order nilpotencies in this case, i.e., with

$$(A - \omega I)^r h = 0 \quad \text{but} \quad (A - \omega I)^{r-1} h \neq 0, \tag{12.33}$$

where $r > 1$, and this is the reason for the absence of factors of j^p in (12.6) with $p > 1$. We would have higher-order nilpotencies like (12.33) if the chain of loops in our graph H had length larger than 2, and this is consistent with the faster rate of growth for the N_i's which would then occur, as in Proposition 5.13.

Let us now consider another class of examples, in which there can be several loops which meet at a single vertex. Specifically, we shall work with an oriented graph M that consists of k distinct oriented loops $L_1, L_2, \ldots, L_k$ that meet at a single vertex v. Each loop L_i should have ℓ_i vertices $z(i,1)$, $z(i,2)$, $\ldots, z(i,\ell_i)$, and there should be exactly one edge in L_i going from $z(i,j)$ to $z(i,j+1)$ when $1 \leq j < \ell_i$, and exactly one edge going from $z(i,\ell_i)$ to $z(i,1)$. We ask that all of the vertices and edges in the different L_i's be distinct from each other, except that each $z(i,1)$ should be the same vertex v. We take M to be the graph which is the union of the L_i's, with this one common vertex v. Thus M contains exactly

$$1 + \sum_{i=1}^{k} (\ell_i - 1) \tag{12.34}$$

vertices.

We allow each ℓ_i to be any positive integer, with $\ell_i = 1$ included, in which case L_i contains only one vertex and one edge. We also allow the number of loops k to be arbitrary, except that we require that $k \geq 2$, so that we do not fall back into the first case discussed in this section.

In this situation, we can describe the adjacency transformation $A = A_M$ as follows. If f is any function on the vertices of M, then

$$\begin{aligned} Af(z(i,j)) &= f(z(i,j-1)) \qquad \text{when } j > 1, \\ Af(v) &= \sum_{i=1}^{k} f(z(i,\ell_i)). \end{aligned} \tag{12.35}$$

(Remember that $z(i,1) = v$ for all i.)

What can we say about the spectral properties of A in this case? If $\ell_i = 1$ for each i, then M contains only the single vertex v, and the vector space of f's has dimension 1. In this case, A simply acts by multiplication by the number k.

We shall assume from now on that $\ell_i > 1$ for at least one choice of i. For simplicity, we also ask that ℓ_k be maximal among the ℓ_i's (which can easily be arranged by relabelling the L_i's).

It is easy to account for the nilpotent vectors of A, as follows. Given $i < k$ and $1 < j \leq \ell_i$, consider the function $g_{i,j}$ which satisfies

$$g_{i,j}(z(i,j)) = 1 \quad \text{and} \quad g_{i,j}(z(k, \ell_k - (\ell_j - j)) = -1 \tag{12.36}$$

and takes the value 0 at all other vertices. One can check that

$$A^{\ell_j - j+1} g_{i,j} = 0. \tag{12.37}$$

Indeed, with each application of A, the 1 on L_i and the -1 on L_k take one step around the loop, and the indices have been chosen so that they both reach v at exactly the same moment, where they cancel each other out. This uses the assumption above that ℓ_k be maximal among the ℓ_i's, to know that the -1 on L_k does not reach v before the 1 on L_i does. (If it did, then we would have a problem, because the definition of A would cause this -1 to spread to the other loops.)

In particular, this choice of $g_{i,j}$ lies in the kernel of A when $j = \ell_i$. It is not hard to show that we get the whole kernel of A in this way. In other words, a vector h lies in the kernel of A if and only if

$$h(z(i,p)) = 0 \qquad \text{when } 1 \leq i \leq k \text{ and } p < \ell_i \tag{12.38}$$

and

$$\sum_{i=1}^{k} h(z(i, \ell_i)) = 0. \tag{12.39}$$

This characterization of the kernel of A can be verified directly from the definitions, and it is clear that the $g_{i,j}$'s with $j = \ell_i$ span this space. Once the kernel of A is pinned down in this way, it is not difficult to work backwards to see that the nilpotent vectors $g_{i,j}$ with $i < k$ and $1 < j \leq \ell_i$ account for all of the vectors which lie in the kernel of some positive power of A. This uses the observation that the kernel of A^2 is generated by the kernel of A together with *single* choices of preimages under A of each element of the kernel of A, and similarly for higher powers of A.

Notice that the $g_{i,j}$'s with $i < k$ and $1 < j \leq \ell_i$ span a vector space of dimension

$$\sum_{i=1}^{k-1} (\ell_i - 1). \tag{12.40}$$

The dimension of our whole vector space is the same as the number of vertices in M, which is given by (12.34). Thus there are ℓ_k more dimensions in our vector space for which to account.

Now let us look for nonzero eigenvalues of A, i.e., nonzero complex numbers λ for which there exist nontrivial vectors f such that $Af = \lambda f$. We can rewrite this equation as $f = \lambda^{-1} Af$, which is the same as

$$f(z(i,j)) = \lambda^{-1} f(z(i,j-1)) \qquad \text{when } j > 1, \tag{12.41}$$

$$f(v) = \lambda^{-1} \sum_{i=1}^{k} f(z(i,\ell_i)),$$

by (12.35). If such an f exists, it must satisfy

$$f(z(i,j)) = \lambda^{-(j-1)} f(v) \tag{12.42}$$

for each $1 \leq i \leq k$ and $1 < j \leq \ell_i$, simply by iterating the first equation above. Conversely, given any nonzero complex number λ, we can always define f through (12.42) and a choice of value at the vertex v, and this will determine an eigenvector for A exactly when the second equation in (12.41) is also satisfied, i.e., when

$$f(v) = \lambda^{-1} \sum_{i=1}^{k} f(z(i,\ell_i)) = \sum_{i=1}^{k} \lambda^{-\ell_i} f(v). \tag{12.43}$$

We should also restrict ourselves to functions f which do not vanish everywhere, which is equivalent to requiring that $f(v) \neq 0$, because of (12.42). Using this, we can convert (12.43) into

$$1 = \sum_{i=1}^{k} \lambda^{-\ell_i}. \tag{12.44}$$

To summarize, a nonzero complex number λ is an eigenvalue for A if and only if (12.44) holds, in which case the corresponding eigenvectors must satisfy (12.42). In particular, the eigenvectors are unique up to scalar multiples.

How many of these nonzero eigenvalues are there? Consider the polynomial P given by

$$P(\alpha) = \sum_{i=1}^{k} \alpha^{\ell_i} - 1. \tag{12.45}$$

The roots of this polynomial are exactly the reciprocals of the nonzero eigenvalues of A, by the preceding argument. The degree of P is ℓ_k, since we are assuming that ℓ_k is maximal among the ℓ_i's, and this implies that P has ℓ_k roots, if we count multiplicities.

In general, multiple roots can correspond to eigenvalues for which the corresponding space of eigenvectors has dimension larger than 1. This does not happen here, since the eigenvector is uniquely determined by the eigenvalue, up to a scalar multiple. However, we are missing exactly ℓ_k dimensions in our vector space from the earlier analysis of nilpotent vectors for A. If P has multiple roots, so that there are fewer than ℓ_k nonzero eigenvalues for A, then there must be some additional "nondiagonalizable" parts of the Jordan canonical form for A.

We can see this concretely as follows. Fix a nonzero complex number λ which satisfies (12.44), which is the same as saying that $P(\lambda^{-1}) = 0$. Suppose that λ^{-1} is not a *simple* zero for P, so that

$$P'(\lambda^{-1}) = \sum_{i=1}^{k} \ell_i \, \lambda^{-(\ell_i - 1)} = 0. \tag{12.46}$$

Define a new function ϕ on the vertices of our graph M by

$$\phi(z(i,j)) = \lambda^{-(j-1)} \, j \tag{12.47}$$

for $1 \leq i \leq k$ and $1 \leq j \leq \ell_i$. Notice that this formula gives the same value to $\phi(z(i,1))$ for all i, as it should, since the $z(i,1)$'s all represent v. Let us compute $(A - \lambda I)\phi$, where I denotes the identity transformation, as usual. When $j > 1$, we have that

$$\begin{aligned}(A - \lambda I)\phi(z(i,j)) &= \phi(z(i,j-1)) - \lambda\, \phi(z(i,j)) \\ &= \lambda^{-(j-2)}\,(j-1) - \lambda \cdot \lambda^{-(j-1)}\, j \\ &= -\lambda^{-(j-2)}.\end{aligned} \tag{12.48}$$

(Remember that the action of A is given by (12.35).) For $j = 1$, we have that

$$\begin{aligned}(A - \lambda I)\phi(v) &= \sum_{i=1}^{k} \phi(z(i,\ell_i)) - \lambda\, \phi(v) \\ &= \sum_{i=1}^{k} \lambda^{-(\ell_i - 1)}\, \ell_i - \lambda\end{aligned} \tag{12.49}$$

since $\phi(v) = 1$, by the formula in (12.47). The double root condition (12.46) implies that the sum over i is equal to zero, and hence that

$$(A - \lambda I)\phi(v) = -\lambda. \tag{12.50}$$

Combining this with (12.48), we see that $(A - \lambda I)\phi$ satisfies the condition (12.42), and hence is an eigenvector of A with eigenvalue λ. (This also uses the fact that λ satisfies (12.44), by assumption, to get (12.43).)

The conclusion of all of this is that if the reciprocal of λ is a double root of P, then we can construct a vector ϕ which is *not* an eigenvector of A, but which does lie in the kernel of $(A - \lambda I)^2$. (Strictly speaking, the fact that ϕ is not an eigenvector uses the assumption from long ago that $\ell_k > 1$.) Thus double roots of P lead to nontrivial contributions to the Jordan canonical form in exactly the manner that one would expect, and indeed we can write down these nilpotent vectors for $A - \lambda I$ in a very simple way.

To account for the whole Jordan canonical decomposition, one should also look at roots of P of higher order, but we shall not pursue that here.

We should also point out that multiple roots for P can indeed occur for the class of graphs under consideration. For instance, in order to have a double zero of $P(\alpha)$ at $\alpha = -1$, the ℓ_i's should satisfy

$$P(-1) = P'(-1) = 0, \tag{12.51}$$

which is the same as saying that

$$\sum_{i=1}^{k} (-1)^{\ell_i} - 1 = \sum_{i=1}^{k} \ell_i \, (-1)^{\ell_i} = 0. \tag{12.52}$$

This can be arranged by taking $k = 5$ and choosing the ℓ_i's so that three of them are equal to 2 and two of them are equal to 3, so that

$$P(\alpha) = 2\,\alpha^3 + 3\,\alpha^2 - 1 \tag{12.53}$$

and $P(-1) = P'(-1) = 0$.

How does the spectral theory of A correspond to the behavior of the visibility of M for this class of examples? The visibility of M necessarily grows exponentially in this situation, because we have multiple loops which intersect. (See Proposition 5.1 in Section 5.2.) The powers of A must grow exponentially as well, because of the formula (12.1) in Section 12.2. We also have a special form for the sequence $\{N_j\}$ which governs the growth of the visibility in terms of the spectral properties of A, coming from Lemma 12.3 in Section 12.2 and the discussion just after the proof of Lemma 12.3.

The leading behavior of $\{N_j\}$ as $j \to \infty$ is controlled by the eigenvalues of A of maximal modulus, and this maximal modulus must be strictly larger than 1 in the present situation, in order to accommodate the exponential growth of the N_j's. We can analyse the eigenvalues of maximum modulus directly through the polynomial P in (12.45), as follows.

Nonzero eigenvalues of A correspond to the reciprocals of roots of P, and so we would like to find a root of P of modulus less than 1. From the definition (12.45) of P, we have that $P(0) = -1$ and $P(1) \geq 1$ (since we are assuming here that $k \geq 2$). This implies that P has a real root r between 0 and 1, because P takes real values on the real axis.

In fact r is the *only* root that P has among the positive real numbers, because P is strictly increasing on the positive real numbers. We also have that r must be a simple root of P, because $P'(t) > 0$ for any positive real number t, as one can see from the definition of P.

Let us check that if α is any other root of P in the complex plane, then

$$|\alpha| \geq r. \tag{12.54}$$

Indeed, if $|\alpha| < r$, then

$$|P(\alpha) + 1| = \left| \sum_{i=1}^{k} \alpha^{\ell_i} \right| \leq \sum_{i=1}^{k} |\alpha|^{\ell_i} < \sum_{i=1}^{k} r^{\ell_i} = 1. \tag{12.55}$$

For the last equality, we used the fact that $P(r) = 0$, by construction. This implies that $P(\alpha) \neq 0$ when $|\alpha| < r$, and hence that (12.54) holds when α is a root of P.

From these observations, we conclude that r^{-1} is a positive eigenvalue of A, which is at least as large as the modulus of every complex eigenvalue of A. It might happen that there are other eigenvalues with modulus r^{-1}, however. This is the same as saying that there might be roots α of P which are different from r but which satisfy $|\alpha| = r$, and we can analyze this possibility as follows. Assume first that there is a positive integer $s > 1$ such that s divides ℓ_i for each i. For if s enjoys this property, then we can rewrite the polynomial $P(\zeta)$ as $Q(\zeta^s)$, where Q is given by

$$Q(\zeta) = \sum_{i=1}^{k} \zeta^{\ell_i/s} - 1. \tag{12.56}$$

Not only is r a root of P in this case, but the product of r with any root of unity of order s is as well.

Conversely, if α is a root of P which has the same modulus as r does, then α is equal to the product of r with a root of unity θ, and the product of r with any power of θ must also be a root of P. Indeed, if $|\alpha| = r$, then we have that

$$\left|\sum_{i=1}^{k} \alpha^{\ell_i}\right| = \sum_{i=1}^{k} |\alpha|^{\ell_i}, \tag{12.57}$$

because

$$\sum_{i=1}^{k} \alpha^{\ell_i} = P(\alpha) + 1 = 1 \tag{12.58}$$

and

$$\sum_{i=1}^{k} |\alpha|^{\ell_i} = \sum_{i=1}^{k} r^{\ell_i} = P(r) + 1 = 1. \tag{12.59}$$

Thus (12.57) puts us in the case of equality for the triangle inequality. This implies that the complex numbers α^{ℓ_i} are all positive multiples of a single complex number β. In fact, β is a positive real number, since the sum of the α^{ℓ_i}'s equals 1, and this shows that $\alpha^{\ell_i} > 0$ for each i. We can write α as $r \cdot \theta$, where θ is a complex number with modulus 1, and the positivity of the α^{ℓ_i}'s implies that θ must be a root of unity. Moreover, the minimal integer s such that $\theta^s = 1$ must divide all of the ℓ_i's. This brings us back to the same situation as before. In particular, the product of r with any root of unity of order s must also be a root of P in this case.

Notice that the positivity of the α^{ℓ_i}'s implies that α *cannot* be a double root of P in this situation. This is because a double root should satisfy $P'(\alpha) = 0$, while here we have that

$$\alpha P'(\alpha) = \sum_{i=1}^{k} \ell_i \alpha^{\ell_i} > 0. \tag{12.60}$$

Thus we have a fairly complete picture for the roots of P of minimal modulus, and hence for the eigenvalues of A of maximum modulus. These types of observations are quite standard, and fit within a more general and systematic theory. See [IM76], for instance. See also Theorem 4.11 on p167 of [RS80].

12.4 Representation problems

Consider the following question. Given a collection of nonnegative integers N_j, $1 \leq j \leq n$, and another positive integer k, under what conditions can one find a linear mapping $A : \mathbf{R}^k \to \mathbf{R}^k$ such that each entry of the corresponding matrix is a nonnegative integer, and such that

$$N_j = \sum_{l=1}^{k} \langle A^j(e_1), e_l \rangle, \tag{12.61}$$

where e_l denotes the lth standard unit basis vector in $\mathbf{R}^k$?

There are many natural variations to this question. One might specify the N_j's only for some choices of j, for instance, or loosen the restriction to matrix entries which are nonnegative integers. One could even permit the matrix entries to be more general algebraic objects, like elements of a ring. One could add restrictions on A, such as bounds on the sizes of its matrix entries, or on the sizes of the row sums or column sums of the matrix. Instead of the sum of $\langle A^j(e_i), e_l \rangle$ on the right side of (12.61), one could consider more general expressions, with coefficients, in particular.

A related matter is to look at infinite sequences, and representations as above for $j \in \mathbf{Z}_+$. See [RS80] for a number of questions and results about these. This includes the notions of *N-rational* and *Z-rational* functions, and their relationships with each other, and with growth functions for D0L-systems.

If the entries of the matrix associated to A are nonnegative integers, then A can be interpreted as the adjacency transformation associated to an oriented graph on k vertices. Bounds on the row sums or column sums of the matrix then correspond to bounds on the number of edges entering or departing from each vertex. Our basic question above can be viewed as providing purely algebraic versions of some of our earlier questions concerning the possibility of representing a given tree as the visibility of an oriented graph with a prescribed number of vertices. (See Sections 9.2 and 10.7.)

As in the geometric problems for graphs, one can consider more difficult versions of the same basic question, in which the numbers N_j are given only *implicitly*. For instance, one might assume from the start that one has a representation of the N_j's of the form

$$N_j = \sum_{l=1}^{p} \langle B^j(e_1), e_l \rangle, \tag{12.62}$$

where $B : \mathbf{R}^p \to \mathbf{R}^p$ is a linear transformation which is given to us, and then ask if there is another such representation (12.61) with more restrictive properties, e.g., with a smaller dimension k instead of p. This is analogous to asking whether a given rooted tree (T, b) can be represented as the visibility of a graph of a given size, but with (T, b) provided initially as the visibility of some other graph.

These issues are also closely connected to some of the ones in Chapter 7 and Sections 9.5 and 10.15, concerning the existence and equivalence of implicit descriptions of objects of interest by means of feasibility graphs, formal proofs, etc. (See also [CS].) In some situations, the constructions performed by feasibility graphs can be represented by "adjacency transformations" in a natural way (but not necessarily acting on vector spaces). For instance, let R be a ring (or a semiring), and imagine that we are working with a feasibility graph G over R, in which focusing vertices represent sums in R, and for which the unary transformations associated to edges are given by multiplications by specific elements in R. From G one can make a matrix over R whose powers represent computations performed by G, in the same way that the powers of the ordinary adjacency matrix computes numbers of paths in G. (See also Section 16.9.)

12.5 Mappings and matrices

Let G and G' be finite oriented graphs, and let $\phi : G \to G'$ be a mapping between them. This induces a mapping $\phi_* : \mathcal{F}(G) \to \mathcal{F}(G')$ between the vector spaces of functions on vertices in G and G' (as in Section 12.1), the *push-forward mapping*, as follows. Given $f \in \mathcal{F}(G)$, we define $f' = \phi_*(f)$ by taking $f'(u)$ to be the sum of $f(w)$ over all vertices w in G which are mapped to u by ϕ, where u is any vertex in G'. If there are no such w's, then we take $f'(u)$ to be 0. It is easy to see that this defines $\phi_* : \mathcal{F}(G) \to \mathcal{F}(G')$ as a *linear* mapping.

Let $A : \mathcal{F}(G) \to \mathcal{F}(G)$ and $A' : \mathcal{F}(G') \to \mathcal{F}(G')$ be the adjacency transformations associated to G and G', as in Section 12.1. If f and g are functions in $\mathcal{F}(G)$, we shall write $f \geq g$ to mean that $f(v) - g(v) \geq 0$ for every vertex v in G, and similarly for functions in $\mathcal{F}(G')$.

Lemma 12.4 *Notations and assumptions as above. If $\phi : G \to G'$ is a local $+$-isomorphism (Definition 10.8), then we have that*

$$(A')^j \circ \phi_* = \phi_* \circ A^j \tag{12.63}$$

for all positive integers j. If ϕ is a local $+$-injection (Definition 10.12), then we have that

$$((A')^j \circ \phi_*)(f) \geq (\phi_* \circ A^j)(f) \tag{12.64}$$

for all $f \in \mathcal{F}(G)$ which satisfy $f \geq 0$, and for all positive integers j. If ϕ is a local $+$-surjection (Definition 10.12), then

$$((A')^j \circ \phi_*)(f) \leq (\phi_* \circ A^j)(f) \tag{12.65}$$

for all $f \in \mathcal{F}(G)$ such that $f \geq 0$, and for all positive integers j.

Proof In each case it suffices to consider only $j = 1$, as the assertion for $j > 1$ follows from repeated application of the one for $j = 1$. This uses the fact that A and A' preserve nonnegativity of functions and inequalities between functions, since they are represented by nonnegative matrices in the standard basis.

We may also restrict ourselves to functions $f \in \mathcal{F}(G)$ of the form $e_u(v)$, where u is a vertex in G, and $e_u(v)$ is the function that equals 1 when $v = u$ and vanishes at all other vertices in G. This follows from the linearity of the transformations involved.

With these reductions, the matter becomes purely local, and one can derive the assertions easily from the definitions of ϕ_* and the adjacency transformations A, A'. We omit the details. □

This lemma can be seen as an algebraic counterpart to some of our earlier observations about induced mappings between visibilities (such as Lemmas 10.9 and 10.14). For instance, both types of statements contain information about comparing numbers of oriented paths in graphs G and G' in terms of mappings from G to G'.

One can also look at Lemma 12.4 as a variation on the themes of Section 11.5, where we discussed mappings between feasibility graphs and their effect on the computations represented by the graphs. (Compare with the comments about feasibility graphs and adjacency transformations made at the end of Section 12.4.)

Similarly, in place of some of our earlier questions about the existence of mappings between graphs with particular properties (as in Chapters 10 and 11), one can consider algebraic versions, e.g., concerning the existence of linear mappings from one vector space to another which behave properly in terms of interwining relations with other operations (as in Lemma 12.4).

13

DUALITY AND NP-COMPLETENESS

13.1 The visibility mapping problem

Definition 13.1 *Let G, H, and K be oriented graphs, and assume that G and H have no nontrivial oriented cycles. Let $g : G \to K$ and $h : H \to K$ be orientation-preserving mappings. Fix vertices v in G and w in H, and assume that $g(v) = h(w)$. Set $z = g(v) = h(w)$, and let*

$$\widehat{g} : \mathcal{V}_+(v, G) \to \mathcal{V}_+(z, K), \qquad \widehat{h} : \mathcal{V}_+(w, H) \to \mathcal{V}_+(z, K), \tag{13.1}$$

be the induced mappings between visibilities, as in Section 10.1. Given these data, the visibility mapping problem *asks whether the images of $\widehat{g}$ and $\widehat{h}$ inside $\mathcal{V}_+(z, K)$ are* not *the same.*

When we speak of the "image" of $\mathcal{V}_+(v, G)$ under $\widehat{g}$ (or similarly for h), we mean the image as a subgraph of $\mathcal{V}_+(z, K)$. It is easy to see that the image as a subgraph is determined by the image of the set of vertices in $\mathcal{V}_+(v, G)$, and so we do not really need to worry about what happens to the edges.

Note that the requirement that G and H contain no nontrivial oriented cycles can be verified in polynomial time. This follows from the existence of polynomial-time solutions to the "reachability" problem (as in [Pap94]). This fact, and variants of it, came up before, in Section 9.1.

Proposition 13.2 *The visibility mapping problem is NP-complete.*

In other words, the problem of determining whether the images of $\widehat{g}$ and $\widehat{h}$ inside $\mathcal{V}_+(z, K)$ are the same is co-NP-complete.

This proposition is really just a small variation of well-known results concerning *regular expression inequivalence* and *finite-state automaton inequivalence.* In the first case, one is given two *regular expressions* (as in Section 1.1) which do *not* use star operations, and one is asked to decide whether the expressions determine different languages. The NP-completeness of this problem is mentioned on p267 of [GJ79] and in item (d) on p504 of [Pap94]. Finite-state automaton inequivalence is nearly the same, but with the languages in question represented by automata. (We shall review the way that this works in Chapter 14.) In that case, one requires that the languages be finite to get an NP-complete problem. See p265 of [GJ79]. If one allows infinite languages, or if one allows regular expressions which use the star operation in the first situation, then the questions become PSPACE-complete, as discussed in the references mentioned above. There are further results which have been established, with different levels of complexity, if one permits other operations, or considers different conditions on the languages, regular expressions, or automata. See [GJ79, Pap94] for more information (and references).

The visibility mapping problem is essentially the same as these questions about formal language theory, but adjusted slightly to obtain a purely geometric formulation. (Compare with Section 14.2.) The proof of NP-completeness is also practically the same as for these other situations, but we shall go through it in some detail for the sake of completeness.

We now have three different general categories of computational questions about mappings between graphs. In the present situation, we are given both the relevant graphs and the mappings between them, and we ask about their behavior at the level of the visibility. By contrast, in Sections 10.2, 10.5, 10.14, and 11.3, the graphs were given, and it was the *existence* of a mapping between them with certain properties which was in question. (This seems to be the more common paradigm for mappings between graphs in complexity theory.) In Sections 10.7, 10.9, 10.11, 10.13, and 11.1, only the domains of the mappings were given, and it was both the graphs in the ranges and the mappings themselves whose existence were under consideration. (Of course, there can be modest variants of one type of problem which fit into another group.)

For the computational problems in Chapters 9 and 10 concerning minimal representations, isomorphic equivalence of visibility graphs, and minimal folding graphs, we had polynomial-time solutions when the given graphs were free of nontrivial oriented cycles, as in Propositions 9.11 and 9.17, Lemma 10.59, Corollaries 10.84 and 10.89, and Remarks 10.42 and 10.90. For the local +-isomorphism problem, the presence or absence of nontrivial oriented cycles did not matter, as in Proposition 10.92 in Section 10.14. The same construction for showing that oriented cycles do not matter also works for a number of other problems in the second category above, in which one is given a pair of graphs, and asked about the existence of a mapping from one to the other with certain properties. For the visibility mapping problem, the presence of nontrivial oriented cycles does matter, and corresponds roughly to allowing the use of the star-operation in the regular-expression inequivalence problem, or to the version of the finite-state automaton inequivalence problem in which the language are permitted to be infinite.

Proof We shall follow closely the discussion in item (d) on p504 of [Pap94].

We must first show that the visibility mapping problem is NP. Let G, H, v, w, $g : G \to K$, and $h : H \to K$ be given as above, and suppose that

$$\widehat{g}(\mathcal{V}_+(v, G)) \neq \widehat{h}(\mathcal{V}_+(w, H)). \tag{13.2}$$

We might as well assume that there is a vertex in $\widehat{g}(\mathcal{V}_+(v, G))$ which does not lie in $\widehat{h}(\mathcal{V}_+(w, H))$.

Concretely, this means that there is an oriented path α in G which begins at v whose image $\beta = g(\alpha)$ in K cannot be realized as the image $h(\gamma)$ of an oriented path γ in H which begins at w. We want to show that the validity of this property for a given path α can be verified effectively from the data of the problem (in polynomial time).

Our assumption that G contain no (nontrivial) oriented cycle implies that α cannot go through any vertex of G more than once. In particular, its length (number of edges traversed) is bounded by the number of vertices in G, which is important for the NP property.

For the non-existence of γ, we should be a bit careful. Let k denote the *length* of α. For $i = 0, 1, 2, \ldots, k$, let U_i denote the set of vertices u in H for which there is an oriented path δ in H from w to u such that δ has length i and $h(\delta)$ is the same as the initial subpath of $g(\alpha)$ of length i. Thus U_0 consists of w alone, and the nonexistence of of a path γ as above is the same as saying that $U_k = \emptyset$. We want to check that this can be determined in polynomial time.

In fact, one can compute every U_i, $1 \leq i \leq k$, in polynomial time. To see this, it suffices to show that U_i can be computed in polynomial time (as a function of the size of our initial data, which includes H and the mappings $g : G \to K$, $h : H \to K$), given the knowledge of U_{i-1}. This is very easy to do, since a vertex p lies in U_i if and only if there is a vertex q in U_{i-1} and an (oriented) edge e from q to p such that $h(e)$ is the ith edge in the path $g(\alpha)$.

Thus the U_i's for $1 \leq i \leq k$ can all be computed in polynomial time, and in particular one can decide whether U_k is empty or not in polynomial time. This finishes the proof that the visibility mapping problem lies in NP.

To show NP-completeness, we shall make a reduction from 3SAT, i.e., the satisfiability problem for formulae in conjunctive normal form and with 3 literals in each clause. For this, we shall need only very special instances of the visibility mapping problem.

Let $x_1, x_2, \ldots, x_n$ be some collection of n Boolean variables, and suppose that we have m clauses $C_1, C_2, \ldots, C_m$, where each C_j is of the form

$$C_j = \lambda_{j,1} \vee \lambda_{j,2} \vee \lambda_{j,3}. \tag{13.3}$$

Here $\lambda_{j,l}$ should either be an x_i or the negation of an x_i for all choices of j and l. To solve the 3SAT problem, one is supposed to determine whether there are choices of truth values for the x_i's so that each of the clauses $C_1, C_2, \ldots, C_m$ becomes "true". We want to show that this problem can be encoded into an instance of the visibility mapping problem.

We choose K so that it consists of a single vertex z and two edges which both begin and end at z. We think of these two edges as being labelled 0 and 1.

For H we do something very similar, except that we have to take care to avoid oriented cycles. We choose H so that it has $n+1$ vertices, which we denote by $w_0, w_1, \ldots, w_n$. We ask that there be exactly two edges which go from w_i to w_{i+1} for $i = 0, 1, 2, \ldots, n-1$. We think of these two edges as being labelled by 0 and 1, and we add no other edges to H.

Note that we could just as well take K to be the same as H, to avoid oriented cycles in K, for instance. This choice would require only modest changes to the argument that is given below. (See also the proof of Proposition 13.6 in Section 13.3.)

We set $w = w_0$, and we define $h : H \to K$ in the obvious way. That is, h maps all of the w_i's to z, and it maps the edges of H to the two edges of K according to the labelling by 0's and 1's mentioned above.

The definition of G is more complicated, and is based on the clauses $C_1, C_2, \ldots, C_m$. Fix j with $1 \leq j \leq m$, and define an oriented graph G_j as follows. We give to G_j exactly $n+1$ vertices, which we denote $v(j,0), v(j,1), \ldots, v(j,n)$. We have to decide how to connect these vertices by edges. As before, we shall only put edges from a vertex $v(j,i)$ to its successor $v(j,i+1)$, but we need to do this more carefully now, using the clause C_j. Basically, we only allow edges when they are compatible with *not* satisfying C_j.

Let us be more precise. Given i with $1 \leq i \leq n$, we attach two edges going from $v(j,i-1)$ to $v(j,i)$ if the variable x_i is *not* involved in the literals $\lambda_{j,1}$, $\lambda_{j,2}$, $\lambda_{j,3}$. We label one of these edges with 0 and the other by 1. If one of the literals $\lambda_{j,1}$, $\lambda_{j,2}$, $\lambda_{j,3}$ is equal to x_i, then we do *not* put an edge from $v(j,i-1)$ to $v(j,i)$ which is labelled by 1. Similarly, if one of the literals $\lambda_{j,1}$, $\lambda_{j,2}$, $\lambda_{j,3}$ is equal to $\neg x_i$, then we do *not* add an edge from $v(j,i-1)$ to $v(j,i)$ which is labelled by 0. If both x_i and $\neg x_i$ appear among $\lambda_{j,1}$, $\lambda_{j,2}$, $\lambda_{j,3}$, then we do not add *any* edge from $v(j,i-1)$ to $v(j,i)$, and we add exactly one edge when only one of x_i and $\neg x_i$ appears among $\lambda_{j,1}$, $\lambda_{j,2}$, $\lambda_{j,3}$.

This defines the graph G_j. That is, we put in edges exactly as in the manner just described, and we do not put in any others.

We do this for each $j = 1, 2, \ldots, m$. To define G, we take the disjoint union of the G_j's, and then identify the first vertices $v(1,0), v(2,0), \ldots, v(m,0)$ of each of them into a single vertex, which we take for our special vertex v. See Fig. 13.1 for an example.

We define $g : G \to K$ in the obvious manner. All vertices in G are sent to the unique vertex z in K, and we map the edges in G to the two edges in K according to the labels 0 and 1.

What happens in the visibilities? The vertices in $\mathcal{V}_+(z, K)$ represent oriented paths in K, and they can be described in an obvious way by arbitrary finite words over 0 and 1. The image of $\mathcal{V}_+(w, H)$ under $\widehat{h}$ corresponds exactly to the set of all binary strings of length at most n.

The image of $\mathcal{V}_+(v, G)$ under $\widehat{g}$ is more interesting. It consists of binary strings of length at most n, and the strings of length exactly n are precisely the ones which *fail* to satisfy at least one of the C_j's.

This is easy to check. Each vertex in $\mathcal{V}_+(v, G)$ represents an oriented path in G which begins at v, and every such path must be inherited from a path in one of the G_j's. The paths that come from a fixed G_j yield precisely those truth assignments which fail to satisfy C_j (if there are any).

If we could solve the visibility mapping problem by a polynomial-time algorithm, then we would be able to decide in polynomial time whether the image of $\mathcal{V}_+(v, G)$ under $\widehat{g}$ is the same as the image of $\mathcal{V}_+(w, H)$ under $\widehat{h}$. If we knew

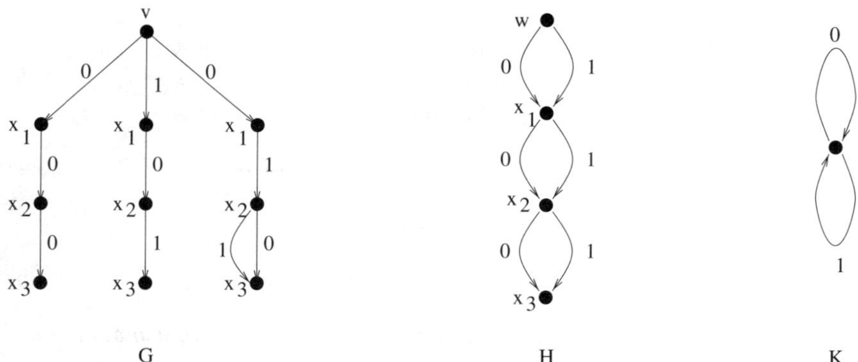

FIG. 13.1. This picture shows the graphs G, H, and K in the case where there are three Boolean variables x_1, x_2, x_3, and three clauses C_1, C_2, and C_3, with $C_1 = x_1 \vee x_2 \vee x_3$, $C_2 = \neg x_1 \vee x_2 \vee \neg x_3$, and $C_3 = x_1 \vee \neg x_2 \vee \neg x_2$. The three "strands" in G correspond to these three clauses, as in the definition of G. The numbers of vertices in G and H correspond to the number of Boolean variables that we have, and the relationship between the vertices and the variables is also indicated in the picture.

this, then we could solve the question of the satisfiability of the collection of clauses $C_1, C_2, \ldots, C_m$. Indeed, if the image of $\mathcal{V}_+(v, G)$ under $\widehat{g}$ is the same as the image of $\mathcal{V}_+(w, H)$ under $\widehat{h}$, then it means that all truth assignments are *non-satisfying*, and we are finished. Conversely, suppose that the two images are *not* the same. By construction, the image of $\widehat{g}$ is *contained* in the image of $\widehat{h}$. If we do not have equality, then it means that there is an element in the image of $\widehat{h}$ which is not in the image of $\widehat{g}$. A priori, this element of the image of $\widehat{h}$ could correspond to a path (or a binary string) of length less than n, but in that case we could extend it to one of length equal to n which could not lie in the image of $\widehat{g}$ either. (This is easy to check from the definitions.) This means that there is a truth assignment which is not in the "failure" set for *any* of the C_j's, which says exactly that we have a satisfying truth assignment.

This completes the proof of the reduction to 3SAT, and the proposition follows. □

The term "duality" in the title of this chapter refers to the fact that in the visibility mapping problem we do not work directly with the graphs G and H themselves, but with "measurements" through "functions" defined on them. This is a very common in mathematics, with Boolean algebras being somewhat special in this regard, because of the natural "self"-dualities and identifications between basic objects and measurements (i.e., sets and indicator functions).

13.2 Monotonicity and stability properties

Let us pause a moment to look at the monotonicity and stability properties enjoyed by the visibility mapping problem. Let G, G_0, H, H_0, K, and K_1 be oriented graphs, all of which are free of oriented cycles, and let v, v_0, w, w_0, z, and z_1 be vertices in these graphs (in that order). Let $g : G \to K$ and $h : H \to K$ be orientation-preserving mappings, with $g(v) = h(w) = z$, and let

$$\alpha : G_0 \to G,\ \beta : H_0 \to H,\ \text{and } \gamma : K \to K_1 \tag{13.4}$$

be orientation-preserving mappings which satisfy

$$\alpha(v_0) = v,\ \beta(w_0) = w,\ \text{and } \gamma(z) = z_1. \tag{13.5}$$

Lemma 13.3 *Notations and assumptions as above. If the induced mappings*

$$\widehat{\alpha} : \mathcal{V}_+(v_0, G_0) \to \mathcal{V}_+(v, G), \quad \widehat{\beta} : \mathcal{V}_+(w_0, H_0) \to \mathcal{V}_+(w, H) \tag{13.6}$$

are surjections, and if the induced mapping

$$\widehat{\gamma} : \mathcal{V}_+(z, K) \to \mathcal{V}_+(z_1, K_1) \tag{13.7}$$

is an injection, then the "answer" to the visibility mapping problem for

$$\gamma \circ g \circ \alpha,\ \gamma \circ h \circ \beta,\ v_0,\ w_0 \tag{13.8}$$

is the same as the answer for

$$g,\ h,\ v,\ w. \tag{13.9}$$

Proof This is easy to check, using the definition of the visibility mapping problem (Definition 13.1) and the "homomorphism property" ((10.2) in Section 10.1) for the induced mapping between visibilities. □

Note that the assumptions on α, β, and γ in Lemma 13.3 are satisfied if α and β are local +-surjections, and γ is a local +-injection, because of Lemma 10.14. (Recall that the notions of "local +-surjection" and "local +-injection" were defined in Definition 10.12 in Section 10.3.)

In the next lemma, we give a slightly more general monotonicity property for the visibility mapping problem.

Lemma 13.4 *Notation and assumptions as above. Suppose now that we only ask that the induced mapping*

$$\widehat{\alpha} : \mathcal{V}_+(v_0, G_0) \to \mathcal{V}_+(v, G) \tag{13.10}$$

be surjective (and do not put conditions on $\widehat{\beta}$ and $\widehat{\gamma}$). If the image of the visibility $\mathcal{V}_+(w_0, H_0)$ in $\mathcal{V}_+(z_1, K_1)$ under the mapping induced by $\gamma \circ h \circ \beta$ contains a vertex that does not lie in the image of the visibility $\mathcal{V}_+(v_0, G_0)$ in $\mathcal{V}_+(z_1, K_1)$ under the mapping induced by $\gamma \circ g \circ \alpha$, then the same is true for g, h, G, H, and K, i.e., the image of $\mathcal{V}_+(w, H)$ in $\mathcal{V}_+(z, K)$ under the mapping induced by h contains a vertex that does not lie in the image of $\mathcal{V}_+(v, G)$ in $\mathcal{V}_+(z, K)$ under the mapping induced by g. (In particular, the images of $\mathcal{V}_+(w, H)$ and $\mathcal{V}_+(v, G)$ under the mappings induced by g and h are distinct in this situation, as in Definition 13.1.)

It is not necessarily true here that the conclusion for g and h implies the assumption about $\gamma \circ h \circ \beta$ and $\gamma \circ g \circ \alpha$, because there can be loss of information coming from β and γ.

Proof This is again easy to verify from the definitions and the homomorphism property (10.2). □

13.3 The visibility surjection problem

Definition 13.5 *Let G and K be oriented graphs without nontrivial oriented cycles, and let $g : G \to K$ be a mapping between them. Fix a vertex v in g, and let*

$$\widehat{g} : \mathcal{V}_+(v, G) \to \mathcal{V}_+(g(v), K) \tag{13.11}$$

be the induced mapping between visibilities (as in Section 10.1). The visibility surjection problem *is to decide whether $\widehat{g}$ is a surjection, i.e., whether the image of $\mathcal{V}_+(v, G)$ under $\widehat{g}$ is all of $\mathcal{V}_+(g(v), K)$.*

As before, the requirement that G and K be free of nontrivial oriented cycles can be checked in polynomial time.

Proposition 13.6 *The visibility surjection problem is co-NP-complete.*

Proof It suffices to show that the problem of deciding whether $\widehat{g}$ is *not* a surjection is NP-complete. (See Proposition 10.1 on p220 of [Pap94].) This is nearly the same as Proposition 13.2. Indeed, the problem of deciding whether $\widehat{g}$ is not a surjection can be viewed as a special case of the visibility mapping problem, with $H = K$ and with $h : H \to K$ taken to be the identity, and this permits us to derive membership in NP from Proposition 13.2. To prove NP-completeness, one can make a reduction from 3SAT in practically the same manner as before, but with the following modifications. One can use the same graphs G and H as in the proof of Proposition 13.2, but the graph K should now be taken to be the same as the graph H (instead of the earlier graph with one vertex and two edges). The previous choice of $h : H \to K$ should be replaced with the identity mapping, and $g : G \to K$ should be modified so that g maps the vertex $v(j, i)$ in G to the vertex w_i in $K = H$, for all choices of j and i. (The action of g on edges should still respect the labelling by 0 and 1.) With these changes, the rest of the argument is practically the same as in the proof of Proposition 13.2, and we omit the details. □

Problem 13.7 *Is there a natural "proof system" for the visibility surjection problem?*

In other words, one would like to have a notion of "derivation" which can be verified effectively and which would guarantee the surjectivity of $\widehat{g}$ for a given mapping $g : G \to K$, in the same way that a formal proof guarantees the validity of a logical formula. One would like for this notion to be "complete", in the sense that a derivation exists whenever $g : G \to K$ has the property that $\widehat{g}$ is a surjection. If one could also show that every mapping $g : G \to K$ for which $\widehat{g}$ is a surjection admits a derivation of *polynomial size*, then one would be able to conclude that the visibility surjection problem lies in NP. This would imply that NP = co-NP, since the visibility surjection problem is co-NP complete. (See Proposition 10.2 on p220 of [Pap94]. This is analogous to the situation for valid formulae and proof systems in formal logic, as in Section 1.5, and similar issues have come up in other mathematical contexts.)

One would like to have proof systems for the visibility surjection problem which are as geometric as possible. The next lemma provides a basic ingredient for this.

Lemma 13.8 *Let G_1, G_2, G_3 be oriented graphs, and let v_1, v_2, v_3 be vertices, with $v_i \in G_i$. Let $g_1 : G_1 \to G_2$ and $g_2 : G_2 \to G_3$ be orientation-preserving mappings such that $g_1(v_1) = v_2$ and $g_2(v_2) = v_3$. Write h for the composition $g_2 \circ g_1$, and let*

$$\widehat{g}_i : \mathcal{V}_+(v_i, G_i) \to \mathcal{V}_+(v_{i+1}, G_{i+1}), \qquad i = 1, 2, \tag{13.12}$$

$$\widehat{h} : \mathcal{V}_+(v_1, G_1) \to \mathcal{V}_+(v_3, G_3) \tag{13.13}$$

be the induced mappings between visibilities (Section 10.1). If $\widehat{g}_1$ and $\widehat{g}_2$ are surjections, then the same is true of $\widehat{h}$. Conversely, if $\widehat{h}$ is a surjection, then $\widehat{g}_2$ is as well.

Note the similarity with Sections 10.17 and 13.2.

Proof This follows easily from the fact that $\widehat{h} = \widehat{g}_2 \circ \widehat{g}_1$, as in (10.2). □

Lemma 13.8 shows that in order to "prove" that a given mapping $g : G \to K$ induces a surjection between visibility graphs, it is enough to find a "derivation" of it through compositions and de-compositions of mappings which are already known to induce surjections on visibility graphs. (By a *de-composition* we mean a passage as from h to g_2 in Lemma 13.8.) This is a nice feature of the visibility surjection problem, that it cooperates well with compositions of mappings.

In order to use this to make a viable proof system for the visibility surjection problem, one needs to have a sufficiently rich supply of mappings which are known to induce surjections between visibility graphs, or for which this surjectivity can be easily verified. One such class of mappings is provided by the local +-surjections (Definition 10.12), since they always induce surjections between visibility graphs, as in Lemma 10.14 in Section 10.3.

The local +-surjectivity property is *not* necessary for having a surjection between visibility graphs. This is very different from the situation for injections and isomorphisms, which will be discussed in Section 13.4. Here is a simple example. Let K be the oriented graph with three vertices 1, 2, 3, with two edges going from 1 to 2 and two edges going from 2 to 3, and let G be the graph with six vertices a, b, c, d, e, f, with an edge going from a to each of b, c, d, and e, and an edge going from each of these four vertices to f. Thus there are exactly four oriented paths in G which go from a to f, and four oriented paths in K that go from 1 to 3. Consider now the mappings $g : G \to K$ that satisfy

$$g(a) = 1,\ \ g(b) = g(c) = g(d) = g(e) = 2,\ \text{and}\ g(f) = 3. \tag{13.14}$$

There are a number of ways that such a mapping might be defined on edges, but it is easy to see that one can choose g in such a way that the induced mapping

$$\widehat{g} : \mathcal{V}_+(a, G) \to \mathcal{V}_+(g(a), K) \tag{13.15}$$

is a surjection. No mapping of this type will ever be a local +-surjection, since there is only one edge in G which comes out of each of b, c, d, and e.

One can make more complicated examples by gluing together many copies of modest configurations like this one, or by attaching a small piece like this to a local +-surjection. In this way, one can have mappings which are not local +-surjections, but which do induce surjections between visibility graphs, and for which there are short "proofs" of this fact.

To make this idea more systematic, one can specify gluing operations that can be used to combine different mappings between graphs, and which preserve the surjectivity of induced mappings between visibility graphs. For this purpose, it will be helpful to describe first some collapsing operations for mappings between graphs.

Definition 13.9 (First collapsing operation) *Let G, K be oriented graphs, and let $g : G \to K$ be an orientation-preserving mapping between them. Let A be a set of vertices in G which are all mapped by g to the same vertex in K, and let G_0 denote the graph obtained from G by collapsing all of the vertices in A to a single point and leaving all other vertices in G intact, as well as the edges. Let $g_0 : G_0 \to K$ denote the mapping which is induced from $g : G \to K$ in the obvious way. We say that $g_0 : G_0 \to K$ was produced from $g : G \to K$ by the* first collapsing operation.

More precisely, we have a canonical mapping

$$\tau : G \to G_0 \tag{13.16}$$

which represents the effect of the collapsing of A to a single vertex, and the mapping $g_0 : G_0 \to K$ satisfies

$$g = g_0 \circ \tau. \tag{13.17}$$

Lemma 13.10 *Let G and K be oriented graphs, let $g : G \to K$ be an orientation-preserving mapping between them, and suppose that $g_0 : G_0 \to K$ was obtained from $g : G \to K$ by the first collapsing operation. Fix a vertex v in G, and let v_0 denote the corresponding vertex in G_0. If the induced mapping*

$$\widehat{g} : \mathcal{V}_+(v, G) \to \mathcal{V}_+(g(v), K) \tag{13.18}$$

is a surjection, then the same is true for

$$\widehat{g}_0 : \mathcal{V}_+(v_0, G_0) \to \mathcal{V}_+(g_0(v_0), K). \tag{13.19}$$

Note that the converse is not true; it could easily happen that $\widehat{g}_0$ is a surjection but $\widehat{g}$ is not. (Consider the case where $G_0 = K$ and g_0 is the identity mapping.)

Proof This follows from Lemma 13.8 and (13.17) (and is easy to verify directly anyway). □

Definition 13.11 (Second collapsing operation) *Let G and K be oriented graphs, and let $g : G \to K$ be an orientation-preserving mapping between them. Let x, y be two vertices in G, and let G' denote the oriented graph obtained from G by identifying x with y but leaving all other vertices alone, and also leaving the edges alone. Define K' in the same way, but using the vertices $g(x)$, $g(y)$. Let $g' : G' \to K'$ denote the mapping obtained from $g : G \to K$ by following these identifications. Assume also that*

$$\begin{aligned}&x \textit{ is the only vertex in } G \textit{ which is mapped by } g \textit{ to } g(x), \\ &\textit{and } y \textit{ is the only vertex in } G \textit{ which is mapped to } g(y).\end{aligned} \tag{13.20}$$

Under these conditions, we say that $g' : G' \to K'$ is obtained from $g : G \to K$ by the second collapsing operation.

In this case, we have mappings

$$\tau' : G \to G' \quad \text{and} \quad \kappa' : K \to K' \tag{13.21}$$

which represent the contractions of the vertices x, y in G and $g(x)$, $g(y)$ in K, and $g' : G' \to K'$ is related to $g : G \to K$ by the equation

$$g' \circ \tau' = \kappa' \circ g. \tag{13.22}$$

Remark 13.12 The assumption (13.20) is not too serious, since one can reduce to that case through the first collapsing operation.

Definition 13.13 (Stable collapsing operations) *Let*

$$G,\, K,\, g : G \to K,\, x, \textit{ and } y \tag{13.23}$$

be as in Definition 13.11. We say that the second collapsing operation is stable *if the induced mappings*

$$\widehat{g}_x : \mathcal{V}_+(x, G) \to \mathcal{V}_+(g(x), K), \quad \widehat{g}_y : \mathcal{V}_+(y, G) \to \mathcal{V}_+(g(y), K) \tag{13.24}$$

between visibilities are both surjective.

The next lemma provides a basic criterion for the second collapsing operation to produce a mapping which induces a surjection between visibility graphs.

Lemma 13.14 *Let G and K be oriented graphs, and let $g : G \to K$ be an orientation-preserving mapping between them. Suppose that $g' : G' \to K'$ is obtained from a second collapsing operation which is stable. Fix a vertex v in G,* *and let v' denote the corresponding vertex in G'. (That is, $v' = \tau'(v)$, where τ' is as in (13.21).) If*

$$\widehat{g}_v : \mathcal{V}_+(v, G) \to \mathcal{V}_+(g(v), K) \tag{13.25}$$

is a surjection, then the same is true of

$$\widehat{g'}_{v'} : \mathcal{V}_+(v', G') \to \mathcal{V}_+(g'(v'), K'). \tag{13.26}$$

Proof Let $g : G \to K$, etc., be as above, and assume that $\widehat{g}_v$ is surjective. The following gives a more concrete formulation of the surjectivity of $\widehat{g'}_{v'}$.

Claim 13.15 *If β' is any oriented path in K' which begins at $g'(v')$, then there is an oriented path α' in G' which begins at v' and which is mapped to β' by g'.*

To prove Claim 13.15, let a path β' in K' be given as above. Let x and y be the vertices in G which are identified with each other to produce G', as in Definition 13.11. Write w' for the vertex in G' which was obtained by identifying x and y in G, and write z' for the vertex in K' which was obtained by identifying $g(x)$ with $g(y)$ in K. If β' never passes through z', then we are finished, because β' can then be viewed as an oriented path in K which begins at $g(v)$, and the assumption that $\widehat{g}_v$ be surjective implies the existence of a suitable path α' in G'. If instead β' does pass through z', then the argument becomes a bit more complicated, and we have to use our assumption of stability of the second collapsing operation.

If β' does pass through z', then we can break β' up into a sequence of subpaths $\beta'_1, \beta'_2, \ldots, \beta'_n$ with the following properties: β' is the same as the concatenation of $\beta'_1, \beta'_2, \ldots, \beta'_n$; β'_i begins at z' when $2 \le i \le n$; β'_i ends at z' when $1 \le i \le n-1$; and no β'_i passes through z' at any time except at the endpoints. This is easy to check.

Since the β'_i's do not pass through z' except at the endpoints, we can find oriented paths $\beta_1, \beta_2, \ldots, \beta_n$ in K with the following properties: (a) β'_i is obtained from β_i by projecting it into K' (through the mapping $\kappa' : K \to K'$ in (13.21)); (b) β_1 begins at $g(v)$; (c) β_i begins at either $g(x)$ or $g(y)$ when $2 \le i \le n$; and (d) β_i ends at $g(x)$ or $g(y)$ when $1 \le i \le n-1$. This follows from the corresponding properties for the β'_i's. (Note that we do not say *which* of $g(x)$ or $g(y)$ is the starting or ending point for β_i in (c), (d).)

Our assumption that $\widehat{g}_v$ be surjective implies that there is an oriented path α_1 in G which begins at v and which is mapped onto β_1 by g. Similarly, for each $i > 1$, there is an oriented path α_i in G which begins at either x or y (as appropriate) and which is mapped to β_i by g. This follows from our "stability" assumption, which ensures the surjectivity of the mappings in (13.24).

We also know that α_i must end at either x or y when $i \le n-1$, because of property (d) of the β_i's above and the condition (13.20) in Definition 13.11. Let α'_i be the image of α_i down in G', i.e., under the mapping $\tau' : G \to G'$ from (13.21). For each $i \le n-1$, we have that the endpoint of α'_i and the starting point of α'_{i+1} are both equal to the same vertex w' (where $w' = \tau'(x) = \tau'(y)$, as above). This permits us to combine the α'_i's into a single oriented path α' in G.

Note that α' begins at v', since α_1 begins at v. The image of α' under g' is equal to β', because the image of α'_i under g' is equal to β'_i for each i, by construction. (This is not hard to verify, and it basically comes down to (13.22).)

This finishes the proof of Claim 13.15. Using the claim, it is easy to see that $\widehat{g'}_{v'}$ must be surjective, and Lemma 13.14 follows. □

The second collapsing operation provides a mechanism by which to glue mappings together, and control what happens to the induced mappings between visibilities. More precisely, if $g_1 : G_1 \to K_1$ and $g_2 : G_2 \to K_2$ are two mappings that one wishes to "glue" together, one can begin by combining them into a single mapping $g : G \to K$ in a trivial way, by taking G to be the disjoint union of G_1 and G_2, doing the same for K_1 and K_2, and for g_1 and g_2. This puts the initial mappings into a common package, and to make genuine "gluings" one can identify vertices as in the second collapsing operation.

We shall see some concrete examples of gluing procedures of this nature in Chapter 15, especially Sections 15.2, 15.3, and 15.4. Note that there are some natural situations in which the assumption (13.20) in Definition 13.11 may not hold, but Lemma 13.14 works anyway, and through roughly the same method. This happens for the "union" operation in Section 15.4, for instance.

If $g_1 : G_1 \to K$, $g_2 : G_2 \to K$ are two orientation-preserving mappings with the *same* target graph K, then one can also combine g_1 and g_2 into a single orientation-preserving mapping $\phi : P \to K$ through the operation of "fiber product". (See Section 15.6.) This operation is compatible with surjectivity of the induced mappings between visibilities as well, as in Lemma 15.8.

Thus there are a number of different types of operations which one could utilize in a proof system for the visibility surjection problem. Remember that we have compositions and de-compositions of mappings too, as in Lemma 13.8. In connection with the NP = co-NP problem, one would like to ask whether every mapping $g : G \to K$ which induces a surjection between visibility graphs can be developed in a reasonably concise way from simpler mappings through operations like these. If one does not believe that NP = co-NP should be true, then one should perhaps not ask for too much here, but it does seem plausible that there could be a definite relationship between the geometry of G and K when there is a mapping $g : G \to K$ which induces a surjection between the visibility graphs and when the visibility graphs are much larger than G and K themselves. There might be a relationship which could be expressed in terms of the existence of a suitable "derivation" of the mapping $g : G \to K$, in particular. (It also seems plausible that such a relationship might exist in a form which would be sufficiently unregulated so as not to be useful for computational issues related to NP and co-NP, etc.)

13.4 The visibility injection problem

Lemma 13.16 *Let G and K be oriented graphs, and let $g : G \to K$ be an orientation-preserving mapping. Fix a vertex v in G, and assume that every* *vertex and edge in G can be reached by an oriented path which begins at v. Then the induced mapping*

$$\widehat{g} : \mathcal{V}_+(v, G) \to \mathcal{V}_+(g(v), K) \tag{13.27}$$

between visibilities is an injection if and only if g is a local $+$-injection (Definition 10.12).

In other words, the "visibility injection problem" is characterized by a local condition which is very easy to check. Thus we do not get an interesting problem for NP or co-NP, as we had before, for surjections.

Proof If g is a local $+$-injection, then we know already from Lemma 10.14 that $\widehat{g}$ must be an injection. Conversely, suppose that g is not a local $+$-injection. This means that there is a vertex x in G and a pair of edges e_1, e_2 in G that go out of x such that g maps e_1 and e_2 to the same edge in K. Let α be an oriented path in G that goes from v to x (whose existence is guaranteed by our hypotheses), and let α_1 and α_2 be the extensions of α obtained by adding the edges e_1 and e_2, respectively. Then g maps α_1 and α_2 to the same oriented path in K, and this is the same as saying that $\widehat{g}$ sends the vertices in $\mathcal{V}_+(v, G)$ that correspond to α_1, α_2 to the same vertex in $\mathcal{V}_+(g(v), K)$. Thus $\widehat{g}$ is not injective, and the lemma follows. □

There is a similar result for isomorphisms between visibility graphs.

Lemma 13.17 *Let G and K be oriented graphs, and let $g : G \to K$ be an orientation-preserving mapping. Fix a vertex v in G, and assume that every vertex and edge in G can be reached by an oriented path which begins at v. Then the induced mapping*

$$\widehat{g} : \mathcal{V}_+(v, G) \to \mathcal{V}_+(g(v), K) \tag{13.28}$$

between visibilities is an isomorphism if and only if g is a local $+$-isomorphism (Definition 10.8).

Proof If g is a local $+$-isomorphism, then the induced mapping between visibilities is an isomorphism, as in Lemma 10.9. Conversely, assume that $\widehat{g}$ does define an isomorphism between the visibility graphs. From Lemma 13.16 we know that g must be a local $+$-injection. Assume, for the sake of a contradiction, that g is not a local $+$-surjection, so that there is a vertex x in G and an edge e in K such that e flows out of $g(x)$, but no outgoing edge at x in G is mapped to e by g. Let α be an oriented path in G which goes from v to x, and let β be the oriented path in K which is the image of α under g. Let β^* denote the oriented path in K obtained by adding e to the end of β. The point now is that there is no oriented path γ^* in G which begins at v and which is mapped to β^* by g. For suppose that there were such a path γ^*, and let γ denote the initial subpath of γ^* which includes all of γ^* except for the last step, i.e., the last vertex and edge. The local $+$-injectivity of g ensures that $\gamma = \alpha$, as in Lemma 10.14. This means that γ^* should be obtained from α by adding an edge to the end of it, but this is impossible, since we are assuming there is no outgoing edge at x (the endpoint of

α) which is mapped to the final edge e of β^*. This proves that g must be a local $+$-surjection as well as a local $+$-injection, and hence a local $+$-isomorphism, as desired. □

14

FINITE AUTOMATA AND REGULAR LANGUAGES

Finite automata (defined below) are basic objects from theoretical computer science which can be used to characterize certain collections of words over an alphabet, namely, *regular languages.* They provide another mechanism by which to make implicit descriptions, and with a degree of implicitness which is tightly controlled. They are connected to the themes of this book in a number of ways, and we review some of their features in the present chapter. General references include [HU79, vL90b]. The first chapter of [ECH$^+$92] provides a nice introductory treatment as well.

14.1 Definitions and the subset construction

A *finite-state automaton* (or *deterministic finite-state automaton*) is, formally, a 5-tuple $(Q, \Sigma, \delta, q_0, F)$. Let us approach the definition slowly. The set Q is a finite set of *states.* Σ is an "alphabet", which means a finite set of "letters" that one uses to make *words* (strings of letters). In the end, the automaton determines a particular set of words over Σ, sometimes called the *language accepted by the automaton.*

The *transitions* of the automaton are governed by δ, which is a mapping from $Q \times \Sigma$ into Q. If q is an element in Q (and so a "state" of the automaton), and a is a letter in Σ, then $\delta(q, a)$ specifies the state in Q that "comes next" under these conditions. In this way, a word w over Σ provides instructions for moving around in Q. If one starts at a state q, then one should first move from there to $q' = \delta(q, a_1)$, where a_1 is the first letter in w, and then from q' to $q'' = \delta(q', a_2)$, where a_2 is the second letter in w, and so on.

The last two ingredients in our 5-tuple $(Q, \Sigma, \delta, q_0, F)$ tell us where to begin and end in Q. Specifically, q_0 is an element of Q called the *initial state*, and F is a subset of Q of *final states*, or *accept states.* To see whether a given word w is "accepted" by the automaton, one starts at the initial state q_0, and moves from there to new states q', q'', etc., according to the letters in w (as in the previous paragraph). If $\overline{q}$ denotes the last state in the chain, reached after using all of the letters in w, then w is accepted by the automaton when $\overline{q}$ lies in F, and otherwise not.

The transitions of the automaton can be encoded into an oriented graph G, called the *transition graph* associated to the automaton, which is defined as follows. For the vertices of G we use the elements of Q. Given q, q' in Q, we attach an edge from q to q' for every letter $a \in \Sigma$ such that $\delta(q, a) = q'$. Thus the total number of outgoing edges at each vertex is the same as the total number of letters in Σ. Each edge should be viewed as being "labelled" by the corresponding element of Σ. Note that there may be multiple edges from q to q', associated to different letters in Σ.

A sequence of transitions in the automaton corresponds exactly to an oriented path in the graph G. A word w over Σ is accepted by the automaton if it can be read off from an oriented path in G that begins at q_0 and ends in F.

An example of a deterministic automaton is shown in Fig. 14.1, with the alphabet $\Sigma = \{a, b, A, B\}$. It includes six states, an initial state 0 (which is also allowed as a final state), four other final states 1, 2, 3, and 4, and a failure state 5 (which is not a final state). This example is nearly the same as ones in [ECH$^+$92, Far92], and we shall come back to this in a moment.

The language L which is recognized by this automaton consists of all words over Σ for which there are never adjacent occurrences of a and A, or of b and B. This language arises naturally if one is interested in the free group with two generators a, b, and with A and B interpreted as representing the inverses of a and b. Thus the words that arise are the so-called *reduced* words, in which all expressions of the form aA, Aa, bB, and Bb have been "cancelled". This ensures that no two words represent the same group element. (Note that L includes the empty word, since the initial state 0 is also a final state in this example.)

This example is motivated in part by the notion of *automatic groups*, to which we shall return in Chapter 17. Basic references for this are [ECH$^+$92, Far92].

Let Σ^* denote the set of all (finite) words over Σ, including the empty word. A *language* over Σ is just a subset of Σ^*. A language is called *regular* if it is recognized by a finite-state automaton in the manner described above. This class of languages turns out to be the same as the one defined through the notion of *regular expressions*, as in Section 1.1. (See [HU79].)

There is also a *nondeterministic* version of the concept of an automaton. For this, one starts with a set of states Q and an alphabet Σ as before, but now if one is given a state q in Q and a letter a in Σ there may be 0, 1, or more transitions to elements q' of Q that are associated to q and a. Thus one may not have a transition *function* from $Q \times \Sigma$ into Q, but instead a "pseudomapping" which is allowed both to take multiple values, or no value (for a given pair q, a). One also specifies an initial state q_0 and a set of final states F, and the language accepted by the automaton consists of the words w which can be read off from sequences of transitions from q_0 to an element of F. One can define an oriented graph G associated to the system in the same manner as before, and again the words accepted by the automaton are the ones which come from oriented paths in G which begin at q_0 and end at an element of F.

If G is the transition graph of a *deterministic* automaton, then for every word w in Σ^* there is a unique oriented path in G which begins at q_0 and follows the transitions represented by the letters in w. For nondeterministic automata, neither existence nor uniqueness of such a path has to hold.

An example of a nondeterministic automaton is shown in Fig. 14.2. In this example, 0 is the initial state, and 3 is the only accept state. The language accepted by this automaton is the one consisting of all words in a and b which

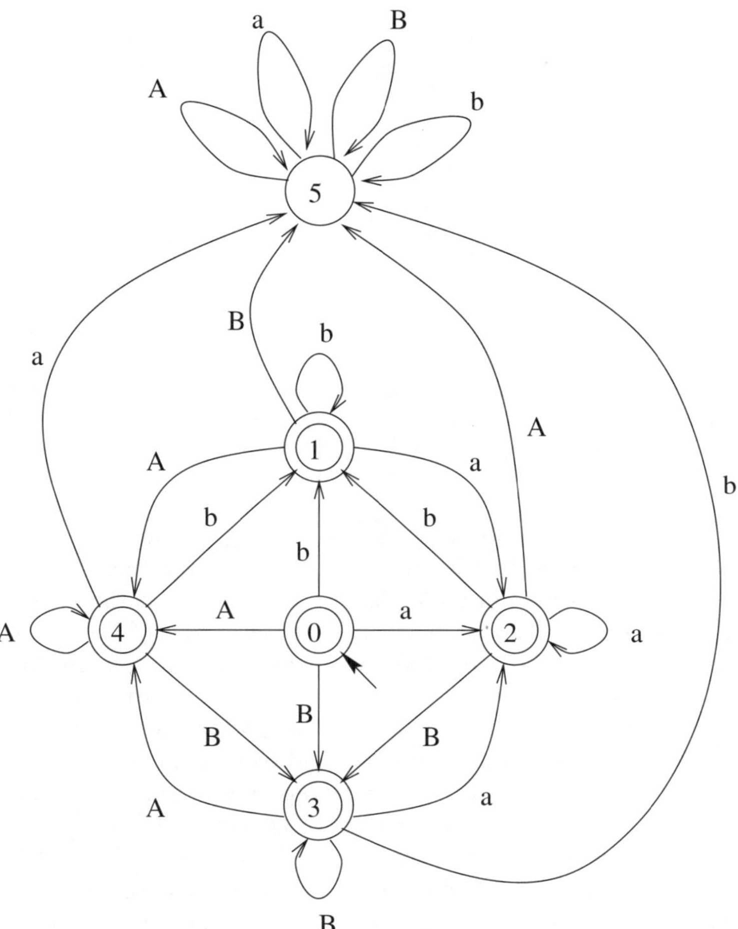

FIG. 14.1. This picture shows a deterministic automaton that recognizes the set of reduced words in a free group with two generators. The symbols a and b denote the generators of the group, and A and B are their inverses. The states in the automaton are indicated by circles, the final states by double circles, and the initial state by an arrow.

end with *bab*. This language is represented by the regular expression $(a+b)^*bab$.

There is a universal construction for converting nondeterministic automata into deterministic ones while preserving the associated language. This is called the *subset construction*. The idea is simple enough; one replaces states in the original automaton with *sets of states* in order to get rid of the possible multiplicities or absences of transitions. This can be described concretely as follows. Let Q denote the collection of states in the original automaton, and let q_0 be the initial state. Given a letter a in the associated alphabet, we look at the collection

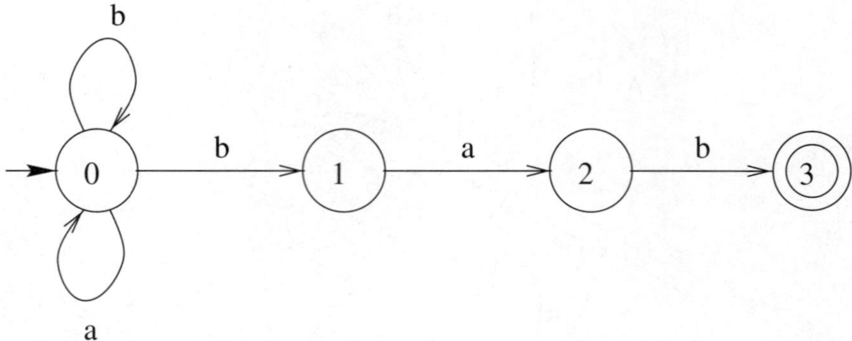

FIG. 14.2. This picture shows a nondeterministic automaton. There are two ways to read the letter b at the initial state 0, and no way to read b at the states 1, 3, or to read a at the states 2, 3.

of *all* states in Q which can be reached by q_0 through a transition labelled by a. We repeat this process, so that in general we start with a collection $\{q_1, \ldots, q_n\}$ of states in Q and a letter b in the alphabet, and we look at the set $\{p_1, \ldots, p_r\}$ of all states in Q which can be reached by at least one of the q_i's through a transition labelled by b. We may generate an empty set of states in this way, but we do not mind. We take our new set of states Q' to consist of $\{q_0\}$ together with all the collections of states $\{p_1, \ldots, p_r\}$ in Q which are eventually obtained from $\{q_0\}$ by applying this procedure over and over again, using all letters, and in any order. (One can also be less parsimonious and take *all* collections of states in Q.) Thus the elements of Q' are *subsets* of Q. We take $\{q_0\}$ to be the initial state of our new automaton, and for the *final* states we take the elements of Q' which contain one of the final states of Q as an element. With transitions between elements of Q' defined as above, we get now a *deterministic* automaton, as one can easily check. It is not hard to verify that it accepts the same language as the original automaton too.

In the case of the nondeterministic automaton shown in Fig. 14.2, this construction leads to the deterministic automaton shown in Fig. 14.3.

See [HU79] for more information about the subset construction. Note that the passage from a nondeterministic automaton to a deterministic can require exponential expansion in the number of states.

Notice that there are really two reasons why an automaton might not be deterministic. There can be states and letters for which more than one transition is defined, as well as states and letters where no transition is defined. The latter possibility is easily fixed by adding a single failure state to which the missing transitions can be made. This state can also admit transitions to itself, to avoid the creation of new missing transitions. By adding no transitions to the other states, one guarantees that the language accepted by the automaton is not changed. In the subset construction, the empty set provides exactly this kind of failure state. The exponential expansion in the subset construction comes

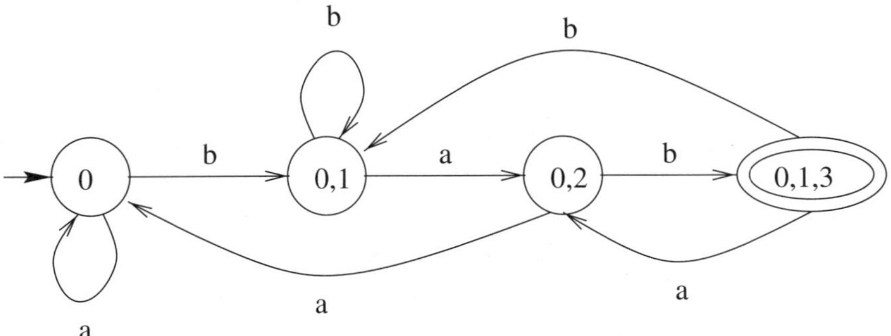

FIG. 14.3. The "determinization" of the automaton shown in Fig. 14.2

from the resolution of the first problem, concerning the possibility of multiple transitions associated to a single choice of state and letter.

The state 5 in the deterministic automaton shown in Fig. 14.1 has exactly the role of this kind of failure state, and its removal would have little effect beyond taking the automaton out of compliance with the details of the definition of a deterministic automaton.

There is another extension of finite automata, in which one allows "ϵ-moves", in addition to ordinary transitions as above. More precisely, an ϵ-move is a transition between states which is not labelled by a letter in the alphabet, or which one might view as being labelled by the empty word ϵ. The language accepted by such an automaton is defined in the same way as before, through finite sequences of transitions and the words that they generate, with no letters used for the ϵ-moves. If a language is accepted by a nondeterministic automaton which accepts ϵ-moves, then the language is also accepted by one which does not. See [HU79].

Two examples of finite automata with ϵ-moves are shown in Fig. 14.4. The language accepted by the first example is represented by the regular expression $ab^* + ba^*$, and the language accepted by the second example is represented by b^*a+a^*b. In the first example, one could convert the automaton into one without ϵ-moves, and which recognizes the same language, simply by taking out the ϵ-moves. That is, the states 0, 1, and 2 would be combined into a single state in the new automaton, and this would be the initial state. One would keep the other states as they are, and use the same transitions as before. In the second example, if one merely took out the ϵ-moves in the same manner, and combined the states 0, 1, and 2, then the language recognized by the automaton would change. One would have to do something different to get an automaton without ϵ-moves which recognizes the same language.

We should emphasize that finite automata are much more restrictive than Turing machines in general. This is made precise by the fact that they require only "bounded space". Conversely, if a language is recognized by a Turing machine that needs only bounded space, then the language is regular, and can be

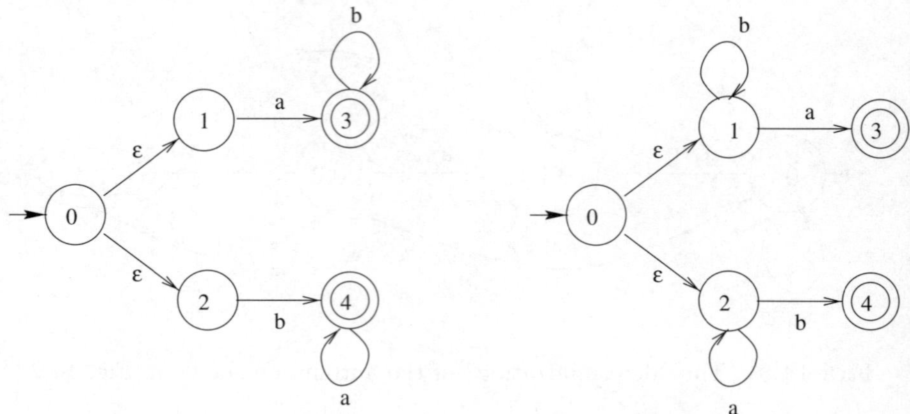

FIG. 14.4. Two finite automata with ϵ-moves

recognized by a finite automaton. (See p54-5 of [Pap94] for more information.) There is an enormous range between Turing machines that use bounded space and Turing machines in general, with many complexity classes and extensions of automata coming up between the two extremes.

In this chapter, we shall take a more geometric view of finite state automata, in terms of mappings between oriented graphs, and the induced mappings between their visibility graphs.

14.2 Geometric reformulations

Let us begin by rephrasing the information contained within a finite automaton in more geometric terms. As before, we can associate to an automaton an *oriented graph* G, in which the set Q of states are the vertices and edges represent transitions between the states. The initial state q_0 and the collection F of finite states simply reflect certain designations of vertices, which one might view as "boundary conditions" for G.

This graph G comes with extra information, namely an assignment of a letter in an alphabet Σ to each edge in graph G. Let us rephrase this as follows. To an alphabet Σ we associate an *alphabet graph* $\Gamma = \Gamma(\Sigma)$ as follows. We give Γ exactly one vertex. To this vertex we attach exactly one edge (with both endpoints at the single vertex) for each letter in Σ. (See Fig. 14.5.) We think of these edges as being labelled by the letters in Σ.

We can reformulate the idea of a (nondeterministic) automaton in the following manner. It consists of an oriented graph G, a choice of initial vertex q_0, a collection of final vertices F, and a mapping from G to the alphabet graph $\Gamma = \Gamma(\Sigma)$ associated to some alphabet Σ. This collection of objects contains exactly the same amount of information as a nondeterministic finite automaton; the mapping g from G to $\Gamma(\Sigma)$ is just another way of saying that we are associating a letter in Σ to each edge in G. The mapping on vertices contains no information, because $\Gamma(\Sigma)$ has only one vertex.

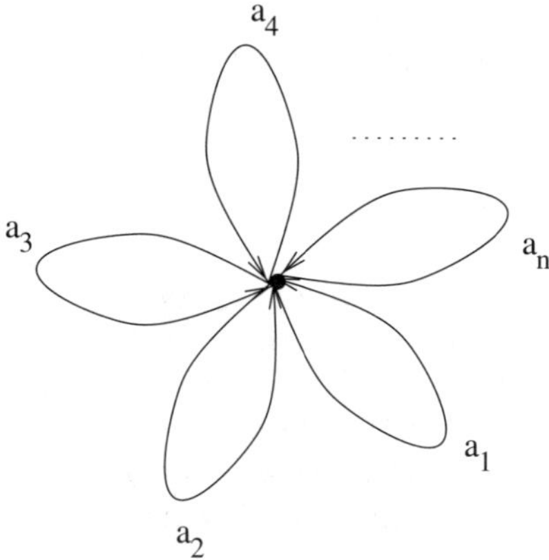

FIG. 14.5. An alphabet graph

It is sometimes convenient to think of alphabet graphs as being *oriented*, and to speak of *orientation-preserving* mappings between them, even if this is somewhat meaningless at the combinatorial level. (If one thinks of graphs *topologically*, so that edges are like actual segments, or intervals of points in the real line, then the orientation is more significant.)

In this reformulation, a *deterministic* automaton corresponds exactly to the same collection of objects, but with the extra requirement that the mapping $g : G \to \Gamma(\Sigma)$ should be a *local +-isomorphism* (Definition 10.8 in Section 10.3). This is not hard to check from the definitions. Similarly, a *nondeterministic finite automaton with ϵ-moves* corresponds to the same collection of objects, except that we allow $g : G \to \Gamma(\Sigma)$ to be a *weak mapping* (Definition 10.2 in Section 10.1), instead of a mapping.

We can also describe the language accepted by an automaton geometrically using the visibilities of our graphs. Normally one reads words from certain paths in the graph G, namely, the oriented paths which begin at q_0 and end at an element of F. These paths correspond to the vertices in the visibility $\mathcal{V}_+(q_0, G)$ which lie in $\pi^{-1}(F)$, where $\pi : \mathcal{V}_+(q_0, G) \to G$ is the canonical projection (defined in Section 4.5). The reading of words from these vertices corresponds to taking the image of $\pi^{-1}(F)$ under the mapping (or weak mapping) $\widehat{g} : \mathcal{V}_+(q_0, G) \to \mathcal{V}_+(o, \Gamma(\Sigma))$ between visibility graphs, where $\widehat{g}$ is induced from $g : G \to \Gamma(\Sigma)$ in the manner of Section 10.1, and where o denotes the unique vertex of the alphabet graph $\Gamma(\Sigma)$. This reinterpretation employs the fact that the vertices in the visibility $\mathcal{V}_+(o, \Gamma(\Sigma))$ are in a natural one-to-one correspondence with the set Σ^* of all words over the alphabet Σ (including the empty word). That is, vertices in $\mathcal{V}_+(o, \Gamma(\Sigma))$ represent oriented paths in the alphabet graph $\Gamma(\Sigma)$ which begin at o, and these paths are exactly characterized by words over Σ, since o is the only vertex in $\Gamma(\Sigma)$, and the edges in $\Gamma(\Sigma)$ are in one-to-one correspondence with the letters in Σ.

In short, we can think of the set $\widehat{g}(\pi^{-1}(F))$ of vertices in the visibility

$\mathcal{V}_+(o, \Gamma(\Sigma))$ as being an equivalent representation of the language accepted by the automaton.

14.3 An extended view

The preceding discussion suggests a more geometric view of regular languages and the finite automata which produce them. Instead of using an alphabet graph $\Gamma(\Sigma)$ as above, we can use any graph H and any mapping (or weak mapping) $g : G \to H$. The analogue of the *language accepted by the automaton* would then be the collection of all paths in H which arise as the image under g of an oriented path in G that begins at the initial vertex q_0 and ends at a final vertex (i.e., an element of F).

For simplicity, let us restrict ourselves to the situation where H is an oriented graph, and where the mapping $g : G \to H$ preserves orientations. Thus we have the induced mapping (or weak mapping) $\widehat{g} : \mathcal{V}_+(q_0, G) \to \mathcal{V}_+(g(q_0), H)$ between visibility graphs, which permits us to represent the aforementioned collection of paths as the set $\widehat{g}(\pi^{-1}(F))$ of vertices in the visibility $\mathcal{V}_+(g(q_0), H)$.

This extended notion of automata could be coded into the usual version, by collapsing the image graph H down to an alphabet graph, identifying all of the vertices to a single vertex, but keeping the edges intact and distinct. (Note that the visibility itself of any finite oriented graph can be coded as a regular language, by letting the graph represent an automaton in which the edges are labelled with distinct letters. The initial state of the automaton would be the vertex in the graph from which one takes the visibility, and all vertices would be used as final states.) Instead, we want to try to keep and use as much of the available geometry as possible. For instance, the formulation in terms of mappings between graphs cooperates well with *compositions* of mappings.

One of the reasons for defining this extension of automata is to have more flexibility in making implicit descriptions of geometric structures. We shall return to this theme in a more concrete way beginning in Section 17.6.

Many standard results about automata extend to this setting with little trouble. The "pumping lemma" (as on p56 of [HU79]) amounts to the statement that if the language associated to $g : G \to H$ includes a path in H of sufficiently large length, then there has to be an oriented cycle in the domain G, and, more precisely, a cycle which is accessible from q_0 and from which one can reach a final vertex. (Compare with Section 4.7.) By traversing this cycle a greater or fewer number of times, one can get infinitely many other "accepted" paths in the image (and which are related to the original one in a simple way). (If $g : G \to H$ is a weak mapping, then one should be a bit careful, and get a cycle in G which is not simply collapsed to a single vertex in H by g.)

Similarly, one has a natural dichotomy between polynomial and exponential growth of the language associated to a mapping $g : G \to H$. This corresponds to results for regular languages, as in Section 1.3 of [ECH$^+$92], and the material in Chapter 5.

Unions and intersections of regular languages are regular, and similar results hold for other operations, such as concatenation and Kleene closure. (See [HU79].) For general graphs H, one can make suitable constructions directly at a geometric level, as in Chapter 15.

This extended notion also helps to emphasize the following geometric point about automata. Let $g : G \to H$ be an orientation-preserving mapping between oriented graphs. As above, we assume that G comes equipped with a designated initial vertex $q_0(G)$ and a set $F(G)$ of final vertices. Let us assume for the moment that $g : G \to H$ is a *local* $+$*-isomorphism* (Definition 10.8), which corresponds to the notion of *deterministic automata.* In this case, the induced mapping

$$\widehat{g} : \mathcal{V}_+(q_0, G) \to \mathcal{V}_+(g(q_0(G)), H) \tag{14.1}$$

is an *isomorphism*, by Lemma 10.9. Thus the two visibility graphs $\mathcal{V}_+(q_0, G)$ and $\mathcal{V}_+(g(q_0(G)), H)$ both represent the same tree T, but they might do so in different ways.

In particular, a set of vertices in T might be simpler to describe in one representation than in another. If we think of T as being the visibility of G, then it is very easy to understand the set $\pi^{-1}(F(G))$ of vertices which project down to final vertices in G through the usual projection $\pi : \mathcal{V}_+(q_0, G) \to G$ (from Section 4.5). This set may not be as easy to understand when we think of T as being the visibility of H. This is because a vertex w in H may have several preimages under the mapping $g : G \to H$, and it may be that some of these preimages lie in $F(G)$, while others do not. (See Fig. 14.6 for an example.)

14.4 Markov languages

In Chapter 9 of [Gd90], there is discussed the notions of *Markov grammars* and *Markov languages.* A Markov grammar amounts to the same thing as nondeterministic finite automata (without ϵ-moves), in which all states are final states, and no state is unreachable by the initial state. A Markov language is a language which is accepted by such an automaton.

The relationship between Markov grammars and nondeterministic automata in general is made clearer with the notion of a *live state.* If one is working with a transition graph G with initial vertex $q_0(G)$ and set of final vertices $F(G)$, then a *live state* is a vertex in G which is contained in an oriented path that goes from $q_0(G)$ to a final vertex. All states are live in the context of a Markov grammar. This is not true for automata, but we can reduce to that case by eliminating all states in a given automaton that are not live, and transitions to them from live states, and transitions from them. This will not effect the language accepted by the automaton, as one can check.

Given any nondeterministic automaton, we can make a Markov grammar by first reducing to the case where all states are live, and then using all states for final states. The Markov language accepted by the resulting automaton is the *prefix closure* of the original one, i.e., the set of words which arise as initial subwords of words in the original language.

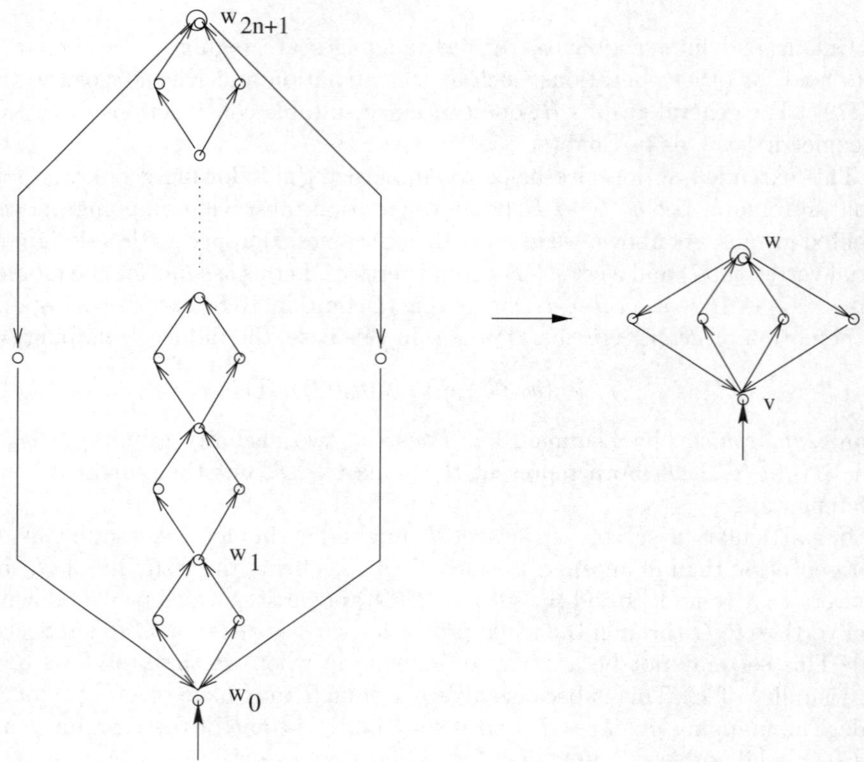

FIG. 14.6. A local +-isomorphism that sends w_{2i} to v and w_{2i+1} to w for every i. We take w_{2n+1} to be the only final vertex on the left, and w to be the only final vertex on the right, so that w has many preimages on the left which are not final vertices. One could make more complicated examples, with multiple "chains" of different lengths inside the perimeter on the left, for instance.

These observations are standard, and similar matters arise in [ECH$^+$92]. They have a useful geometric consequence which we would like to discuss briefly (and which also comes up in [ECH$^+$92]).

Lemma 14.1 *Let G be an oriented graph, with a specified initial vertex $q_0(G)$ and set of final vertices $F(G)$. Assume that all vertices in G are "live", in the sense that each one lies on an oriented path from $q_0(G)$ to an element of $F(G)$. Then there is a nonnegative integer κ such that for each vertex s in the visibility $\mathcal{V}_+(q_0(G), G)$ there is another vertex t in $\pi^{-1}(F(G))$ such that t can be reached by an oriented path (in $\mathcal{V}_+(q_0(G), G)$) from s of length at most κ. One can take κ to be less than the number of vertices in G.*

Thus every vertex in $\mathcal{V}_+(q_0(G), G)$ lies within a bounded distance of an element of $\pi^{-1}(F(G))$ in these circumstances, so that the entire visibility is approximately "filled" by $\pi^{-1}(F(G))$. The assumption about all vertices being live is needed and reasonable, and for automata one can easily reduce to that case, as indicated above.

Proof For each vertex u in G, there is an oriented path γ in G which begins at u and ends in $F(G)$, because of the requirement that all vertices be live. We may assume that γ does not pass through any vertex in G more than once, by removing unnecessary loops. This ensures that the length of γ be strictly less than the total number of vertices in G.

Given any oriented path α in G which begins at $q_0(G)$, we can extend it to an oriented path β which ends in $F(G)$, by adding to α a path γ chosen as in the previous paragraph. In particular, the length of the extension can be taken to be strictly less than the number of vertices in G. Lemma 14.1 follows easily from this assertion by lifting paths to the visibility of G, with s corresponding to α, and t to β. □

15

CONSTRUCTIONS WITH GRAPHS

We have seen in Sections 14.2 and 14.3 how finite automata and regular languages are very close to oriented graphs and their visibilities, and how the notion of automata can be extended by allowing arbitrary graphs in the image instead of only alphabet graphs. In this chapter, we discuss this extension more formally, and describe some constructions for graphs which correspond to familiar properties of regular languages, and which work in the extended setting. These constructions are also related to the topics of Section 13.3.

15.1 Mappings and automata

Let us begin by making some of the material from Section 14.3 a bit more formal.

Definition 15.1 (Marked graphs) *A* marked graph *will mean an oriented graph G together with a choice of "initial vertex" $q_0(G)$ and a collection of "final vertices" $F(G)$.*

If G is a marked graph, G' is another oriented graph, and $g : G \to G'$ is a mapping which preserves orientations, then we can think of these data as representing a kind of generalized automaton, as in Section 14.3. For the analogue of the *language accepted by an automaton*, we can use the visibility, as follows.

Definition 15.2 *Suppose that G is a marked graph, G' is a graph which is oriented, and that $g : G \to G'$ is a mapping (or weak mapping) between them which preserves orientations. Define $F(\mathcal{V}_+(q_0(G), G))$, the set of final vertices in the visibility $\mathcal{V}_+(q_0(G), G)$, by*

$$F(\mathcal{V}_+(q_0(G), G)) = \pi^{-1}(F(G)), \tag{15.1}$$

where $\pi : \mathcal{V}_+(q_0(G), G) \to G$ is the canonical projection (from Section 4.5). The language associated to $g : G \to G'$ *is the subset of the set of vertices in the visibility $\mathcal{V}_+(g(q_0(G)), G')$ which is the image of $F(\mathcal{V}_+(q_0(G), G))$ under the mapping $\widehat{g} : \mathcal{V}_+(q_0(G), G) \to \mathcal{V}_+(g(q_0(G)), G')$ between visibilities that is induced by g (as in Section 10.1). In other words, the language associated to $g : G \to G'$ corresponds to the set of oriented paths in G' which arise as images of oriented paths in G that begin at $q_0(G)$ and end at an element of $F(G)$.*

We can think of the visibility graph $\mathcal{V}_+(q_0(G), G)$ as being a marked graph itself, with its standard basepoint as initial vertex, and with the set of final vertices chosen as in (15.1). We shall sometimes write $\mathcal{V}_+(G)$ for this visibility graph as a marked graph when a particular choice of marking for G is understood.

At times, it will be convenient to think of G' and its visibility graph as being marked graphs too. This adds some symmetry to the general discussion, and it can be useful in dealing with compositions of mappings. There is also a natural notion of *mappings between marked graphs.*

Definition 15.3 *If G and G' are marked graphs, then a mapping (or a weak mapping) $g : G \to G'$ is* compatible with the markings *if it preserves orientations, and if $g(q_0(G)) = q_0(G')$ and $g(F(G)) \subseteq F(G')$.*

If G is a marked graph, and $\mathcal{V}_+(G)$ is the visibility of G as a marked graph (as above), then the canonical projection $\pi : \mathcal{V}_+(G) \to G$ is automatically compatible with the markings.

We can always ask that G' be a marked graph, and that $g : G \to G'$ be compatible with the markings, without any loss of generality. For if G' is not marked at the beginning, we can always give it a marking by setting $q_0(G') = g(q_0(G))$ and taking $F(G')$ to be any set of vertices which contains $g(F(G))$, and even the set of all vertices in G'. The choice of $F(G')$ does not play a role in Definition 15.2.

For alphabet graphs, there is never any trouble with the marking, since there is only one vertex in the graph. We shall always take it to be the initial vertex, as well as the only final vertex. (The only other option would be to have no final vertices.)

From now on, when we speak of generalized automata (in the spirit of Section 14.3), we shall typically do so in terms of mappings between marked graphs which preserve the markings. The reader should feel free to choose the set of final vertices for the image graph to be the entire set of its vertices if that is convenient.

Let us mention a couple of small observations about mappings between marked graphs.

Lemma 15.4 *(a) If G and G' are marked graphs, and $g : G \to G'$ is a mapping (or a weak mapping) which is compatible with the markings, then the lifting $\widehat{g} : \mathcal{V}_+(G) \to \mathcal{V}_+(G')$ (in the sense of Section 10.1) is compatible with the markings.*

(b) If G'' is another marked graph, and $g_2 : G' \to G''$ is also compatible with the markings, then $g_2 \circ g : G \to G''$ is compatible with the markings.

This is easy to derive from the definitions.

15.2 Cartesian products and concatenation

Let G and H be marked graphs. We define a new marked graph K from G and H as follows. For the vertices of K, we take the disjoint union of sets of vertices in G and H. For the edges, we take all of the edges in G and H, and we also add an edge going from each vertex in $F(G)$ to $q_0(H)$. This defines K as an oriented graph. For the marking we take $q_0(K)$ to be $q_0(G)$ and $F(K)$ to be $F(H)$.

This construction for marked graphs provides a natural representation for operations of *concatenation* and *Cartesian product* for formal languages. To see this, we begin at the level of the visibility. Since K is a marked graph, it has a marked visibility graph $\mathcal{V}_+(K)$ with a special set of final vertices $F(\mathcal{V}_+(K))$ (defined in Section 15.1). There is a natural one-to-one correspondence between this set of final vertices and ordered pairs of "final" vertices in the visibility graphs $\mathcal{V}_+(G)$, $\mathcal{V}_+(H)$. This comes down to the fact that every oriented path in K which begins at $q_0(G)$ and ends at an element of $F(H)$ determines a pair of paths in G and H, where the path in G begins at $q_0(G)$ and ends in $F(G)$, and the path in H begins at $q_0(H)$ and ends at $F(H)$. Conversely, any pair of paths in G and H with these properties can be combined (together with one of the new edges in K that go from G to $q_0(H)$) to make a path in K that begins at $q_0(G)$ and ends in $F(H)$.

Although we might think in terms of Cartesian products here, we are actually closer to the idea of *concatenations* in terms of what is happening geometrically with the paths which are being represented by vertices in the visibilities. For this one can think of the extra edges in K (going into $q_0(H)$) as being like ϵ-moves. One can also think of them as providing explicit markers for the transition between the pieces coming from G and H (which brings one closer to the idea of a Cartesian product).

Now let us look at *mappings* between marked graphs. Suppose that G' and H' are two more marked graphs, and that $g : G \to G'$ and $h : H \to H'$ are mappings (or weak mappings) which are compatible with the markings. Let K be as above, and let K' be the marked graph obtained from G' and H' in the same manner. We can combine g and h to get a mapping (or a weak mapping) $k : K \to K'$ in an obvious way.

According to Definition 15.2, the language associated to k is the given by the image of $F(\mathcal{V}_+(K))$ in $F(\mathcal{V}_+(K'))$ under the mapping $\widehat{k} : \mathcal{V}_+(K) \to \mathcal{V}_+(K')$ between visibilities that is induced by k as in Section 10.1. It is easy to see that there is a natural one-to-one correspondence between $\widehat{k}(F(\mathcal{V}_+(K)))$ and the set of ordered pairs of elements in the "languages" associated to $g : G \to G'$ and $h : H \to H'$ (as in Definition 15.2). Again this corresponds geometrically to a kind of concatenation between paths in K', except for the presence of additional edges which both facilitate and mark the transition from the end of one path to the beginning of the next.

In the special case in which G' and H' are alphabet graphs, we do not quite get an alphabet graph for K', but instead we obtain a graph as in Fig. 15.1.

The standard notion of concatenation for languages corresponds better to the situation where we have mappings $g : G \to M$ and $h : H \to M$ into the *same* marked graph M. In this case, we should also ask that

$$g \text{ maps every vertex in } F(G) \text{ to } h(q_0(H)). \tag{15.2}$$

This permits us to combine g and h into a *weak* mapping $\kappa : K \to M$, in which κ is left undefined on the edges which join vertices in $F(G)$ to $q_0(H)$ inside K. Note that (15.2) is automatic when M is an alphabet graph, since it then has only one vertex.

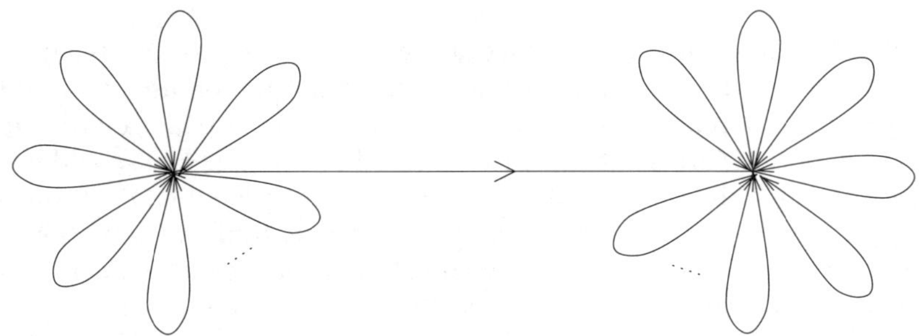

FIG. 15.1. The case where G' and H' are alphabet graphs

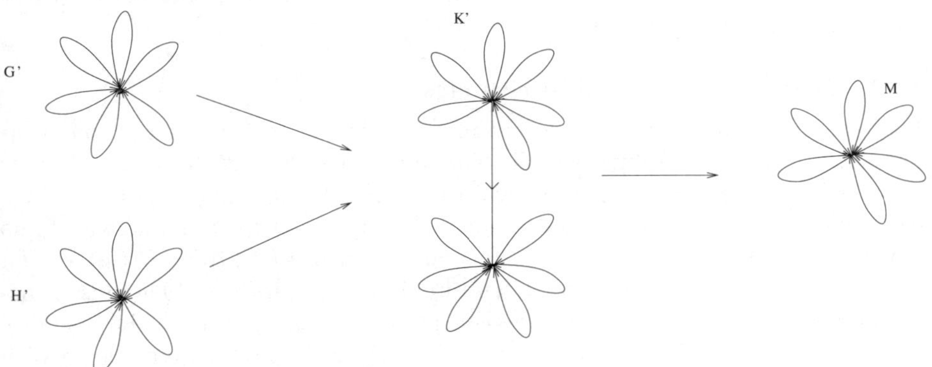

FIG. 15.2. From G' and H' to K' and then to M

This weak mapping κ is compatible with the markings when g and h are, and, more generally, when $g(q_0(G)) = q_0(M)$ and $h(F(H)) \subseteq F(M)$.

It is easy to see that the language associated to $\kappa : K \to M$ as in Definition 15.2 corresponds exactly to the set of paths in M obtained by concatenating the paths coming from the languages associated to g and h. In this situation, there are no longer any auxiliary edges in M which separate the parts of the paths coming from g and those coming from h.

If M is an alphabet graph, then we are back to the setting of standard automata, and the language (in the ordinary sense of words) recognized by $\kappa : K \to M$ is precisely the concatenation of the languages recognized by $g : G \to M$ and $h : H \to M$. Of course, it is well-known that concatenations of regular languages are again regular, and through this kind of procedure.

If G' and H' are distinct alphabet graphs, then we can map them directly into a single alphabet graph M, by combining the underlying alphabets. We can also think of the graph K' defined above as an intermediate step in the passage from G', H' to the combined graph M, as in Fig. 15.2. Note that G' and H' might correspond to alphabets which are not disjoint, so that some loops from G' and H' might be identified in M.

The notion of a Cartesian product above — in essence a concatenation with a marker between the individual pieces — should not be confused with ordinary automata which happen to use alphabets in which the letters are ordered pairs of other symbols (or are identified with such ordered pairs). This would lead to a language in which the words could be interpreted as ordered pairs of words over a more primitive alphabet, but this is very different from the situation above, in which the words are written in a completely independent manner, first one, and then the other. On the other hand, there is also a notion of *asynchronous automata*, in which one writes to pairs of words, but not necessarily at the same rate of speed. We shall encounter some of these other ways of making pairs of words later (in Chapters 17 and 18), in connection with *automatic groups* [ECH+92] and their generalizations.

15.3 Free products and positive closure

Let marked graphs G and H be given. We define a new marked graph N as follows. We begin by taking the disjoint union of G and H. We add edges as before, going from each element of $F(G)$ to $q_0(H)$. We also add edges from each element of $F(H)$ to $q_0(G)$. These are all of the edges that we add, and we add no new vertices. We take $q_0(N)$ to be $q_0(G)$, and we take $F(N)$ to be $F(G) \cup F(H)$.

This construction plays the role of a kind of "free product". There is a one-to-one correspondence between the elements of $F(\mathcal{V}_+(N))$ (inside the visibility of N) and arbitrary finite strings over $F(\mathcal{V}_+(G)) \cup F(\mathcal{V}_+(H))$ which begin with an element of $F(\mathcal{V}_+(G))$, continue with an element of $F(\mathcal{V}_+(H))$, and constantly alternate between the two. This is not hard to see; each element of $F(\mathcal{V}_+(N))$ represents an oriented path in N which begins at $q_0(N) = q_0(G)$ and ends at an element of $F(N) = F(G) \cup F(H)$, and, by construction, any such path arises from a string of paths which alternate between G and H. Each constituent path in this alternation must either begin at $q_0(G)$ and end in $F(G)$, or begin at $q_0(H)$ and end in $F(H)$.

The difference between this construction and the one in the previous section is that we allow ourselves to return to G after going to H, from which the process can be repeated. As before, we do not really concatenate paths from G and H directly, but instead we have additional edges between them.

This construction can also be applied to *mappings* between marked graphs. Suppose that G', H' are additional marked graphs, and that $g : G \to G'$ and $h : H \to H'$ are compatible with the markings. We can construct a marked graph N' from G' and H' in the same manner as before, and we can combine g and h to get $n : N \to N'$ in an obvious way. One can check that n will be compatible with the markings on N and N'. The "language" associated to $n : N \to N'$ in the manner of Definition 15.2 is a kind of "free product" of the languages associated to g and h, in the same manner as above. In terms of paths, the language associated to n is obtained by taking paths from the languages associated to $g : G \to G'$, $h : H \to H'$ and combining them, always alternating from G' to H' and then back again, etc.

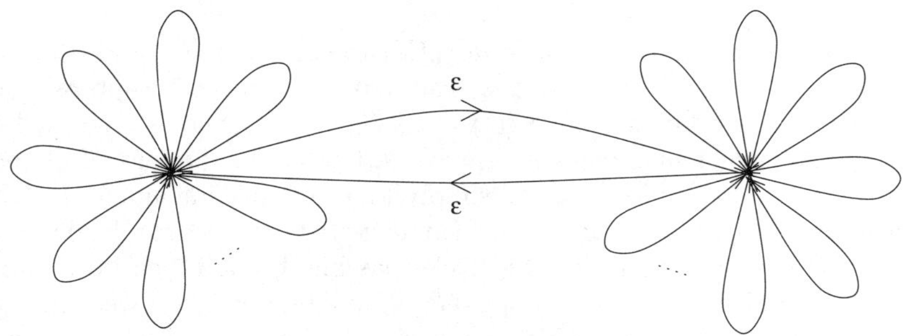

FIG. 15.3. The combination of alphabet graphs

Notice that N' is not an alphabet graph when G' and H' are. See Fig. 15.3. In particular, this concatenation of paths is slightly different from the one for words, because of the extra edges between G' and H' in N'.

These nuisances disappear when we have mappings (or weak mappings) $g : G \to M$ and $h : H \to M$ into the *same* marked graph M. For this we assume also that

$$\begin{aligned} &g \text{ sends every element of } F(G) \text{ to } q_0(M), \text{ and} \\ &h \text{ sends every element of } F(H) \text{ to } q_0(M). \end{aligned} \tag{15.3}$$

This permits us to combine g and h to get a weak mapping $\nu : N \to M$. Note that ν is not defined on the edges between G and H in N; this is compatible with the definition of a weak mapping, because of our assumptions (15.3). The language associated to ν is given in terms of concatenations of paths from $g : G \to M$ and $h : H \to M$, without auxiliary edges in between.

Now suppose that G and H are the same graph, and that $g : G \to M$ and $h : H \to M$ are the same mapping. In this case, we get a kind of "positive closure" of a language, consisting of arbitrary concatenations of elements of the original language. This is similar to the Kleene closure, except that one does not include the concatenation of zero elements of the original language, which would correspond to the path at $q_0(M)$ that traverses no edges. (This might be contained in the final result anyway, by having $q_0(G) \in F(G)$.) One could consider adjustments for this, and there are similar issues for the situation where G and H (or g and h) are different.

Instead of combining two copies of G to get the graph N, one can obtain the same effects by taking a single copy of G, and adding edges from the elements of $F(G)$ to $q_0(G)$. This is similar to taking H to consist of a single vertex, and no edges. At any rate, there are some variations on these themes, and also when G and H are different.

15.4 Unions and intersections

Let G and H be marked graphs, and let us make a new marked graph U in the following manner. We start with the disjoint union of G and H. To this we add a new vertex $q_0(U)$, and an edge from $q_0(U)$ to each of $q_0(G)$ and $q_0(H)$. This defines U as an oriented graph, and we can make it a marked graph using $q_0(U)$ as the initial vertex, and setting $F(U) = F(G) \cup F(H)$. The new marked graph U provides a convenient geometric representation for the "union" of G and H.

It would be a little nicer here to simply take the disjoint union of G and H and then identify $q_0(G)$ and $q_0(H)$ to a single vertex, but this can lead to trouble when G or H have nontrivial oriented cycles passing through their initial vertices.

If we have additional marked graphs G', H' and mappings (or weak mappings) $g : G \to G'$ and $h : H \to H'$ which are compatible with the markings, then we can apply the same construction to G' and H' to get a marked graph U', and we can combine g and h into a mapping $u : U \to U'$. The language associated to u as in Definition 15.2 corresponds naturally to the union of the languages associated to g and h. This is very easy to see from the definitions.

The presence of the auxiliary edges is a bit of a nuisance. In particular, if G' and H' are alphabet graphs, then this construction does not give back an alphabet graph. This problem can be avoided by using the simpler "union" operation in the image (without the auxiliary edges), but *not* for the domains (for which the possibility of cycles passing through the initial vertices is more troublesome). That is, we can take U' to be the graph obtained from the disjoint union of G' and H' and identifying $q_0(G')$ and $q_0(H')$ to a single vertex $q_0(U')$. If we do this in the image but not in the domain, then $u : U \to U'$ would be a weak mapping, undefined on the edges coming out of $q_0(U)$.

Now suppose that G and H already take values in the same graph M, and that $g(q_0(G)) = h(q_0(H))$. We can combine $g : G \to M$ and $h : H \to M$ into a weak mapping $\mu : U \to M$ which satisfies $\mu(q_0(U)) = g(q_0(G)) = h(q_0(H))$, and which is undefined on the edges in U that go from $q_0(U)$ to $q_0(G)$ and $q_0(H)$. The language associated to this weak mapping is the union of the languages associated to g and h. As usual, this corresponds to standard constructions for ordinary automata.

In the next sections, we discuss an operation of "fiber products" for graphs, which provides an alternate approach to the union that applies also to intersections. In fact, it will work somewhat better for intersections than for unions. (On the other hand, the constructions above for unions are already very simple anyway, and simpler than fiber products.) For standard automata, one sometimes deals with intersections by taking complements to reduce to the case of unions, as in Theorems 3.3 and 3.2 on p59 of [HU79]. To handle complements, one uses representations in terms of deterministic automata. A nice feature of fiber products is that one can work directly with nondeterministic automata.

The use of fiber products for the intersection of languages is natural, in that the operation of intersection often behaves like a kind of product anyway. This came up (at least implicitly) in Chapter 3, especially Sections 3.2 and 3.6, and we shall return to similar issues in Section 16.13. In particular, intersections can behave like products in terms of "complexity", and the fiber product also behaves like a product in terms of complexity, by definition.

15.5 Fiber products (in general)

The notion of a *fiber product* is very basic in mathematics, and appears in many contexts. While it can be described abstractly in terms of category theory, we shall follow a more concrete approach. We discuss it in this section in various contexts, and then specialize to graphs in Section 15.6. We give some interpretations for regular languages and automata in Section 15.7.

Suppose that we have sets A, B, and C, and mappings $g : A \to C$ and $h : B \to C$. The *fiber product* consists of a set P and a mapping ϕ defined as follows. P is the subset of $A \times B$ given by

$$P = \{(a, b) \in A \times B : g(a) = h(b)\}. \tag{15.4}$$

We define $\phi : P \to C$ by $\phi(a, b) = g(a) = h(b)$.

Note that we have canonical projections $pr_1 : P \to A$ and $pr_2 : P \to B$ defined by $p_1(a, b) = a$, $p_2(a, b) = b$. These projections are related to the other mappings through the identities

$$g \circ pr_1 = h \circ pr_2 = \phi. \tag{15.5}$$

As an example, let us consider the case where $A = C \times D$, $B = C \times E$, and g, h are the obvious projections onto C, i.e., $g(c, d) = c$ and $h(c, e) = c$ for all $c \in C$, $d \in D$, and $e \in E$. Then P becomes the set

$$\{(c, d, c, e) \in C \times D \times C \times E : c \in C, d \in D, e \in E \text{ are arbitrary}\}, \tag{15.6}$$

and ϕ, pr_1, and pr_2 are given by

$$\phi(c, d, c, e) = c, \quad pr_1(c, d, c, e) = (c, d), \quad pr_2(c, d, c, e) = (c, e). \tag{15.7}$$

One can think of P as being equivalent to $C \times D \times E$ in this case.

In general, if $c \in C$ has m preimages in A under g and n preimages in B under h, then c has $m \cdot n$ preimages in P under ϕ. This follows from the definitions. In the special case that we just considered, one might say that these preimages were all "parallel" to each other, but in other situations there can be more twisting, and the behavior of the fiber product is more complicated.

This defines the basic notion for sets, but in fact it cooperates well with many different kinds of structure that one might have on A, B, and C. If A, B, and C are groups, for instance, and the mappings g and h are group homomorphisms, then P inherits a natural group structure, and ϕ, pr_1, and pr_2 are group homomorphisms.

Similarly if A, B, and C are topological spaces, and g and h are continuous mappings, then the fiber product P has a natural topology in which ϕ, pr_1, and pr_2 are continuous mappings. If g and h are *covering mappings* — local homeomorphisms — then ϕ, pr_1, and pr_2 will be too.

Let us consider a concrete example in some detail. Let j be an integer, and let S_j denote the real numbers modulo j, also written $\mathbf{R}/j\,\mathbf{Z}$. We can think of this as a group, or a topological space (which is homeomorphic to the unit circle in $\mathbf{R}^2$), or both (a topological group). If j divides k, then we have a natural mapping $g : S_k \to S_j$, in which we take a real number modulo k, and reduce it to a real number modulo j. We can do the same again with $h : S_l \to S_j$ when j divides l. How does the fiber product P look in this case? By definition

$$P = \{(x, y) \in S_k \times S_l : x \equiv y \bmod j\}, \tag{15.8}$$

but what does this really mean?

Suppose first that $k = l$. Then it makes sense to talk about $x - y$, because they live in the same place, and so we can write P as

$$\{(x, y) \in S_k \times S_k : x - y \equiv 0 \bmod j\}. \tag{15.9}$$

This is the same as

$$\left\{(x, x + i \cdot j) \in S_k \times S_k : i \in \mathbf{Z},\, 0 \leq i < \frac{k}{j}\right\}. \tag{15.10}$$

Thus, in effect P is the same thing as the Cartesian product of S_k with $k\,\mathbf{Z}/j\,\mathbf{Z}$. This is true as a group, and as a topological space. In particular, P is not connected as a topological space. This can be seen directly, because the map $(x, y) \mapsto x - y$ is well-defined and continuous, and maps P onto a discrete set.

Let us now consider an opposite extreme, and assume that j is the greatest common divisor of k and l, so that $\frac{k}{j}$ and $\frac{l}{j}$ are relatively prime. In this case, we cannot talk about $x - y$ directly, when $x \in S_k$ and $y \in S_l$, and in fact there is a completely different description of P. Let m denote the least common multiple of k and l, so that

$$m = j\,\frac{k}{j}\,\frac{l}{j}. \tag{15.11}$$

Since both k and l divide m, we have natural mappings $\psi_1 : S_m \to S_k$ and $\psi_2 : S_m \to S_l$, in which we simply take a point z in S_m, and reduce it modulo k or l, as appropriate. Define $\psi : S_m \to S_k \times S_l$ by $\psi(z) = (\psi_1(z), \psi_2(z))$. Note that

$$\psi_1(z) \equiv \psi_2(z) \bmod j \tag{15.12}$$

for every $z \in S_m$, since $\psi_1(z)$ and $\psi_2(z)$ are both equal to $z \bmod j$. This implies that $\psi(z)$ actually lies in P, so that ψ defines a mapping from S_m into P. It is not hard to show that $\psi : S_m \to P$ is continuous and a group homomorphism, and in fact that it is a bijection, a homeomorphism, and a group isomorphism. This is very different from the previous case, where P was not even connected, while here it is.

Let us put this into a more general context. Let M be a set on which a group Γ acts freely. This means that we have a mapping from Γ into the set of bijections on M which is a group homomorphism, and that when γ is an element of Γ which is distinct from the identity element, then the bijection on M associated to γ

has *no* fixed points. This group action defines an equivalence relation on M, in which two points x and y are considered to be equivalent if there is an element γ of Γ such that $\gamma(x) = y$. Here we permit γ to denote both an element of Γ and the bijection on M to which it is associated. That this is an equivalence relation comes from the assumption that Γ be a group, and the equivalence classes are normally referred to as orbits under Γ. The space of Γ-orbits in M is usually written as M/Γ, and it will be convenient for us to denote it by C.

In practice, M often has extra structure which is preserved by Γ. For instance, M could be a topological space, and the bijections associated to elements of Γ could be homeomorphisms. Under modest conditions, C will inherit a reasonable topology from M, and the canonical projection from M to C will be a *covering map*, i.e., a local homeomorphism.

Suppose that Γ_1 and Γ_2 are two subgroups of Γ, and let $A = M/\Gamma_1$ and $B = M/\Gamma_2$ be the corresponding orbit spaces. This leads to natural mappings $g : A \to C$, $h : B \to C$, which take an orbit in M under Γ_1, Γ_2 (respectively) and associate to it the larger orbit under Γ. How does the fiber product P behave? By definition, P consists of the ordered pairs of Γ_1-orbits and Γ_2-orbits in M which are contained in the same Γ-orbit.

Set $\Gamma_0 = \Gamma_1 \cap \Gamma_2$. This is another subgroup of Γ. We have natural mappings $\mu_1 : M/\Gamma_0 \to M/\Gamma_1$, $\mu_2 : M/\Gamma_0 \to M/\Gamma_2$ which take an orbit under Γ_0 and associate it to the orbits of Γ_1, Γ_2 which include it. We can also define a mapping $\mu : M/\Gamma_0 \to M/\Gamma$ in the same manner. It is not hard to see that

$$\mu = g \circ \mu_1 = h \circ \mu_2, \tag{15.13}$$

just from the definitions. Specifically, the first equality says that the Γ orbit that contains a given Γ_0 orbit is the same as the Γ orbit that contains the Γ_1 orbit that contains the given Γ_0 orbit, and this holds because $\Gamma_0 \subseteq \Gamma_1 \subseteq \Gamma$. The second inequality is similar, but with Γ_2 in place of Γ_1. These equalities permit us to define $\nu : M/\Gamma_0 \to P$ by $\nu = (\mu_1, \mu_2)$.

This mapping ν is always injective. In other words, a Γ_0-orbit in M is uniquely determined by the knowledge of the Γ_1-orbit and the Γ_2-orbit which contain it. To see this, suppose that p and q are elements of M which both lie in the same Γ_1 orbit and in the same Γ_2 orbit, and let us show that they lie in the same Γ_0 orbit. By assumption, there exist $\gamma_1 \in \Gamma_1$ and $\gamma_2 \in \Gamma_2$ such that $\gamma_1(p) = q$ and $\gamma_2(p) = q$. Thus $\gamma_2^{-1} \circ \gamma_1$ maps p to itself, from which we conclude that $\gamma_2^{-1} \circ \gamma_1$ is the identity element of the group, since we are assuming that Γ is acting on M without fixed points. This implies that $\gamma_1 = \gamma_2$ lies in $\Gamma_0 = \Gamma_1 \cap \Gamma_2$, and hence that p and q lie in the same Γ_0 orbit in M. This shows that ν is automatically injective.

Note that ν cooperates well with the mappings $\phi : P \to M/\Gamma$, $pr_1 : P \to M/\Gamma_1$ and $pr_2 : P \to M/\Gamma_2$ which are associated to the fiber product, in the sense that

$$\mu = \phi \circ \nu, \;\; \mu_1 = pr_1 \circ \nu, \;\; \mu_2 = pr_2 \circ \nu. \tag{15.14}$$

This follows from (15.13) and the definitions.

It may or may not be true that ν is a bijection, depending on Γ and its subgroups Γ_1, Γ_2, $\Gamma_0 = \Gamma_1 \cap \Gamma_2$. The action on M does not matter for this, because the matter works orbit by orbit. The question of whether ν is a bijection can be reformulated as follows. Consider the coset spaces $H_1 = \Gamma/\Gamma_1$, $H_2 = \Gamma/\Gamma_2$, $H_0 = \Gamma/\Gamma_0$. These are merely *sets* at the moment, since our various subgroups are not assumed to be normal. Let $\theta_1 : H_0 \to H_1$, $\theta_2 : H_0 \to H_2$ denote the mappings which take a Γ_0-coset and replace it with the Γ_1-coset or Γ_2-coset that contains it, and define $\theta : H_0 \to H_1 \times H_2$ by $\theta = (\theta_1, \theta_2)$. One can check that θ is always an injection, and that it is a bijection if and only if $\nu : M/\Gamma_0 \to P$ is. The reason for the latter is that the question of surjectivity for ν depends only on what happens in each individual Γ-orbit, and the behavior of ν on a Γ-orbit is the same as the behavior of θ. This uses the assumption that the action of Γ on M be fixed-point free.

At the level of *sets*, there is not much point in looking at the action of Γ, Γ_i on M instead of merely the coset spaces as above, but M can carry extra structure, like a topology, which is not included in this. For instance, in this situation, it may or may not be that P is connected, even if A, B, and C are, as we saw before, in the case of the circle.

Keep in mind that the groups Γ_i are not required to be abelian now, as they were in the previous situation. This allows for more "twisting". Consider the special case where $\Gamma_1 = \Gamma_2$, but they are proper subgroups of Γ. Then $\theta : H_0 \to H_1 \times H_2$ is certainly not a surjection, since $H_1 = H_2$ and θ maps onto the diagonal

$$\{(x, y) \in H_1 \times H_2 : x = y\}. \tag{15.15}$$

To analyze the part of $H_1 \times H_2$ that remains, one would like to have a mapping from $H_1 \times H_2$ to H_1 which would represent the "difference" between x and y, as we did before. If $\Gamma_1 = \Gamma_2$ were a *normal* subgroup of Γ, then one would be able to do this, because the quotient space would have the structure of a group, but in general this does not work.

15.6 Fiber products of graphs

Let us now specialize to the case of graphs. Let G, H, and K be graphs, and suppose that $g : G \to K$ and $h : H \to K$ are mappings between them. Assume also that G, H, and K are oriented, and that g and h preserve orientations (which will be our context of main concern). We want to define the fiber product P as an oriented graph, and $\phi : P \to K$ as an orientation-preserving mapping between graphs.

Instead of ordinary mappings, one could also consider weak mappings between graphs. These could be treated in much the same way, with some modest adjustments. We shall not pursue this here, for the sake of simplicity, although a version of this will be discussed in Section 16.13 (which covers the main changes). In the context of automata, as in Section 15.7, weak mappings correspond to the use of ϵ-moves.

Now let us define P. For the set of vertices of P, we take the fiber product of the sets of vertices of G and H, in the same manner as in Section 15.5. Thus a vertex in P is given by an ordered pair (u, v), where u is a vertex in G, v is a vertex in H, and $g(u) = h(v)$. For the set of edges in P, we use the fiber product of the sets of edges in G and H. To be explicit, suppose that α is an edge in G which goes from a vertex u_1 to a vertex u_2, and that β is an edge in H which goes from a vertex v_1 to a vertex v_2. If $g(\alpha) = h(\beta)$, then we must have that $g(u_1) = h(v_1)$ and $g(u_2) = h(v_2)$, since $g(\alpha)$ is an edge in K which goes from $g(u_1)$ to $g(u_2)$, and $h(\beta)$ is an edge which goes from $h(v_1)$ to $h(v_2)$. We take (α, β) to be an edge in P which goes from (u_1, v_1) to (u_2, v_2) (which are vertices in P, as above). This gives the set of edges in P, and the way that the edges are attached to vertices in P.

This defines P as an oriented graph. As in Section 15.5, we define $\phi : P \to K$ by setting $\phi(u, v) = g(u) = h(v)$ when (u, v) is a vertex in P, and $\phi(\alpha, \beta) = g(\alpha) = h(\beta)$ when (α, β) is an edge in P. It is easy to check that this defines ϕ as a mapping between graphs (i.e., that the usual compatibility conditions for the vertices and edges are satisfied), and that ϕ preserves orientations.

One can also define projections $pr_1 : P \to G$ and $pr_2 : P \to H$ in the same manner as in Section 15.5. Specifically, $pr_1(u, v) = u$ and $pr_1(\alpha, \beta) = \alpha$ for vertices (u, v) and edges (α, β) in P, and similarly for pr_2. It is again easy to check that pr_1 and pr_2 do give mappings between graphs (with the compatibility conditions for vertices and edges), and that they preserve orientations. As in Section 15.5, we have the identities

$$g \circ pr_1 = h \circ pr_2 = \phi \tag{15.16}$$

for these mappings between graphs.

This construction is compatible with the visibility, as we now explain.

Lemma 15.5 *Let G, H, and K be oriented graphs, and let $g : G \to K$ and $h : H \to K$ be orientation-preserving mappings between them. Fix vertices $u \in G$, $v \in H$, and $w \in K$ with $g(u) = h(v) = w$, so that $z = (u, v)$ defines a vertex in the (oriented) fiber product P described above. Let $pr_1 : P \to G$, $pr_2 : P \to H$, and $\phi : P \to K$ be the mappings associated to P as above.*

Consider the corresponding visibility graphs

$$\mathcal{V}_+(u, G),\ \mathcal{V}_+(v, H),\ \mathcal{V}_+(w, K),\ \text{ and } \mathcal{V}_+(z, P) \tag{15.17}$$

and the associated mappings

$$\widehat{g} : \mathcal{V}_+(u, G) \to \mathcal{V}_+(w, K),\ \ \widehat{h} : \mathcal{V}_+(v, H) \to \mathcal{V}_+(w, K), \tag{15.18}$$

$$\widehat{pr_1} : \mathcal{V}_+(z, P) \to \mathcal{V}_+(u, G),\ \ \widehat{pr_2} : \mathcal{V}_+(z, P) \to \mathcal{V}_+(v, H), \tag{15.19}$$

and

$$\widehat{\phi} : \mathcal{V}_+(z, P) \to \mathcal{V}_+(w, K). \tag{15.20}$$

Then

$$\widehat{\phi} = \widehat{g} \circ \widehat{pr_1} = \widehat{h} \circ \widehat{pr_2}. \tag{15.21}$$

Moreover, if we take the fiber product of $\mathcal{V}_+(u, G)$ *and* $\mathcal{V}_+(v, H)$ *over* $\mathcal{V}_+(w, K)$ *using the mappings* $\widehat{g}$ *and* $\widehat{h}$ *mentioned above, then there is a canonical isomorphism between the resulting oriented graph and* $\mathcal{V}_+(z, P)$*. The basepoint in* $\mathcal{V}_+(z, P)$ *corresponds under this isomorphism to the ordered pair of basepoints in* $\mathcal{V}_+(u, G)$ *and* $\mathcal{V}_+(v, H)$*. The mappings* $\widehat{\phi}$*,* $\widehat{pr_1}$*, and* $\widehat{pr_2}$ *all correspond under the isomorphism to the mappings provided by the fiber product construction.*

In short, the visibility of the fiber product is practically the same as the fiber product of the visibilities.

Proof This is basically a matter of chasing definitions. Notice first that (15.21) follows from the usual identity (15.5) for fiber products, and the general result (10.2) for visibilities and compositions.

Let us look now at the visibilities. What does the visibility $\mathcal{V}_+(z, P)$ of the fiber product P look like? A vertex in this visibility represents an oriented path ξ in P which begin at z. This path ξ projects down to oriented paths ξ_1 and ξ_2 in G and H which start at u and v, respectively, using the mappings pr_1 and pr_2. By taking images under g and h, we get oriented paths $g(\xi_1)$ and $h(\xi_2)$ in K which start at w. In fact, these must be the same path, because of the way that the fiber product is defined. One can use (15.16) to see this, and to show that $g(\xi_1)$ and $h(\xi_2)$ are equal to the image of ξ under ϕ.

We can think of ξ_1 and ξ_2 as representing vertices in the visibilities $\mathcal{V}_+(u, G)$, $\mathcal{V}_+(v, H)$, while $g(\xi_1) = h(\xi_2)$ represents a vertex in the visibility $\mathcal{V}_+(w, K)$. Furthermore, (ξ_1, ξ_2) determines a vertex in the fiber product of $\mathcal{V}_+(u, G)$ and $\mathcal{V}_+(v, H)$ defined with respect to the mappings $\widehat{g}$ and $\widehat{h}$ into $\mathcal{V}_+(w, K)$, since $g(\xi_1)$ and $h(\xi_2)$ coincide as paths in K, and hence as vertices in $\mathcal{V}_+(w, K)$.

Roughly speaking, one can say that the difference between vertices in $\mathcal{V}_+(z, P)$ and vertices in the fiber product of $\mathcal{V}_+(u, G)$ and $\mathcal{V}_+(v, H)$ is like the difference between paths of ordered pairs and ordered pairs of paths. There is no real difference, except in "formatting".

To be more formal, the discussion above describes a mapping from the vertices of $\mathcal{V}_+(z, P)$ to the vertices of the fiber product of $\mathcal{V}_+(u, G)$ and $\mathcal{V}_+(v, H)$ over $\mathcal{V}_+(w, K)$. It is not hard to see that this process can be reversed, so that we actually have a bijection between the sets of vertices. Similarly, one can go through the definitions to get a one-to-one correspondence between edges which is compatible with the one for vertices, and which is also compatible with the orientations. Thus $\mathcal{V}_+(z, P)$ is isomorphic to the fiber product of $\mathcal{V}_+(u, G)$ and $\mathcal{V}_+(v, H)$ over $\mathcal{V}_+(w, K)$. It is easy to see that the basepoints match up too, as in the statement of the lemma.

That the mappings $\widehat{\phi}$, $\widehat{pr_1}$, and $\widehat{pr_2}$ correspond to the ones provided by the construction of the fiber product of the visibility graphs is again a matter of rearranging the "formats", between paths of ordered pairs and ordered pairs of paths, and we omit the details. This completes the proof of Lemma 15.5. □

The notion of fiber products also cooperates fairly well with the property of being a local +-isomorphism, as in the next result.

Lemma 15.6 *Let G, H, and K be oriented graphs, and let $g : G \to K$ and $h : H \to K$ be orientation-preserving mappings. Let P denote the fiber product, with the usual mappings $pr_1 : P \to G$, $pr_2 : P \to H$, and $\phi : P \to K$, as before.*

If g is a local +-isomorphism, then pr_2 is too. If h is a local +-isomorphism, then pr_1 is also. If both g and h are local +-isomorphisms, then ϕ is one as well.

The analogous statements for local +-injections and local +-surjections also hold.

Proof This is not hard to check from the definitions, but let us be a bit careful. Assume that g is a local +-isomorphism. We want to say that pr_2 is one too. Fix a vertex (u, v) in P. We want to show that pr_2 induces a one-to-one correspondence between edges in P that flow away from (u, v) and edges in H which flow away from v. If β is an outgoing edge at v in H, then $h(\beta)$ is an outgoing edge at $h(v)$ in K, and the definition of the fiber product ensures that $h(v) = g(u)$. The assumption that g be a local +-isomorphism implies that g induces a one-to-one correspondence between outgoing edges at u in G and outgoing edges at $g(u)$ in K. In particular, there is exactly one edge α in G which flows away from u and satisfies $g(\alpha) = h(\beta)$. Thus we have an edge (α, β) in P which flows away from (u, v). It is the *only* such edge which is projected to β by pr_2, as one can verify from the uniqueness of α. This implies that pr_2 is a local +-isomorphism.

The argument for h and pr_1 is the same. One can use the same method when both g and h are local +-isomorphisms to conclude that ϕ is one too, or one can use (15.16) and Lemma 10.11 in Section 10.3.

The statements for local +-injections and local +-surjections can be established by the same arguments. This proves the lemma. □

Let us record a couple of minor variations on these themes. The first was used in Section 10.16, in the proof of Proposition 10.101.

Lemma 15.7 *Let G, H, and K be oriented graphs, and let $g : G \to K$ and $h : H \to K$ be orientation-preserving mappings which are surjective (on both vertices and edges, as in Remark 10.16 in Section 10.3). Let P denote the fiber product, with the usual mappings $pr_1 : P \to G$, $pr_2 : P \to H$, and $\phi : P \to K$ as above. Then each of the mappings pr_1, pr_2, and ϕ is also surjective.*

The proof is an easy exercise.

The next observation came up in Section 13.3.

Lemma 15.8 *Let G, H, and K be oriented graphs, and let $g : G \to K$ and $h : H \to K$ be orientation-preserving mappings between them. Fix vertices $u \in G$, $v \in H$, $w \in K$, and assume that $g(u) = h(v) = w$, so that $z = (u, v)$ defines a vertex in the fiber product P of G and H over K. If the induced mappings*

$$\widehat{g} : \mathcal{V}_+(u, G) \to \mathcal{V}_+(w, K), \qquad \widehat{h} : \mathcal{V}_+(v, H) \to \mathcal{V}_+(w, K) \tag{15.22}$$

between visibilities are surjections, then the same is true of

$$\widehat{\phi} : \mathcal{V}_+(z, P) \to \mathcal{V}_+(w, K), \tag{15.23}$$

where $\phi : P \to K$ is the usual mapping associated to the fiber product.

Proof One could actually derive this from Lemmas 15.5 and 15.7, but one can also see the matter directly. (One could also view this as a special case of Lemma 15.9 below.) The main point is the following. Let γ be an oriented path in K which begins at w. The assumption of surjectivity for the mappings in (15.22) implies the existence of oriented paths α, β in G, H which begin at u, v and which are mapped to γ by g and h, respectively. These two paths can then be "combined" to produce an oriented path δ in P which begins at z and which is mapped by ϕ to γ. This is easy to check from the definitions, and it implies the surjectivity of (15.23) as a mapping between vertices. It is easy to see that (15.23) is also surjective on edges, and the lemma follows. □

15.7 Interpretations for automata

In the context of automata, it is natural to consider fiber products obtained from *marked* graphs G, H, K and mappings $g : G \to K$, $h : H \to K$ which are compatible with the markings. Let P be the fiber product graph, as in Section 15.6, so that P already has the structure of an oriented graph. There is an obvious choice of initial vertex for P, namely $q_0(P) = (q_0(G), q_0(H))$. For the set $F(P)$ of final vertices in P, there is some flexibility, and one can choose it in different ways depending on the situation.

Lemma 15.9 *Let G, H, K, $g : G \to K$, $h : H \to K$, and the fiber product P be as above. Set*

$$F(P) = \{(u, v) \in P : u \in F(G) \text{ and } v \in F(H)\}, \tag{15.24}$$

so that P becomes a marked graph. If $\phi : P \to K$ is as in Section 15.6, then ϕ is compatible with the markings, and the induced mappings $\widehat{g} : \mathcal{V}_+(G) \to \mathcal{V}_+(K)$, $\widehat{h} : \mathcal{V}_+(H) \to \mathcal{V}_+(K)$, and $\widehat{\phi} : \mathcal{V}_+(P) \to \mathcal{V}_+(K)$ between the associated visibility graphs satisfy

$$\widehat{\phi}(F(\mathcal{V}_+(P))) = \widehat{g}(F(\mathcal{V}_+(G))) \cap \widehat{h}(F(\mathcal{V}_+(H))). \tag{15.25}$$

Recall that $\mathcal{V}_+(G)$, $\mathcal{V}_+(H)$, and $\mathcal{V}_+(K)$ are the visibility graphs associated to G, H, and K as marked graphs, and with markings coming from the markings of G, H, and K, as in the comments just after Definition 15.2 in Section 15.1. In particular, the sets of final vertices is specified.

One can rephrase Lemma 15.9 as saying that, with $F(P)$ given by (15.24), the language associated to $\phi : P \to K$ (Definition 15.2) is equal to the intersection of the languages associated to $g : G \to K$ and $h : H \to K$. If K is an alphabet graph, then this corresponds to the well-known fact that the intersection of two regular languages is again regular. See also Sections 15.4 and 16.13.

Proof This is mostly a question of unwinding definitions. Notice first that ϕ is indeed compatible with the markings. This is easy to check from the definitions, and the hypothesis that g and h be compatible with the markings. It remains to prove (15.25).

A vertex in $\widehat{\phi}(F(\mathcal{V}_+(P)))$ represents an oriented path δ in K which arises as the image under ϕ of an oriented path γ in P that begins at the initial vertex $q_0(P) = (q_0(G), q_0(H))$ in P and ends at an element of $F(P)$. Let $pr_1 : P \to G$ and $pr_2 : P \to H$ be as in Section 15.6. Using these mappings, one can project γ to oriented paths α in G and β in H. It is easy to see from the definitions that α begins at $q_0(G)$ and ends at an element of $F(G)$, and that β begins at $q_0(H)$ and ends at an element of $F(H)$. Thus the images of α and β in K under g and h represent elements of $\widehat{g}(F(\mathcal{V}_+(G)))$ and $\widehat{h}(F(\mathcal{V}_+(H)))$, respectively. The images of α, β, and γ in K under g, h, and ϕ are all the same; this follows from the definitions, and one could also use (15.16). This shows that every element of $\widehat{\phi}(F(\mathcal{V}_+(P)))$ also lies in the intersection of $\widehat{g}(F(\mathcal{V}_+(G)))$ and $\widehat{h}(F(\mathcal{V}_+(H)))$, which gives us one inclusion for (15.25).

For the other inclusion, one argues in the same way, but going backwards. An element of the right side of (15.25) represents an oriented path δ in K for which there are oriented paths α in G and β in H such that α begins at $q_0(G)$ and ends at an element of $F(G)$, β begins at $q_0(H)$ and ends at an element of $F(H)$, g maps α to δ in K, and h maps β to δ in K. Because α and β project down to the same path in K, it is not hard to see that they can be combined to give a path γ in P. This is also part of Lemma 15.5. This path γ is mapped to δ in K by ϕ as well, and γ ends at an element of $F(P)$ in P, because of (15.24) and the corresponding properties of α and β. The latter implies that γ represents an element of $F(\mathcal{V}_+(P))$, and we conclude that every element of the intersection on the right side of (15.25) also lies in the left side of (15.25). This proves the other inclusion in (15.25), and the lemma follows. □

Now let us consider the situation for unions.

Lemma 15.10 *Let G, H, K, $g : G \to K$, $h : H \to K$, and the fiber product P be as above. Set*

$$F(P) = \{(u, v) \in P : u \in F(G) \text{ or } v \in F(H)\}, \tag{15.26}$$

so that P again becomes a marked graph. Then $\phi : P \to K$ is compatible with the markings, and

$$\widehat{\phi}(F(\mathcal{V}_+(P))) \subseteq \widehat{g}(F(\mathcal{V}_+(G))) \cup \widehat{h}(F(\mathcal{V}_+(H))), \tag{15.27}$$

where $\widehat{\phi}$, $\widehat{g}$, and $\widehat{h}$ are the liftings of the ϕ, g, and h to mappings between the visibilities, as usual.

In general, one should not expect to have equality in (15.27). We shall discuss this further after the proof.

Proof The argument is very similar to the one used for Lemma 15.9. The compatibility of ϕ with the markings follows easily from the corresponding assumption about g and h. It remains to verify (15.27).

A vertex in the left side of (15.27) represents a path δ in K which is the image of an oriented path γ in P that begins at $q_0(P) = (q_0(G), q_0(H))$ and ends at an element of the set $F(P)$ in (15.26). As in the proof of Lemma 15.9, we can project γ to paths α, β in G and H, where α and β have the same image δ in K under the mappings g and h. We also have that α begins at $q_0(G)$, and that β begins at $q_0(H)$. The definition of $F(P)$ in this case ensures that at least one of α and β ends at an element of $F(G)$ or $F(H)$. This implies exactly that our vertex in the left side of (15.27) lies in at least one of the two parts on the right side, which is what we wanted. □

What about the opposite inclusion? Suppose that δ is an oriented path in K which determines a vertex in the visibility $\mathcal{V}_+(K)$, and which lies in the right side of (15.27). Assume for the sake of definiteness that this vertex lies in $\widehat{g}(F(\mathcal{V}_+(G)))$, so that there is an oriented path α in G which begins at $q_0(G)$, ends at an element of $F(G)$, and projects down to δ in K. To prove the opposite inclusion in (15.27), we would like to be able to "lift" α to a path in P which begins at $q_0(P)$. In general, we cannot do this, unless we know that H contains an oriented path that begins at $q_0(H)$ and which also projects down to δ.

Lemma 15.11 *Notation and assumptions as in Lemma 15.10. Suppose also that $g : G \to K$ and $h : H \to K$ have the property that the induced mappings $\widehat{g} : \mathcal{V}_+(G) \to \mathcal{V}_+(K)$, $\widehat{h} : \mathcal{V}_+(H) \to \mathcal{V}_+(K)$ between visibilities are surjections. Then we have equality in (15.27), i.e.,*

$$\widehat{\phi}(F(\mathcal{V}_+(P))) = \widehat{g}(F(\mathcal{V}_+(G))) \cup \widehat{h}(F(\mathcal{V}_+(H))). \tag{15.28}$$

Note that the requirement that $\widehat{g} : \mathcal{V}_+(G) \to \mathcal{V}_+(K)$, $\widehat{h} : \mathcal{V}_+(H) \to \mathcal{V}_+(K)$ be surjections is satisfied if g and h are local $+$-surjections (Definition 10.12), as in Lemma 10.14. Compare also with Section 13.3.

Proof The hypothesis of surjectivity for the induced mappings between the visibilities permits us to complete the argument which was indicated before the statement of Lemma 15.11. Specifically, if δ and α are oriented paths in K and G with the properties described above, then the surjectivity of $\widehat{h}$ implies that we can find an oriented path β in H which begins at $q_0(H)$ and which projects down to δ in K. At this stage, we are in essentially the same situation as before, in the second part of the proof of Lemma 15.9. The paths α and β can be combined to give an oriented path γ in P which begins at $q_0(P)$ and projects down to δ in K, because of the corresponding properties for α and β. This path γ ends in $F(P)$, since α ends in $F(G)$. This implies that the vertex on the right side of (15.28) determined by δ also lies in the left side of (15.28), which is what we wanted. Thus we get the equality in (15.28). □

16

STRONGER FORMS OF RECURSION

So far in this book, we have mostly considered only modest amounts of "implicitness", corresponding roughly to at most one level of exponential expansion in the passage to explicit constructions. In terms of formal proofs, we have been working in the realm in which only propositional rules of inference are used. In this chapter, we shall look at combinatorial structures whose behavior is more like that of formal proofs in which quantifier rules are allowed.

One of the nice features of formal proofs is that there is a simple way in which to separate amounts of implicitness which can occur into different levels, in terms of the total number of alternations of quantifiers. Roughly speaking, each new level of implicitness (as measured by the alternations of quantifiers) brings an additional exponential in the possible complexity of the objects being described. We shall see a concrete example of this in Section 16.1, and in Section 16.2, we shall look at one of the basic mechanisms for implicitness in formal proofs through a comparison with transformations on functions and sets. In this analogy, each new level of alternation of quantifiers in a formal proof corresponds to a transition from working with objects in a particular set X to working with functions on X, and this can be done over and over again.

Beginning in Section 16.3, we shall explain how similar effects can be obtained by allowing feasibility graphs to be used to construct other feasibility graphs. The possibility for exactly one new degree of exponentiation in complexity for each new level of implicitness comes out particularly clearly in this setting, and we shall examine it in some specific contexts, including ordinary numbers, formal expressions, and sets of words. We shall also see how additional levels of implicitness can correspond to stronger forms of symmetry in the objects being described.

These constructions with feasibility graphs can be seen as providing an infinite hierarchy of measurements of information content which are similar to the measurements used in Kolmogorov complexity and algorithmic information theory, but with fairly precise restrictions on the type of recursions allowed. As one moves within the hierarchy, one trades between concreteness of the representations involved and the amount of compression which they can achieve. This is similar to the situation for formal proofs, and the "trades" that one makes by increasing or decreasing the extent to which alternations of quantifiers are allowed.

16.1 Feasible numbers

Let us begin with examples of implicit constructions in formal proofs that use quantifiers. For this purpose, it is convenient to return to the concept of *feasible numbers*, as in Section 4.8. We shall follow the treatment in [Car00].

As before, we work in the context of arithmetic, but with an extra unary predicate F. The intended meaning of $F(x)$ is that "x is feasible", and part of the idea is that a proof of $F(t)$ for some term t will include some kind of recipe for building up t from primitive objects.

More precisely, the predicate F is subject to the rules that 0 is feasible, and that feasibility is preserved by addition, multiplication, the successor function (which represents addition by the number 1), and equality of terms. We do not permit induction to be used over formulae containing F, because otherwise one could prove $\forall x F(x)$ immediately, and a proof of $F(t)$ for a particular term t would not have to contain any information about t itself. (One might say that induction represents an *infinite* process, while we shall restrict ourselves to finite processes here. This is a well-known aspect of arithmetic which appears in various guises.)

For simplicity, we shall permit ourselves to use exponential functions and their basic properties freely. Normally one would build these up directly from more primitive objects in arithmetic, but we shall not bother with this here. It is important, however, that we do *not* include compatibility with exponentiation among the basic properties of the feasibility predicate F. Thus any such compatibility must be *proved* in the particular context.

The choice of operations which the feasibility predicate is required to respect is a bit arbitrary and not too important for the present purposes. The main point is to be clear about what comes for free and what reflects more substantial issues of complexity and implicitness.

For instance, the present conventions permit one to make formal proofs of $F(2^n)$ in $O(n)$ lines, simply using the rule for multiplications repeatedly. This does not reflect a nontrivial exponential effect, but simply the basic choices for the notion of feasibility.

A more interesting exponential effect can be obtained using cuts and contractions, as discussed in Section 4.8. The main point is to prove

$$F(2^{2^j}) \to F(2^{2^{j+1}}) \tag{16.1}$$

for an arbitrary nonnegative integer j, and then to combine a series of these proofs using cuts to get a proof of $F(2^{2^n})$ in $O(n)$ lines.

For our purposes, it will be a little better to think of (16.1) as being a version of

$$F(x) \to F(x^2). \tag{16.2}$$

That is, we understand first the basic rule of squaring, which we can then repeat over and over again.

We can get a stronger exponential effect using quantifiers, in the following manner. Our basic building block will now be

$$\forall x(F(x) \supset F(x^k)) \to \forall x(F(x) \supset F(x^{k^2})). \tag{16.3}$$

Keep in mind that this is supposed to be a *sequent*, and not a formula. Each of

$$\forall x(F(x) \supset F(x^k)) \quad \text{and} \quad \forall x(F(x) \supset F(x^{k^2})) \tag{16.4}$$

is a *formula*, in which $\supset$ is used to denote the connective of implication. The sequent arrow $\to$ is *not* a connective, but a special symbol used to divide the formulae which appear on the left and right sides of the sequent from each other. (While the sequent arrow $\to$ is interpreted as saying that the *conjunction* of the formulae on the left side implies the *disjunction* of the formulae on the right side, its precise role in the rules of sequent calculus is substantially different from that of the connectives. See Appendix A for a brief review of sequent calculus.)

It is easy to give a proof of (16.3) in just a few steps. The idea is that if one knows that $F(x) \supset F(x^k)$ holds for any x, then one can apply it with x replaced by x^k to get $F(x^k) \supset F(x^{k^2})$. Combining this with $F(x) \supset F(x^k)$, one can get $F(x) \supset F(x^{k^2})$, and this does the job, because x is arbitrary.

It is not difficult to convert this into a formal proof, but we shall not bother to do this now. We would like to emphasize two points, however. The first is that the proof requires just a few lines, with *the number of lines not depending on* k. The second point is that the formal version of the argument sketched above leads first to a proof of

$$\forall x(F(x) \supset F(x^k)), \forall x(F(x) \supset F(x^k)) \to \forall x(F(x) \supset F(x^{k^2})), \tag{16.5}$$

from which (16.3) is derived by applying the contraction rule to the left-hand side. This reflects the fact that we really used the "hypothesis" $\forall x(F(x) \supset F(x^k))$ *twice* in order to obtain the conclusion $\forall x(F(x) \supset F(x^{k^2}))$ (just once) in the informal argument described before.

As soon as we have (16.3), we can get

$$\forall x(F(x) \supset F(x^2)) \to \forall x(F(x) \supset F(x^{2^{2^n}})) \tag{16.6}$$

in $O(n)$ lines, using the cut rule to combine a series of copies of (16.3) One can also prove

$$\to \forall x(F(x) \supset F(x^2)) \tag{16.7}$$

quite easily, using (16.2). This can be combined with (16.6) using the cut rule to get

$$\to \forall x(F(x) \supset F(x^{2^{2^n}})). \tag{16.8}$$

Using the quantifier rules one can convert this into

$$\to F(2) \supset F(2^{2^{2^n}}). \tag{16.9}$$

This leads to a proof of

$$\to F(2^{2^{2^n}}) \tag{16.10}$$

in a total of $O(n)$ lines, using a proof of $\to F(2)$ and a few more simple steps.

This proof of the feasibility of $2^{2^{2^n}}$ in $O(n)$ lines uses a single layer of quantifiers, and without alternations. This should be compared with the earlier proofs of the feasibility of 2^{2^n} in $O(n)$ lines that did not use quantifiers at all. In the other direction, there is a construction due to Solovay which uses many levels of nested quantifiers to give proofs of the feasibility of towers of exponentials (with 2's, say) in a small number of lines. See [Car00] for details.

What were really the "operations" used in the above proof of the feasibility of $2^{2^{2^n}}$? A basic point is that the main step (16.3) did not really rely on *any* of the special properties of the feasibility predicate F or the underlying arithmetic operations. Instead, it reflects a very general mechanism for making substitutions in predicate logic (that we shall discuss further in Section 16.2). In the remaining portions of the proof, there were only two places where the particular nature of F and the underlying arithmetic operations played a significant role. The first was in the proof of (16.7), which used the rule which says that the product of feasible numbers is again feasible. The second place came at the end, where we needed to have the feasibility of the number 2.

If these were the only places where the arithmetic structure is used in a serious way, then how do we get a proof of the feasibility of $2^{2^{2^n}}$ by the end? This is part of the power of the cut rule. If one traces carefully the computations that are made implicitly in the proof, then one finds oneself returning back to (16.7) many times, even though it appears only once in the proof. This is reflected geometrically in the presence of numerous cycles in the logical flow graph of the proof. This is discussed in [Car00].

The many trips back to (16.7) also emerges clearly when one simplifies the applications of the cut rule in the proof over the contractions. We shall discuss this further in Section 16.16, and we shall discuss the effect of the simplification of the cuts on the logical flow graph of the proof there as well.

In the short proofs of the feasibility of towers of exponentials mentioned above, there is a much richer structure of cycles in the underlying logical flow graphs, with more complicated patterns of nesting of cycles reflecting the more intricate substitutions which occur in the proof. See [Car00, Cara] for more information.

16.2 Combinatorial interpretations

In order to understand what is really going on in the proofs with quantifiers described in Section 16.1, it is helpful to formulate an auxiliary notion of feasibility for *mappings* and to look at compositions of feasible mappings with themselves.

Let ϕ be a mapping that takes nonnegative integers into themselves. We shall express the feasibility of ϕ by the logical formula

$$\forall x(F(x) \supset F(\phi(x))). \tag{16.11}$$

In other words, a proof of the feasibility of ϕ should show that ϕ preserves the feasibility of any given x. A basic example is provided by the function ϕ defined by $\phi(x) = x^2$, whose feasibility in this sense follows from the rule that says that the product of feasible numbers is again feasible.

Strictly speaking, we ought to be more careful about what we mean by a "mapping", or how it might be defined in arithmetic. However, these technical matters are not central to the present discussion, and we shall ignore them for the sake of simplicity.

It is easy to see that the feasibility of a function ϕ automatically implies the feasibility of the composition of ϕ with itself. This assertion is represented by the sequent

$$\forall x(F(x) \supset F(\phi(x))) \to \forall x(F(x) \supset F(\phi(\phi(x)))). \tag{16.12}$$

An informal argument would proceed by saying that if x is chosen arbitrarily so that $F(x)$ holds, then we can use the feasibility of ϕ to obtain that $F(\phi(x))$ is also true, and then again to get $F(\phi(\phi(x)))$. It is not difficult to convert this into a formal proof of a few lines.

Let us write ϕ^j for the j-fold composition of ϕ. The sequent

$$\forall x(F(x) \supset F(\phi(x))) \to \forall x(F(x) \supset F(\phi^{2^n}(x))) \tag{16.13}$$

expresses the idea that the feasibility of ϕ implies that of ϕ^{2^n}. It can be proved in $O(n)$ lines, by combining n proofs like the one for (16.12) using the cut rule.

This is practically the same as what we did in Section 16.1, in order to establish (16.6). This is because the mapping

$$x \mapsto x^{2^{2^n}} \tag{16.14}$$

is the same as the 2^n-fold composition of $\phi(x) = x^2$. Thus (16.6) is really just a particular version of (16.13). Similarly, (16.3) is the same as (16.12), but for functions of the form

$$x \mapsto x^k. \tag{16.15}$$

Thus every step in the discussion above about feasibility for functions has a clear counterpart in Section 16.1, and this accounts for most of the proof of (16.10) described in Section 16.1.

This general language about functions is not at all needed for the proof of (16.10), and indeed for making formal proofs it is better to simply write things out directly, as in Section 16.1. However, the general language of functions does help to make clearer what is really going on in the proof with quantifiers. A basic point is that at the level of functions, the heart of the proof is completely analogous to the earlier proof of $\to F(2^{2^n})$ which did not use quantifiers (described in Section 4.8). That is, the proof of (16.13) proceeds by exactly the same kind of repeated "squaring" that we did before, even if "squaring" refers now to compositions of mappings instead of multiplication of numbers. In the present context, we are using quantifiers to formulate the idea of feasibility for mappings, and to show that it is preserved by "squaring", but otherwise they do not really play a role in the "global" structure of the proof of (16.13). Instead, the large-scale structure relies on the use of cuts and contractions, and in this respect it is exactly like the proof for numbers discussed in Section 4.8.

One might say that the proof described in Section 16.1 uses a layer of quantifiers in order to encode "propositional" properties of *functions* without speaking about functions directly.

As mentioned in Section 16.1, there is a method of Solovay that uses multiple layers of alternating quantifiers to obtain short proofs of feasibility for terms which represent towers of exponentials. In this case, the alternation of quantifiers leads not only to feasibility properties of mappings, but also to "feasibility" of an increasingly complicated hierarchy of transformations on functions.

Let us be more precise. Given a set X, let $\mathcal{F}(X)$ denote the set of all functions which map X into itself. Define $\mathcal{F}_j(X)$ recursively by taking $\mathcal{F}_1(X) = \mathcal{F}(X)$ and $\mathcal{F}_{j+1}(X) = \mathcal{F}(\mathcal{F}_j(X))$ for each $j \geq 1$. Thus $\mathcal{F}_{j+1}(X)$ is the set of functions which act on $\mathcal{F}_j(X)$, which is a collection of functions in its own right, etc.

If we have a notion of "feasibility" for the set X, then this leads to a notion of feasibility for $\mathcal{F}(X)$, just as in (16.11). Similarly, we can define notions of feasibility on every $\mathcal{F}_j(X)$, $j \geq 1$, by recursion.

Given $\Phi \in \mathcal{F}_j(X)$, $j \geq 1$, it is not hard to show that the feasibility of Φ implies that of $\Phi \circ \Phi$. This is just like (16.12). We can formulate this in a different way as follows. Define an element S_{j+1} of $\mathcal{F}_{j+1}(X)$ by

$$S_{j+1}(\Phi) = \Phi \circ \Phi \quad \text{for all} \quad \Phi \in \mathcal{F}_j(X). \tag{16.16}$$

Thus S_{j+1} represents the operation of "squaring" on $\mathcal{F}_j(X)$. It is automatically feasible as an element of $\mathcal{F}_{j+1}(X)$, since the feasibility of $\Phi \in \mathcal{F}_j(X)$ implies the feasibility of $\Phi \circ \Phi$, as noted above.

From this we obtain in particular that there are always nontrivial elements of $\mathcal{F}_k(X)$ which are feasible when $k \geq 2$, namely the S_k's. (Of course, one could also take cubes, etc.) This is completely general, and does not require any special information about the original notion of feasibility on X itself. In terms of formal logic, it is like pure predicate logic, with no special axioms or rules of inference. However, the feasibility of these "universal" transformations is not very useful unless we have some nontrivial feasible elements of $\mathcal{F}_1(X)$, and for this we do need to know something about the original notion of feasibility on X. In the context of feasible numbers, for instance, we have nontrivial feasible elements of $\mathcal{F}_1(X)$, like $\phi(x) = x^2$.

Notice that

$$S_l(T^a) = (T^a)^2 = T^{2a} \tag{16.17}$$

for all $T \in \mathcal{F}_{l-1}(X)$, $a \in \mathbf{Z}_+$, and $l \geq 2$. As usual, we are using exponents here to denote multiple compositions of transformations, so that T^a represents the a-fold composition of T. By iterating this we get that

$$S_l^k(T) = T^{2^k} \tag{16.18}$$

for all $k \geq 1$ and $l \geq 2$. In particular, we have that

$$S_l^k(S_{l-1}) = (S_{l-1})^{2^k} \tag{16.19}$$

for all $k \geq 1$ and $l \geq 3$.

In other words, each step in the hierarchy of $\mathcal{F}_j(X)$'s can be traded for another exponential in the level of complexity of the operations that we define on them. By using several layers in the hierarchy, we can obtain towers of exponentials. Imagine that we start with a large value of j, for instance, and with the fact that S_j and S_{j-1} define feasible elements of $\mathcal{F}_j(X)$ and $\mathcal{F}_{j-1}(X)$, respectively. By definitions this implies the feasibility of $S_j(S_{j-1})$ in $\mathcal{F}_{j-1}(X)$, which is to say the feasibility of $(S_{j-1})^2$. Using this and the feasibility of S_{j-2} in $\mathcal{F}_{j-2}(X)$ we obtain the feasibility of $S_{j-1}^2(S_{j-2}) = S_{j-2}^{2^2}$ in $\mathcal{F}_{j-2}(X)$. We then apply this transformation to S_{j-3} in $\mathcal{F}_{j-3}(X)$, and so forth. In the end we obtain the feasibility of S_2^k, where k is a tower of exponentials (of 2's) of height $j-2$.

This part of the construction is completely universal, and does not require any special information about the original notion of feasibility on X. If, as in the context of feasible numbers, we actually have a nontrivial element ϕ of $\mathcal{F}(X) = \mathcal{F}_1(X)$ which we know to be feasible, then we can obtain the feasibility of ϕ^r, where r is given by a tower of exponentials of height $j-1$. In the context of feasible numbers, we could take $\phi(x) = x^2$, and then use the feasibility of ϕ^r as an element of $\mathcal{F}(X)$ to obtain the feasibility of $\phi^r(2)$ as a number, and this number represents a tower of exponentials of height j.

Roughly speaking, this combinatorial discussion corresponds closely to what happens in the formal proofs of Solovay for feasibility of numbers mentioned before. Solovay's construction does not rely on the set-theoretic language of functions and transformations on them, but instead one expresses everything directly in terms of arithmetic. There is a counterpart of the higher notions of feasibility for the $\mathcal{F}_j(X)$'s, but it is employed in a more specific way, and realized concretely through logical formulae with j layers of alternating quantifiers. It is exactly this use of alternating quantifiers which permits one to avoid the need for such a rich language, in which one has transformations on functions, transformations on other transformations of functions, etc.

16.3 Feasibility graphs for feasibility graphs

One can also accommodate stronger forms of recursion and implicitness using feasibility graphs. This can be accomplished by working with feasibility graphs that operate on objects which are themselves feasibility graphs. We shall describe a basic mechanism for doing this in the present section, and illustrate it with more concrete examples afterwards.

Before we talk about feasibility graphs that operate on other kinds of feasibility graphs, we should look at ordinary graphs, without feasibility structures. For our purposes, it will be convenient to restrict ourselves to the following special class of graphs.

Definition 16.1 (IO graphs) *By an* IO graph *we mean an oriented graph which contains exactly one input vertex and one output vertex. (Recall that a vertex in an oriented graph is called an* input *vertex if it has no incoming edges, and it is called an* output *vertex if it has no outgoing edges.) We shall sometimes write* $\mathcal{IO}$ *for the class of all IO graphs. (One can also work with suitable collections of representatives up to isomorphism.)*

Note that the graph which consists of a single vertex and no edges is considered admissible as an IO graph, with the single vertex being both an input vertex and an output vertex. Otherwise, the input and output vertices will be distinct (unless the graph has a single vertex as a connected component, and the other components have nontrivial oriented cycles).

In practice, we shall often be concerned with IO graphs that come from feasibility graphs, and which contain no nontrivial oriented cycles in particular (as in (7.1) in Section 7.1). Although this restriction is not necessary for the basic concepts, it does have some benefits, as in the following observation.

Lemma 16.2 *Suppose that G is an IO graph which does not contain nontrivial oriented cycles, and let u be any vertex in G. Then there is at least one oriented path in G which begins at the input vertex and ends at u, and there is at least one oriented path which begins at u and ends at the output vertex.*

If oriented cycles were allowed, then G could be disconnected, with the input vertex feeding into a cycle, and the output vertex being reachable from a different cycle. One could also have components with cycles which are not accessible by either the input or the output vertex.

In Lemma 16.2, we are relying on our convention (from the beginning of Chapter 4) that the word "graph" mean "finite graph", and indeed the lemma is not true in general for infinite graphs.

Proof Let u be any vertex in G, and let us show that it can be reached by an oriented path which begins at the input vertex. If u is the input vertex, then there is nothing to do. Otherwise u has at least one incoming edge. Let u_{-1} be a vertex in G with an edge that goes to u. Either u_{-1} is an input vertex, in which case we are finished, or it has a predecessor u_{-2}. Repeating this process indefinitely, we either reach the input vertex, which is fine, or we have to repeat a vertex, since G is *finite* by assumption. In the latter case, we would have a nontrivial oriented cycle, which is not permitted. Thus every vertex can be reached by an oriented path that begins at the input vertex, and a similar argument shows that there is always an oriented path to the output vertex. This proves Lemma 16.2. □

Next we want to define some basic operations on IO graphs.

Definition 16.3 (Operations on IO graphs) *Let A and B be arbitrary IO graphs.*

The sum *of A and B is defined to be the graph obtained by taking the disjoint union of A and B, and then identifying the input vertices with each other, and also identifying the output vertices with each other. The identified input vertices become the unique input vertex of the sum, and similarly for the output vertices.*

The product *of A and B is defined by taking the disjoint union of A and B, and then identifying the output vertex of A with the input vertex of B. In this case, the input vertex of A becomes the unique input vertex of the product, while the output vertex of B becomes the unique output vertex of the product.*

The successor *of A is defined as follows. One starts with A and adds exactly one new vertex (which we denote by o) and one new edge, which goes from the output vertex of A to o. Thus o becomes the new unique output vertex for the graph that results, while the unique input vertex remains the same as for A.*

In all three cases, the resulting graph is taken to be an oriented *graph, using the orientations which are naturally induced from A and B. To be precise, it is perhaps better to say that the resulting graph is defined only up to isomorphic equivalence (in each case).*

It is easy to check that sums, products, and successors of IO graphs are always IO graphs.

For the record, when we speak of the "disjoint union" of A and B, we mean that one should take isomorphic copies of A and B, if necessary, to ensure that their vertex and edge sets are disjoint from each other, if they were not disjoint already. In particular, this should be done whenever one takes the sum or product of a graph with itself.

Note that the notion of product for IO graphs defined above is not commutative. Let us also mention the following, which will be useful in the context of feasibility graphs.

Lemma 16.4 *Let A and B be IO graphs, and assume that A and B are both free of nontrivial oriented cycles. Then the same is true of the sum and product of A and B, and of the successor of A.*

Proof This is not hard to verify, directly from the definitions. □

Once we have defined these three operations on IO graphs, we can also talk about feasibility graphs which describe constructions over the class $\mathcal{IO}$ of all IO graphs, in the same manner as in Chapter 7. Thus a feasibility graph for the class of IO graphs would consist of an oriented graph G without nontrivial oriented cycles, together with certain designations, as follows. Each focusing branch point of G would either be associated to sums of IO graphs or products of IO graphs.

One can restrict oneself to graphs with at most two incoming edges at each vertex, or allow repeated sums or products at focusing branch points with more incoming vertices. In the case of products, one would have to specify the order in which the product would be performed. Each edge in G would either represent the successor operation on IO graphs, or no operation.

Given these designations, we can employ the notion of *value functions* as before (from Section 7.1). Specifically, a value function on G would be a mapping f from the vertices of G to IO graphs such that the value of f at a particular vertex v is given (up to isomorphic equivalence) in terms of the values at the preceding vertices through the operations designated as above. Value functions on G are uniquely determined (up to isomorphic equivalence) by their values at the input vertices of G, and arbitrary values at the input vertices can be used, as in Lemma 7.2.

Keep in mind that G itself is *not* required to be an IO graph here. In some circumstances one might wish to impose this condition, but there is no reason to do this in general.

In order to talk about *normalized value functions* for G, one should also chose a notion of "zero element" or initial element for the class of IO graphs. The most obvious choice is the graph (up to isomorphic equivalence) with only a single vertex and no edges, but in some contexts it will be better to take the graph which consists of a single oriented edge with two distinct endpoints.

Remark 16.5 If G is a feasibility graph for constructing IO graphs, and if A is an IO graph which is one of the values of the normalized value function for G (based on either type of "zero element"), then the size of A (total number of vertices and edges) is at most exponential compared to the size of G, and exponential expansion can occur. One can obtain exponential expansion with feasibility graphs G which look like the graph pictured in Fig. 4.2 in Section 4.3, in the same manner that we have seen before. To see that at most exponential expansion can occur, one can argue as follows. Define an integer-valued function f on the vertices of G by taking $f(v)$ to be the size (total number of vertices and edges) of the IO graph A_v which is the value of the normalized value function on G at the vertex v. If $I(v)$ denotes the set of incoming edges to v, then it is not hard to check that

$$f(v) \leq 1 + \sum_{e \in I(v)} (1 + f(\iota(e))), \tag{16.20}$$

where $\iota(e)$ denotes the vertex in G at which the edge e begins. This inequality accounts for both the sizes of the graphs constructed at the preceding stage, and the additions which can come from applications of the successor operation. It also takes into account the way that some vertices are identified in the sum and product operations, so that the number of vertices is reduced in the process.

Let G_v denote the subgraph of G which consists of all vertices and edges which come "before" v in G, i.e., which are contained in an oriented path that ends at v. Using the inequality (16.20), it is not difficult to obtain exponential bounds for the size of $f(v)$ (as a number) in terms of the size of G_v. One can also get estimates in terms of the negative visibility, as follows. (This is similar to what happens in Sections 7.4 and 7.5, although we are phrasing it somewhat differently here.) Let $b(v)$ denote the size of the negative visibility

$$\mathcal{V}_-(v, G) = \mathcal{V}_-(v, G_v) \tag{16.21}$$

(i.e., counting both vertices and edges). Then

$$f(v) \leq b(v) \tag{16.22}$$

for all vertices v in G. Indeed, both $f(v)$ and $b(v)$ take the value 1 at input vertices of G, assuming that we use single vertices for the zero elements of the class of IO graphs. (If not, one could make modest modifications to adjust for this.) The step-by-step growth of $f(v)$ is controlled by (16.20), and $b(v)$ satisfies the analogous relation with the inequality replaced by an equality. The latter is easy to verify, just from the definitions. Using these observations, one can derive (16.22) through the same type of arguments as in the proof of Lemma 7.2 in Section 7.1. Thus one can derive bounds for $f(v)$ from bounds for the visibility.

Constructions with feasibility graphs

The preceding discussion provides a way to define feasibility graphs over the class of IO graphs, and to use these feasibility graphs to describe constructions of IO graphs through the notions of value functions and normalized value functions. Of course, one could work with other classes of graphs or operations on them, but these definitions serve fairly well in illustrating the main points, and accommodating some basic examples. (See also Section 16.13.)

Even with these particular definitions, the story becomes more intricate if we allow our IO graphs to come equipped with additional structure, like some kind of labellings. Let us simply concentrate for now on the situation where the IO graphs are feasibility graphs in their own right, which is the primary case of concern for the present chapter.

Let X be a fixed collection of objects, and let $\mathcal{C}$ be a set of operations defined on X. We shall sometimes refer to the pair $X, \mathcal{C}$ as a *structural system*. X might consist of words over some alphabet, for instance, or nonnegative numbers. In any case, as in Chapter 7, a feasibility graph over $X, \mathcal{C}$ means an oriented graph G without nontrivial oriented cycles such that the focusing branch points in G with j incoming edges are associated to j-ary operations in $\mathcal{C}$, and for which the edges in G may be associated to unary operations in $\mathcal{C}$, or to no operation. For focusing branch points, one may have to specify an ordering on the incoming edges, as discussed in Section 7.1. Note that we could also allow unary operations to be associated to the vertices in G with only 1 incoming edge, as mentioned in Section 7.2. For the sake of simplicity, let us ignore this possibility for the moment, and utilize the edges for unary operations.

Given IO graphs A and B which are feasibility graphs over X, $\mathcal{C}$, how might we define the sums or products of A and B, or successors of A, as feasibility graphs over X, $\mathcal{C}$? They are defined already as IO graphs, but we have to decide how to add suitable designations to them to make them into feasibility graphs over X, $\mathcal{C}$.

For the product of A and B, there is no trouble or ambiguity. One simply keeps all of the old designations from A and B, and the operation of product does not create any new edges or focusing branch points, nor does it change the number or arrangement of incoming branch points. This is not the case for the sum and successor, which require more care.

In general, the successor operation on ordinary IO graphs will lead to several operations on IO graphs which are feasibility graphs. Specifically, if $\mathcal{C}$ contains k *unary* operations on X, then there are $k+1$ versions of the successor operation for IO graphs which are feasibility graphs over X, $\mathcal{C}$. Each of these $k+1$ operations follows the same basic recipe. One starts with an IO graph A which is a feasibility graph over X, $\mathcal{C}$, to which one adds a new edge and a new vertex to A, just as in Definition 16.3. On the portion of the new graph that comes from A, one keeps all of the same designations for the feasibility structure as for A itself. To the new edge, one either associates one of the unary operations in $\mathcal{C}$, or no operation at all. It is only in the last step that there is any freedom in making choices, and we have exactly $k+1$ choices available.

For sums of IO graphs, there can be a bit more trouble than that, and it is often better to use the following variant of the notion of sums of IO graphs.

Definition 16.6 (Modified sum of IO graphs) *Let A and B be arbitrary IO graphs. We define the* modified sum *of A and B to be the sums of the successors of A and B.*

Thus the modified sum of a pair of IO graphs is still an IO graph, and it does not have nontrivial oriented cycles if the original graphs did not either. The main change is that we have better control on the output vertex of the resulting graph, which now has exactly two incoming edges, with one from each of A and B.

Coming back to the story of feasibility graphs, if there are m binary operations in $\mathcal{C}$, then there are exactly m natural versions of the modified sum operation for IO graphs which are feasibility graphs over X, $\mathcal{C}$. These m versions of the modified sum are defined in practically the same manner as before: given IO graphs A and B which are also feasibility graphs over X, $\mathcal{C}$, we start by taking their modified sum to get a new IO graph D (which is not yet a feasibility graph); on the part of D that comes from A or B, we keep exactly the same designations for the feasibility structure as we had for them; for the two new edges in D (coming from the use of the successor operation in Definition 16.6), we associate no operation on X (so that they are like ϵ-moves in an automaton); and to the new output vertex in D, we associate one of the m binary operations in $\mathcal{C}$.

For this last step, we should add a bit of extra structure to the modified sum of A and B. Namely, the two edges in D which arrive at the new output vertex should be *ordered*, with the edge coming from A viewed as arriving first (for the sake of definiteness). This is needed to accommodate the fact that binary operations in $\mathcal{C}$ may not be commutative.

There is no reason to stop at the level of *binary* operations. Let us first extend the definition of modified sums as follows.

Definition 16.7 (j-fold modified sums) *Let j be an arbitrary positive integer, and let $A_1, \ldots, A_j$ be IO graphs. The* modified sum *of $A_1, \ldots, A_j$ is defined as follows. We begin by taking the successors of the A_i's to get IO graphs $A'_1, \ldots, A'_j$. We then take the "sum" of the A'_i's, in the same manner as in Definition 16.3. More precisely, we first take the disjoint union of all of the A'_i's. Afterwards, we identify all of the input vertices of the individual A'_i's to a single vertex, and we identify all of the output vertices of the A'_i's to a single vertex as well. The resulting graph H is the modified sum of $A_1, \ldots, A_j$.*

The modified sum H is considered to be an oriented graph, with the orientation which is naturally induced from the A_i's. As usual, one should perhaps view it as being defined only up to isomorphic equivalence.

When the A_i's come from feasibility graphs, one should (normally) consider the j incoming edges at the unique output vertex of H to be ordered, *with the edge coming from A_1 arriving first, and then the edges coming from A_2, etc. Thus the modified sum is* not *commutative in this situation.*

Note that we allow j to be 1 here, so that the "family" of modified sums now includes the successor operation as well.

As usual, we have the following.

Lemma 16.8 *The modified sum of IO graphs is again an IO graph. If the summands $A_1, \ldots, A_j$ contain no nontrivial oriented cycles, then the same is true for the modified sum of $A_1, \ldots, A_j$.*

The proof is straightforward, and we omit the details.

The modified sum permits one to *lift* operations in $\mathcal{C}$ to operations on feasibility graphs which are IO graphs, in the manner that was indicated before. The next definition makes this more formal.

Definition 16.9 (Canonical lifting of operations) *Let X and $\mathcal{C}$ be as above, i.e., a collection of objects together with a set of operations defined on them. Fix an operation T in $\mathcal{C}$, and let $j \in \mathbf{Z}_+$ be the* arity *of T (i.e., the number of its arguments). The* canonical lifting of *T is the j-ary operation $\widetilde{T}$ that acts on the class of IO graphs that are feasibility graphs over X, $\mathcal{C}$, and which is defined as follows.*

Let $A_1, \ldots, A_j$ be any sequence of j IO graphs which are feasibility graphs over X, $\mathcal{C}$, and let H denote their modified sum, initially as only an IO graph. To make H into a feasibility graph, we use all of the same designations on the part of H that come from an A_i as we have on A_i itself. If $j = 1$, so that T

is a unary operation, and H is the successor of A_1 as an IO graph, then we associate to the new edge in H the operation T in $\mathcal{C}$, and this defines H as a feasibility graph. If $j > 1$, then we do not assign any (unary) operations to the new edges in H (which occur because we took the successors of the A_i's in Definition 16.7). In this case, the output vertex of H is a focusing branch point, and to it we assign the j-ary operation T in $\mathcal{C}$. We also use the ordering on the incoming edges at the output vertex which comes from the initial ordering on the A_i's. This completes the description of H as a feasibility graph over X, $\mathcal{C}$. (Remember that the definition of a feasibility graph from Section 7.1 requires that there be no oriented cycles present, which is true of H since it must also be true of $A_1, \ldots, A_j$.)

We now set $\widetilde{T}(A_1, \ldots, A_j) = H$. This completes the definition of $\widetilde{T}$, since $A_1, \ldots, A_j$ can be chosen arbitrarily.

In our previous discussion of successor operations for feasibility graphs (before Definition 16.6), we also included the possibility of assigning no operation to the new edge in the successor of a given graph A. We have omitted this here, but one could easily incorporate it again. One can also think of it as corresponding to the unary operation which is the identity mapping (which may or may not be included in $\mathcal{C}$).

With these operations defined on IO graphs that are also feasibility graphs, we can now talk about feasibility graphs which describe constructions of other feasibility graphs. The following definition provides a convenient setting in which to do this.

Definition 16.10 (Canonical lifting of structural systems) *Let $X, \mathcal{C}$ be as above, so that X is a collection of objects and $\mathcal{C}$ is a set of operations defined on them. By the* canonical lifting *of X, $\mathcal{C}$ we mean the pair X', $\mathcal{C}'_*$ which is defined as follows. For X' we take the class of IO graphs which are also feasibility graphs over X, $\mathcal{C}$. (One can get a* set *here by thinking of feasibility graphs as being defined only up to isomorphic equivalence, and choosing X' so that all isomorphism classes are represented.) Let $\mathcal{C}'$ denote the family of canonical liftings of the operations in $\mathcal{C}$ to X', as in Definition 16.9. For $\mathcal{C}'_*$ we take $\mathcal{C}'$ together with the binary operation of* product *of IO graphs which are feasibility graphs, described earlier in this section (shortly before Definition 16.6).*

Strictly speaking, we should probably call X', $\mathcal{C}'$ the canonical lifting of X, $\mathcal{C}$, and then call X', $\mathcal{C}'_*$ something like the "augmented canonical lifting" of X, $\mathcal{C}$. The terminology above will be a bit simpler in practice, and so we shall follow it instead.

As before, one might wish to include the unary operation on feasibility graphs given by the successor, with no unary operation from $\mathcal{C}$ associated to the additional edge in the successor.

The canonical lifting X', $\mathcal{C}'_*$ provides a reasonable setting in which to talk about feasibility graphs that describe constructions of other feasibility graphs. To understand better what this really means, it is helpful to think about feasibility graphs over $X, \mathcal{C}$ as defining functions on X. We shall discuss this correspondence with functions in some detail in Section 16.4, and then pursue this and other interpretations of the canonical lifting more informally in Sections 16.5 and 16.6. We shall look at some concrete examples afterwards, beginning in Section 16.7.

16.4 Correspondence with functions

Feasibility graphs which are IO graphs give rise to functions on the underlying set of objects in a fairly straightforward manner, through the existence and uniqueness of "value functions" on feasibility graphs with prescribed values at the input vertices. We shall describe this more precisely below, and also the way that operations on graphs correspond to operations on functions. In particular, *products* of graphs lead to *compositions* of functions. Before we get to that, we set some notation concerning the operations on functions that we shall use, and concerning the lifting of operations from a set X to functions on X.

Definition 16.11 (Canonical lifting to functions) *Let X, $\mathcal{C}$ be a collection of objects together with a set of operations defined on it, and let $\mathcal{F}(X)$ denote the collection of all functions from X into itself. If T is a j-ary operation on X, then we can* lift *T to a j-ary operation $\widehat{T}$ on $\mathcal{F}(X)$ through the following (straightforward) recipe: given j functions $f_1, \ldots, f_j$ in $\mathcal{F}(X)$, we take $\widehat{T}(f_1, \ldots, f_j)$ to be the function given by*

$$\widehat{T}(f_1, \ldots, f_j)(x) = T(f_1(x), \ldots, f_j(x)) \tag{16.23}$$

for all $x \in X$.

We denote by $\mathcal{F}(\mathcal{C})$ the collection of all operations $\widehat{T}$ on $\mathcal{F}(X)$ which arise from elements T of $\mathcal{C}$ in the manner just described. We write $\mathcal{F}_*(\mathcal{C})$ for the collection of operations on $\mathcal{F}(X)$ consisting of the elements of $\mathcal{F}(\mathcal{C})$ together with the binary operation of composition of functions.

To be precise, we use here the "backward" composition operation on $\mathcal{F}(X)$, which takes functions F, G in $\mathcal{F}(X)$ and gives back

$$G \circ F. \tag{16.24}$$

In other words, if we think of applying our functions to actual elements in X, then we do so in the same order in which our functions arrive.

If A is an IO graph which is also a feasibility graph over X, $\mathcal{C}$, then we can associate to it a function $F_A : X \to X$ using the notion of "value functions" from Section 7.1. More precisely, given any element x of X, there is a unique "value function" g_x on A which takes the value x at the input vertex of A, as in Lemma 7.2 in Section 7.1. By evaluating g_x at the output vertex of A, we get a new element y of X. This transformation from x to y is exactly the function F_A on X that we to associate to A.

Thus we have a correspondence between IO graphs which are feasibility graphs over X, $\mathcal{C}$ and functions on X. We can reformulate this as a mapping from X' into $\mathcal{F}(X)$, as in the next definition.

Definition 16.12 (The canonical correspondence) *Let $X, \mathcal{C}$ be a set of objects together with a collection of operations on it, and let X', $\mathcal{C}'_*$ and $\mathcal{F}(X)$, $\mathcal{F}_*(\mathcal{C})$ be as in Definitions 16.10 and 16.11. The* canonical correspondence *between X', $\mathcal{C}'_*$ and $\mathcal{F}(X)$, $\mathcal{F}_*(\mathcal{C})$ consists of the mapping from X' to $\mathcal{F}(X)$ defined above, and the "obvious" one-to-one correspondence between $\mathcal{C}'_*$ and $\mathcal{F}_*(\mathcal{C})$. Specifically, the latter is defined by associating the product operation for graphs in $\mathcal{C}'_*$ to the operation of composition of functions in $\mathcal{F}_*(\mathcal{C})$, and by using the the fact that each element of $\mathcal{C}'$ and $\mathcal{F}(\mathcal{C})$ comes from an element of $\mathcal{C}$ (i.e., we already have one-to-one correspondences between $\mathcal{C}'$ and $\mathcal{C}$, and between $\mathcal{F}(\mathcal{C})$ and $\mathcal{C}$).*

In other words, if T is an operation in $\mathcal{C}$, then it is associated to an operation $\widetilde{T}$ in $\mathcal{C}'$ as in Definition 16.9, and also to an operation $\widehat{T}$ in $\mathcal{F}(\mathcal{C})$ as in Definition 16.11. The correspondence between $\mathcal{C}'$ and $\mathcal{F}(\mathcal{C})$ is defined by associating $\widetilde{T}$ to $\widehat{T}$.

Lemma 16.13 *The canonical correspondence from X', $\mathcal{C}'_*$ to $\mathcal{F}(X)$, $\mathcal{F}_*(\mathcal{C})$ is a homomorphism (in the sense of Section 7.12).*

This is just a fancy way of saying that if one performs some operation from $\mathcal{C}'_*$ on graphs in X', and then converts the resulting graph in X' into a function on X as above, then one obtains the same result as if one made the conversion from graphs in X' to functions first, and then applied the corresponding operation in $\mathcal{F}_*(\mathcal{C})$ to the functions that came from the graphs. The proof proceeds by a straightforward unwinding of the definitions, but for the sake of clarity and completeness, we shall go through it in some detail.

Proof Suppose first that T is some operation from $\mathcal{C}$, a binary operation for instance, and let $\widetilde{T}$ be the corresponding operation in $\mathcal{C}'$. Thus $\widetilde{T}$ acts on X', and we also have a similar operator $\widehat{T}$ on $\mathcal{F}(X)$.

Let A, B be a pair of fixed elements of X', and set $D = \widetilde{T}(A, B)$. Let F_A, F_B, and F_D denote the functions on X associated to A, B, and D. We are supposed to show that

$$F_D = \widehat{T}(F_A, F_B), \tag{16.25}$$

which is the same as saying that

$$F_D(x) = T(F_A(x), F_B(x)) \tag{16.26}$$

for all $x \in X$.

By definitions, A, B, and D are all IO graphs which are also feasibility graphs over $X, \mathcal{C}$. Fix $x \in X$, and let g_x, h_x be the (unique) value functions on A and B (respectively) which take the value x at the input vertices of A, B, as in Lemma 7.2. Thus g_x and h_x are mappings from vertices in A and B into X, whose values are compatible with the operations on X designated by A and B in the usual way. If o_A and o_B denote the output vertices of A and B, then we have that

$$F_A(x) = g_x(o_A) \quad \text{and} \quad F_B(x) = h_x(o_B) \tag{16.27}$$

by the definition of F_A and F_B.

To compute $F_D(x)$, we want to build a value function j_x on D which takes the value x at the input vertex of D. It does not matter how we produce such a value function on D, as long as it is a value function, because of the uniqueness assertion in Lemma 7.2. We shall in fact build j_x from g_x and h_x.

Specifically, we define $j_x(v)$ for any vertex v in D in the following manner. If v comes from a vertex in A, then we give j_x the same value as g_x at the corresponding vertex in A, and we do the same for vertices that come from B. The initial vertex of D comes from both A and B, but one has the same value there from both g_x and h_x, namely x itself. This accounts for all vertices in D except for the output vertex o_D of D, where we set

$$j_x(o_D) = T(g_x(o_A), h_x(o_B)). \tag{16.28}$$

This defines j_x unambiguously at all vertices in D, and it is easy to see that j_x is a value function on D, i.e., that the value of j_x at any given vertex v is related to the values of j_x at the predecessors of v in the usual way. This comes from the corresponding properties of g_x and h_x, and from the way that we defined j_x at the output vertex of D (which is consistent with the definition of D as $\widetilde{T}(A, B)$, as in Definition 16.9).

Because of the uniqueness of value functions with prescribed values at the input vertices, we may conclude that

$$F_D(x) = j_x(o_D). \tag{16.29}$$

This combined with (16.27) and (16.28) yields (16.26), as desired.

The same argument works for any operation T in $\mathcal{C}$ of any arity, and so we proceed now the case of products and compositions. Fix elements A and B of X', as before, and let us now take $D \in X'$ to be the product of A and B (in the sense of Section 16.3, and in that order). In this case, we want to show that

$$F_D = F_B \circ F_A \tag{16.30}$$

(as in (16.24)), or, equivalently, that

$$F_D(x) = F_B(F_A(x)) \tag{16.31}$$

for all $x \in X$.

Fix an arbitrary $x \in X$, and let g_x be the value function on A which takes the value x at the input vertex of A (as above). This time we take h to be the value function on B which takes the value

$$g_x(o_A) = F_A(x) \tag{16.32}$$

at the input vertex. Thus we have

$$h(o_B) = F_B(F_A(x)), \tag{16.33}$$

by the definition of F_B.

We define now a function j on the vertices of D simply by taking j to agree with g_x on the part of D that comes from A, and taking j to be the same as h on the part of D that comes from B. The output vertex of A and the input vertex of B correspond to the same vertex in D, by the definition of the product operation for IO graphs, and the values of g_x and h at these vertices coincide with each other, by construction.

Thus we have a well-defined function j on the vertices of D. One can check that j is a value function on D. The value of j at the input vertex of D is equal to x, because it is the same as the value of g_x at the input vertex of A, by construction. From the uniqueness of value functions, we obtain that

$$F_D(x) = j(o_D). \tag{16.34}$$

On the other hand, $j(o_D) = h(o_B)$, by construction, and so (16.33) and (16.34) imply (16.31), as desired. This completes the proof of the lemma. □

Remark 16.14 Feasibility graphs which are not IO graphs can also be interpreted as representing functions, but functions which have multiple arguments or values, as many as there are input or output vertices. With suitable labellings, one can think of these as mappings between Cartesian products of X, where the number of factors of X in the domain and range need not be the same. In this extended setting, the notions of products, compositions, and other operations for graphs and functions become slightly more complicated, since one has to say how the operations are performed (i.e., how the various inputs and outputs should be matched up). The basic constructions in the proof of Lemma 16.13 (for combining value functions from different graphs) still apply, however.

16.5 Implicit representations of functions

Let X, $\mathcal{C}$ be a collection of objects together with a set of operations on them, and let X', $\mathcal{C}'_*$ be the canonical lifting of X, $\mathcal{C}$ to feasibility graphs which are IO graphs, as in Definition 16.10. We can think of each element of X' as representing a function on X, in the manner described in Section 16.4.

In fact, elements of X' represent not only functions on X, but also specific recipes by which to compute their values from given inputs. Feasibility graphs over X', $\mathcal{C}'_*$ in turn describe recipes for building elements of X', and hence functions on X.

Let us be more precise. There is a natural "zero element" for X', namely a graph Z with only one vertex and no edges (and none of the usual designations associated to feasibility graphs, since there are no edges or focusing branch points). This graph corresponds (in the sense of Definition 16.12) to the identity mapping on X. Using Z as the "zero element" for X', we can talk about "normalized value functions" on feasibility graphs over X', $\mathcal{C}'_*$, as in Section 7.1, and these normalized value functions give the elements of X' which are "constructed" by a given feasibility graph over X', $\mathcal{C}'_*$.

Imagine that G is a feasibility graph over X', $\mathcal{C}'_*$ which is also an IO graph, and let A denote the element of X' which is the value of the normalized value function on G at the output vertex of G. Thus A represents a function F_A on X, as in Section 16.4, and G represents a construction of A. One can say that G provides a kind of implicit description of F_A. Indeed, we cannot go directly from G to the computation of values of F_A at given points in X, but G provides a recipe for making a recipe for computing the values of F_A at particular points. Specifically, we can execute G to obtain A, and then we can execute A to obtain actual values of F_A.

Each of these "executions" can involve exponential expansion, in the usual way. In passing from G to actual values of F_A, we can have double exponential expansion, and we shall see this concretely in the examples given after the next section.

Remember that the basic source of implicitness in a feasibility graph comes from the fact that one can have defocusing branch points, which permit the result of a partial computation to be used many times, without repeating the supporting computation. In the present context of feasibility graphs over X', $\mathcal{C}'_*$, this can have the effect of duplicating whole functions or recipes, rather than just individual values (like numbers or words, for instance).

Of course, we do not have to stop at the level of X', $\mathcal{C}'_*$ either. We can start from there and repeat the whole construction, to get the collection X'' of feasibility graphs over X', $\mathcal{C}'_*$ which are also IO graphs, together with certain operations over X'', and so on. This is like the earlier discussion from Section 16.2, where we started with a set X, and then looked at the space $\mathcal{F}(X) = \mathcal{F}_1(X)$ of functions on X, and then the space $\mathcal{F}_2(X)$ of functions on $\mathcal{F}(X)$, and so forth.

A feature of this story is that the functions which occur at each stage are always represented in a concretely computable way, through feasibility graphs. Each layer of implicitness brings another exponential in the possible level of complexity, but no more, since the execution of a feasibility graph entails at most a (single) exponential expansion in size. (Compare with Section 7.4 and Remark 16.5.)

16.6 Functions and points

Let $X, \mathcal{C}$ be a set of objects and a collection of operations on them, as usual, and fix some element ξ of X. In some situations, one might want to take ξ to be a particular "zero element" or initial element of X, as in the context of normalized value functions. For the purposes of this section, ξ could be any element of X, and in some circumstances, one may wish to use more than one element of X for ξ, or all elements of X.

Let X', $\mathcal{C}'_*$ be the canonical lifting of X, $\mathcal{C}$, as in Definition 16.10. In Section 16.4, we defined a mapping from X' to *functions* on X, but now we want to define a mapping $p : X' \to X$, to actual "points" in X. We do this in the following manner. Let A be an element of X', so that A is a feasibility graph over X, $\mathcal{C}$ which is also an IO graph. If F_A is the function on X associated to A as in Section 16.4, then we set

$$p(A) = F_A(\xi). \tag{16.35}$$

If ξ is designated as the "zero element" of X, then $F_A(\xi)$ is the same as the element of X which is the value of the (unique) normalized value function on A at the output vertex of A.

Using this mapping from X' into X, one can think of feasibility graphs over X', $\mathcal{C}'_*$ as representing constructions in X, and not just in X'. This is quite different from making constructions in X through feasibility graphs over X, $\mathcal{C}$. In X', $\mathcal{C}'_*$, we have a new operation that we did not have before, namely, the product operation on IO graphs which are feasibility graphs. By working at the level of X', one has the possibility for making other operations that we do not have directly at the level of X. We shall discuss an example of this in Section 16.13.

Recall from Definition 16.10 that $\mathcal{C}'$ denotes the collection of operations on X' which are lifted directly from the operations $\mathcal{C}$ on X'. If one restricts oneself to $\mathcal{C}'$, then X' is not so different from X, in terms of representing constructions in X, as in the following lemma.

Lemma 16.15 *The mapping $p : X' \to X$ defined above together with the natural one-to-one correspondence between elements of $\mathcal{C}'$ and $\mathcal{C}$ (that comes from the definition of $\mathcal{C}'$) defines a homomorphism from X', $\mathcal{C}'$ to X, $\mathcal{C}$ (in the sense of Section 7.12).*

This is an easy consequence of Lemma 16.13, and we omit the details. Note that it works for any choice of ξ in X.

From Lemma 16.15, we have that every feasibility graph G' over X', $\mathcal{C}'$ can be reinterpreted as a feasibility graph G over X, $\mathcal{C}$, and that the "answers" given by G' and G are compatible with each other. (See Section 7.12.) More precisely, for this correspondence, one should use compatible choices of values for value functions at the input vertices of feasibility graphs over X', $\mathcal{C}'$ and X, $\mathcal{C}$, respectively. The simplest choices for this are $\xi \in X$ for feasibility graphs over X, $\mathcal{C}$, and the graph Z with one vertex and no edges as an element of X' for feasibility graphs over X', $\mathcal{C}'$ (as in Section 16.5). These two choices match up automatically here, because the function F_Z on X associated to Z as a feasibility graph over X, $\mathcal{C}$ is the same as the identity mapping on X.

This correspondence between feasibility graphs over X', $\mathcal{C}'$ and X, $\mathcal{C}$ would not work for $\mathcal{C}'_*$ instead of $\mathcal{C}'$ on X', for which there is the extra operation of products of feasibility graphs. In general, the product operation need not even be definable directly at the level of X. If the product operation (or other operation) on X' does correspond in a simple way to an operation on X (and we shall see examples of this in the next sections), then one can get a correspondence between feasibility graphs over X', $\mathcal{C}'$ and feasibility graphs over X in the same way as above, but now with a new operation on X, in addition to the ones in $\mathcal{C}$. (Compare with Section 16.14, as well as with the concrete examples in the next sections.)

One can always "simulate" feasibility graphs over X', $\mathcal{C}'_*$ by feasibility graphs over X, $\mathcal{C}$, by "executing" the feasibility graphs over X', $\mathcal{C}'_*$, and this can entail exponential expansion in the size of the graph. By contrast, if one can simply reinterpret a feasibility structure on a graph, as in the case where one has a homomorphism (as in Lemma 16.15), then the underlying graph does not change, and in particular its size does not change.

One can also look at these matters in terms of functions on X, rather than feasibility graphs. Let $\mathcal{F}(X)$ denote the space of functions on X, and let $\mathcal{F}(\mathcal{C})$ denote the canonical lifting of the collection $\mathcal{C}$ of operations on X to a collection of operations on $\mathcal{F}(X)$, as in Definition 16.11. Using ξ in X, we obtain a mapping from $\mathcal{F}(X)$ to X by evaluation at ξ, i.e.,

$$F \mapsto F(\xi), \qquad F \in \mathcal{F}(X). \tag{16.36}$$

If one makes computations at the level of functions using only the operations in $\mathcal{F}(\mathcal{C})$, and projects back down into X at the end using the evaluation mapping (16.36), then one does not get anything more than what could have been done directly on X, using the operations in $\mathcal{C}$. If instead one also permits the use of compositions of functions, or other operations on functions, then one can get much more than what was originally available with X, $\mathcal{C}$. There are possibilities to make much more concise representations of individual elements of X in particular.

One can keep going with this, working with functions on $\mathcal{F}(X)$ (and not just on X), and so forth, as in Section 16.2. With each new level, one gains a new form of "composition", and this permits one to make more and more concise descriptions of objects which are increasingly large or complicated.

Feasibility graphs provide a framework in which these general ideas are realized effectively. One has a fairly precise delineation of different levels of implicitness, and concrete ways to analyze the effect of individual layers of implicitness (using tools related to visibility graphs, for instance).

With each new level of implicitness comes stronger forms of internal symmetry, and this appears clearly in the basic examples. We shall discuss a few concrete settings for these ideas in the next sections, and mention some visual interpretations in Section 16.11.

16.7 Graphs and numbers

Let X be the set of positive integers, equipped with only the binary operation $+$ of addition. For the moment, it will be convenient to use the number 1 as the "zero element" or initial element of X (i.e., for defining normalized value functions, as in Section 7.1).

Let A be a feasibility graph over $X, +$ which is also an IO graph. In this case, the edges in A have no interpretation in terms of operations over X (since we are not employing any unary operations on X), and the focusing branch points all represent sums.

Lemma 16.16 *Let X and A be as above, and suppose that f is a value function on A (as in Section 7.1). Let i_A denote the initial vertex of A, and let v be any other vertex in A. Then $f(v)$ is the same as the product of $f(i_A)$ and the number of distinct oriented paths in A which go from i_A to v.*

Proof This lemma is just a minor variation on the theme of "counting functions" from Section 9.1, but let us quickly indicate a proof.

Note that every vertex v in A is accessible by an oriented path which begins at the input vertex of A. (See Lemma 16.2 in Section 16.3. Remember that A should be free of nontrivial oriented cycles, since it is a feasibility graph, as in (7.1) in Section 7.1.)

Let g be the function on the vertices of A mentioned in the conclusion of Lemma 16.16, i.e., $g(u)$ is the product of $f(i_A)$ and the number of distinct oriented paths in A which go from i_A to u, where u is any vertex in A. Thus f and g agree at the input vertex i_A, by construction. It suffices to show that g is itself a value function, because the uniqueness of value functions (as in Lemma 7.2) would then imply that $f = g$ everywhere, as desired.

In other words, we need to know that the value of g at any vertex w in A is the same as the sum of the values of g at the immediate predecessors of w. This is easy to verify directly from the definition of g. This completes the proof of Lemma 16.16. □

Let X' denote the set of all feasibility graphs over $X, +$ which are also IO graphs, and let $p : X' \to X$ be the mapping described in Section 16.6, with $\xi = 1$. Thus p takes a feasibility graph A over $X, +$ which is an IO graph, and associates to it the positive integer which is the value of its normalized value function at the output vertex. This means that $p(A)$ is equal to the number of oriented paths in A which go from the input vertex to the output vertex, because of Lemma 16.16.

In the present situation, our set $\mathcal{C}$ of operations over X consists only of the binary operation of addition, and when we make the lifting to X' (as in Definition 16.10), we get a set $\mathcal{C}'_*$ of operations on X'. The first operation is the canonical lifting of $+$ to an operation on graphs, which also incorporates the notion of "modified sum" of feasibility graphs from Definition 16.6. (See Definition 16.9.) The second operation on X' comes from the product operation for feasibility graphs, as in Definitions 16.3 and 16.10.

What do these operations on X' mean at the level of X? Let A and B be arbitrary elements of X'. If C is the element of X' which corresponds to the "sum" of A and B (through the lifting of $+$ on X to a binary operation on X'), then

$$p(C) = p(A) + p(B). \tag{16.37}$$

This equation can be seen as a consequence of Lemma 16.15, but it is easy to check directly too. If D denotes the element of X' which is the product of A and B, then

$$p(D) = p(A) \cdot p(B), \tag{16.38}$$

where the right-hand side refers to the ordinary product for positive integers. This is not hard to check, using Lemma 16.16, which tells us exactly how the value functions on these graphs behave.

One can also look at the product operation for these graphs in the following way. For each of our graphs A, B, and D in X' we get functions F_A, F_B, and F_D on X, as discussed just before Definition 16.12 in Section 16.4. Thus F_A is the function that takes in a number x, uses that for the input value of a value function f_x on A, and then gives back the number y which is the value of f_x at the output vertex of A. In particular, $F_A(1) = p(A)$, and we get that

$$F_A(x) = x \cdot p(A) \tag{16.39}$$

for all positive integers x, because of Lemma 16.16. Of course, we have analogous formulae for F_B and F_D. On the other hand, Lemma 16.13 implies that F_D is given by the composition of F_A and F_B. (Remember also (16.24).) In this case, the composition rule contains exactly the same information as the product formula (16.38).

To summarize, when we start with the set X of positive integers equipped with the binary operation of addition, our general constructions lead naturally to *multiplications* of numbers.

In this context of numbers, the exponential expansion that can come with feasibility graphs and the canonical lifting to X' is particularly clear. Given a feasibility graph G over X, $+$, the elements of X (i.e., numbers) which arise as values of the normalized value function for G at the output vertices can be of at most exponential size compared to the size of G, and this type of exponential expansion can occur. (This corresponds closely to the discussion in the first subsection of Section 7.6.) If instead one uses feasibility graphs over the canonical lifting X', $\mathcal{C}'_*$ of X, $+$, then one can reach numbers of double-exponential size, but no further. Let us be more precise. If H is a feasibility graph over X', $\mathcal{C}'_*$, then H describes the construction of some elements of X', namely the values of its normalized value function at the output vertices of H. As usual, these elements of X' will have at most exponential size compared to the size of H for the usual reasons. (See Remark 16.5, and also Remark 16.17 below.) If one converts these elements of X' into numbers using the mapping $p : X' \to X$, one gets another exponential level of expansion, but no more. This can be derived from Lemma 16.16 and our usual bounds for the size of the visibility, as in Section 8.8.

Remark 16.17 Strictly speaking, the estimates indicated in Remark 16.5 should be modified slightly to account for the use of "modified sums" (Definition 16.6), as opposed to the original notion of sums of IO graphs from Definition 16.3. In the present setting, the difference between the two kinds of sums does not really matter, because the larger size of the modified sum is exactly balanced by the fact that we are not using the successor operation for IO graphs here, and they were included in Remark 16.5.

In any case, the basic matter of exponential bounds is not so delicate as to be disturbed much by the change from ordinary sums of IO graphs to modified sums. One could go much further and multiply the right-hand side of the stepwise inequality (16.20) by a bounded constant and still obtain roughly the same kind of exponential estimate in the end, even if the rate of the exponential growth would change. That is, one always gets estimates like C^n, where n is the size of the feasibility graph, and not like n^n, or anything like that. (See also Section 8.8.)

Remark 16.18 There is a basic feature of the product formula (16.38) that makes sense in general, but which is not normally true. Namely, the result of the transformation $p : X' \to X$ applied to the product graph D depends only on the values of p at the constituent graphs A and B, so that the product operation on X' "descends" to a well-defined binary operation on X itself. We shall encounter some other special situations in which this also happens, but typically one should expect $p(D)$ to depend not only on $p(A)$ and $p(B)$, but also on the internal behavior of A and B.

16.8 Graphs and numbers, continued

Let $Y, \cdot$ be the set of positive integers equipped with the binary operation of multiplication. In this context, it is convenient to use the number 2 for our "zero element", which is used for making normalized value functions.

If $X, +$ is as in Section 16.7, then we have a natural "homomorphism" from X, $+$ into $Y, \cdot$, in which a positive integer x is replaced by 2^x. This homomorphism is also compatible with the "zero elements" that we have chosen in these two cases, i.e., the number 1 for X and the number 2 for Y.

One can make for $Y, \cdot$ roughly the same types of computations and constructions as we made for $X, +$ is as in Section 16.7. Let us start with is the following analogue of Lemma 16.16.

Lemma 16.19 *Let $Y, \cdot$ be as above, and let A be a feasibility graph over $Y, \cdot$ which is also an IO graph. Let f be a value function on A (as in Section 7.1), let i_A denote the initial vertex of A, and let v be any other vertex in A. Then $f(v)$ is the same as $f(i_A)^{\alpha(v)}$, where $\alpha(v)$ is the number of distinct oriented paths in A which go from i_A to v.*

In this case, the edges in A have no interpretation in terms of operations over Y, and the focusing branch points in A all represent multiplications.

Proof This follows from exactly the same kind of argument as we used to prove Lemma 16.16. (One could also use the idea of "homomorphisms" to make comparisons between the two situations, as in Section 7.12, but for this one ought to extend the discussion from integers to a larger set of numbers in the domain, to get a homomorphism whose image includes all positive integers.) □

Let Y' denote the set of IO graphs which are also feasibility graphs over Y, $\cdot$, and let $p_Y : Y' \to Y$ be the mapping described in Section 16.6, with $\xi = 2$. Thus p_Y takes a feasibility graph A over Y, $\cdot$ which is an IO graph and assigns to it the positive integer which is the value of its normalized value function at the output vertex. This means that $p_Y(A)$ is equal to 2^α, where α denote the number of oriented paths in A which go from the input vertex to the output vertex of A, as in Lemma 16.19.

As before, we have natural operations of "sum" and "product" on the feasibility graphs in Y', and these have simple arithmetic interpretations at the level of numbers. Specifically, if A and B are two elements of Y', and C denotes their "sum", then

$$p_Y(C) = p_Y(A) \cdot p_Y(B). \tag{16.40}$$

This follows from Lemma 16.15, but it is easy to check directly as well. If D denotes the "product" of A and B, then

$$p_Y(D) = 2^{(\log_2 p_Y(A))(\log_2 p_Y(B))}. \tag{16.41}$$

This is not hard to verify, using Lemma 16.19. (Compare also with Lemma 16.13 in Section 16.4.) Note that these formulae are compatible with (16.37) and (16.38) in Section 16.7, in terms of the homomorphism from X, $+$ to Y, $\cdot$ mentioned earlier (which sends $x \in X$ to 2^x).

If A is an element of Y' which represents a graph of size n, then $p_Y(A)$ can be of at most *double-exponential* size in n, and double-exponential expansion can occur. Because of Lemma 16.19, this is equivalent to saying that the number of oriented paths going from the input vertex in A to the output vertex is at most exponentially-larger than the size of A (as in Sections 4.7 and 8.8), and that the exponential increase can take place. (Recall the example shown in Fig. 4.2 in Section 4.3, and see also Section 8.8.)

On the other hand, we can just as well view Y' as a collection of objects in its own right, with the binary operations of sum and product (of IO graphs which are feasibility graphs over Y, $\cdot$). This permits us to use feasibility graphs to describe the construction of elements of Y', which we can then reinterpret as numbers, through the mapping $p_Y : Y' \to Y$. This leads to another exponential in expansion, i.e., a feasibility graph of size n over Y' can be used to construct elements of Y' of exponential size, and these elements of Y' can then describe constructions of numbers of triple-exponential size compared to n.

All of these statements about Y, $\cdot$ are completely analogous to the ones for X, $+$ in Section 16.7, and indeed they are also compatible with the homomorphism from X, $+$ to Y, $\cdot$ that we have through the mapping $x \mapsto 2^x$. Instead of comparing Y with X, however, let us compare it with the set X' of IO graphs which are feasibility graphs over X, $+$. Every element of X' determines a number through the mapping $p : X' \to X$ defined in Section 16.7, and the product operation on feasibility graphs in X' corresponds to the usual product of numbers, as in (16.38).

On X' we also have the operation of sum for feasibility graphs, and this corresponds to usual addition of numbers, as in (16.37). Thus if we think of X' as being equipped with both the sum and product operations for feasibility graphs, then our mapping from X' to numbers can be viewed as a "homomorphism" into $Y, \{+, \cdot\}$.

If we allow both addition and multiplication to be used on Y, then we lose some of the simplicity that we had before. Lemma 16.19 no longer remains valid, and indeed the behavior of the value function on a feasibility graph A over Y, $\{+, \cdot\}$ now depends on the way that the operations $+$ and $\cdot$ are assigned to the focusing branch points. For the same reason, the product operation on feasibility graphs over Y, $\{+, \cdot\}$ no longer represents a single binary operation on numbers, as it did before (in (16.41)). Instead, the number associated to the product D of two feasibility graphs A and B now depends on the internal structure of A and B, and not just on the numbers associated to them.

The inclusion of the operation $+$ on Y does not really change the magnitudes of the numbers which can be constructed by feasibility graphs, however. A feasibility graph A over Y, $\{+, \cdot\}$ of size n can describe the construction of numbers of at most double-exponential size in n, and double-exponential expansion can occur. This is the same as for Y, $\cdot$ (without $+$), and indeed $+$ plays little role here, since it is easier to make numbers large through multiplication rather than addition.

There is a minor technical point here, which is that when we include $+$ and think of Y, $\{+, \cdot\}$ as receiving a homomorphism from the set X' equipped with the operations of sum and product for feasibility graphs, then we should also go back to using the number 1 as our "zero element" for the purposes of defining normalized value functions, rather than the number 2 as we did previously. Thus, to make large numbers with feasibility graphs, one should first perform some additions to get numbers which are larger than 1, before continuing with only multiplications to make numbers which are as large as possible.

Our homomorphism from X' (equipped with the operations of sum and product for feasibility graphs which are IO graphs) into Y, $\{+, \cdot\}$ permits us to reinterpret a feasibility graph over X' (with the operations just mentioned) as a feasibility graph over Y, $\{+, \cdot\}$, and to say that the constructions made by these two types of feasibility graphs are compatible with each other. (See Section 7.12.) The double-exponential expansion which occurs in using feasibility graphs over Y, $\{+, \cdot\}$ to describe numbers can be seen as nothing but another manifestation of the double-exponential expansion involved in using feasibility graphs to construct elements of X' (which are themselves feasibility graphs), and then converting these elements of X' into numbers (elements of X), as discussed in Section 16.7. Similarly, the triple-exponential expansion that comes from using feasibility graphs to construct elements of Y' (feasibility graphs over Y) which are then converted into numbers (elements of Y) reflects the expansion involved in using feasibility graphs which describe the construction of feasibility graphs over X', and then converting these into feasibility graphs over X', and then into elements of X', and then into numbers.

Of course this process can be repeated indefinitely, with exactly 1 new level of exponentiation occurring with each stage. This is what one should expect in general, but the case of arithmetic is especially crisp and easy to make explicit.

16.9 Rings and semirings

Let us pause a moment to consider the basic algebraic ideas of addition and multiplication in the broader context of rings, or semirings.

A semiring is like a ring but without additive inverses. More precisely, let us say that a set S equipped with operations $+$, $\cdot$ is a *semiring* if S, $+$ is a commutative semigroup, S, $\cdot$ is a semigroup which may or may not be commutative, and if the two operations $+$ and $\cdot$ are related by the usual distributive laws

$$x \cdot (y + z) = x \cdot y + x \cdot z, \qquad (y + z) \cdot w = y \cdot w + z \cdot w \tag{16.42}$$

for all $x, y, z, w \in S$. It will be convenient for us to require also that S have both additive and multiplicative identity elements.

A basic example of a semiring is given by the set of nonnegative integers equipped with the usual operations of addition and multiplication. For another example, let Σ be any given set, which we think of as an "alphabet", and let S_Σ denote the semiring which is freely generated by Σ. To define this in a precise way, let Σ^* denote the set of all (finite) words over Σ, including the empty word ϵ, and let S_Σ consist of all (finite) formal sums of elements of Σ^*, with suitable identifications so that the operation of addition becomes commutative and associative. We also include in S_Σ a 0 element, which one can view as representing the empty formal sum. This gives us S_Σ as a set, together with the operation $+$ of addition. To define multiplication, we use the operation of concatenation of words in Σ^*, and extend it to all of S_Σ through the distributive laws, with 0 times anything giving 0 back again. Note that the empty word $\epsilon \in \Sigma^*$ represents a nonzero element of S_Σ, which provides a multiplicative identity element.

This semiring S_Σ is very close to the idea of "regular expressions" over Σ (Section 1.1), but there are some differences. Consider the quantities

$$a(b + c) \qquad \text{and} \qquad ab + ac, \tag{16.43}$$

where a, b, c are elements of Σ, say. These two quantities represent the same element of S_Σ, but as regular expressions, they may be treated as (syntactically) distinct. They are distinct but "equivalent", in the sense that they define the same language (through the method described in Section 1.1). It is easy to check that regular expressions over Σ are always equivalent when they represent the same element of S_Σ, i.e., that the laws of commutativity (for addition), associativity, and distributivity preserve the underlying language. It is not true that equivalent regular expressions necessarily define the same element of S_Σ, because

$$a \qquad \text{and} \qquad a + a \tag{16.44}$$

define distinct elements of S_Σ, but are equivalent as regular expressions.

We are implicitly restricting ourselves here to regular expressions which do not use the star operation $*$. Concerning versions of the star operation in broader settings, see [Eil74, KS86].

Given an arbitrary semiring $S, +, \cdot$, we can define feasibility graphs for making constructions in S in the usual way. One might also include a set $\mathcal{U}$ of unary operations on S, which can represent additions or multiplications by fixed elements of S. For the purpose of defining normalized value functions (as in Section 7.1), one should choose a "zero element" or initial element. The additive identity element 0 is a natural choice, or perhaps the multiplicative identity element, depending on the circumstances.

To make the idea of feasibility graphs more concrete in this context, let us give the following analogue of Lemma 16.16 from Section 16.7.

Lemma 16.20 *Let $S, +, \cdot$ be a semiring, and let $\mathcal{U}$ be a collection of unary operations on S, each of which is given by right-multiplication by a fixed element of S. Let A be an IO graph which is also a feasibility graph over $S, \mathcal{U} \cup \{+\}$ (and without the (binary) operation of multiplication $\cdot$). Let i_A denote the unique input vertex in A, and let f be a value function on A (in the sense of Section 7.1). If v is any vertex in A, then $f(v)$ can be written as $f(i_A) \cdot h(v)$, where $h(v)$ is the element of S obtained in the following manner.*

If p is any oriented path in A, then the edges traversed by p give rise to a succession of unary operations of $\mathcal{U}$. (Some edges may correspond to no operation, in which event they are skipped for this purpose. This is equivalent to having these edges correspond to right-multiplication by the multiplicative identity element in S.) We apply these unary operations (in order) to the multiplicative identity element of S, to get an element $\eta(p)$ of S, which reflects the total effect of these unary operations along the path p. For $h(v)$, we take the sum of $\eta(p)$ over all oriented paths p in A which go from i_A to v.

Although this statement is somewhat more complicated than that of Lemma 16.16, they both amount to roughly the same thing. One can think of Lemma 16.16 as a special case of Lemma 16.20, with S taken to be the semiring of nonnegative integers, and with $\mathcal{U}$ consisting of only the unary operation of multiplication by 1 (or with no unary operations at all). In general, $h(v)$ is a more complicated expression resulting from the collection of oriented paths going from i_A to v than merely their total number.

Note that there may be no elements of $\mathcal{U}$ associated to a path p, either because no unary operations are associated to edges traversed by p, or because p is the degenerate path which contains only the vertex i_A, and crosses no edges. In this case, $\eta(p)$ is equal to the multiplicative identity element in S.

Proof Lemma 16.20 can be proved using the same kind of arguments as for Lemma 16.16. We begin by defining a function g on the vertices of A by

$$g(v) = f(i_A) \cdot h(v). \tag{16.45}$$

The main point is that $g(v)$ is a value function for A which has the same value as f at the (unique) input vertex i_A. This is not difficult to check from the definition of h. The uniqueness of value functions (Lemma 7.2 in Section 7.1) permits us to conclude that $f(v) = g(v)$ at all vertices v in A, which is exactly what we want. □

Of course, there are analogous statements for left-multiplication.

Corollary 16.21 *Let S, $+$, $\cdot$ and $\mathcal{U}$ be the same as in Lemma 16.20, and let A and B be IO graphs which are also feasibility graphs over S, $\mathcal{U} \cup \{+\}$. Let D be the product of A and B (as in Section 16.3), and let x, y, and z denote the values of the normalized value functions of A, B, and D at their respective output vertices. Then*

$$z = xy. \tag{16.46}$$

This is not hard to verify, using Lemma 16.20. (Remember also Lemma 16.13 in Section 16.4.) The issue is practically the same as for ordinary numbers, as in Section 16.7 (especially (16.38)). One should be a bit more careful here, though, since the multiplication may not be commutative.

We end this section with a couple of examples, to illustrate the nature of "exponential expansion" as it arises in this context. For these examples, we shall work in the semiring S_Σ, using an alphabet Σ which consists of exactly two letters a and b, and we shall take $\mathcal{U}$ to be the pair of unary operations defined by right-multiplication by a and b, respectively.

Let A be the IO graph with $n+1$ vertices $v_0, v_1, \ldots, v_n$ and exactly two edges from v_j to v_{j+1} for $0 \le j \le n-1$. We denote these two edges as $e_j(a)$ and $e_j(b)$, and we want to think of A as defining a feasibility graph over S_Σ, $\mathcal{U} \cup \{+\}$, where all the focusing vertices represent additions, and the edges $e_j(a)$ and $e_j(b)$ represent right-multiplications by a and b, respectively.

Define a function f on the vertices of A by setting

$$f(v_j) = (a+b)^j. \tag{16.47}$$

It is easy to see that this is a value function for A. It is the the normalized value function for A, if we use the empty word ϵ as the designated "zero element" or initial element for S_Σ. (If instead one prefers to use the additive identity element 0 as the starting value for normalized value functions, then one should modify A slightly, and add a unary operation of addition by ϵ to $\mathcal{U}$, for instance.)

The value of f at the output vertex n is

$$(a+b)^n. \tag{16.48}$$

This has exponential size in n, in the sense that if it is written out explicitly as a sum of words in a and b, then this sum will have 2^n terms.

Conversely, let B be any IO graph which is a feasibility graph over S_Σ, $\mathcal{U} \cup \{+\}$, and suppose that the value of the normalized value function of B at the output vertex o_B is $(a+b)^n$ (where again we use the empty word ϵ for the value of the normalized value function at the input vertex i_B). Then there must be at least 2^n different oriented paths in B which go from i_B to o_B. This can be derived Lemma 16.20. In other words, there must be at least one oriented path from i_B to o_B for every word of length n over a and b. This implies that the negative visibility $\mathcal{V}_-(o_B, B)$ must contain at least 2^n vertices. In particular, B must contain fairly long chains of focal pairs, if it is substantially smaller in size than that, as in Theorem 8.9 in Section 8.4. This provides a concrete example for the comments in Remark 7.6 in Section 7.5.

If we also allow the binary operation of multiplication $\cdot$ to be used in our feasibility graphs, then we could produce expressions like

$$(a+b)^{2^n} \tag{16.49}$$

by feasibility graphs of size $O(n)$. The same effect could be achieved by using feasibility graphs which themselves describe constructions of feasibility graphs over S_Σ, $\mathcal{U} \cup \{+\}$, since the product operation for feasibility graphs can be used to simulate multiplications at the level of S_Σ, as in Corollary 16.21.

This is all very much analogous to the situation for numbers described in Sections 16.7 and 16.8. Part of the point of working with more general semirings, though, is that they can be used to represent more complicated structures than only quantity. A nice example of this is provided by the way that regular expressions can be used to represent sets, and we shall take this up next.

16.10 Feasibility of sets

Fix an alphabet Σ, and let Σ^* denote the set of all words over Σ. We can use feasibility graphs to describe subsets of Σ^*, in the following way.

Let X denote the collection of all finite subsets W of Σ^*. For each element a of Σ, we can get a unary operation R_a on X, which is defined by the (concatenation) rule

$$R_a(W) = \{wa : w \in W\}. \tag{16.50}$$

In other words, we take every element of W, and multiply it on the right by a, in order to get $R_a(W)$. Let $\mathcal{C}$ denote the set of operations on X which consists of the unary operations R_a, $a \in \Sigma$, and the binary operation of union (of sets).

Finite automata

Feasibility graphs over X, $\mathcal{C}$ are practically the same thing as finite (nondeterministic) automata with ϵ-moves and no nontrivial oriented cycles. This is easy to see, but let us be somewhat precise. For the purpose of defining normalized value functions (as in Section 7.1), we shall use the set which contains only the empty word as our "zero element" or initial element of X. (This is *not* the same as the empty set of words, which can be another natural choice of zero element of X in slightly different circumstances.)

Lemma 16.22 *Fix a feasibility graph A over X, $\mathcal{C}$, and assume that A is an IO graph (for simplicity). Let f denote the normalized value function for A. If*

v is a vertex in A, then set of words in Σ^* determined by $f(v)$ is the same as the set of words obtained by reinterpreting A as a nondeterministic automaton with ϵ-moves over Σ, with the input vertex i_A of A as the initial state of the automaton, and with v as the final state. The marking of the transitions in the automaton by letters in Σ, or by ϵ-moves, follows the designation of edges in A by unary operators in $\mathcal{C}$, or by no operation.

The empty word may be in the set of words determined by $f(v)$, or in the set of words accepted by the corresponding automaton, because of paths which do not cross any edges with unary operators or letters associated to them. This includes the path at i_A which traverses no edges.

Lemma 16.22 can be proved in the usual way, by observing that the function f as described above does indeed define a normalized value function, and then using uniqueness of normalized value functions, as in Lemma 7.2 in Section 7.1. This is also very similar to Lemma 16.20 in Section 16.9, since the language accepted by the automaton associated to A and the vertex v as above is the same as the set of words in Σ^* which can be "read" from oriented paths in A which go from i_A to v. Note that focusing branch points in A are always associated to unions, when A is viewed as a feasibility graph over X, $\mathcal{C}$, and something similar happens automatically for automata.

Regular expressions

One can also look at constructions of sets of words through feasibility graphs in terms of regular expressions. (See Section 1.1 for the definition of a regular expression.) Let Y denote the collection of all regular expressions over Σ which do not use the star operation $*$. For the moment, let us think of Y as being equipped with the binary operation of addition $+$, and the collection $\mathcal{U}$ of unary operations which represent right-multiplication by elements of Σ.

Each regular expression r in Y represents a set of words in Σ^*, as reviewed in Section 1.1. We can think of this representation as defining a mapping from Y to the set X defined above, and we have a natural correspondence between the operations on Y and those on X. Specifically, $+$ on Y corresponds to the operation of union on elements of X, and right-multiplication by a on Y corresponds to R_a on X. These operations are compatible with the mapping from Y to X, so that we have a homomorphism from $Y, \mathcal{U} \cup \{+\}$ to $X, \mathcal{C}$, in the sense of Section 7.12. If we use the regular expression ϵ as our "zero element" or initial element for Y, then this corresponds (under the mapping from Y to X) to the one for X that we specified before, i.e., the set which contains only the empty word.

Using these correspondences, we can convert feasibility graphs over $Y, \mathcal{U} \cup \{+\}$ to feasibility graphs over $X, \mathcal{C}$, and vice-versa, as in Section 7.12. The normalized value functions on these two types of feasibility graphs match up in the correct manner as well. The next lemma gives a more precise statement of this.

Lemma 16.23 *Let A be an IO graph which is also a feasibility graph over X, $\mathcal{C}$, and let f be the normalized value function on A. Let $\widetilde{A}$ be the feasibility graph* *over Y, $\mathcal{U} \cup \{+\}$ which is obtained by using the same underlying IO graph as A, but converting the operations in $\mathcal{C}$ to their counterparts in $\mathcal{U} \cup \{+\}$. Denote by $\widetilde{f}$ the normalized value function on $\widetilde{A}$. If v is any vertex in A, then the subset of Σ^* determined by $f(v)$ is the same as the one represented by the regular expression $\widetilde{f}(v)$.*

This follows from the general discussion in Section 7.12, and it can be checked more directly too.

The effect of products of regular expressions can be given by products of feasibility graphs over Y, $\mathcal{U} \cup \{+\}$, as in the next lemma.

Lemma 16.24 *Let A and B be IO graphs which are feasibility graphs over Y, $\mathcal{U} \cup \{+\}$, and let D denote the product of A and B (as in Section 16.3). Let g, h, and k denote the normalized value functions on A, B, and D, respectively, and let r, s, and t denote the regular expressions which are the values of g, h, and k at the output vertices of A, B, and D. Then*

$$t = rs \tag{16.51}$$

as elements of the free semiring S_Σ (defined in Section 16.9). In particular, t and rs define the same collections of words in Σ^*.

In other words, t and rs will be the same up to rearrangements permitted by the commutative (for addition), associative, and distributive laws, as in Section 16.9, even if they may not be given by the same expressions.

Proof Let us simply reinterpret A, B, and D as feasibility graphs over S_Σ, with the analogous operations (i.e., addition and right-multiplication by elements of Σ). The normalized value functions for this new interpretation are essentially the same as the old ones, with the regular expressions reinterpreted as elements of S_Σ. In the context of S_Σ, we do have the equation (16.51), as in Corollary 16.21. This gives the first part of Lemma 16.24. The last part, about t and rs representing the same sets of words, is an easy consequence of the fact that they are equal as elements of S_Σ. (We also mentioned this in Section 16.9, as part of the general discussion of the free semiring S_Σ.) □

Corollary 16.25 *Let z be any nonempty regular expression over Σ (which does not use the $*$ operation). Then there is an IO graph A which is a feasibility graph over Y, $\mathcal{U} \cup \{+\}$ and which has the following properties:* (1) *the size of A is bounded by a constant multiple of the size of z;* (2) *if r is the regular expression which is the value of the normalized value function of A at the output vertex of A, then r and z are the same as elements of S_Σ, and they represent the same collection of words in Σ^*.*

This is an straightforward consequence of Lemma 16.24 and the definitions, and indeed it is the same in essence as the standard conversion from regular expressions to nondeterministic finite-state automata with ϵ-moves, as discussed on p30 of [HU79].

Behavior under products

Let X, $\mathcal{C}$ be as defined at the very beginning of this section, so that X consists of finite subsets of Σ^*, and $\mathcal{C}$ consists of the binary operation of union and the unary operations R_a of concatenation on the right by a letter a in Σ, as in (16.50). Let X' denote the set of all IO graphs which are also feasibility graphs over X, $\mathcal{C}$, and let $p : X' \to X$ be as in Section 16.6, with $\xi \in X$ taken to be the set which consists of only the empty word (our choice of initial element in X). If $A \in X'$, then $p(A) \in X$ is equal to the value of the normalized value function on A at the output vertex o_A of A. Suppose that B is another IO graph which is a feasibility graph over X, $\mathcal{C}$, and let D denote the product of A and B (in the sense of Section 16.3). Thus D is again an IO graph and a feasibility graph over X, $\mathcal{C}$.

Lemma 16.26 *Notations and assumptions as above. The set of words in Σ^* given by $p(D)$ is the same as the concatenation of the sets of words given by $p(A)$ and $p(B)$, i.e.,*

$$p(D) = \{w \in \Sigma^* : w = xy,\ x \in p(A),\ y \in p(B)\}. \tag{16.52}$$

This can be seen as a consequence of Lemma 16.24, but let us indicate a direct proof, using the following.

Sublemma 16.27 *Let A be an IO graph over X, $\mathcal{C}$ which is a feasibility graph, and let f denote the normalized value function of A. Let W be any set of words in Σ^*, and let g be the value function on A such that $g(i_A) = W$, where i_A denotes the input vertex of A. If v is any vertex in A, then the set of words in Σ^* given by $g(v)$ is the same as the concatenation of W with the set of words given by $f(v)$.*

In other words,

$$g(v) = \{z \in \Sigma^* : z = wy,\ w \in W,\ y \in f(v)\}. \tag{16.53}$$

Proof (Sublemma 16.27) This is analogous to Lemmas 16.16 and 16.20, but let us review the argument. Define a function g' from vertices in A into X by taking $g'(v)$ to be the set of words in Σ^* given by the concatenation of W and $f(v)$, as in (16.53). We want to show that $g = g'$. To this end, we observe that g and g' have the same value at the input vertex of A, by construction, and that g' is a value function on A, since f is. This is not hard to check, and it implies that g and g' must be the same, since value functions are uniquely determined by their restrictions to the input vertices (as in Lemma 7.2 in Section 7.1). □

Using Sublemma 16.27, it is is easy to derive Lemma 16.26 from the definition of the product of feasibility graphs. This mirrors the discussion of products in the previous three sections. One can also see Lemma 16.26 as a reformulation of standard constructions for combining automata in order to represent concatenations of languages (as in Section 15.2).

Feasibility graphs of feasibility graphs

Let X, $\mathcal{C}$ be as before, and let X' again denote the collection of IO graphs which are feasibility graphs over X, $\mathcal{C}$. As usual, we can also employ feasibility graphs to make constructions over X', using operations corresponding to ones in $\mathcal{C}$, as well as products of graphs in X. This follows the definition of the "canonical lifting", in Definition 16.9 in Section 16.3.

As in Section 16.6, this gives a more "implicit" way of making constructions of elements of X, i.e., of sets of words in Σ^*. In effect, these more implicit constructions in X correspond to feasibility graphs over X itself, but with the extra operation of concatenation added to the others. This is because of Lemma 16.26, and it is analogous to the way that multiplications arose from feasibility graphs based on addition in Sections 16.7 and 16.9.

In this context of sets of words, we can get a more geometric view of what these different levels of implicitness mean. If W is any finite set of words in Σ^*, then a completely explicit description of W would simply be a listing of all of its elements. If W_n is the set of all words of length n in the letters a and b, $a \neq b$, then we get a more succinct representation of W_n through the regular expression

$$(a+b)^n, \tag{16.54}$$

or through finite-state automata, or feasibility graphs over X, $\mathcal{C}$. The "implicitness" in these descriptions is reflected in the fact that W_n has 2^n elements, while these descriptions can be given with size $O(n)$. For W_{2^n}, this first level of implicitness is not sufficient to make a description of less than exponential size, but this is easy to achieve if we go to the "second" level of implicitness, in which feasibility graphs are allowed to be used to construct other feasibility graphs. This is analogous to the examples discussed at the end of Section 16.9, since W_{2^n} is represented by the regular expression

$$(a+b)^{2^n}. \tag{16.55}$$

(See (16.49).) As in that situation, the regular expression (16.55) can be realized by a feasibility graph of size $O(n)$, if one allows multiplications in the feasibility graph, in addition to sums and the unary operations of multiplication by a single letter.

These two levels of implicitness reflect two different kinds of symmetry. The first type of implicitness would work just as well for any set W which can be represented by a regular expression of the form

$$(a_1+b_1)(a_2+b_2)\cdots(a_n+b_n), \tag{16.56}$$

no matter how the a_i's and b_i's behave as a function of i. This is not true for the second level of implicitness, which requires some kind of regularity among the a_i's and b_i's in a product of the form

$$(a_1+b_1)(a_2+b_2)\cdots(a_{2^n}+b_{2^n}). \tag{16.57}$$

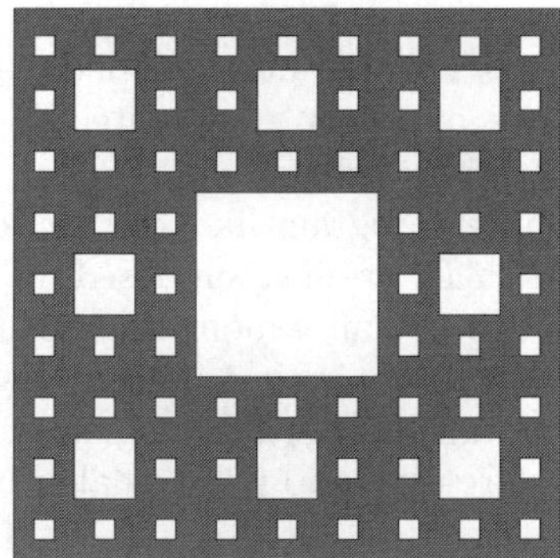

FIG. 16.1. The Sierpinski carpet

Notice that if Σ contains at least 3 elements, then the number of ways of choosing the a_i's and b_i's for $1 \leq i \leq 2^n$ is of double-exponential size in n, while the number of implicit descriptions of size k (through feasibility graphs which construct feasibility graphs, for instance) is around a single exponential in k (or, more precisely, in $k \log k$). Analogous points were discussed in Sections 7.6 and 7.9.

On the other hand, sets described by regular expressions of the form (16.56) are already very special within the broader collection $\mathcal{E}$ of subsets E of Σ^* consisting of 2^n words of length n. One can make the same kind of counting as before: the number of different regular expressions of the form (16.56) is clearly bounded by a single exponential in n, while the number of elements of $\mathcal{E}$ is of double-exponential size in n, at least if Σ contains more than 2 letters.

16.11 Visual interpretations

The idea of different levels of implicitness can be considered in very concrete and even visual ways.

In this connection, let us look at the *Sierpinski carpet*, which is the well-known fractal set pictured in Fig. 16.1. This set can be constructed as follows. One starts by taking a square (like the unit square $[0,1] \times [0,1]$) in the Euclidean plane $\mathbf{R}^2$. One then subdivides it into 9 (closed) squares of size one-third that of the original, in the usual way. One throws away the square in the center, and keeps the remaining 8 squares. In each of these, one repeats the construction, to get 8 new squares, in which one repeats the construction again, etc.

In general, there are 8^j squares of size (sidelength) 3^{-j} at the jth level of the construction. The unit square from which we started is treated as the 0th stage of the construction. The union E_j of these squares is a closed subset of the plane, and the (decreasing) intersection of all the E_j's is the Sierpinski carpet.

For making a picture like Fig. 16.1, one has to stop at a finite level of the construction, of course. How exactly might one make a copy of E_j by hand, and for reasonably large values of j?

The most naive answer is also the most "explicit" one, in which one makes each of the 8^j little squares of size 3^{-j} separately, and arranges them in the correct way. This could be pretty exhausting.

If one is using a computer program like xfig, or if one is allowed to use a photocopier (and the holes in E_j are just supposed to be shaded in black), then one can use "duplication" to get an exponential speed-up. One starts with a single little box of size 3^{-j}. One copies it 7 times to have 8 little squares, which one can then arrange to lie in a slightly larger square of size 3^{-j+1}, with all but the center of the larger square covered by the smaller ones. This gives a slightly larger box with a hole in it.

This box with the hole can again be copied 7 times to make 8 of them, and then these 8 can be arranged inside a box of size 3^{-j+2}. Again one omits the middle of the new box, and places previous sets in each of the remaining 8 locations inside the bigger box.

This gives a copy of E_2 inside of a box of size 3^{-j+2}. Again one can make copies to have 8 of them, which can then be arranged to make a copy of E_3 in a box of size 3^{-j+3}. (Figure 16.1 is in fact a copy of E_3. Note that $8^3 = 2^9 = 512$.)

Of course one can repeat this indefinitely. This kind of construction corresponds roughly to the "first level of implicitness". What about the second level?

For that, there is a concrete manifestation as follows. Suppose that one has a copy of E_j for some j already, and that one wants to make a copy of E_{2j}. To do this, one can start with a fixed copy of E_j, which has 8^j little squares of size 3^{-j}, and replace each of these little squares with a copy of E_j instead, shrunk by a factor of 3^{-j} so that it fits properly into the larger picture. This gives E_{2j}. If one repeats this, then one gets constructions of E_{2^k}, $k \in \mathbf{Z}_+$.

This type of construction uses more of the symmetry in the Sierpinski carpet. It uses the fact that the Sierpinski carpet behaves in the same way at different scales, as well as at different locations. The previous type of construction used the symmetry of the Sierpinski carpet in terms of the common behavior at different locations, but not at different scales. In fact, that kind of construction would also work for sets in which different rules or patterns are used at different scales, as long as they are used uniformly over all locations, for each individual scale.

Similar matters came up in Section 1.1, and we shall continue with this in the next section.

16.12 Codings and sets

There is a simple (and well-known) relationship between the Sierpinski carpet and the earlier story about regular expressions. Let S be a square in the plane (with sides parallel to the axes), and suppose that we cut it up into 9 subsquares, each with sidelength equal to one-third that of S. These 9 subsquares are called the "children" of S. We can assign labels to them in a natural way, to designate which subsquare lies in the middle of S, which lies in the upper-left corner, etc. These labels make sense for all squares S at once, and we shall use the same set Σ of 9 labels for all choices of S.

Now suppose that we fix a square S_0 and cut it up into 9 smaller squares, then 9^2 even smaller squares, etc., with 9^j squares at the jth generation. The size of each of the 9^j squares at the jth level is equal to 3^{-j} times the size of S_0. Let $\mathcal{S}_j$ denote the collection of these 9^j subsquares of S_0. If T lies in $\mathcal{S}_j$, then we can assign to T a word $w(T)$ in Σ^* of length j which represents the history of T in the subdivision of S_0. Namely, the first letter in $w(T)$ should specify which of the 9 children T_1 of S contains T, the second letter should specify which of the 9 children T_2 of T_1 contains T, and so on, until we have done this for all j generations. It is easy to see that this defines a one-to-one correspondence between the 9^j squares in $\mathcal{S}_j$ and the set of words in Σ^* of length j.

Let Σ_1 be the subset of Σ which consists of the 8 labels that do not correspond to middle squares. Fix j, and let $\mathcal{E}_j$ denote the collection of 8^j squares T in $\mathcal{S}_j$ such that the word $w(T)$ only involves letters from Σ_1. The union

$$\bigcup_{T \in \mathcal{E}_j} T \tag{16.58}$$

is then the same as the set E_j described in Section 16.11.

Of course, we can take any set $\mathcal{A}$ of words over Σ, and convert it into a subset of S_0 by taking the union

$$A = \bigcup_{T \in \mathcal{A}} T. \tag{16.59}$$

If $\mathcal{A}$ consists of 8^j words of length j, then this set A will have the same area in the plane as E_j does, but in general it will not be nearly as symmetric as E_j is.

Analogous codings and representations were discussed in Section 1.1, in the context of the Sierpinski gasket. As there, the symmetry and structure of the set A depends on the behavior of the set $\mathcal{A}$ of words from which it is defined. A basic point is that if $\mathcal{A}$ is described by a regular expression of the form

$$B_1 \cdot B_2 \cdot B_3 \cdots B_j, \tag{16.60}$$

where each B_j represents a sum of letters from Σ, then A will enjoy a lot of homogeneity, in the sense that the local structure of A looks the same everywhere, and at any fixed scale. The behavior at different scales will normally be completely different from each other, since there need not be any pattern to the B_i's. In the case of the sets E_j in the construction of the Sierpinski carpet, the B_i's are all the same, and this leads to the similarity in the patterns which occur at different scales.

The process that goes from E_j to E_{2j} mentioned at the end of Section 16.11 corresponds at the level of words over Σ to taking the set of words $\mathcal{E}_j$ which is associated to E_j, and then concatenating two copies of it. One can also look at this in terms of multiplying regular expressions. Related matters arose in the last part of Section 16.10.

16.13 Other operations

We have emphasized the operations of sum and product on IO graphs so far in part because of their universal applicability, in connection with lifting of operations and compositions of functions. One can of course consider other types of operations, including operations which make sense only in special settings. In this section, we shall discuss another kind of product operation on graphs, which can be used to represent intersections in the context of sets of words.

Let Σ be a finite collection of letters, and let Σ^* denote the set of all words over Σ. As in Section 16.10, we write X for the collection of all finite sets W of words in Σ^*, and we denote by $\mathcal{C}$ the collection of operations on X consisting of the binary operation of union together with the unary operations R_a, $a \in \Sigma$, which represent concatenation by the letter a on the right (as in (16.50)). For the present purposes, it is convenient to permit ourselves to take unions of several sets at once, and not just two sets at a time, and so we add to $\mathcal{C}$ union operations of arbitrary arity.

Let A and B be IO graphs which are also feasibility graphs X, $\mathcal{C}$. We want to define a kind of product feasibility graph C which represents intersections of sets. We could do this through the use of fiber products (as in Sections 15.6 and 15.7), but for the sake of concreteness, let us describe the construction directly. (There are some adjustments to be made anyway.)

We start by defining a "preliminary" graph P in the following manner. The vertices of P consist of all ordered pairs of the form (x, y), where x is a vertex in A and y is a vertex in B. Let (x, y) and (z, w) be vertices in P, and let a be an element of Σ. If there is an edge e in A that goes from x to z and an edge f in B which goes from y to w, and if e and f are both associated to the unary operator R_a on X, then we attach an edge to P that goes from (x, y) to (z, w), and which is associated to the unary operation R_a. We do this for all choices of a in Σ, and for all pairs of edges e, f in A and B, respectively.

We treat edges which are associated to no unary operator R_a in a slightly different manner. If e is an edge in A that goes from x to z and does not have any unary operator assigned to it, then we attach an edge to P that goes from (x, y) to (z, y) and is not associated to a unary operator, for every vertex y in B. Similarly, if there is an edge f in B that goes from y to w and which is not associated to any unary operation R_a, then we attach an edge to P that goes from (x, y) to (x, w) and does not have a unary operation associated to it, for every vertex x in A.

This defines P as an oriented graph. To relate this to the earlier discussion of fiber products, one can think of A and B as coming equipped with (weak) mappings into the alphabet graph over Σ, where these weak mappings collapse all of the vertices to a single point, and represent the assignments of unary operators R_a to the edges, when there is such an assignment. (In the earlier discussion, we only considered mappings between graphs, rather than *weak* mappings, but this is not a serious issue.)

This graph P is not quite the one that we want, however. Let i_A, o_A, i_B, and

o_B denote the input and output vertices of A and B, respectively, and let i_P and o_P denote the vertices (i_A, i_B) and (o_A, o_B) in P. The graph C that we want is the subgraph of P consisting of all vertices and edges which are contained in an oriented path which goes from i_P to o_P, if there are any. If not, we simply take C to consist of the vertex i_P, and no edges.

Lemma 16.28 *C is an IO graph which contains no nontrivial oriented cycles.*

Proof This is trivial if C consists only of the single vertex i_P, and so we assume that there is an oriented path in P which goes from i_P to o_P.

It is easy to see that i_P is necessarily an input vertex of C, i.e., that i_P admits no incoming edges. Indeed, i_P is in fact an input vertex of P, since i_A and i_B are input vertices of A and B. Similarly, o_P is automatically an output vertex of P, and hence of C. These are the only possible input and output vertices in C, since, by construction, every other vertex lies on an oriented path in C which goes from i_P to o_P. Thus C is an IO graph.

As to the nonexistence of nontrivial oriented cycles in C, let γ be any non-degenerate oriented path in C (or in P for that matter), and let us show that it cannot begin and end at the same vertex.

To do this, we shall use weak mappings $pr_1 : P \to A$ and $pr_2 : P \to B$, which are defined in nearly the same manner as in Section 15.6. Specifically, if (x, y) is a vertex in P, then we set $pr_1(x, y) = x$ and $pr_2(x, y) = y$. If (x, y) and (z, w) are two vertices in P, and if g is an edge in P which goes from (x, y) to (z, w), then either g is associated to edges e and f in A and B which go from x to z and from y to w, respectively, or we are in the more degenerate situation where we have only one of the edges e or f, and where either $x = z$ or $y = w$, as appropriate. In the first case, we simply set $pr_1(g) = e$ and $pr_2(g) = f$. If g is associated to an edge e in A but not to an edge f in B, then $y = w$, and we set $pr_1(g) = e$ but leave $pr_2(g)$ undefined. This is consistent with the definition of a weak mapping (Definition 10.2). Similarly, if g is associated an edge f in B but not to an edge e in A, then we set $pr_2(g) = f$, and leave $pr_1(g)$ undefined. Note that pr_1 and pr_2 preserve orientations, by construction.

Let γ_A and γ_B be the oriented paths in A and B which are the images of γ under pr_1 and pr_2, respectively. If γ begins and ends at the same vertex, then the same must be true of γ_A and γ_B. On the other hand, A and B were assumed to be feasibility graphs, and hence free of nontrivial oriented cycles. (See (7.1) in Section 7.1.) This implies that γ_A and γ_B should both be degenerate paths, i.e., paths which traverse no edges. This is impossible, since γ is assumed to be nondegenerate, and since it never happens that the weak mappings pr_1 and pr_2 are both undefined at the same edge.

This proves that P (and hence C) does not contain any nontrivial oriented cycles, and the lemma follows. □

The construction of C also permits us to view it as a feasibility graph over $X, \mathcal{C}$. This is because the edges in C (as well as P) inherit designations by unary operations R_a, or by no operation, from the corresponding designations for A

and B. (The focusing branch points are all treated as representing unions. Note that the number of incoming edges at a vertex in C can be larger than in A or B.)

Let W_A, W_B, and W_C denote the elements of X (subsets of Σ^*) whose construction is described by the feasibility graphs A, B, and C. In other words, W_A should be the value of the normalized value function on A at the output vertex o_A, and similarly for B and C. As in Section 16.10, we use the set consisting of only the empty word for the "zero element" or initial element of X needed for defining normalized value functions.

Lemma 16.29 $W_C = W_A \cap W_B$.

Proof Given an oriented path α in A, let us write $w(\alpha)$ for the word in Σ^* which is obtained by interpreting the edges traversed by α as letters in Σ. That is, we write the letter u when α traverses an edge associated to the unary operation R_u, and we write no letter when we traverse an edge which is not associated to any of the unary operations R_q. We take $w(\alpha)$ to be the empty word when α passes through no edges which are associated to a unary operation R_q.

From Lemma 16.22 in Section 16.10, we know that W_A is the same as the set of words $w(\alpha)$ which arise from oriented paths α in A that go from i_A to o_A. The analogous statements for B and C hold as well.

Let γ be an oriented path in C, and let γ_A and γ_B be its projections in A and B, as in the proof of Lemma 16.28. It is not hard to check that

$$w(\gamma) = w(\gamma_A) = w(\gamma_B), \tag{16.61}$$

just using the definitions. (The graph P was constructed so that this would happen.)

If γ begins at the input vertex of C and ends at the output vertex, then the analogous statements are true for γ_A and γ_B. This and (16.61) imply that

$$W_C \subseteq W_A \cap W_B. \tag{16.62}$$

To get the opposite inclusion we need the following.

Claim 16.30 *Let α and β be oriented paths in A and B, respectively, and assume that α begins at i_A and ends at o_A, and similarly for β. If $w(\alpha) = w(\beta)$, then there is an oriented path γ in C which goes from the input vertex to the output vertex and which has the property that $\alpha = \gamma_A$ and $\beta = \gamma_B$, where γ_A, γ_B are the paths in A and B associated to γ as above.*

This is not hard to verify from the construction of C, and we omit the details. (One has to be slightly careful about the edges for which there is no associated unary operation, but this is the only tricky point.)

From Claim 16.30 and (16.61) we get that

$$W_C \supseteq W_A \cap W_B, \tag{16.63}$$

and this completes the proof of the lemma. □

Thus the operation of intersections of sets of words can be simulated at the level of feasibility graphs over $X, \mathcal{C}$. In some situations, one might prefer to think of intersections as providing "extractions" from a given set.

By allowing intersections of sets, we have the possibility of obtaining the empty set by the end, and indeed one is sometimes more interested in determining the emptiness or nonemptiness of the final result than anything else. For instance, suppose we are working in the set $\mathcal{B}_n$ of binary strings of length n, and let $C_{i,j}$ be a collection of "1-cells" in $\mathcal{B}_n$. In other words, each $C_{i,j}$ should be defined by specifying a particular entry in the word (e.g., "the second entry is always 0", or "the fourth entry is always 1"), and leaving the rest free. The problem of deciding whether an intersection of the form

$$\bigcap_{i=1}^{k} (\bigcup_{j=1}^{l} C_{i,j}) \tag{16.64}$$

is nonempty is the same as the satisfiability problem for Boolean formulae which are in conjunctive normal form (i.e., which are conjunctions of disjunctions of Boolean variables and their negations). This problem is NP-complete, as is well-known. One can also think about this in terms of validity and provability. Compare with Chapter 3, and Section 3.6 in particular.

16.14 Simulations and conversions

We have seen a number of examples in this chapter which illustrate the way that operations on feasibility graphs can represent operations on more basic objects, such as numbers, words, sets of words, etc. We would like to look at this a slightly different way now, starting with a feasibility graph A, and transforming it into a larger graph $\mathbf{A}$ which represents an equivalent computation, but with fewer or more primitive operations.

The considerations in this section are similar to ones in previous sections, but with emphasis now going in a somewhat different direction from before. (One might compare especially with the discussions in Sections 16.5 and 16.6.)

For the sake of concreteness, we shall restrict ourselves to working with the set X of positive integers equipped with the binary operations $+, \cdot$ of ordinary addition and multiplication. We shall assume that A is an IO graph which is also a feasibility graph over $X, \{+, \cdot\}$, and we shall seek to replace it with a feasibility graph $\mathbf{A}$ over $X, +$. The basic principles involved in doing this are quite general, but this special case will serve well in indicating the key points in simple terms.

Let X' denote the collection of IO graphs which are also feasibility graphs over $X, +$. As in Section 16.3, we can think of X' as being equipped with graph-theoretic operations of (modified) sum and product, where the operation of sum on X' is the "lifting" of the operation $+$ on X (in the sense of Definition 16.9).

Let A' be the feasibility graph over X' which is obtained by reinterpreting A in the obvious way. More precisely, the feasibility structure on A simply amounts to an assignment of $+$ or $\cdot$ to each vertex in A which is a focusing branch point, and to get A', we keep the same underlying graph, and convert these assignments for A to assignments of the operations of modified sum and product on X'.

Since A' is a feasibility graph over X', it can be viewed as describing the construction of an element $\mathbf{A}$ of X', in the following manner. As usual, let us take the graph Z with one vertex and no edges as the "zero element" or initial element of X'. Once we have this, we can get a unique normalized value function f' on A', as in Lemma 7.2 in Section 7.1. Let $\mathbf{A}$ be the value of f' at the unique output vertex of A'. Thus $\mathbf{A}$ is an element of X'.

This concludes the construction of $\mathbf{A}$. The next lemma makes precise the idea that $\mathbf{A}$ and A represent equivalent constructions on X.

Lemma 16.31 *Notation and assumptions as above. Let f and $\mathbf{f}$ denote the normalized value functions on A and $\mathbf{A}$, respectively, where we use the number 1 as the initial element of X for the purpose of defining normalized value functions in both cases. If o_A and $o_{\mathbf{A}}$ denote the output vertices of A and $\mathbf{A}$, then*

$$f(o_A) = \mathbf{f}(o_{\mathbf{A}}). \tag{16.65}$$

Proof Let $p : X' \to X$ be the mapping defined as in Section 16.6, with $\xi = 1$. Thus, if B is an element of X', and hence an IO graph which is a feasibility graph over X, $+$, then $p(B)$ is defined by taking the normalized value function on B and evaluating it at the output vertex of B.

The principal point is that $p : X' \to X$ is a "homomorphism", in the sense that sums of graphs in X' correspond to sums of numbers in X, and products of graphs in X' correspond to products of numbers in X. See (16.37) and (16.38) in Section 16.7. This gives a correspondence between the computations in X' and X at each step.

We also have that the "zero elements" for X' and X match up under p, i.e., that

$$p(Z) = 1. \tag{16.66}$$

This follows from unwinding the definitions in a straightforward manner.

Let A' denote the feasibility graph over X' obtained from A as described before the statement of the lemma, and let f' be the normalized value function for A'. Thus if v is any vertex in A, then $f'(v)$ represents an element of X' (a feasibility graph over X, $+$), while $f(v)$ represents an element of X (i.e., a number). The homomorphism property for p implies that

$$p(f'(v)) = f(v) \tag{16.67}$$

for every vertex v in A. This follows from the general discussion in Section 7.12. (Specifically, one can show that $p(f'(v))$ defines a normalized value function on A, because of the analogous property for f' on A', and then conclude that $p(f'(v))$ is equal to $f(v)$ at all vertices, because of the uniqueness of normalized value functions, as in Lemma 7.2 in Section 7.1.)

We now apply (16.67) with $v = o_A$. On the other hand, $\mathbf{A} = f'(o_A)$, by definition of $\mathbf{A}$, and $p(\mathbf{A}) = \mathbf{f}(o_{\mathbf{A}})$, by definition of p. Thus $p(f'(o_A)) = \mathbf{f}(o_{\mathbf{A}})$, and (16.65) follows from this and (16.67). This proves Lemma 16.31. □

To summarize, we have shown how a feasibility graph A over X, $\{+, \cdot\}$ which is an IO graph can be effectively transformed into a feasibility graph $\mathbf{A}$ over X, $+$ which is also an IO graph, in such a way that A and $\mathbf{A}$ represent the same construction on X (in terms of the normalized value function). The choice of operations involved was not important for the basic construction, except for the fact that the operation that we wanted to simplify (multiplication of numbers in this case) could be simulated by an operation on feasibility graphs (i.e., products).

16.15 Sums and visibility graphs

There is a slightly annoying feature of the construction in Section 16.14, which is that if the original feasibility graph A over X, $\{+, \cdot\}$ did not use the operation $\cdot$ of multiplication, then the new graph $\mathbf{A}$ could still be very different from A, and indeed much larger than A, despite the fact that A itself would already satisfy the properties of $\mathbf{A}$ that we were seeking (of not using multiplications).

Roughly speaking, the construction in Section 16.14 has not only the effect of eliminating the use of multiplication in A, but also of simplifying the structure of A in a way which is similar to that of the visibility. To make this more precise, we shall consider a slightly simpler geometric formulation of the issue.

Let G be an oriented graph which is free of nontrivial oriented cycles. We can interpret G as a feasibility graph for describing constructions of IO graphs by interpreting each focusing branch point of G as representing the operation of sums for IO graphs, and each edge as representing the operation of successor (Definition 16.3). Here we really mean "sums" in the sense of Definition 16.3, rather than "modified sums" in the sense of Definition 16.6; for working with feasibility graphs we have taken modified sums as the standard (for the reasons discussed in Section 16.3), but for the moment it will be more convenient to use the original notion of sums. (We shall return to this point later in the section.)

Again we shall use the graph Z with one vertex and no edges as the "zero element" for the collection of all IO graphs.

Lemma 16.32 *Let g denote the normalized value function for G, viewed as a feasibility graph over IO graphs as in the preceding discussion. If v is any vertex in G, then $g(v)$ is isomorphic as an IO graph to the graph $B(v)$ defined by taking the negative visibility $\mathcal{V}_-(v, G)$ of G at v, and identifying all of the vertices in $\mathcal{V}_-(v, G)$ which have no incoming edges to a single point.*

Proof This is really just a modest variant of the discussion in Section 7.10. It suffices to check that $B(v)$ is a normalized value function for G, because of the uniqueness of normalized value functions Lemma 7.2 in Section 7.1. If v is an input vertex of G, then $\mathcal{V}_-(v, G)$ consists of only its basepoint, without any edges or other vertices, and this says exactly that $B(v)$ agrees with the zero element Z for the IO graphs (up to isomorphism). Thus $B(v)$ is properly "normalized", and to say that it is a value function means that $B(v)$ is obtained from the values of B at the vertices immediately before v through the rules of the feasibility graph

G when v is not an input vertex. This can easily be checked from the definitions, and we omit the details. □

If we used modified sums of IO graphs instead of ordinary sums in the interpretation of G as a feasibility graph over IO graphs, then the story would be roughly the same, except that the comparison with negative visibility graphs would be complicated by the possibility of some "stretching", through the addition of extra edges. These extra edges would not affect the total amount of branching involved, however (or the "topology" of the result). On the other hand, if some of the edges in G are not used to represent the successor operation on IO graphs, then we also have some contraction in the comparison with the visibility. The two effects cancel out when none of the edges going into focusing branch points are used to represent successor operations on IO graphs, and all other edges are used to represent successor operations.

Note that the graphs $B(v)$ which arise as in Lemma 16.32 are always *steady* graphs, in the sense of Definition 6.2 in Section 6.12. This is because the negative visibility graphs $\mathcal{V}_-(v, G)$ can contain only focusing vertices (as in Lemma 4.3 for the usual (forward) visibility graphs), so that the input vertex of $B(v)$ is the only defocusing branch point.

The evolution of the graphs $B(v)$ fits better with some of the phenomena that we saw in Chapter 6, concerning cut elimination, than the ordinary visibility does. With the graphs $B(v)$, we do not simply split apart all of the defocusing branch points, as we do for the negative visibility, but instead we push them back to the beginning, at the input vertex of $B(v)$. This has the effect of preserving the focal pairs (Definition 4.15), even if it disrupts all of the chains of focal pairs (Definition 4.17) of length at least 2. Similar matters came up in Section 7.4. In making comparisons with other situations, one might keep in mind that here the underlying feasibility graph G is an IO graph. In general, instead of identifying all of the input vertices in the negative visibility, as occurs with $B(v)$, one might identify them in groups. This can be more compatible with what happens with cut elimination, or in the context of Section 7.4, depending on the circumstances.

In connection with cut elimination, it can be better to replace ordinary sums of IO graphs with a variant of modified sums, in which extra incoming edges are attached at the input vertices, in addition to the extra outgoing edges attached at the output vertices as in Definition 16.6. (The extra incoming edges would be attached in essentially the same manner as the extra outgoing edges were before.) This prevents the number of outgoing edges at any vertex from ever becoming too large, and it is similar in spirit to the way that edges and vertices are added to the logical flow graph of a formal proof from the contractions that are used at the "bottom" of the duplication of subproofs (as in (6.5) in Section 6.2). In particular, it leads to evolutions of graphs like the one in Fig. 16.2, evolutions which produce graphs like the one called H in Section 4.3 (Fig. 4.4).

Remember that logical flow graphs are always *optical* graphs, with never more than two incoming edges or two outgoing edges at any vertex. The original operation of sums of IO graphs does not preserve the class of optical graphs,

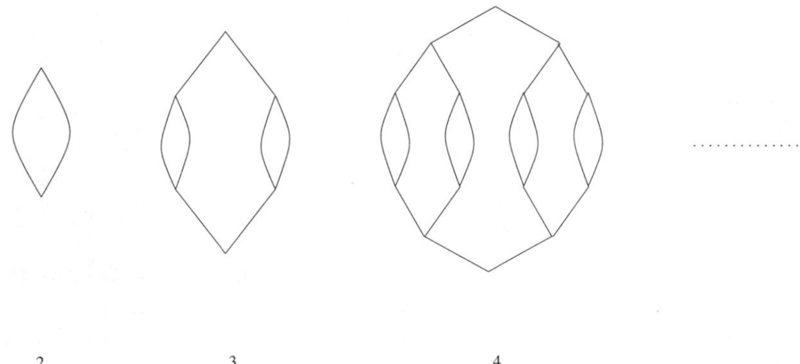

FIG. 16.2. An evolution of graphs

nor does the modified sum, but the variant of modified sums mentioned in the preceding paragraph does enjoy this property. This is not hard to verify.

Remark 16.33 If we apply the process described in Section 16.14 to a feasibility graph A over X, $\{+, \cdot\}$, and if this feasibility graph does *not* use the operation $\cdot$ of multiplication, then the geometric effects in the production of the graph **A** in Section 16.14 follow the discussion above. In particular, one gets a steady graph in the end (in the sense of Definition 6.2), with no chains of focal pairs of length at least 2. (Compare with Lemma 6.5 in Section 6.12.) If multiplications are employed by A, so that the product operation on IO graphs is used in the construction of **A**, then **A** can behave very differently, with plenty of long chains of focal pairs coming from the use of the product operation on IO graphs.

16.16 Back to formal proofs

Let us return now to the formal proof of feasibility of $2^{2^{2^n}}$ discussed in Section 16.1, and look at what happens when we simplify the cuts in order to derive an explicit construction from the proof. We shall see that this conversion behaves a lot like the ones that have arisen in this chapter in the context of feasibility graphs.

We begin by reviewing the broad structure of the proof. The basic step was given by

$$\forall x(F(x) \supset F(x^k)) \to \forall x(F(x) \supset F(x^{k^2})). \tag{16.68}$$

This can be proved in a few lines for each value of k. By combining a sequence of these proofs using cuts, one can obtain a proof of

$$\forall x(F(x) \supset F(x^2)) \to \forall x(F(x) \supset F(x^{2^{2^n}})). \tag{16.69}$$

More precisely, one combines proofs of (16.68) for $k = 2^{2^j}$, $j = 0, 1, \ldots, n-1$ in order to get a proof of (16.69). One then gets

$$\to \forall x(F(x) \supset F(x^{2^{2^n}})) \tag{16.70}$$

by combining (16.69) with a proof of

$$\to \forall x(F(x) \supset F(x^2)) \tag{16.71}$$

using a cut. Once one has (16.70), it is easy to get a proof of $\to F(2^{2^{2^n}})$ using a proof of $\to F(2)$.

The building blocks (16.68) and (16.71) both use contractions in their proofs, and the main point in deriving an explicit construction from the proof as a whole is to simplify the cuts over these contractions. The contractions involved in (16.68) and (16.71) are quite different from each other, though. The contractions used in the proof of (16.68) occur *after* the quantifiers are already present, while the contractions in the proof of (16.71) are employed *before* the quantifier $\forall$ is added. This difference turns out to play an important role.

The contraction in the proof of (16.68) comes at the last step, in deriving (16.68) from

$$\forall x(F(x) \supset F(x^k)), \forall x(F(x) \supset F(x^k)) \to \forall x(F(x) \supset F(x^{k^2})). \tag{16.72}$$

We can eliminate these contractions from the proof of $\to F(2^{2^{2^n}})$ as a whole by duplicating the supporting subproofs in a way that we have seen several times before, as in Section 3.3 (especially the part that begins around (3.42)), and in Section 4.8.

Specifically, the proof of (16.69) sketched above uses a single proof of (16.68) for each choice of $k = 2^{2^j}$, $j = 0, 1, \ldots, n-1$. The new proof would need 2^{n-1-j} copies of the proof of (16.72) when $k = 2^{2^j}$, $j = 0, 1, \ldots, n-1$. That is, it would use one proof of (16.72) when $k = 2^{2^j}$, $j = n-1$, two proofs of (16.72) when $k = 2^{2^j}$, $j = n-2$, and so on. With each step backwards in j, one doubles the number of proofs of (16.72). This reflects the ratio of two formulae to one between the left and right sides of the sequent arrow in (16.72), as opposed to the ratio of one-for-one in (16.68). One would still combine the proofs of (16.72) at successive levels through cuts, as before, but now the process of combining the proofs of (16.72) behaves like a binary tree, instead of a linear sequence.

In the end, one would not get a proof of (16.69), as before, but rather a proof of

$$\Gamma_n \to \forall x(F(x) \supset F(x^{2^{2^n}})), \tag{16.73}$$

where Γ_n consists of 2^n copies of the formula

$$\forall x(F(x) \supset F(x^2)). \tag{16.74}$$

To get a proof of

$$\to \forall x(F(x) \supset F(x^{2^{2^n}})), \tag{16.75}$$

one would combine the proof of (16.73) with 2^n proofs of $\to \forall x(F(x) \supset F(x^2))$, using 2^n applications of the cut rule.

This completes the first stage in the simplification of the proof of $\to F(2^{2^{2^n}})$. We eliminated the contractions used in the proof of (16.68), at the expense of a (single) exponential expansion in the number of lines in the proof.

In the second stage of the simplification of the proof, one goes back and eliminates all of the quantifiers from the proof, substituting explicit terms for the various occurrences of the free variable x. This is an application of the part of the standard method for cut elimination which deals with cut-formulae that have quantifiers. For instance, in the proof of $\to F(2^{2^{2^n}})$ as it now stands, there is 1 copy of (16.72) being used when $k = 2^{2^{n-1}}$, and this one copy would be replaced with

$$F(2) \supset F(2^{2^{2^{n-1}}}), F(2^{2^{2^{n-1}}}) \supset F(2^{2^{2^n}}) \to F(2) \supset F(2^{2^{2^n}}). \tag{16.76}$$

In general, each of the 2^{n-1-j} copies of (16.72) with $k = 2^{2^j}$ would be replaced with sequents of the form

$$F(t) \supset F(t^k), F(t^k) \supset F(t^{k^2}) \to F(t) \supset F(t^{k^2}), \tag{16.77}$$

where the t's are explicit terms which will be different for the different sequents. Similarly, each of the 2^n proofs of $\to \forall x(F(x) \supset F(x^2))$ will now be replaced with 2^n proofs of sequents of the form

$$\to F(s) \supset F(s^2) \tag{16.78}$$

for various explicit terms s.

Of course it is important to choose all of these terms in the right way, so that the various subproofs can still be combined with cuts to make a correct proof of $\to F(2^{2^{2^n}})$. This was the reason for choosing the terms as we did in the top level (16.76). For the next-to-top level, with $k = 2^{2^{j-1}}$, we would need two versions of (16.77), one with $t = 2$, the other with $t = 2^{2^{2^{n-1}}}$. Working backwards, one can easily choose the terms t and s in (16.77) and (16.78) so that the various subproofs fit together properly.

We should mention that, as in Section 16.1, we are ignoring some technical issues about arithmetic manipulation of terms, especially with regard to exponentials. This helps to make clearer the basic combinatorial patterns of substitutions which are encoded into the proof.

In the end, we get a proof of $\to F(2^{2^{2^n}})$ which uses cuts and contractions but not quantifiers, and for which the number of lines is exponential in n. This finishes the second stage in the simplification of the original proof. Notice that it would not have been possible to eliminate the quantifiers directly from the original proof, i.e., if we had not passed through the first stage of simplification beforehand. This is because the explicit versions (16.76) and (16.77) of (16.72) really use two different choices of terms for the variable x on the left-hand side. In the more implicit version with quantifiers, the two formulae on the left side of (16.72) are exactly the same, and hence they can be contracted together (as in (16.68)). The two formulae on the left side of the sequent arrow are no longer identical when we work with explicit terms, and we cannot contract the formulae together.

Thus we see here a concrete example of a standard phenomenon in predicate logic, which is that the use of quantifiers can lead to more "uniformity" in a formal proof, and hence to shorter proofs, through the use of contractions and cuts. Conversely, we also see how the extraction of explicit computations may entail many duplications in order to accommodate the variety of explicit terms which are implicitly being manipulated in a single formula with quantifiers.

The proof without quantifiers that we obtain from these first two stages of simplification of the original proof of $\to F(2^{2^{2^n}})$ still uses cuts and contractions in a significant way, however. This is reflected in the fact that we have a proof of the feasibility of $2^{2^{2^n}}$ which requires only exponentially-many lines (as a function of n), whereas a more direct proof (based simply on repeated multiplications) would entail a double-exponential number of lines. Indeed, the proof that we obtain from these first two stages of simplification is similar to the one described in Section 4.8, with the n there replaced with 2^n. The 2^n versions of (16.78) needed here correspond to the main steps

$$F(2^{2^j}) \to F(2^{2^{j+1}}) \tag{16.79}$$

in Section 4.8, for instance.

There are some minor differences in the details of these two proofs of feasibility, in that the current proof contains formulae of the form $F(v) \supset F(w)$, while the proof in Section 4.8 used only atomic formulae of the form $F(u)$. This is not too significant for the way that the implicit computations are being performed in the two proofs, and indeed the logical flow graphs have basically the same structure in the two situations. In each case, the logical flow graph consists of a long chain of focal pairs coming from the repeated usages of (16.78) and (16.79), with focal pairs not arising from anything else in the proof. This chain is wrapped around the present proof in a more complicated way than before, however.

The process of simplifying the original proof of $\to F(2^{2^{2^n}})$ with quantifiers that we have described here follows closely some of the earlier constructions with feasibility graphs. Let Y denote the set of positive integers, equipped with the operation $\cdot$ of multiplication. Proofs of feasibility that use cuts and contractions but not quantifier rules are roughly like feasibility graphs over Y, $\cdot$, as we have discussed before (in Chapter 7). The proof with quantifiers is more like making a feasibility graph which itself constructs feasibility graphs over Y, $\cdot$. If Y' denotes the set of feasibility graphs over Y, $\cdot$ which are IO graphs, then the first two stages of simplification of the proof with quantifiers played a role like that of "executing" a feasibility graph over Y' in order to get a more explicit feasibility graph over Y itself. At a geometric level, this is particularly true of the first stage in the simplification of the original proof with quantifiers, in which the elimination of the contractions from the proofs of (16.68) (for various values of k) behaved like the execution of the product operation for feasibility graphs in Y'. In both cases, there was roughly the same effect in terms of exponential expansion, and in terms of the geometric structures that result (i.e., in the logical flow graph of the proof without quantifiers, or in the feasibility graphs over Y, which are being constructed by the feasibility graph over Y').

The relationship between these two constructions is not surprising, since the product operation of feasibility graphs which are IO graphs encodes compositions of (unary) functions, and since the implicit representation of compositions of functions is a basic feature of formal proofs in predicate logic in general, and of these proofs of feasibility in particular. For the present proof with quantifiers, one can see the "compositions" taking place in the proofs of the building blocks (16.3), as in Section 16.1.

We have discussed these general themes before, in Sections 16.2, 16.3, and 16.4. It is helpful to look at examples like these, to see concretely and explicitly how the different steps in the two types of constructions can correspond to each other. For that matter, it is also helpful to look at examples and see what happens in them, and how they work, on their own. We have tried to include a number of examples in this chapter for these reasons.

We have focused on IO graphs and compositions of unary functions in this chapter largely because they are easier to manage, and accommodate a number of basic situations of interest, but one could just as well consider more complicated graphs and compositions, as in Remark 16.14 in Section 16.4.

17

GROUPS AND GRAPHS

A fundamental class of examples of implicit descriptions of large objects in mathematics is provided by *finitely-presented groups*. A finite presentation is given by a finite set of generators and a finite set of relations, but then the resulting group can be an infinite object which may be quite complicated. In particular, the word problem — deciding whether a particular expression in the generators represents the identity element in the group — is algorithmically undecidable for many finitely-presented groups. (See [Man77], for instance.)

If we think of a finite presentation as being an implicit description, then the complete explicit object is given by the group itself, with the underlying group law, the identity element, and the function which gives inverses of elements of the group. One can also think geometrically, in terms of the *Cayley graph* and *word metric* (discussed in Section 17.1) that the presentation defines. These objects automatically have a great deal of symmetry, because of the group invariance.

In groups, one has a particular context in which to consider problems related to working with objects and their implicit descriptions. One of the main topics of this chapter will be the notion of an *automatic group* from [ECH+92]. This provides another kind of implicit description of a group, which is more restrictive than a finite presentation, and which can be much better in terms of computational effectiveness, and in representing global geometry.

From the setting of groups, one can see notions that make sense more generally. We shall mention analogous structures for the implicit representation of graphs which do not come from groups, for instance.

17.1 Cayley graphs and the word metric

Let G be a finitely-generated group, and let S be a finite generating set for G. Thus every element of G can be expressed in terms of products of elements of S and their inverses. With this data, one can define a graph, called the *Cayley graph* associated to G and S, as follows. For the vertices, one simply takes the elements of G. Between two elements g and h of G, one attaches an edge for each generator $a \in S$ such that $g = ha$ (as in Fig. 17.1).

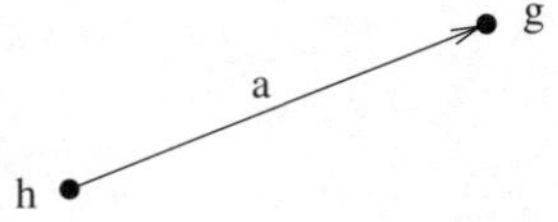

FIG. 17.1. An edge in the Cayley graph, where $g = ha$

The *word metric* $d(g_1, g_2)$ on G can be defined as the length of the shortest path between g_1 and g_2 in the Cayley graph. One can also define $d(g_1, g_2)$ as the size of the smallest word in the generators in S and their inverses which represents $g_1^{-1} g_2$, with the empty word corresponding to the identity element in G. It is easy to see that this is indeed a metric (i.e., it is nonnegative and vanishes exactly when $g_1 = g_2$, it is symmetric in g_1, g_2, and the triangle inequality holds).

Instead of thinking of the Cayley graph of G in combinatorial terms, one can also work with the corresponding 1-dimensional topological space, in which the edges are topologically equivalent to intervals in the real line. One can define a natural metric on this topological space too, in such a way that each edge is isometrically equivalent to the standard unit interval $[0, 1]$. (For this we should forbid the use of the identity element in the generating set, to prevent edges from having both endpoints at the same vertex.) One could still determine distances in this topological version of the Cayley graph by taking the lengths of shortest paths, but now the paths would be allowed to begin and end inside edges, and not just at their endpoints.

If one thinks in terms of representing elements of G by words, then this metric need not be at all easy to compute. Just knowing when two words represent the same element can be algorithmically undecidable in the context of finitely-*presented* groups, and this is the same as deciding when the distance between two words is equal to 0. Keep in mind that a group can easily be finitely-generated but not finitely-presented. (See [Man77] for the related topic of *recursive* groups.)

Note that the Cayley graph is homogeneous. That is, if we let G act on itself by *left* multiplication, then this action extends to an action on the Cayley graph. It preserves distances as defined by the word metric.

Free groups provide an interesting class of examples. The Cayley graph of a free group with its standard generators is a homogeneous tree. At the opposite end are the free abelian groups $\mathbf{Z}^k$, for which the Cayley graph can be seen as a lattice in $\mathbf{R}^k$. (See Fig. 17.2 for the $k = 2$ case of each.)

If G is generated by a set S with n elements, then it is a simple and well-known fact that G can be realized as a quotient of the free group F_n on n generators. The quotient mapping simply takes elements of F_n and reinterprets them as elements of G by replacing the standard generators for F_n with the given generators for G. Of course there can be many cancellations in this mapping, coming from the relations of G.

This homomorphism from F_n to G also induces a mapping from the Cayley graph of F_n onto the Cayley graph of G. One can again view this as an aspect of simplicity or symmetry of G and its Cayley graph, since the Cayley graph of F_n is just a tree and the mapping comes from a group homomorphism. However, one should be careful here; quotients in mathematics can be very subtle. That is part of the point that we would like to explore, in the context of combinatorial geometry.

The theme of groups as geometric objects is one that has been especially emphasized in recent years in work of M. Gromov. Related references include

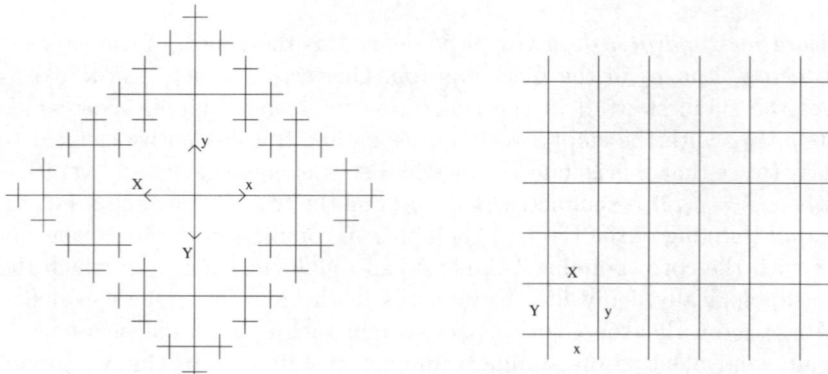

FIG. 17.2. This picture shows the Cayley graphs of the free group and free abelian group, each with two (standard) generators x, y. If X and Y denote the inverses of x and y, then one can think of them as indicating directions opposite to those of x and y. In the Cayley graph of the free group, on the left, the directions indicated by x, y, X, and Y lead to separate parts of the graph, which are not connected to each other except through the point at which one starts. In the free abelian group, one can go in the direction x, then the direction y, then the direction X, and then the direction Y, and end up at the place where one started, because $xyXY$ is equal to the identity element in the group.

[Gro81a, Gro84, Gro87, Gro93, Gd90, ECH+92, Far92, FM98].

17.2 Pause for some definitions

When one thinks about groups and mappings between, them one might typically think of homomorphisms, isomorphisms, etc. Once one has the word metric, one can also look at geometric conditions on mappings. Let us take a moment to review some definitions concerning mappings between metric spaces.

Let $(M, d(x,y))$ and $(N, \rho(u,v))$ be two metric spaces. For the record, a *metric* on a set M is a nonnegative function on $M \times M$ which is symmetric, vanishes exactly on the diagonal, and satisfies the triangle inequality.

Let f be a mapping from M to N. One says that f is *Lipschitz* if there is a constant $C > 0$ so that

$$\rho(f(x), f(y)) \leq C\, d(x,y) \tag{17.1}$$

for all $x, y \in M$. In other words, the Lipschitz condition asks that distances not be increased by the mapping by more than a bounded factor.

One calls f *bilipschitz* if there is a constant $C > 0$ so that

$$C^{-1}\, d(x,y) \leq \rho(f(x), f(y)) \leq C\, d(x,y) \tag{17.2}$$

for all $x, y \in M$. Thus distances should neither be expanded nor contracted by f by more than a bounded factor for this condition.

Two metric spaces M and N are *bilipschitz equivalent* if there is a bilipschitz mapping from M *onto* N. The inverse of the mapping is also bilipschitz in this case. This provides a way to say that two metric spaces are roughly equivalent even if the distances are not quite the same. A basic example for this notion is provided by the word metric for finitely-generated groups. The precise values of the word metric depends on the choice of a finite generating set S, but any two such sets S, S' lead to metrics which are bounded above and below by constant multiples of each other. This is a well-known observation, and it is not hard to check, using the fact that an element of one generating set can be expressed as a finite word in the elements of the other generating set. One can reinterpret this as saying that the identity mapping on the group is a bilipschitz mapping with respect to the two metrics.

In short, while the word metric itself depends on the choice of the generating set S, the approximate geometry does not.

Sometimes it is useful to work with a weaker condition, as follows. A mapping $f : M \to N$ is called a *quasi-isometry* if there is a constant $C > 0$ so that

$$C^{-1}\, d(x,y) - C \leq \rho(f(x), f(y)) \leq C\, d(x,y) + C \tag{17.3}$$

for all $x, y \in M$. This means that f distorts distances by only a bounded factor at large scales, but at small scales this condition tolerates more confusion. One says that $f : M \to N$ is a *quasi-isometric equivalence* if it is a quasi-isometry, and if there is a constant $C' > 0$ so that for every $v \in N$ there is an $x \in M$ such that $\rho(v, f(x)) \leq C'$.

Bilipschitz equivalences are always bijections, while quasi-isometric equivalences need not be either injective or surjective. The latter do behave like bijections at large scales, in the sense that the discrepancy from being injective or surjective is bounded.

For instance, suppose that G is a finitely-generated group for which we have two different generating sets S_1 and S_2. The corresponding word metrics on G are bilipschitz-equivalent, but if we take the topological version of the Cayley graphs (with edges included as point sets in their own right, as discussed in Section 17.1), then they may not be. The Cayley graphs will always be quasi-isometrically equivalent to each other, however, and also to G itself (as one can easily verify). This reflects the fact that G and its Cayley graphs look roughly the same at large scales, even if they are very different locally (since G is discrete and its Cayley graphs are not). See Fig. 17.3 for a simple example.

Another class of examples of quasi-isometric equivalences is provided by finite-index subgroups of finitely-generated groups G. An isomorphism between two finitely-generated groups defines a bilipschitz equivalence with respect to the word metrics, while isomorphisms between finite-index subgroups lead to quasi-isometric equivalences between the original groups.

The standard inclusion of $\mathbf{Z}^n$ inside $\mathbf{R}^n$ is a quasi-isometric equivalence (with respect to the Euclidean metrics). There are numerous other situations where discrete groups sit inside continuous groups in a similar manner, and with a lot

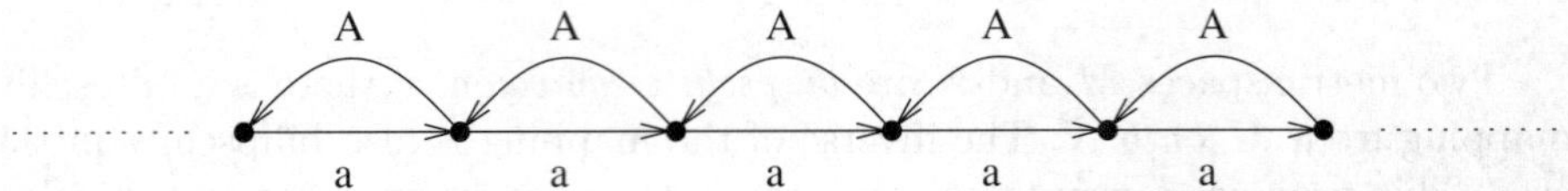

FIG. 17.3. For the group **Z** of integers, suppose that a is the standard generator (i.e., 1), and A is its inverse. If one defines the Cayley graph of **Z** using both a and A as separate elements of a generating set, then the graph looks like the one shown in the picture. If one only uses a, then the Cayley graph looks like an ordinary line.

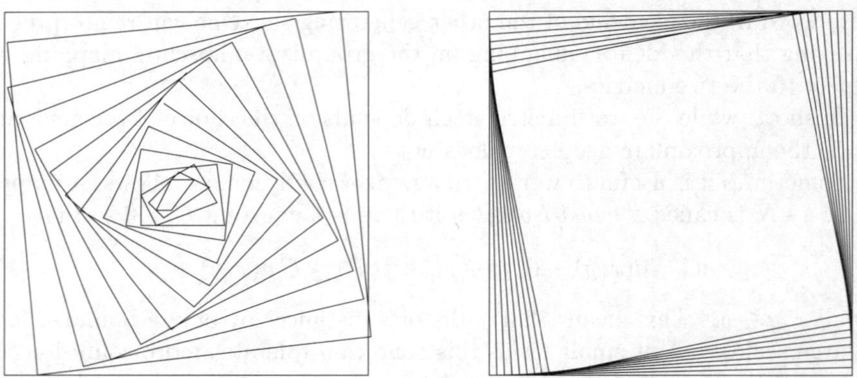

FIG. 17.4. Vertigo: bilipschitz mappings can entail substantial spiralling, even on the Euclidean plane

of interesting structure related to this. See [Rag72, GP91], for instance, and the references therein.

In general, bilipschitz and quasi-isometric mappings are free to twist around in strong ways, even on a relatively simple space like the Euclidean plane. (This is illustrated in Fig. 17.4.) Mappings that come from isomorphisms between groups are much more rigid than that. In the case of $\mathbf{R}^n$, they would simply correspond to *linear* mappings.

There are situations in which it is known that bilipschitz or quasi-isometric equivalence between finitely-generated groups implies something like isomorphic equivalence of subgroups of finite index, however (and perhaps special properties of the original mappings). The study of finitely-generated groups up to quasi-isometry was initiated by Gromov [Gro84, Gro93]. See [FM98] for some recent results and references to the literature. Some related topics are discussed in [GP91].

17.3 The Heisenberg groups

Let us look now at a specific family of examples, the *Heisenberg groups*. These groups have the interesting feature that they are very simple to define (algorithmically as well as mathematically), while the geometry that they induce is remarkably subtle, and indeed quite strange in comparison with Euclidean geometry and with many self-similar fractals (of which the continuous version of the Heisenberg group can be viewed as an example). We begin with their basic properties in this section, and discuss their geometry in Section 17.4.

Like $\mathbf{R}^n$ and $\mathbf{Z}^n$, these groups have natural continuous versions as well as discrete versions, and for the definition, we begin with the former.

In its continuous version, the Heisenberg group can be described as follows. Let n be an arbitrary positive integer. As a set, we take H_n to be $\mathbf{R}^n \times \mathbf{R}^n \times \mathbf{R}$. We write x and y for elements of $\mathbf{R}^n$, and t for an element of $\mathbf{R}$. The group operation is given by

$$(x, y, t) \cdot (x', y', t') = \Big(x + x',\, y + y',\, t + t' + \sum_{j=1}^{n} (y_j\, x'_j - x_j\, y'_j)\Big). \tag{17.4}$$

One can check that this defines a noncommutative group which is *nilpotent.* Elements of the form $(0, 0, t)$, $t \in \mathbf{R}$, lie in the center of H_n, and the quotient of H_n by its center is just $\mathbf{R}^n \times \mathbf{R}^n$ with the standard abelian structure (given by ordinary addition of the components).

If (x, y, t) is an element of H_n, then its inverse is given by $(-x, -y, -t)$.

Define $H_n(\mathbf{Z})$ by

$$H_n(\mathbf{Z}) = \{(x, y, t) \in H_n : x, y \in \mathbf{Z}^n, t \in \mathbf{Z}\}. \tag{17.5}$$

It is easy to see that this defines a subgroup of H_n. Also set

$$\widetilde{H}_n(\mathbf{Z}) = \Big\{(x, y, t) \in H_n(\mathbf{Z}) : t \equiv \sum_{j=1}^{n} x_j\, y_j \bmod 2\Big\}. \tag{17.6}$$

This defines a subgroup of $H_n(\mathbf{Z})$. Indeed, the inverse of an element of $\widetilde{H}_n(\mathbf{Z})$ clearly lies in $\widetilde{H}_n(\mathbf{Z})$. To show that $\widetilde{H}_n(\mathbf{Z})$ is closed under the group operation, one should check that

$$t + t' + \sum_{j=1}^{n} (y_j\, x'_j - x_j\, y'_j) \equiv \sum_{j=1}^{n} (x_j + x'_j)\,(y_j + y'_j) \mod 2 \tag{17.7}$$

when $(x, y, t), (x', y', t') \in \widetilde{H}_n(\mathbf{Z})$. This is easy to do. (Keep in mind that the difference between $+$ and $-$ does not matter when one is working modulo 2.)

The group element $(0, 0, 1)$ lies in $H_n(\mathbf{Z})$ but not in $\widetilde{H}_n(\mathbf{Z})$. On the other hand, every element of $H_n(\mathbf{Z})$ is either in $\widetilde{H}_n(\mathbf{Z})$ or can be written as a product (in the group structure) of an element of $\widetilde{H}_n(\mathbf{Z})$ and $(0, 0, 1)$. Thus $\widetilde{H}_n(\mathbf{Z})$ is a subgroup of $H_n(\mathbf{Z})$ of index 2.

The group $\widetilde{H}_n(\mathbf{Z})$ will be a bit more convenient for us as a discrete version of the Heisenberg group than $H_n(\mathbf{Z})$. In this regard, let us write down a set of generators for $\widetilde{H}_n(\mathbf{Z})$, and the relations that they satisfy. Let $e_1, \ldots, e_n$ denote the standard basis vectors in $\mathbf{R}^n$, so that e_j has jth component equal to 1 and all other components equal to 0. Define $g_1, \ldots, g_n, h_1, \ldots, h_n$ in $\widetilde{H}_n(\mathbf{Z})$ by $g_i = (e_i, 0, 0)$, $h_j = (0, e_j, 0)$, where the first and second entries in these ordered triples refer to elements of $\mathbf{R}^n$, reflecting the fact that H_n is $\mathbf{R}^n \times \mathbf{R}^n \times \mathbf{R}$ as a set. From the definition of the group law, we have that

$$g_i g_j = g_j g_i, \;\; h_i h_j = h_j h_i, \;\; g_i h_j = h_j g_i \tag{17.8}$$

for all i, j with $i \neq j$ (where this last condition is a serious restriction only for the third equation, with g_i and h_j together). When $i = j$, we have that

$$h_i g_i = g_i h_i \zeta, \qquad \zeta = (0, 0, 2). \tag{17.9}$$

We also have that ζ commutes with all of the g_i's and h_j's. Using these properties, it is easy to see that the g_i's and the h_j's generate $\widetilde{H}_n(\mathbf{Z})$; every element of $\widetilde{H}_n(\mathbf{Z})$ can be realized as a product of powers of the g_i's, times a product of powers of the h_j's, times a power of ζ. To define a set of relations just in terms of the g_i's and h_j's, one can take the relations indicated in (17.8), together with ones that assert that the commutators $h_k g_k h_k^{-1} g_k^{-1}$ are all equal to each other, $1 \leq k \leq n$, and commute with all of the g_i's and h_j's. (The latter is a consequence of the former and (17.8) if $n > 1$.) It is not hard to show that this gives a presentation for $\widetilde{H}_n(\mathbf{Z})$, i.e., it is a complete set of relations.

One often starts with a presentation like this to define the discrete version of the Heisenberg group. Let us also mention that there are slightly different but equivalent ways to define the continuous version of the Heisenberg group. The definition above has some convenient features, including the fact that the inverse of (x, y, t) is given by $(-x, -y, -t)$.

If $n = 1$, then $\widetilde{H}_1(\mathbf{Z})$ is generated by $g = (1, 0, 0)$ and $h = (0, 1, 0)$. If we write $\zeta = (0, 0, 2)$, then every element of $\widetilde{H}_n(\mathbf{Z})$ can be written as $g^j h^k \zeta^l$ for integers j, k, l, where all triples (j, k, l) in $\mathbf{Z}^3$ occur exactly once. Using these "coordinates", the Cayley graph of $\widetilde{H}_1(\mathbf{Z})$ is indicated in Fig. 17.5. Note that we are now using g, h, and ζ as a set of generators, rather than just g and h. (This gives a simpler picture, and a simpler representation of the vertices.)

Let us come back now to the continuous version of the Heisenberg groups, and return to the discrete groups later. A basic feature of H_n is that it admits a natural family of *dilations*. On Euclidean spaces, one has the dilations defined by multiplication by positive real numbers, and these define group homomorphisms. For the Heisenberg group H_n, one can define the dilations $\delta_r : H_n \to H_n$ for $r \in \mathbf{R}$, $r > 0$, by

$$\delta_r(x, y, t) = (r\, x, r\, y, r^2\, t). \tag{17.10}$$

These dilations preserve the group structure (17.4), as one can easily check. The r^2 in the t-component here is very important, to match with the terms of degree 2 in x, y, x', y' in the t-component of the right side of (17.4).

If r is a positive integer, then δ_r maps $H_n(\mathbf{Z})$ and $\widetilde{H}_n(\mathbf{Z})$ into themselves. This is analogous to the situation with $\mathbf{R}^n$ and $\mathbf{Z}^n$, and ordinary dilations there.

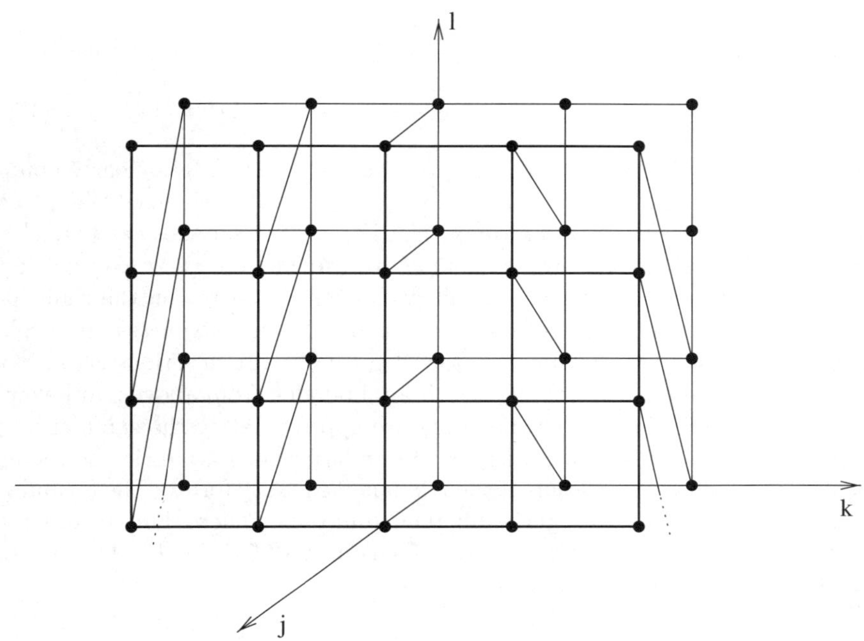

FIG. 17.5. The Cayley graph for the Heisenberg group $\widetilde{H}_1(\mathbf{Z})$. Redrawn with the permission of Birkhäuser from the work *Sur les Groupes Hyperboliques d'après Mikhael Gromov*, Etienne Ghys and Pierre de la Harpe, editors.

This additional symmetry of dilations plays an important role in the geometry of the Heisenberg groups. In this regard, let us define a "homogeneous norm"

$$\|(x, y, t)\| = (|x|^2 + |y|^2 + |t|)^{\frac{1}{2}} \tag{17.11}$$

on H_n. Here we write $|x|$ for the usual Euclidean norm on $\mathbf{R}^n$, and one can think of (17.11) as being analogous to it on H_n. The homogeneous norm has the property that

$$\|\delta_r(x, y, t)\| = r\,\|(x, y, t)\| \tag{17.12}$$

for all $(x, y, t) \in H_n$ and $r > 0$, just as in the Euclidean case. This was the reason for using $|t|$ in (17.11), rather than $|t|^2$, as in a Euclidean norm. Similarly,

$$\|(-x, -y, -t)\| = \|(x, y, t)\| \tag{17.13}$$

for all $(x, y, t) \in H_n$, i.e., the norm of a group element is the same as the norm of its inverse. It is not hard to show that there is a constant $C > 0$ so that

$$\|(x, y, t) \cdot (x', y', t')\| \leq C(\|(x, y, t)\| + \|(x', y', t')\|) \tag{17.14}$$

for all $(x, y, t), (x', y', t') \in H_n$. This is like the usual subadditivity property for quasinorms on vector spaces. (We shall say more about this in a moment.)

Next we set

$$d((x, y, t), (x', y', t')) = \|(-x, -y, -t) \cdot (x', y', t')\|. \tag{17.15}$$

This is symmetric in (x, y, t), (x', y', t'), because of (17.13). It is clearly nonnegative, and it vanishes exactly when $(x, y, t) = (x', y', t')$. We would like to say that $d(\cdot, \cdot)$ actually defines a *metric* on H_n, i.e., that it satisfies the triangle inequality. The subadditivity property in (17.14) provides only a weaker version of the triangle inequality, in which one allows a constant factor on the right-hand side. With this property, one often says that $d(\cdot, \cdot)$ is a *quasimetric*, and this is adequate for some purposes. In this case, this is not really a serious issue, in the sense that there are metrics on H_n which are bounded from above and below by constant multiples of $d(\cdot, \cdot)$, so that they are approximately the same as $d(\cdot, \cdot)$. See [KR85] (Section 1.F) for an explicit formula for such a metric (of a similar type as above), and a derivation of the triangle inequality for it. (The formula for the group law in [KR85] is slightly different from the one here, but the difference does not cause serious trouble. Also, the focus in [KR85] is on H_1, but this part extends to H_n for all n without difficulty.)

If $u, v \in H_n$ and $r > 0$, then

$$d(\delta_r(u), \delta_r(v)) = r\, d(u, v). \tag{17.16}$$

This is again analogous to the situation for Euclidean geometry. Also, we have the *left invariance* property that

$$d(w \cdot u, w \cdot v) = d(u, v) \tag{17.17}$$

for all $u, v, w \in H_n$. This follows easily from the definitions (and since the inverse of $w \cdot u$ is the same as the product of the inverses of w and u, but in the *opposite* order).

As in the case of $\mathbf{R}^n$ and $\mathbf{Z}^n$, there is a constant C_n so that every element (x, y, t) of H_n lies at distance $\leq C_n$ of an element of $\widetilde{H}_n(\mathbf{Z})$. This is not hard to check. (One should be a bit careful, as the combination of the integer parts of x, y, and t may not give a good approximation to (x, y, t) in H_n.) Thus the embedding of $\widetilde{H}_n(\mathbf{Z})$ inside H_n is a quasi-isometric equivalence with respect to $d(\cdot, \cdot)$ as defined above.

On the other hand, using the system of generators for $\widetilde{H}_n(\mathbf{Z})$ discussed earlier, we get a word metric for $\widetilde{H}_n(\mathbf{Z})$, as in Section 17.1. It turns out that there is a constant $C_n' > 0$ so that distance from any element $u \in \widetilde{H}_n(\mathbf{Z})$ to the identity element $(0, 0, 0)$ in the word metric is bounded by $C_n' \|u\|$. This is not hard to check, but one should be a bit careful. The main point is that if $t \in \mathbf{Z}$, then the distance from $(0, 0, t)$ to $(0, 0, 0)$ in the word metric is bounded by a constant multiple of $\sqrt{|t|}$ (rather than $|t|$, as one might get more naively). This can be derived from the group law and simple calculations. Once one has this, one can get the desired bound for all elements of $\widetilde{H}_n(\mathbf{Z})$ without much trouble.

This inequality implies that if $u, v \in \widetilde{H}_n(\mathbf{Z})$, then the distance between u and v as measured by the word metric is at most $C'_n\, d(u,v)$. This is because one can reduce to the case where $v = (0,0,0)$, using the invariance of both distance functions under left translations, and when $v = (0,0,0)$, the inequality is the same as the one in the preceding paragraph.

If fact, we also have that $d(\cdot,\cdot)$ is bounded from *above* by a constant multiple of the word metric on $\widetilde{H}_n(\mathbf{Z})$. To understand why this is true, let us observe first that

$$d(u, u \cdot f) \leq 1 \tag{17.18}$$

for all $u \in \widetilde{H}_n(\mathbf{Z})$ and all generators f, i.e., for $f = g_j$ or h_j, $j = 1, \ldots, n$. Indeed, by left invariance we have that $d(u, u \cdot f) = d(0, f)$, where 0 means the identity element $(0,0,0)$, and it is easy to see that $d(0,f) = 1$ for each generator f. Now, if we knew that $d(\cdot,\cdot)$ were an actual metric, and satisfied the triangle inequality, then we would be able to use (17.18) to conclude immediately that $d(u,v)$ is less than or equal to the distance between u and v in the word metric for all $u, v \in \widetilde{H}_n(\mathbf{Z})$. As it is, the triangle inequality for $d(\cdot,\cdot)$ does not directly work well enough for this, but because $d(\cdot,\cdot)$ is bounded from above and below by constant multiples of a true metric (as mentioned above), one can use the same reasoning (applied to that metric) to obtain that $d(u,v)$ is bounded by a constant multiple of the distance between u and v in the word metric on $\widetilde{H}_n(\mathbf{Z})$, for all $u, v \in \widetilde{H}_n(\mathbf{Z})$. (One could also make more direct computations.)

In summary, $d(\cdot,\cdot)$ and the word metric on $\widetilde{H}_n(\mathbf{Z})$ are each bounded by constant multiples of the other. Thus they each define approximately the same geometry on $\widetilde{H}_n(\mathbf{Z})$. Notice that the analogous assertions hold for $\mathbf{Z}^n$ inside $\mathbf{R}^n$, i.e., the Euclidean metric and the word metric on $\mathbf{Z}^n$ are each bounded by constant multiples of the other. This is easy to see.

We shall discuss the geometry of the Heisenberg groups further in the next section. One of the main points is that it is already very subtle by itself, even if we forget about the underlying group structure.

17.4 Geometry of Heisenberg groups

How do the Heisenberg groups look geometrically?

A basic point is to estimate the *volume growth.* Fix a positive integer n, and let $V_n(r)$ denote the number of elements of $\widetilde{H}_n(\mathbf{Z})$ which lie within distance r of the identity element with respect to the word metric. Then there is a constant $C_n > 0$ so that

$$C_n^{-1}\, r^{2n+2} \leq V_n(r) \leq C_n\, r^{2n+2} \tag{17.19}$$

for all $r \geq 1$. To see this, one can use the fact that the word metric and $d(\cdot,\cdot)$ from (17.15) are each bounded by constant multiples of the other. Because of this, it suffices to show that the number of elements u of $\widetilde{H}_n(\mathbf{Z})$ which satisfy $||u|| \leq t$ is bounded above and below by constant multiples of t^{2n+2} when $t \geq 1$, where $||\cdot||$ is defined by (17.11). This assertion is immediate from (17.11).

In general, a finitely-generated group G is said to have *polynomial growth* if there exist constants C, d so that the number of elements of G which lie within distance r of the indentity element (with respect to the word metric) is less than or equal to $C\,r^d$ for all $r \geq 1$. This property does not depend on the choice of the generators, but the constant C can change if one changes the generating set. (The power d would remain the same.) A celebrated theorem of Gromov [Gro81a] states that a finitely-generated group G has polynomial growth only if it is *virtually nilpotent*, i.e., has a nilpotent subgroup of finite index. The converse was known before, and the Heisenberg groups provide basic examples.

There is something a bit strange about the volume growth here. On the one hand, H_n is a topological group (meaning that the group operations are continuous functions), and in fact a Lie group, whose underlying topological space (or manifold in this case) is $\mathbf{R}^{2n+1}$. On the other hand, the rate of growth of the discrete group $\widetilde{H}_n(\mathbf{Z})$ makes it look like it has dimension $2n+2$. What happened? Why do we have the gap?

In this regard, let us look for an analogue of (17.19) in the continuous setting. Instead of counting the number of elements in the discrete group $\widetilde{H}_n(\mathbf{Z})$ inside a given ball, one can compute the *measure* of the ball in H_n. One has to be careful about what measure to use on H_n, but in this case there is a simple answer. As a set, we can think of H_n as being $\mathbf{R}^n \times \mathbf{R}^n \times \mathbf{R}$, on which we have ordinary Lebesgue measure, which we denote by μ. (One could just as well use Riemann integrals and volumes for subsets of $\mathbf{R}^n \times \mathbf{R}^n \times \mathbf{R}$ here.) In fact, Lebesgue measure is *left-invariant* with respect to the Heisenberg group operation. To be precise, define $\lambda_u : H_n \to H_n$ for $u \in H_n$ by

$$\lambda_u(v) = u \cdot v, \tag{17.20}$$

using the Heisenberg group structure on the right side of the equation. Then μ is invariant under λ_u for all $u \in H_n$, i.e.,

$$\mu(\lambda_u(E)) = \mu(E) \tag{17.21}$$

for all measurable subsets of H_n. This can be verified by direct calculation, using the change of variables formula from vector calculus. The main point is that the Jacobian of λ_u is equal to 1 everywhere, because the nonlinearities in λ_u contribute only to parts of the matrix of the differential which lie on one side of the diagonal.

The left-invariance of Lebesgue measure for the Heisenberg group makes it a natural measure to use, and there are general results to the effect that Lebesgue measure is uniquely determined (up to a scale factor) by the left-invariance property (17.21), under reasonable assumptions. For Lebesgue measure, there is a simple version of (17.19), namely

$$\mu(\{u \in H_n : d(u,v) < r\}) = \mu(\{u \in H_n : d(u,0) < 1\}) \cdot r^{2n+2} \tag{17.22}$$

for all $v \in H_n$ and $r > 0$, where $d(\cdot,\cdot)$ is as in (17.15). In other words, the volume of *any* ball in H_n is equal to the volume of the unit ball times r^{2n+2}. To see this, we can argue by symmetry considerations. The left-invariance of μ permits us to reduce to the case where $v = 0$. There is also a natural scaling property for dilations, namely

$$\mu(\delta_r(E)) = r^{2n+2}\,\mu(E) \tag{17.23}$$

for all measurable subsets of H_n and all $r > 0$, where $\delta_r(w)$, $w \in H_n$, is defined as in (17.10). This is easy to check using (17.10). Once we have this, we can reduce (17.22) to the case where $r = 1$. In summary, we can use translations and dilations to reduce (17.22) to the case where $v = 0$ and $r = 1$, for which it is trivial.

If instead we used the ordinary Euclidean metric in (17.22), then we would have a similar identity, but with r^{2n+2} replaced with r^{2n+1}. Thus the power in the formula for the volume for the balls is the same as the topological dimension of the underlying space in the case of Euclidean geometry, while for the Heisenberg geometry we have r^{2n+2}, and the power is 1 greater than the topological dimension. This power $2n + 2$ is also manifested in the concept of *Hausdorff dimension.* (See [Fal90] for the definition.)

Does this mean that we chose the distance function $d(\cdot,\cdot)$ in a bad way? To the contrary, it is practically unavoidable. This measurement of distance enjoys three basic properties. The first is that it is compatible with the usual Euclidean *topology* that underlies H_n. The second is that it is invariant under left translations on H_n, as in (17.17). Third, it scales in the natural way using the dilations $\delta_r : H_n \to H_n$, as in (17.16). It is not hard to check that these three properties determine $d(\cdot,\cdot)$ up to size, i.e., any other candidate which satisfies the same conditions would be bounded from above and below by constant multiples of $d(\cdot,\cdot)$. We also know from the end of Section 17.3 that $d(\cdot,\cdot)$ is equivalent in size to the word metric on $\widetilde{H}_n(\mathbf{Z})$.

This discrepancy between the topological and "geometric" dimensions make the Heisenberg group (with this geometry) a quintessential example of a *fractal* [Fal90]. As such, it has some very special features, however. To see this, it is helpful to look at *Lipschitz functions* from $(H_n, d(\cdot,\cdot))$ into the real line (with its standard metric). Before we do this, let us first review the situation for ordinary Euclidean spaces.

If $f : \mathbf{R}^n \to \mathbf{R}$ is Lipschitz (as in Section 17.2), then f is actually *differentiable* (in the usual sense of calculus) at almost every point in $\mathbf{R}^n$ (with respect to Lebesgue measure). This is a famous fact from real analysis [Fed69, Sem99b, Ste70]. In other words, if one "zooms in" on the behavior of f at a point p in $\mathbf{R}^n$, as though looking through a microscope with increasing levels of magnification, then for most points p, the pictures that appear will begin to look the same after a while, and in fact they will become more and more linear. This is not at all apparent from the definition of the Lipschitz condition, which only seems to guarantee that the pictures remain bounded.

It turns out that similar assertions hold for $(H_n, d(\cdot,\cdot))$. In this case, the "zooming in" process uses the group structure of H_n, namely left translations to slide the microscope over to a given point, and then dilations to play the role of increasing the magnification of the microscope. For a given Lipschitz function $g : H_n \to \mathbf{R}$, it is again true that at almost any point in H_n (with respect to Lebesgue measure), the pictures of g that one sees from the microscope are all approximately the same after a while, and in the limit the picture will be that of a group homomorphism from H_n into $\mathbf{R}$.

See [Pan89b] for a precise formulation of this statement, and also a proof. There are analogous results for Lipschitz mappings between Heisenberg groups, and for mappings between other nilpotent Lie groups as well.

For metric spaces in general, this type of phenomenon does not work, even when the underlying metric space enjoys strong self-similarity properties. Some aspects of this are discussed in [Sem99b]. Very recently, new results pertaining to differentiability of functions on metric spaces have been obtained. See [Che99].

Here is another manifestation of the special nature of the geometry associated to the Heisenberg group.

Proposition 17.1 *For each pair of positive integers n, N, there does* not *exist a bilipschitz mapping from $(H_n, d(\cdot,\cdot))$ into $\mathbf{R}^N$ equipped with the standard Euclidean metric.*

By contrast, the familiar fractals that one sees in the literature (as in [Fal90]) typically live in a Euclidean space by construction.

This proposition was known to Assouad and perhaps to others as well. One can derive it from the differentiation theorem for Lipschitz functions on the Heisenberg group, as in [Pan89b], using also the noncommutativity of the Heisenberg group. See [Sem96] for more details.

Of course, one should not expect just any metric space to admit a bilipschitz embedding into a finite-dimensional Euclidean space. One needs to have some kind of *finite-dimensionality* condition, even for only a topological embedding [HW41]. For bilipschitz embeddings, one needs a more "quantitative" version of finite-dimensionality. If M is a metric space which admits a bilipschitz embedding into some $\mathbf{R}^N$, then M must be *doubling*, which means that each ball B in M should admit a covering by a *bounded* number of balls of half the radius (with a bound that does not depend on the initial choice of B). (See Definition 19.6 in Section 19.4.) This property is satisfied by $(H_n, d(\cdot,\cdot))$, and in fact one can use dilations and left-translations to reduce the question to the single case of the unit ball, which can be handled directly.

Thus the doubling condition is not sufficient for the existence of a bilipschitz embedding from a given metric space into some $\mathbf{R}^N$. See [Ass77, Ass79, Ass83] for some positive results about the existence of certain types of well-behaved embeddings of arbitrary metric spaces which are doubling into finite-dimensional Euclidean spaces. We shall recall a basic result in Section 19.8.

Although this type of invariant geometry for the Heisenberg group (and other nilpotent groups) is quite distinct from that of Euclidean spaces, as in Proposition 17.1 above, spaces like the Heisenberg group nonetheless manage to behave a lot like Euclidean spaces in other respects. For instance, they enjoy much the same kind of "Sobolev" and "Poincaré" inequalities, for controlling the average behavior of functions in terms of the average behavior of their "first derivatives", as one has on Euclidean spaces. The idea of these inequalities can be formulated as soon as one has a metric space equipped with a reasonable measure; one does not need actual "derivatives", or a "gradient" with a vector attached to it, as on Euclidean spaces. Instead one can work with more direct measurements of small-scale or infinitesimal oscillation. In analysis, one can often control the oscillations of functions at large scales through some kind of "integration" of the oscillations at small scales. This is true for ordinary Euclidean geometry, and it also turns out to work for the Heisenberg groups and other nilpotent groups equipped with their natural invariant geometries. See [Hei95] and [Ste93] concerning these and other aspects of analysis related to the Heisenberg group.

This kind of "calculus" does not work on arbitrary metric spaces, but requires substantial structure of a special nature. Until recently, the situations in which it was known to work involved geometry which was either approximately Euclidean or connected to the invariant geometry of nilpotent groups. Striking new examples have been found by Bourdon and Pajot [BP99] and by Laakso [Laa98]. The Hausdorff dimensions of these examples do not have to be integers, and in fact all real numbers greater than or equal to 1 can occur. (These are the all the dimensions that would have a possibility to work.) See [Bou97, Che99, HK96, HK98, HS97, Sem99a] for related information.

In conclusion, the geometry of the Heisenberg group is very interesting and remarkable in a way, and this is true purely at the level of the distance function, without the underlying group structure.

See [DS97] for other aspects of fractal geometry related to symmetry and rigidity.

We have so far emphasized the geometric behavior of $(H_n, d(\cdot,\cdot))$, but what about $\widetilde{H}_n(\mathbf{Z})$? The basic answer is that *all* of the geometry of H_n is encoded in $\widetilde{H}_n(\mathbf{Z})$. If H_n is like the surface of the Earth and we fly high above it in an airplane, then $\widetilde{H}_n(\mathbf{Z})$ and H_n look almost the same. This is easier to understand in the abelian case, i.e., $\mathbf{Z}^n$ and $\mathbf{R}^n$ look almost alike at large scales. To make this more precise, one can proceed as follows. If one understands $(\widetilde{H}_n(\mathbf{Z}), d(\cdot,\cdot))$, then one should know about $(\widetilde{H}_n(\mathbf{Z}), \epsilon \cdot d(\cdot,\cdot))$ for all $\epsilon > 0$. That is, we simply multiply the distance function by a small parameter ϵ, which is like flying high above it in an airplane and looking down (so that houses, cars, and people appear to be very small). As ϵ tends to 0, one can recover the continuous space $(H_n, d(\cdot,\cdot))$. For instance, $(\widetilde{H}_n(\mathbf{Z}), \epsilon \cdot d(\cdot,\cdot))$ is isometrically equivalent to $\delta_\epsilon(\widetilde{H}_n(\mathbf{Z}))$ as a subset of $(H_n, d(\cdot,\cdot))$, because of (17.16). This subset becomes more and more dense in H_n as ϵ tends to 0. One can use this to make precise the idea that $(H_n, d(\cdot,\cdot))$ can be recovered from $(\widetilde{H}_n(\mathbf{Z}), \epsilon \cdot d(\cdot,\cdot))$.

These principles are illustrated by the proof of the following result.

Proposition 17.2 *For each pair of positive integers n, N there does* not *exist a bilipschitz mapping from $\widetilde{H}_n(\mathbf{Z})$ (equipped with the word metric) into $\mathbf{R}^N$ (equipped with the standard Euclidean metric).*

Proof Suppose to the contrary that $f : \widetilde{H}_n(\mathbf{Z}) \to \mathbf{R}^N$ is bilipschitz. We already know that the word metric on $\widetilde{H}_n(\mathbf{Z})$ is bounded from above and below by constant multiples of $d(\cdot,\cdot)$, and so we can think of f as being bilipschitz as a mapping from $(\widetilde{H}_n(\mathbf{Z}), d(\cdot,\cdot))$ into $\mathbf{R}^N$ (with the usual Euclidean metric).

We may assume that f maps the identity element of $\widetilde{H}_n(\mathbf{Z})$ to the origin in $\mathbf{R}^N$, because we can always arrange this to be true by subtracting a constant from f, which would not affect the bilipschitz property.

We can use f to define a family of mappings $f_\epsilon : \delta_\epsilon(\widetilde{H}_n(\mathbf{Z})) \to \mathbf{R}^N$. Namely, we set

$$f_\epsilon(u) = \epsilon \cdot f(\delta_\epsilon^{-1}(u)) \tag{17.24}$$

for each $u \in \delta_\epsilon(\widetilde{H}_n(\mathbf{Z}))$. Each f_ϵ is bilipschitz, as a mapping from $\delta_\epsilon(\widetilde{H}_n(\mathbf{Z}))$ equipped with $d(\cdot,\cdot)$ into $\mathbf{R}^N$ with the standard metric, and a key point is that the f_ϵ's are bilipschitz *with a constant that does not depend on* ϵ. This comes from the way that we made the rescaling, which permits us to use the same constant in the bilipschitz condition for f_ϵ as for f. (See (17.16), and note that one has an analogous property for ordinary dilations and distances on $\mathbf{R}^N$.)

The next point is to say that there exists a sequence $\{\epsilon_j\}$ of positive numbers with $\epsilon_j \to 0$ as $j \to \infty$ such that the corresponding sequence of f_ϵ's converges "uniformly on compact subsets" to a mapping $g : H_n \to \mathbf{R}^N$. This can be obtained from the well known Arzela–Ascoli argument for compactness of equicontinuous families of functions, but one has to be a little more careful than usual because the f_ϵ's are not all defined on the same set. The same type of reasoning applies nonetheless, and in the limit one obtains a mapping g defined on all of H_n, because of the increasing thickness of the subsets $\delta_\epsilon(\widetilde{H}_n(\mathbf{Z}))$ of H_n as $\epsilon \to 0$. (Convergence arguments of this type are commonly used, and are discussed in some detail in [DS97], for instance.)

The last important fact is that g is actually *bilipschitz* as a mapping from $(H_n, d(\cdot,\cdot))$ into $\mathbf{R}^N$. This comes from the uniform bound on the bilipschitz constants of the f_ϵ's, and it is not hard to check. From here we get a contradiction with Proposition 17.1, and Proposition 17.2 follows. □

A conclusion of all of this is that one should think of $\widetilde{H}_n(\mathbf{Z})$ equipped with the word metric as being at least as subtle as a metric space as $(H_n, d(\cdot,\cdot))$. In particular, the Cayley graph of $\widetilde{H}_n(\mathbf{Z})$ must contain quite subtle information, since the word metric on $\widetilde{H}_n(\mathbf{Z})$ can be obtained from the Cayley graph. The $n = 1$ case is already very interesting.

17.5 Automatic groups

A finite presentation for a group implicitly defines the group. It does *not* necessarily provide a good way to list the elements without repetitions. One can represent all the elements as words, but in general there is no algorithm to tell when two words represent the same group element.

In recent years, a more effective notion of *automatic groups* has emerged [ECH$^+$92]. In [Far92], Farb describes how one of the basic ideas behind automatic groups comes from the paper [Can84] of Cannon, about the possibility to "see" Cayley graphs of groups by taking a ball around the identity element (with respect to the word metric) and looking at the way that copies of this ball should be pieced together as one goes out to infinity.

In this regard, notice that the Cayley graph of a finitely-generated group "looks" the same at every point, in the sense that any vertex in the Cayley graph can be moved to any other vertex through an automorphism of the graph, coming from the natural action of the group on its own Cayley graph by left multiplication. However, as one moves around the group, this "frame of reference" may be turning in a significant way. This happens in the case of the Heisenberg group, for instance.

Part of Cannon's idea is to exploit this homogeneity of the Cayley graph to unwrap a local picture of the group (around the identity element) into a global one. A difficulty with this is that shortcuts might suddenly emerge because of the relations in the group; one can follow these local charts for a while, but then suddenly find oneself back at the identity element for global reasons that were not evident from the individual steps. In general, this can be quite tricky, because of the algorithmic unsolvability of the word problem. In automatic groups, this possibility is controlled by something called the "k-fellow traveller property" [ECH$^+$92, Far92]. (We shall return to this in Section 17.10).

In rough terms, the definition of an *automatic structure* for groups works as follows. One starts with a generating set S of the group G, and one treats this set together with the set of inverses of elements in S as an alphabet Σ. (More generally, one can use any set of semigroup generators for the group.) One specifies a regular language L over Σ, i.e., a language defined by a finite state automaton, as in Section 14.1. One assumes that every element of G can be represented by a word in L, but still there may be more than one such representation. Thus one also asks for a finite state automata which generates all ordered pairs of words in L which represent the same element of G. Finally, one asks for finite state automata which generate ordered pairs of elements of L for which the corresponding elements of G are related by multiplication by an element of Σ on the right. (That is, there would be such an automaton for each element of Σ.) These automata describe the local geometry of the Cayley graph at the level of the language L.

See [ECH$^+$92, Far92] for more details about the definition of an automatic structure. An important point is that the two words in an ordered pair are supposed to be treated "synchronously", so that for every letter added to one of the words there is a letter added to the other at the same time. In terms of formatting, one should think of these ordered pair of words as being words over an alphabet whose individual elements are themselves ordered pairs of *letters* from the original alphabet. To deal with pairs of words of different length, one adds to the original alphabet a special "padding" symbol \$ to fill out the shorter word at the end. (See [ECH+92, Far92].)

One of the motivations behind the concept of automatic structures is to be able to deal with the word problem effectively for a wide class of groups of interest. See Theorem 2.3.10 on p50 of [ECH+92] for a quadratic-time algorithm for solving the word problem in groups with automatic structures.

An *automatic group* is simply a group which admits an automatic structure. Among the examples are Finitely-generated abelian groups, fundamental groups of compact Riemannian manifolds of negative curvature, and, more generally, groups which are hyperbolic in Gromov's sense [Gro87]. Nilpotent groups are *not* automatic except when they contain abelian subgroups of finite index. In particular, the Heisenberg groups $\widetilde{H}_n(\mathbf{Z})$ are *not* automatic. (See [ECH+92].)

The failure of automaticity of the Heisenberg groups is very interesting, since they are otherwise quite simple algorithmically. For instance, it is easy to make normal forms for words which represent elements of the Heisenberg group, and to understand how to compose them effectively. Let us be more explicit. The Heisenberg group $\widetilde{H}_n(\mathbf{Z})$ can be generated by elements $g_1, \ldots, g_n, h_1, \ldots, h_n$, with the rules that the g_i's commute with each other, the h_j's commute with each other, g_i commutes with h_j when $i \neq j$, and finally that the commutators $g_i\, h_i\, g_i^{-1}\, h_i^{-1}$ all coincide with a single group element t which commutes with all of the g_i's and h_j's. (See Section 17.3.) Every element in $\widetilde{H}_n(\mathbf{Z})$ can be written in a unique way as

$$g_1^{\alpha_1}\, g_2^{\alpha_2} \cdots g_n^{\alpha_n}\, h_1^{\beta_1}\, h_2^{\beta_2} \cdots h_n^{\beta_n}\, t^{\gamma}, \tag{17.25}$$

where the α_i's, β_j's, and γ are integers. Group multiplication for two such representations is easily computed (simply by adjusting the powers of t to fit the implicit commutations), but the way that increments in the α's and β's produce increments in γ is not compatible with the requirements of the automata. (We shall discuss this further in Section 18.2.)

Of course we also know that the Heisenberg groups are quite subtle geometrically even if they are very simple algorithmically. The nonexistence of automatic structures for the Heisenberg groups (and other nilpotent groups) can be analyzed in geometric terms, through "isoperimetric functions". See [ECH+92] for the first Heisenberg group $\widetilde{H}_1(\mathbf{Z})$, and [Gro93] (Section 5.D, especially parts (5)(e) and (5)(f)) and [Bur96] for the Heisenberg groups in general, and in connection with other nilpotent groups.

17.6 Automatic structures for graphs

Cayley graphs of finitely-generated groups can be quite interesting for purely geometric reasons, apart from the group structure which underlies them. The Heisenberg groups $\widetilde{H}_n(\mathbf{Z})$ are remarkable simply as *metric spaces*, for instance, with the metric induced from the lengths of paths in the Cayley graph.

The notion of an automatic structure is interesting already for graphs, without regard to anything like a group multiplication. The basic idea still applies; one can try to "see" the global structure of a graph implicitly through some kind of machines which create local models and the rules for gluing them together.

One can follow one's nose and extend the notion of automatic groups to graphs in a simple way. To define an "automatic structure for a graph", we start with a regular language L to represent the vertices of the graph. We can allow this representation to have repetitions, as in the representation of group elements through words, in which case we would also like to have a regular language which lists the ordered pairs of representatives in L which denote the same vertex. Finally, we ask for a regular language to decide which ordered pairs of words represent adjacent vertices (including an orientation, if there is one). As in Section 17.5, "ordered pairs of words" should be formatted here as words made up of ordered pairs of letters from the alphabet Σ over which L is defined. To accommodate words of different length, one should use a padding symbol \$ in addition to the letters in Σ, as in [ECH+92, Far92].

What kinds of graphs would this allow? Cayley graphs of automatic groups are certainly included, but one can have automatic structures for graphs which are less homogeneous. Remember that homogeneous trees arise as Cayley graphs of free groups; for automatic graphs one can easily make non-homogeneous examples. Similarly, one can follow the example of free abelian groups and make graphs that follow a kind of rectilinear structure even if it is not quite homogeneous. (See Fig. 17.6 for some illustrations of graphs with some regularity, but not as much homogeneity as Cayley graphs. Part of Fig. 17.6 was inspired by a picture in [Gro93], but in a different role.)

One can formulate some ideas of this type into general mechanisms. If one has already an automatic structure for some graph, like a Cayley graph, then one can get new examples by looking at regular languages which are contained in the one that describes the vertices. This enables one to break the symmetry in certain ways.

Keep in mind also that it makes sense to think about automatic structures for *finite* graphs. The point would be to have automata which are relatively small compared to the graph itself.

Note that the concept of graphs is very close to that of metric spaces. Every (connected) graph defines a metric on its vertex set, by taking the length of the shortest path which connects a given pair of vertices, and, conversely, there are natural ways to associate graphs to metric spaces, as we shall discuss in Chapter 19. Thus one can also see the idea of automatic structures in much broader geometric terms, and this point will be present in a number of the considerations of this chapter.

Let us turn now to some examples.

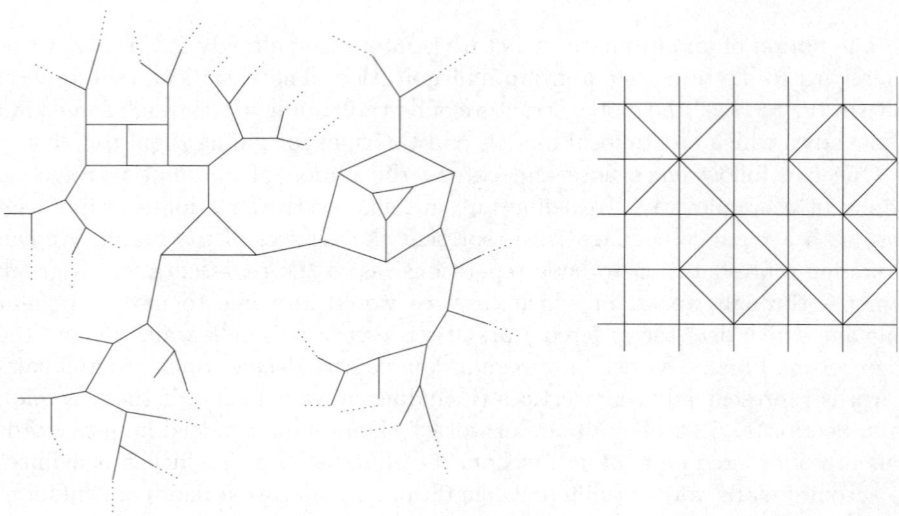

FIG. 17.6. Illustrations of graphs roughly like Cayley graphs, but less symmetric

The configuration graph of a Turing machine

Given a Turing machine M (deterministic or nondeterministic), consider its *configuration graph* as defined on p148 of [Pap94]. In short the *vertices* of the configuration graph represent the possible *configurations* (or "snapshots") of the machine, while the *edges* are oriented and represent the *transitions* that the machine allows to occur. To be precise, if M has k tapes, then a *configuration* of M is represented by a $2k+1$-tuple $(q, w_1, u_1, \ldots, w_k, u_k)$, where q denotes the current state of the machine, and the w_i's and u_i's combine to represent the strings currently found on the k tapes together with the positions of their cursors. Specifically, w_i represents the string to the *left* of the cursor on the ith tape, including the symbol being scanned by the cursor, while u_i is the string to the *right* of the cursor, and which may be empty.

We attach an (oriented) edge from a vertex labelled by the configuration

$$C = (q, w_1, u_1, \ldots, w_k, u_k) \tag{17.26}$$

to the vertex described by $C' = (q', w_1', u_1', \ldots, w_k', u_k')$ exactly when the Turing machine makes a transition from the former to the latter. This possibility is governed by a function δ associated to the Turing machine. Specifically, δ takes in the state of the machine and the symbols currently being scanned by the k cursors, and it responds by telling the machine how to change its state (or perhaps to halt), how to change the symbols being scanned, and where to move the cursors, if at all (by a single step to the left or right).

The set of possible configurations of the machine can easily be realized as a regular language. To make this precise, we should specify how these ordered

$2k+1$-tuples are represented as words. It is not unreasonable to simply concatenate the components of an ordered $2k+1$-tuple into a single word, with a special symbol added to the alphabet to mark the separation between successive components. (One might consider other methods as well.)

For the *transitions* between configurations we should be more careful. We can think of these as being represented by ordered pairs of words, where the individual words represent configurations as in the preceding paragraph. These ordered pairs should be formatted not by concatenation, but "synchronously", as words made up of ordered pairs of letters. This follows the discussions in Section 17.5 and near the beginning of this section. (As before, a special padding symbol \$ should be used to equalize the lengths of the words in an ordered pair.)

It is not very difficult to check that the collection of ordered pairs of words which represent the transitions of the Turing machine M can be realized as a *regular* language. We shall not go through this in detail, but let us mention some of the main points. Since the class of regular languages is closed under finite unions, one can restrict oneself to transitions which correspond to a particular state of the machine, and to particular symbols which are being read. This leads to a division of cases, according to how the cursors are being moved (if they are being moved), and whether the machine moves to a part of a tape that it has not visited before (so that one of the constituent strings becomes longer). The latter possibility can cause two words that represent "adjacent" configurations of the machine to become out of phase with each other. However, this happens only in a very controlled way, and one can verify that it is not incompatible with the property of being a regular language. (More precisely, in the construction of elements in this language, one may, at a given step, need to know what happened in some of the previous steps. However, the amount of memory that one needs remains *bounded*.)

To say that the configuration graph of the Turing machine M admits an automatic structure merely reflects the fact that the individual steps in the execution of M are very simple, with "bounded distortion". The large-scale structure of the configuration graph is another matter, and will be at least as complicated as the execution of M as a whole. In particular, the existence of an oriented path between a given pair of vertices may depend on the possibility of moving very far away from the given vertices (as compared to the sizes of the words which represent them).

See also the discussion of the "reachability method" in Section 7.3 of [Pap94], beginning on p146.

Unrestricted grammars and recursively enumerable languages

Fix a finite alphabet Σ, and let Σ^* be the set of all finite words over Σ. This will be the set of *vertices* for our new graph. Suppose that we have a finite collection of *production rules* of the form $\alpha \to \beta$, where α and β are words over Σ, and α is nonempty. These production rules provide a way to transform some strings in Σ^* to other strings; given words γ and δ, we permit ourselves to transform $\gamma\alpha\delta$

into $\gamma\beta\delta$.

We use these transformations to define *edges* in our graph. That is, we attach an oriented edge from a vertex labelled by $\gamma\alpha\delta$ to the vertex labelled by $\gamma\beta\delta$ whenever $\alpha \to \beta$ is a production rule and γ, δ are arbitrary words over Σ.

This again defines a graph with an automatic structure. The vertices of the graph are words in a regular language by definition (since we take all of Σ^*), while the treatment of the edges is very similar to the case of the configuration graph of a Turing machine. The present setting is a bit simpler, but one has the same phenomenon of words associated to adjacent vertices becoming slightly out of phase when α and β as above have different lengths.

An oriented path in this graph represents the transformation of one word in Σ^* to another through successive applications of the production rules. Normally, one specifies also a special *starting symbol* S from which to begin, and a subset of Σ of "terminal" letters (as opposed to "variables"). The *language generated* by the given grammar (as on p220 of [HU79]) is then the set of all words which involve only terminal letters in Σ and which can be reached from S by repeated use of the production rules.

Note that a language can be generated by an unrestricted grammar if and only if it is recursively enumerable. See Theorems 9.3 and 9.4 on p221-2 of [HU79].

"Linear" graphs

We can also obtain examples of graphs with automatic structures from some standard constructions for self-similar fractals. Before we do that, let us consider the simpler case of *linear graphs*. Fix a nonnegative integer n, and imagine first the finite linear graph L_n, in which the vertices are labelled by the integers $0, 1, 2, \ldots, n$, and we have exactly one edge from vertex j to vertex $j+1$ and no others. It is a simple matter to give this an automatic structure in the sense above. We can use an alphabet Σ with a single letter a, so that the vertices are represented by words involving only the letter a and having length at most $n+1$, and it is easy to characterize the pairs of words which represent adjacent vertices.

This is sort of trivial since the graph is finite, but there are some important points here. The first is that we would still get an automatic structure if we used the *infinite* version L_∞ of this graph. This reflects the fact that we can use *cycles* in automata which describe infinite regular languages.

A second point is that for n finite, we should not really use *unary* representations of the vertices, but *binary* representations instead (or some other base larger than 1). For simplicity of exposition, let us assume that n is of the form $2^k - 1$ for some positive integer k, so that each of the numbers $0, 1, 2, \ldots, n$ has a unique representation by a binary string of length equal to k. (See Fig. 17.7 for a picture, with $k = 3$.) The edges between these vertices can be characterized in terms of binary representations as follows. If u is a binary string (of length k) which ends in 0, then we have an edge from the vertex marked by u to the one marked by u', where u' is the same as u except that the final 0 is replaced with

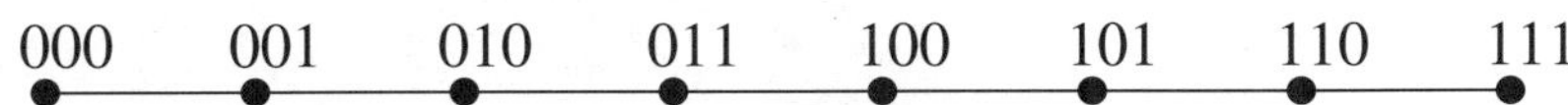

FIG. 17.7. Binary labels for linear graphs

1. More generally, if u is of the form $v01^j$, then we attach an edge from u to $v10^j$. The remaining possibility is that u consists of all 1's, in which case there are no outgoing edges at u.

Using these rules, one can make an automatic structure for L_n which is based on automata of size $O(k)$, rather than $O(n)$, as before. This is not too hard to show. (For the automaton for pairs of adjacent vertices, one should be a bit careful to re-use different parts of the transition graph for the different cases (different j's), to get size $O(k)$ rather than $O(k^2)$. One can have a section of the transition graph which generates two copies of arbitrary binary strings, and another which generates pairs with 1's in the first coordinate, and 0's in the second. One would then have transitions between the two sections, to accommodate the different j's, and to put in the 0 in the first coordinate and the 1 in the second at the moment of the transition.) This illustrates the way that automatic structures can contain nontrivial information even for finite graphs.

In terms of metric spaces, it is nice to think of L_n as a discrete approximation the ordinary unit interval [0,1], but with the Euclidean metric rescaled so that the individual steps have size 1. This fits better with the next example.

The Sierpinski gasket

The *Sierpinski gasket* (Fig. 17.8) is a well-known self-similar fractal set in the plane. This set was described in Section 1.1, but let us review the construction. One starts with a fixed equilateral triangle T. One can divide T into four equilateral triangles, each of which has sides of half the size of the sides of T. Three of these four triangles lie in the corners of T, while the fourth lies in the middle. We forget about the middle triangle, and keep only the other three, which we call T_U, T_L, and T_R. The labels U, L, and R refer to the "upper", "leftmost", and "rightmost" corners of T. To these three triangles we repeat the same process, decomposing them into four, throwing away the one in the center, and keeping only the three in the corners. Each of the three new triangles that we get lies in either the "upper", "leftmost", or "rightmost" corner of the original. We can repeat the process indefinitely, obtaining in the limit a certain (fractal) set in the plane. For the present purposes, we shall use only *finitely* many levels of the construction, to obtain interesting examples of graphs.

At the jth level of the construction, we get a collection $\mathcal{T}_j$ of 3^j triangles. (We allow j to be 0 here, in which case $\mathcal{T}_j$ consists only of T.) Each of these 3^j triangles has its bottom edge parallel to that of T, and can be obtained from our original triangle T by a combination of a translation and a dilation by a factor of 2^{-j}. This is not hard to check. We can code the elements of $\mathcal{T}_j$ by words of length j over the alphabet $\{U, L, R\}$, where the jth letter in the word describes the placement of the given triangle inside its "parent" triangle in $\mathcal{T}_{j-1}$,

FIG. 17.8. The Sierpinski gasket

the second-to-last letter describes the location of the parent triangle inside the grandparent, etc.

Let G_n be the graph whose vertices are the vertices of all the triangles in $\mathcal{T}_n$, and whose edges are the sides of all the triangles in $\mathcal{T}_n$. A nice point about this graph is that it is connected, and that it enjoys strong *uniform* local connectedness properties. Any pair of vertices in G_n can be connected in G_n by a path whose length is bounded by a universal constant times the Euclidean distance between the vertices. For this we measure the lengths of paths in terms of the Euclidean lengths of the edges, which is 2^{-n} times the sidelength of T for the edges in G_n, rather than simply counting each edge as having length 1, as in the word metric for a finitely-generated group. (One could always make a rescaling in the plane so that the edges in G_n do have length 1, though.) This fact about paths is not too difficult to establish. Another nice point is that G_n is "contained" in G_{n+1} for each n, in the sense that every vertex in G_n is also a vertex in G_{n+1}, and every edge in G_n is the union of two edges in G_{n+1}, as segments in the plane. This can be checked through induction, using the way that the boundary of T is contained in the union of the boundaries of its children.

How might we describe this graph algorithmically? We begin by coding the vertices of G_n by words of length $n + 1$ over the alphabet $\{U, L, R\}$, with the first n letters used to determine a triangle S in $\mathcal{T}_n$, and the last letter to specify a vertex in S.

This coding is not unique, but the redundancies can be analyzed as follows. Suppose that we have two different words w_1 and w_2 of length $n + 1$ which describe the same point. Let v be the largest common initial subword of w_1 and w_2. Let j be the length of v, which may be 0, but which is less than or equal to n. This word v determines a triangle R in $\mathcal{T}_j$, and the vertices corresponding to w_1 and w_2 must lie within R (on the boundary or in the interior). Let S_1 and S_2 denote the children of R in $\mathcal{T}_{j+1}$ determined by the letters in the $(j + 1)$th entries in w_1 and w_2. These entries are distinct, because we chose j to be as large as possible. Now, the self-similarity of the construction implies that the position of S_1 and S_2 inside R looks exactly like the arrangement of two of the children of T inside T. In particular, S_1 and S_2 always have disjoint interiors, and their perimeters intersect in exactly one point, a vertex. The vertex of G_n described by w_1 and w_2 must be the same as this common vertex in S_1 and S_2, because w_1 and w_2 must each lie within S_1 and S_2, respectively (including the perimeters of these triangles).

Suppose for instance that S_1 is the upper child of R, and S_2 is the leftmost child of R. Then the $(j+1)$th entries of w_1 and w_2 must be U and L, respectively. Since w_1 and w_2 both represent the common vertex of S_1 and S_2, we must have that the kth entry of w_1 equals L for all $k > j+1$, and that the kth entry of w_1 equals U for all $k > j+1$. In other words, w_1 and w_2 must be of the form vUL^i and vLU^i, where $i = n - j$.

The other cases are similar. One can check that there are never more than *two* distinct codes which determine the same vertex. In particular, the numbers of vertices of G_n is on the order of 3^n.

The main conclusion now is that the ordered pairs of words which represent the same vertex can be described through an automaton of size $O(n)$. This is analogous to the previous situation for the linear graphs L_n. The adjacency problem can be handled in nearly the same manner. Indeed, every edge in G_n comes from an edge in a triangle in $\mathcal{T}_n$. Thus, if two vertices in G_n are adjacent, then they admit representations by words of length $n+1$ which agree in the first n entries and then disagree in exactly the last one. To generate all pairs of words which determine adjacent vertices, one must also take the nonuniqueness into account, and this can be achieved as above.

In short, the graphs G_n admit exponentially-concise implicit descriptions by automatic structures. This is similar to the use of binary representations for the linear graphs L_n.

A fractal tree

Another example is provided by the *fractal tree* shown in Fig. 17.9. In this case, we start with a square in the plane, break it up into 9 pieces of equal size, and then keep the four in the corners and the one in the middle. We repeat the rule as before, and in the limit one can get a fractal set which is a kind of fuzzy tree.

We can also stop at a finite level n of approximation, in which our construction produces a collection of 5^n squares with disjoint interiors, each having sidelength equal to 3^{-n} times that of the original. One can make a graph out of the union of the perimeters of these squares in the same manner as before. These graphs are a bit simpler than the earlier ones, because they behave like a tree except for the finest level of resolution. In any case, one can make codings based on strings using an alphabet with 5 letters, and get an efficient implicit description through automatic structures as before.

One could instead consider the graphs obtained by using the squares at a given level as vertices, with an edge between two vertices when they correspond to squares which intersect. In this case, this graph would be a tree. One could also do this in the setting of the Sierpinski gasket.

FIG. 17.9. A fractal tree

These examples of self-similar sets are quite standard, and they illustrate nicely how something like an automatic structure for graphs can behave in a good way in situations where there may be a lot of symmetry, even if one is not working with a group structure.

One could also consider the Sierpinski carpet, as in Section 16.11. This case is slightly more complicated, because the neighboring squares can have whole edges in common, rather than just vertices. Still, the basic principles apply, so that one can again code vertices in the corresponding graphs by words which represent the positions of squares, and one can analyze when two words represent the same or adjacent vertices in much the same way as before.

There are plenty of other sets along these lines that one could consider as well.

Comparison with semi-Markovian spaces

The preceding discussions of fractal examples are very close in spirit to the notion of "semi-Markovian spaces" introduced by Gromov [Gro87]. This notion applies to *topological* spaces, and (roughly speaking) it asks that a given space be realizable as a quotient of a subset of a Cantor set by a fairly "homogeneous" equivalence relation.

In this context, a Cantor set should be viewed as the space of *infinite* words over a finite alphabet Σ (with at least two elements). This set comes with the standard product topology, which can be described as follows. Let u be any infinite word over Σ, and let j be a positive integer. From these data, one can form the set $N_j(u)$ of all infinite words v over Σ which agree with u at the first j entries. (These sets are sometimes called *cells*.) For fixed u, these sets define a complete system of neighborhoods at u, and an arbitrary set of infinite words is considered to be open if it can be realized as a union of sets of the form $N_j(u)$. It is well-known (and not hard to prove) that the topological spaces produced in this manner are all homeomorphic to each other, and to the standard "middle-thirds" Cantor set in the real line. (We encountered a similar situation in Section 5.5.)

To realize a topological space X as a semi-Markovian space, the first step is to find a way to represent the elements of X as infinite words over an alphabet Σ. For the Sierpinski carpet, for example, every element can be described by an infinite word over the alphabet $\{U, L, R\}$. For the unit interval $[0, 1]$ in the real line, every element can be represented by a binary expansion.

In both cases, there are elements of the topological space which are represented by more than one infinite word, and there are points which are very close to each other, but which have very different representations by words. For real numbers in $[0, 1]$, for instance, the points which are slightly below $1/2$ and the ones which are slightly above $1/2$ have very different binary representations. This is the reason for taking a quotient of the Cantor set, to make suitable identifications.

The use of the quotient here plays a role like that of the collections of ordered pairs of words which determine when two words represent the same or adjacent vertices in the context of automatic structures. Adjacency in a graph is like "closeness" in a topological space, except that it is more rigid and precise.

For the notion of semi-Markovian spaces, one does not allow arbitrary quotients of subsets of Cantor sets, but only ones which are tightly regulated in a certain way. One allows quotients only of a particular class of subsets of the set of all infinite words (over a given alphabet Σ), and this is analogous to the initial choice of a regular language to represent the vertices in a given graph in the context of automatic structures. The type of quotients are restricted in a similar manner, with the quotient being described by a set of pairs of points in the given subset (of which the quotient is being taken). More precisely, this set of pairs can also be viewed as a subset of a Cantor set (the Cartesian product of the original Cantor set with itself), and it is asked to lie in the same general class as in the first step. This is analogous to the condition that the collections of ordered pairs which determine equality and adjacency in an automatic structure be regular languages.

See [Gro87, CP93] for more details. Note that there are only *countably* many different semi-Markovian spaces, up to topological equivalence. This reflects the combinatorial nature of the concept, even if it does deal with fairly complicated topological spaces (and infinite processes).

Semi-Markovian spaces can also arise from finitely-generated groups. If a group Γ is *hyperbolic* in the sense of Gromov [Gro87], then it has a special "space at infinity", defined in [Gro87]. This is a topological space which Gromov has shown to admit a semi-Markovian structure. (See also [CP93].)

For the simplest case, think of a finitely-generated free group. In this case, the "space at infinity" is exactly a Cantor set, and one does not have to pass to a quotient.

In general, one should allow for "relations" in the space at infinity. For instance, the fundamental group of a compact oriented topological surface with no boundary and at least two handles is hyperbolic in Gromov's sense, and its space at infinity is homeomorphic to an ordinary circle. (Surfaces like these were came up before, in Sections 4.10 and 4.11. See Fig. 4.8 in Section 4.10 in particular.)

See [Gro87, CP93] for more information.

17.7 Between Cayley graphs and graphs in general

There are some important differences between automatic structures for groups and the extension to graphs mentioned in the previous section, and we would like to look at some particular aspects of this in this section.

In both cases, one starts with a regular language to describe the vertices of the graph, and in the context of groups the mechanism by which the words are interpreted is very special and relies on the group structure. For graphs in general, we have not restricted the manner in which vertices can be represented by words.

In this direction, there are some intermediate situations. The graph G could have a labelling of its edges by elements of an alphabet Σ, for instance, as there is for the Cayley graph of a finitely-generated group. In analogy with the identity element of the group, one could fix a basepoint in G, and have words over Σ be related to paths in G which begin at the basepoint. Under suitable conditions, one could say that every such word does correspond to at least one such path, or at most one, etc. For Cayley graphs of groups, one has that for any given vertex, there is exactly one edge coming from it which is labelled by a given element of Σ, and that all such edges arise in this manner. This property makes sense for graphs in general, without the presence of a group, and it ensures a good correspondence between paths and words.

Along these lines, one can refine the earlier notion of an automatic structure for graphs, by asking that the description of vertices by words cooperate with a labelling on edges in G. That is, the language would only use letters from the alphabet Σ involved in the labelling, and if a word ω in the language represents a vertex v in the graph, then ω should correspond to a path in the graph from the basepoint to v through the labelling. One could also require a stronger description of adjacency than before, in which for each letter $a \in \Sigma$ there is a separate regular language to determine when an ordered pair of vertices (v, w) is connected by an edge labelled by a. (There are natural variations on these themes, of course.)

These refined notions of automatic structures for graphs are closer to the one for groups. They do not apply to the earlier discussions of the Sierpinski gasket and fractal tree in Section 17.6, or to the binary representations for the linear graph L_n. In those cases, the automatic structure was based on very different geometric considerations, to which we turn next.

17.8 Scales and paths

In the treatment of the Sierpinski gasket and fractal tree in Section 17.6 (and the treatment of linear graphs based on binary representations), we moved around the graphs by *scales* and locations, rather than by paths. Each new letter in a description of a vertex by a word specified an extra degree of resolution in the location, an extra factor of 2 and 3 for the gasket and tree, respectively. We zoomed to a point as through a microscope, rather than driving there in a car.

Here is a related issue. Fix a graph G and a basepoint v_0 (a vertex in G), and consider the rate of *volume growth*, i.e., the number of points in the graph which can be connected to v_0 by a path of length $\leq m$ as a function of m. For the linear graphs, or the graphs associated to the Sierpinski gasket and fractal tree, this rate was always polynomial (and in fact, less than quadratic), with uniform bounds (in the family of graphs).

In general, if one is coding vertices in a graph by words over a finite alphabet, then one has the possibility to code an *exponential* number of vertices by words of a given length. If one restricts oneself to codings by words which act by paths, as in the context of Cayley graphs (or Section 17.7), then the number of vertices that can be coded by words of length less than or equal to m is bounded by the volume growth of the graph at the appropriate basepoint.

Thus the coding through paths will inevitably be inefficient when the number of these vertices grows relatively slowly. If the volume growth is exponential, then it is a different matter.

For finitely-generated groups, the volume growth is often going to be large. A famous theorem of Gromov [Gro81a] states that a finitely-generated group has polynomial growth if and only if it is virtually nilpotent (i.e., has a nilpotent subgroup of finite index). Basic examples of this phenomenon are provided by free abelian groups and the Heisenberg groups. For hyperbolic groups (in the sense of [Gro87]), there is always exponential growth. Note that the Cayley graph of a finitely-generated group never has more than (linear) exponential growth. This is not hard to check, and it is true more generally for graphs in which the number of edges attached to any given vertex is uniformly bounded.

In summary, when the volume growth is large, it makes sense to look at codings by paths, and when it is small (like polynomial), it is often better to try codings by scales, as in the examples above.

17.9 Connections between scales and paths

We have seen two basic ways to move around in a graph so far, by steps along paths and by scales and locations, as through an adjustable microscope. In fact, these two approaches are very closely linked; to move around by scales and locations in one space is often close to moving around by steps in another space. Conversely, in the presence of suitable assumptions of "negative curvature" for a space X, one can define a "boundary" of X, in such a manner that moving around by steps in X is related to moving around in its boundary by scales and locations. (See [Gro87, Gd90].)

A very classical version of this duality occurs with ordinary Euclidean geometry on $\mathbf{R}^n$ and the standard hyperbolic metric on the corresponding upper half-space $\mathbf{R}^n \times \mathbf{R}_+$. Alternatively, as a modestly different form of this picture, one can use the realization of the $(n+1)$-dimensional hyperbolic space based on the unit ball in $\mathbf{R}^{n+1}$, equipped with the standard hyperbolic metric there, instead of the upper half-plane. One then has the unit sphere $\mathbf{S}^n$ at the boundary, in place of $\mathbf{R}^n$. (In the first picture, there is really an extra "point at infinity" which is involved, and which is included more explicitly in the second picture, as a point in $\mathbf{S}^n$.) We shall discuss a general version of this type of construction in Chapter 19.

In Gromov's theory of hyperbolic groups, there is a precise correspondence between moving around the Cayley graph by steps and moving around in an associated "space at infinity" by scales and locations. In the case of free groups, the space at infinity is a Cantor set, and moving from cell to cell in the Cantor set corresponds in a simple way to moving around the Cayley graph by paths. This case arose already in Section 5.5.

Another special case occurs for the so-called *uniform hyperbolic lattices*. These are finitely-generated groups which arise as the fundamental group of compact Riemannian manifolds of constant negative curvature -1 (and without boundary). Their Cayley graphs are approximately the same geometrically as the classical hyperbolic spaces from Riemannian geometry mentioned above (with the upper half-space and ball models), in the sense of "quasi-isometric equivalence". (The definition of this was reviewed in Section 17.2.) This quasi-isometric equivalence is based on mappings from the groups into the hyperbolic spaces which come from the fact that the universal covering of the corresponding compact Riemannian manifold is isometrically equivalent to the hyperbolic space of the same dimension in a natural way. (Compare also with Section 4.11.) For the space at infinity of these groups, one gets a Euclidean sphere of dimension 1 less that of the original Riemannian manifold, at least to within some limited distortion in the geometry. This fits with the way that an ordinary sphere arises as the space at infinity of the hyperbolic spaces. Moving around by steps and paths in these Cayley graphs corresponds to moving around the Euclidean sphere by locations and scales, and in a way that can be made precise and concrete.

There are also *uniform lattices* for *complex* hyperbolic geometry. Here "complex hyperbolic geometry" is as opposed to ordinary hyperbolic geometry, as above, which is sometimes called *real* hyperbolic geometry, to be precise. These uniform lattices are again hyperbolic groups in the sense of [Gro87]. The corresponding spaces at infinity look like the continuous versions of the Heisenberg group with the distance function (17.15). More precisely, one gets a compact space whose local geometry is that of the Heisenberg group, in the same way that the local geometry of a standard Euclidean sphere is approximately the same as that of the Euclidean space of the same dimension. Alternatively, there is an "upper half-space" model for complex hyperbolic geometry, and for this the continuous version of the Heisenberg group comes up as the boundary in a natural way, just as $\mathbf{R}^n$ comes up as the boundary for the upper half-space model of real hyperbolic geometry (as before). (We shall say a bit more about uniform lattices, for real and complex hyperbolic geometry, in a moment.)

Thus, while the discrete Heisenberg groups $\widetilde{H}_n(\mathbf{Z})$ do not cooperate directly with the notion of automatic groups, the relevant geometry is captured by the asymptotic behavior of certain hyperbolic groups (uniform lattices in complex hyperbolic geometry). Here we are using the fact that the *large-scale* structure of $\widetilde{H}_n(\mathbf{Z})$ is very close to the *local* structure of the continuous version H_n. This is brought out by the dilations (17.10) on H_n, and we discussed related points in Section 17.4, especially towards the end. One can look at this in analogy with $\mathbf{Z}^n$ and $\mathbf{R}^n$, for which the comparison between large-scale and local geometry through dilations is more standard. (For instance, consider the case of $n = 1$. One can take the integers between $-L$ and L for some large number L, and rescale them into the interval $[-1, 1]$ through a dilation by a factor of L^{-1}. This gives a discrete "model" for $[-1, 1]$ which "converges" to $[-1, 1]$ as $L \to \infty$. One can make use of this general idea in various ways.)

Although $\mathbf{Z}^n$ is different from $\widetilde{H}_n(\mathbf{Z})$ in the existence of an automatic structure (as a finitely-generated group), this difference is not as great as it might seem at first. As we indicated in Section 17.8, the coding of elements of $\mathbf{Z}^n$ by words over a generating set can never be too efficient, in that words of length k can never represent more than $O(k^n)$ elements of $\mathbf{Z}^n$. This is in contrast to the exponential number of elements in a group which can often be obtained in this manner (in other groups). For both $\mathbf{Z}^n$ and $\widetilde{H}_n(\mathbf{Z})$, the basic geometry is captured more efficiently in terms of the asymptotic behavior of uniform real and complex hyperbolic lattices, respectively. One might think of this as being roughly analogous to the difference between unary and binary representations for natural numbers. (Note that the uniform real and complex hyperbolic lattices always have exponential rates of growth.)

These realizations as spaces at infinity of hyperbolic groups do not represent the group structures of $\mathbf{Z}^n$ and $\widetilde{H}_n(\mathbf{Z})$ in a precise way, even if they do capture the basic geometry.

See [Rag72] for more information about "uniform lattices" for real and complex hyperbolic geometry. Let us be a bit more precise, and say that these lattices are discrete groups of isometries of the real and complex hyperbolic spaces. The groups of isometries of the real and complex hyperbolic spaces can be identified with classical groups of invertible matrices. One can think of the lattices as being discrete subgroups of the matrix groups $O(n, 1)$ and $U(n, 1)$, respectively (for the real and complex case), and these groups are defined explicitly by looking at all invertible matrices which preserve certain quadratic forms. For $O(n, 1)$, one uses matrices with *real* entries, while for $U(n, 1)$, one uses matrices with *complex* entries. See also [GP91].

Let us mention that $\mathbf{Z}^n$ is a uniform lattice in the same sense inside of $\mathbf{R}^n$ (viewed as an abelian group, under addition). Similarly, the discrete Heisenberg group $\widetilde{H}_n(\mathbf{Z})$ is a uniform lattice inside of its continuous version H_n. The groups $O(n, 1)$ and $U(n, 1)$ are quite different from $\mathbf{R}^n$ and H_n, in terms of being more noncommutative, and having exponential growth in a suitable sense.

We should emphasize that the polynomial volume growth for $\mathbf{Z}^n$ and $\widetilde{H}_n(\mathbf{Z})$ has two roles here, both "bad" and "good". It ensures that the coding of group elements by words can never be very efficient, as mentioned above, and, on the other hand, that the large-scale structure of these groups behave like spaces which are *finite-dimensional*. In other words, polynomial volume growth provides

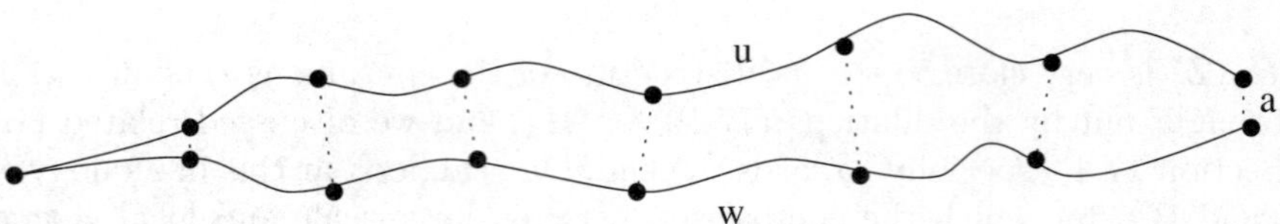

FIG. 17.10. The k-fellow traveller property: the paths stay close together

both a reason why it is better to use codings by scales and locations instead of steps and paths, and a reason why coding by scales and locations can have the possibility to work well. Polynomial volume growth is very much connected to ideas of dilations and scaling too, as in the groups $\mathbf{Z}^n$ and $\widetilde{H}_n(\mathbf{Z})$, and in standard examples of fractals, like the Sierpinski gasket.

For a hyperbolic group, it is the "space at infinity" which has this kind of polynomial behavior in its geometry, and for which notions of scale-invariance are applicable (and tied to the individual steps in the group).

17.10 The k-fellow traveller property

For finitely-generated groups, there is an alternative characterization of the *existence* of an automatic structure which is more directly geometric. As before, one starts by asking for a regular language L of words over a set of generators (and their inverses) such that every element of the group G is represented by at least one word in L. Instead of asking for automata which detect when words represent the same or adjacent elements, this time one asks for the following "k-fellow traveller" property. Let u and w be two words in L which represent elements of G that differ by only an element of the generating set. (As simply *words*, they may differ by much more than that.) Each of these words defines a *path* in the Cayley graph, starting at the identity and traversing edges in accordance with the letters in the words. For the k-fellow traveller property, one requires that the paths corresponding to u and w always lie within distance k of each other with respect to the word metric on the Cayley graph. This means that the jth vertex in the path associated to u should lie within distance k of the jth vertex in the path associated to w for all j. (If the paths have unequal length, then one repeats u or w (as appropriate), when j exceeds the number of letters in the word. One can think in terms of extending the path to be constant after the whole word has been used.) Of course, it is important here that the number k be independent of the choices of u and w. (See Fig. 17.10. The name "k-fellow traveller property" is employed in [Far92], while [ECH$^+$92] uses different terminology.)

See Theorem 2.3.5 on p48 of [ECH$^+$92] and Proposition 1 on p299 of [Far92] for more information about this characterization of automatic groups. In particular, the relationship between k and the number of states in the automata can be seen in the proof of Lemma 2.3.2 near the bottom of p46 in [ECH$^+$92], and in the definition of "standard automata" on p47 in [ECH$^+$92].

Note that the validity of the k-fellow traveller property depends on the choice of the language L, and not just on the group G. For instance, the k-fellow traveller property certainly fails for (infinite) abelian groups if one takes L to be the set of all words over the generating set.

In the proof that the existence of a language L with the k-fellow traveller property is sufficient for the existence of an automatic structure, one uses the underlying group structure in a strong way, to reduce computations concerning two elements at bounded distance apart to computations within a fixed finite part of the group. The argument does not work at a purely geometric level, for graphs, for instance.

However, the idea of the k-fellow traveller property by itself is reflected in the notion of "combability", which applies to any metric space. See Section 3.6 of [ECH$^+$92]. A moderately different definition is given in [Ger92]. Other variants have been considered, as mentioned on p83 of [ECH$^+$92]. Here we would like to mention a version (of the definition in [ECH$^+$92]) for graphs in which one can also use an automaton in a certain way.

Let H be an infinite graph with a fixed basepoint, a vertex v_0 in H. To control the geometry of H we start by asking for a rooted tree (T, b) and a mapping $\alpha : T \to H$ which is surjective on vertices and sends b to v_0. Note that H could easily be far from tree-like, both because of the non-injectivity of α on vertices, and the lack of surjectivity on edges.

We further restrict the behavior of α in the following way. For any vertex t in T there is a *unique* path p_t in T which goes from b to t and which does not cross itself. (For the existence, one can start with any path, and simply remove any cycles which occur. Uniqueness comes from the assumption that T be a tree.) One can formulate a version of the k-fellow traveller property by asking that if s and t are vertices in T such that $\alpha(s)$ and $\alpha(t)$ are the same or adjacent in H, then the paths $\alpha(p_s)$ and $\alpha(p_t)$ should lie within distance k of each other at every step. Here distance between vertices in H is measured by the length of the shortest path between them (as usual). Also, if p_s and p_t do not have the same length, then in the step-by-step comparison, one should continue using the endpoint of the one that runs out first. This is roughly the same as in the definition of combability in [ECH$^+$92], except for the way that we ask for the combing to arise.

By employing trees in this manner, it is easy to add an extra ingredient which is more algorithmic, namely, the requirement that the tree arise as the visibility of a (*finite*) graph G. In this way, the mapping $\alpha : T \to H$ becomes like a listing of vertices in H by a regular language. It goes directly to the transition graph of an automaton (the graph G of which T is the visibility), without the intermediate role of letters and words.

Note that the Cayley graphs of the Heisenberg groups do not satisfy the purely geometric condition of combability. See Chapter 8 of [ECH$^+$92] (and also Theorem 3.6.6 on p86 of [ECH$^+$92]) for the first Heisenberg group, and [Gro93] (parts (5)(e) and (5)(f) in Section 5.D) and [Bur96] for the Heisenberg groups in general (and concerning other nilpotent groups as well).

18

EXTENDED NOTIONS OF AUTOMATA

Heisenberg groups and Baumslag–Solitar groups (described below) provide basic examples of finitely-presented groups which are *not* automatic. They also provide a good setting in which to consider questions about the nature of recursion in broader terms.

Geometry and computation are closely linked. It is often convenient to think of computations geometrically, as in the cases of finite automata, Boolean circuits, Cayley graphs of groups, and the configuration graph of a Turing machine.

Formal proofs have similarly geometric aspects. Through the *logical flow graph* (reviewed in Section A.3), one can attempt to trace the computations which lie below a given proof, and geometric features of the logical flow graph (such as the behavior of oriented *cycles*) can reflect the nature of the recursion which takes place within the proof. In particular, when there are nontrivial oriented cycles in the logical flow graph, it may not be possible to extract the underlying computations directly from the proof. (There can be ambiguities in the way that substitutions should be made, for instance.) Proofs without cuts cannot have nontrivial oriented cycles [Car97], but in proofs with cuts this difficulty can easily occur. One may apply the method of *cut elimination* in order to eliminate oriented cycles, but at possibly great cost of expansion in the proof. An alternative method to eliminating cycles and extracting the underlying computation has been given in [Car00], in connection with matters of *feasibility.*

The precise mechanisms of recursion and symmetry within formal proofs are far from understood. Formal proofs provide a kind of universal setting in which all sorts of mathematical structures can be coded. For instance, one can code derivations of identities in finitely-presented groups into formal proofs, and these have the possibility for intricate structure, through the use of the cut rule.

In view of the difficulty of dealing with formal proofs, and the breadth of structures which they can incorporate or reflect, it is natural to look for simpler combinatorial models for the mathematical structures found within them. This is one of the points of [Carc], and in [Cara] a method is described by which finitely-presented groups are associated to formal proofs in a way that can reflect the recursions within.

In this chapter, we want to look at some of the limitations of standard automata and extensions of them which enjoy different features. We are particularly interested in geometric structures, but we are also motivated by the complicated phenomena which can occur within formal proofs.

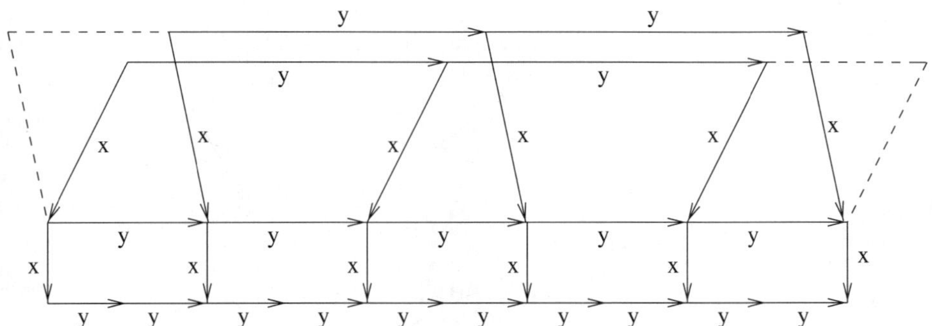

FIG. 18.1. A piece of the Cayley graph of the Baumslag–Solitar group. (Redrawn (with permission) from David Epstein et al., *Word Processing in Groups*, A K Peters Publishers, 1992, p155.)

18.1 Asynchronous automata

A basic example of a finitely-presented group is the Baumslag–Solitar group, defined by taking two generators x and y and prescribing the one relation

$$yx = xy^2. \tag{18.1}$$

This group turns out *not* to be automatic [ECH+92]. To see why a problem should exist, we can look at what goes wrong with a naive attempt to choose an automatic structure for this group.

Before we do that, let us review some basic facts about the structure of this group. The Cayley graph has some nice geometry, as discussed in detail in Section 7.4 of [ECH+92], and illustrated there with several very good pictures. The basic idea from [ECH+92] for picturing the Cayley graph is indicated in Figures 18.1, 18.2, and 18.3. The first picture, Fig. 18.1, shows how the basic "building blocks" for the Cayley graph are patched together, with the individual rectangles reflecting the relation (18.1). In this diagram, one has a kind of splitting of sheets, and this occurs over and over again in the Cayley graph as a whole. Fig. 18.2 gives a side view of the Cayley graph, which shows how the different sheets branch out, as in a binary tree. Fig. 18.3 indicates the way that the Cayley graph would look if one followed only a single sheet in each "vertical" step. (See [ECH+92] for more information.)

There is an obvious homomorphism from the Baumslag–Solitar group onto an infinite cyclic group, in which one takes a word in x and y and throws away all occurrences of y, leaving a power of x. (Thus the word $xyxyxy$ would be transformed into x^3, for instance.) The kernel of this homomorphism consists of elements of the group for which the *total* number of x's (counting powers with *signs*) is zero. To understand the behavior of this kernel, it is helpful to begin by observing that for *any* word in x and y, we can use the relation above to move all of the *positive* powers of x to the extreme left-hand side. For this we are also using the fact that

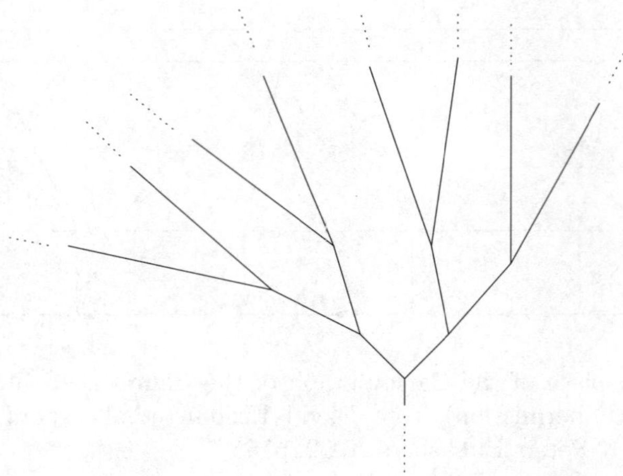

FIG. 18.2. A view from the side. (Redrawn (with permission) from David Epstein et al., *Word Processing in Groups*, A K Peters Publishers, 1992, p156.)

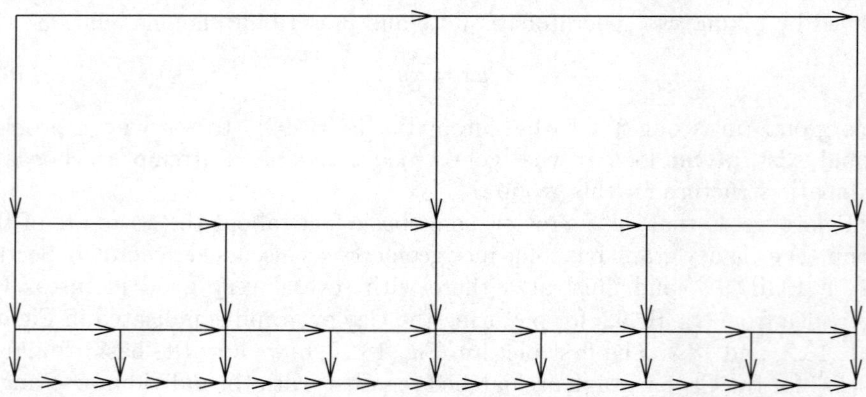

FIG. 18.3. A single "sheet"

$$y^{-1}x = xy^{-2}, \tag{18.2}$$

as one can derive from (18.1). Given a word in the kernel with all of the positive powers of x on the far left-hand side (if there are any), we can write it as a product of words of the form $x^n y^j x^{-n}$, where n is an a nonnegative integer and j is any integer. This is not difficult to verify, simply by regrouping and using the fact that the total number of x's in the word (counting signs) must be zero. We can go a step further and say that every element of the kernel can be written as a single term of the form $x^n y^j x^{-n}$ for some n and j. In other words, the powers of x can be made to match up, because of the identity

$$x^m y^k x^{-m} = x^n y^{k2^{n-m}} x^{-n} \tag{18.3}$$

for $n \geq m$, which can itself be derived easily from (18.1) and (18.2).

Using this normal form for elements of the kernel, we see that the kernel is actually abelian, so that the group as a whole is solvable. The word $x^n y x^{-n}$ provides a 2^n-th root of y inside the group, and the kernel is isomorphic to the additive group of rational numbers for which the denominator is a power of 2. The kernel is *not* finitely generated, but instead more complicated geometrically.

Another point about this group is that it admits *exponential distortion*, in the sense that

$$y^{2^n} = x^{-n} y x^n. \tag{18.4}$$

In words, exponentially large powers of y can be represented as words of *linear* size in n. One can also see this in the Cayley graph, as in Fig. 18.3 (for the first couple of steps). (See [Gro93] for more information about distortion in finitely-presented groups.)

To see concretely the difficulty with automaticity in this case, let us simplify the discussion by looking only at the *semigroup* generated by x and y and satisfying the relation (18.1). There is an obvious normal form for elements of this semigroup, using words of the form $x^j y^k$ for arbitrary nonnegative integers j and k, since (18.1) permits us to move positive powers of x to the left of positive powers of y. This collection of words is recognized by a standard automaton.

We run into trouble when we try to describe pairs of words which are related through right-multiplication by x. The pairs of words $(x^j y^k, x^l y^m)$ such that

$$x^l y^m = (x^j y^k)x = x^{j+1} y^{2k} \tag{18.5}$$

cannot be recognized by a standard automaton, because (roughly speaking) of the "imbalance" between y^k and y^{2k}. This is related to the fact that a standard automaton has only bounded memory.

One should be careful about the "formatting" issues here. In asking about the existence of a description through an automaton, one should view the ordered pairs $(x^j y^k, x^l y^m)$ as being realized as words over an alphabet whose letters are ordered pairs of x, y, as in Section 17.5. (One should also allow an extra padding symbol \$, to equalize the lengths of words in ordered pairs, as mentioned in Section 17.5.)

This problem with representing the set of pairs of words $(x^j y^k, x^l y^m)$ that satisfy (18.5) would disappear though if we were to permit ourselves to use an *asynchronous automaton*. Roughly speaking, this means a machine like a standard automaton, except that it is allowed to read from two tapes separately. The reading of the jth place on one tape need not occur at the same time as the reading of the jth place on the other tape, as would be the case for a standard automaton (over an alphabet of ordered pairs, as above). Instead, one can take two steps on one tape for every step on the other, for instance, as in (18.5).

The Baumslag–Solitar group with which we started is in fact an example of an *asynchronous automatic group*, in the sense of [ECH+92]. The definition of an asynchronous automatic structure for a group is similar to that of a (synchronous) automatic structure, except that one allows asynchronous automata for recognizing the sets of pairs of words which determine the same or adjacent elements of the group. The proof of the existence of an asynchronous automatic structure for the Baumslag–Solitar group is more tricky than the discussion above shows, because of the way that powers of x^{-1} can lead to "fractional powers" of y, as we saw before (just after (18.3)). This complicates even the normal form for elements of the group. See Chapter 7 of [ECH$^+$92] for more information.

The treatment in [ECH$^+$92] applies more generally to the family of Baumslag–Solitar groups $G_{p,q}$, which have two generators x, y and the one relation

$$y^p x = x y^q, \tag{18.6}$$

where p and q are distinct positive integers. See [FM98] for some recent results concerning the large-scale geometry of these groups.

The existence of a polynomial-time algorithm for solving the word problem for asynchronous automatic groups is left open in [ECH$^+$92]. A quadratic-time algorithm is given in [ECH$^+$92] for (synchronous) automatic groups, but in the asynchronous case only an exponential-time algorithm is provided.

18.2 Heisenberg groups

Let us consider now the (discrete) Heisenberg groups. For concreteness, we take the first Heisenberg group, with the presentation provided by the three generators x, y, and z, and the relations

$$yx = xyz,\ xz = zx,\ yz = zy. \tag{18.7}$$

(This is slightly different from the presentation discussed in Section 17.3, but the two are easily seen to be equivalent.) Note that we have similar identities for the inverses, i.e.,

$$yx^{-1} = x^{-1}yz^{-1},\ y^{-1}x = xy^{-1}z^{-1},\ y^{-1}x^{-1} = x^{-1}y^{-1}z. \tag{18.8}$$

This permits us to manipulate both positive and negative powers of the generators in the same way. (This is somewhat different from what happened with the Baumslag–Solitar group.)

This group is *not* automatic, as in [ECH$^+$92]. Let us try to see explicitly what goes wrong when we follow our noses in making a naive construction.

There is a very simple normal form for words in the group, in which every element of the group can be represented uniquely as $x^j y^k z^l$ for some integers j, k, and l. For an automatic structure, one would like to have automata which are able to describe right-multiplications by the generators. There is a problem with right-multiplication by x, which is described by the equation

$$(x^j y^k z^l)x = x^{j+1} y^k z^{l+k}. \tag{18.9}$$

For this one is interested in the set of ordered pairs of the form

$$(x^j y^k z^l, x^{j+1} y^k z^{l+k}), \tag{18.10}$$

where j, k, and l should be arbitrary integers. (Strictly speaking, we should treat x^{-1}, y^{-1}, and z^{-1} as separate generators, and make other adjustments so that only nonnegative integers arise, but this is a relatively minor issue here.)

In fact, the Heisenberg group is *not* even an asynchronous automatic group in the sense of [ECH$^+$92], as in Theorem 8.2.8 on p169 of [ECH$^+$92]. Although there is some similarity with the phenomenon of "doubling" from Section 18.1, there is additional trouble now with nonlocality that was not present before, in the way that information is distributed between the y's and z's. That is, a single increment in the power of y in the first component of (18.10) leads to an increment in both the powers of y and of z in the second component, while the analogous step in Section 18.1 lead to an increase by 2 in the power of y in the second component, and nothing else.

It would be easy to accommodate this type of phenomenon through asynchronous automata with more tapes, i.e., separate tapes for each of x, y, and z. This way of breaking apart the words is not really compatible with the spirit of automatic structures for groups, though. It would rely on special features of this particular situation, rather than providing a natural concept which is applicable in a general way.

One could try to look at the idea of an automatic structure for the Heisenberg group in a different way, allowing not only words over generators as usual, but also use of dilations, as in (17.10). The customary idea for automatic structures of groups (as mentioned in Section 17.5) is that one tries to reconstruct the entire group from a finite piece around the identity element together with simple rules to explain how these pieces should be unwrapped as one moves away from the identity. In this case, one could try to use dilations to take into account rescaled copies of a single finite piece too. This is not quite the same as the notion of "unwrapping" the group structure from a finite piece, but analogous points are entailed. One could also take the view that the dilations are an integral part of the structure, which is simply not present for groups in general. Something roughly similar occurs in the realization of the continuous version of the Heisenberg group in the space at infinity for a uniform lattice for complex hyperbolic geometry, as mentioned in Section 17.9.

18.3 Expanding automata

The study of groups and semigroups provides a interesting backdrop for looking at extensions of automata which allow greater freedom of computation in a controlled manner. In this section, we describe one such extension, called an *expanding automaton.*

A basic example to consider is the collection of ordered pairs

$$(x^j y^k z^l, x^{j+1} y^k z^{l+k}), \tag{18.11}$$

which arose before in (18.10), in the context of the Heisenberg group. For simplicity, we shall restrict ourselves in the present discussion to *nonnegative* values of j, k, and l (rather than allowing all integers, as occur in the Heisenberg group).

With the notion of an expanding automaton, we want to think of describing families of pairs of words in such a way that the first word in the pair is recognized by a standard automaton, while the second word is constructed in a way which is encoded by the first word. The operations that are allowed to be performed on the second word will be more complicated than simply the addition of new letters at the end, so as to accommodate situations like the one presented in (18.11). (Otherwise, we would not get anything beyond what is covered by asynchronous automata.)

Let us be more precise. Fix an alphabet Σ and an automaton A which recognizes a collection of words over Σ. For instance, think of Σ as being $\{x, y, z\}$, and A as being an automaton which recognizes the language

$$\{x^j y^k z^l : j, k, l \geq 0\}. \tag{18.12}$$

To define an *expanding automaton* E_A over this automaton A, one should also specify a set $\mathcal{V}$ of *variables*, a mapping $\theta : \Sigma \times \mathcal{V} \to (\Sigma \cup \mathcal{V})^*$, and an initial element S of $\mathcal{V}$ (the "starting symbol").

We shall give an example for the choice of $\mathcal{V}$ and θ in a moment, but let us first explain the interpretation of an expanding automaton. One can think of an ordinary automaton as being a machine which reads in a word and then "accepts" the word if it corresponds to a sequence of transitions in the automaton which begin at the initial state and end at a final state. The expanding automaton "looks over the shoulder" of the original automaton, reads the same word, letter by letter, and then writes down words of its own, on its own "tape". It performs its own operations, in a fashion that is completely determined by the word which is read in by the original automaton.

In general, one could allow the expanding automaton to perform all sorts of operations on its tape, but for the moment we shall restrict ourselves to some which are analogous to the substitutions performed by a context-free language, or to the "direct derivations" in an L-system. (See [HU79, PL96, RS80], and Section 4.9.)

More precisely, suppose that our automaton A is about to read in a word ω over Σ. Before the automaton begins, the expanding automaton writes down the starting letter S on its own tape. If the first letter in ω is x, for example, then the expanding automaton looks at the word that it presently has, the letter S in this case, and replaces it with the word over $\Sigma \cup \mathcal{V}$ specified by the mapping θ, i.e., $\theta(x, S)$.

Now suppose that the automaton A has operated for several steps, and the expanding automaton E_A with it. This means that we have so far read a certain initial subword μ of ω, and the expanding automaton has a word M over $\Sigma \cup \mathcal{V}$ written on its own tape. Let y be the next letter in ω. The expanding automaton operates on M to produce a new word M' in the following way. It looks at the letters in M, one by one, and when it reaches a variable, i.e., an element V of $\mathcal{V}$, it replaces this occurrence of V with the word $\theta(y, V)$. It then proceeds to the next occurrence of a variable — let us call it V' — ignoring any new variables that might have been introduced in the previous substitution. It replaces V' with $\theta(y, V')$, and continues in the same way until it reaches the end of the word.

By the end, the expanding automaton E_A makes exactly m such substitutions to the word M with which it began in this stage of the process, where m is the number of variables (elements of $\mathcal{V}$) in M. When it is finished, the automaton A reads in the next letter in ω, and the process is repeated. This is somewhat analogous to what happens in a context-free language, except that the execution of the substitutions is regulated by the automaton A. (An example showing the effect of an expanding automaton for a certain word ω is given in (18.16) below.)

In general situations, the number of substitutions performed by the expanding automaton E_A can grow exponentially with the number of letters read in by the original automaton A. This is the same as saying that the number of occurrences of elements of $\mathcal{V}$ in the word written by the expanding automaton can grow exponentially with the number of steps of execution of the automaton A. However, the substitutions performed by the expanding automaton at a fixed step in the execution of A are all independent of each other, involving disjoint parts of the word in question, and one can think of them as being performed in parallel. These transformations are similar to ones used in the context of *L-systems*, as in [PL96, RS80, HU79] and Section 4.9.

Let us give a concrete example now, corresponding to the ordered pairs (18.11). As before, we take $\Sigma = \{x, y, z\}$, and we assume that A is an automaton which recognizes the language given by (18.12). We take $\mathcal{V} = \{S, L, Y, Z, I\}$, and we define $\theta : \Sigma \times \mathcal{V} \to (\Sigma \cup \mathcal{V})^*$ by setting

$$\begin{aligned}
\theta(x, S) &= xL \\
\theta(x, L) &= xL \\
\theta(y, S) &= xyYzZ \\
\theta(y, L) &= xyYzZ \\
\theta(y, Y) &= yIz \\
\theta(y, I) &= yIz \\
\theta(y, Z) &= Z \\
\theta(z, S) &= zZ, \\
\theta(z, L) &= zZ, \\
\theta(z, Z) &= zZ,
\end{aligned} \tag{18.13}$$

and by requiring that θ take the value ϵ (the empty word) in all other cases.

With these rules, our expanding automaton E_A determines a collection of ordered pairs of words which is almost the same as the one given in (18.11). We can make it work slightly better through the following modifications. Let us add a letter \$ to our alphabet Σ, and let us replace (18.12) by the collection of words

$$\{x^j y^k z^l \,\$: j, k, l \geq 0\}. \tag{18.14}$$

In other words, we simply use \$ to mark the end of a word in our basic language (18.14). We keep the same collection $\mathcal{V}$ of variables as before, and we set

$$\theta(\$, S) = x, \tag{18.15}$$

and take $\theta(\$, V)$ to be the empty word ϵ for all other $V \in \mathcal{V}$.

When our basic automaton A reads in a word of the form $x^j y^k z^l \,\$$, the expanding automaton will end up with the word $x^{j+1} y^k z^{l+k}$, as in (18.11). This is not hard to check, and we omit the details. (One should not forget "degenerate" cases, such as $j = 0$, or $j > 0$ but $k = 0$, etc.) To illustrate this, here is an example, corresponding to the word $x^2 y^3 z^2 \$$:

$$\begin{aligned} S &\xrightarrow{x} xL \xrightarrow{x} x^2 L \xrightarrow{y} x^3 y Y z Z \xrightarrow{y} x^3 y^2 I z^2 Z \xrightarrow{y} x^3 y^3 I z^3 Z \\ &\xrightarrow{z} x^3 y^3 z^4 Z \xrightarrow{z} x^3 y^3 z^5 Z \xrightarrow{\$} x^3 y^3 z^5 \end{aligned} \tag{18.16}$$

Each of the transitions shows the effect of the expanding automaton on the string in hand, given the particular letter from the word $x^2 y^3 z^2 \$$ which is being read by the automaton A at that step. (In some steps, this involves more than one substitution by the expanding automaton. For this example, there are no more than two substitutions in any given step.)

We mentioned before that, in general, expanding automata can generate words which contain exponentially-many "variables", and hence they can involve exponentially-many substitutions. This does not happen in the present example of an expanding automaton, for a very simple reason. In order to have exponential expansion, there must be substitution rules which allow a single variable to be replaced by a word which contains more than one variable, and one needs to be able to make substitutions like this over and over again. In the present example, one can replace a single variable by a word with more than one variable, through the substitutions for the variables S and L when the automaton A reads in the letter y. However, the new variables produced in this way (i.e., Y and Z) are ones which are themselves never "expanded" in this way, nor do they lead back to the variables S or L. In fact, none of the words produced by the expanding automaton in this case ever contain more than 2 variables.

To put this into perspective, imagine a situation in which Σ consists only of a single letter a, and our collection of variables $\mathcal{V}$ consists of T, U, and W (with T as the starting symbol). If we had a rule like

$$\theta(a, T) = TT, \tag{18.17}$$

then our expanding automata could generate words with exponentially many T's quite easily. Now suppose we do not have this rule, but instead the rules

$$\theta(a, T) = TU, \qquad \theta(a, U) = U. \tag{18.18}$$

With these rules for making substitutions, our expanding automaton can create words in which the number of variables grows *linearly*, but not more than that. If we replace these rules with

$$\theta(a,T) = TU, \qquad \theta(a,U) = UW, \qquad \theta(a,W) = W, \tag{18.19}$$

then the expanding automaton could generate words in which the number of variables grows *quadratically*, but not more. (More precisely, the number of T's would not grow, the number of U's would grow linearly, and the number of W's would grow quadratically.) If instead we took the rules

$$\theta(a,T) = UW, \qquad \theta(a,U) = W, \qquad \theta(a,W) = U, \tag{18.20}$$

then we would be in a situation more like the one in the earlier example with the rules given in (18.13), in which there were never more than 2 variables in a word constructed by the expanding automaton.

Thus, while the notion of an expanding automaton allows for the possibility of exponential expansion in general, in practice one may be able to show that the amount of expansion is much more limited. In particular, in the context of extended notions of "automatic structures" for Heisenberg groups or other nilpotent groups, the nilpotency of the group leads to the sort of "nilpotency" for the substitution rules that we have seen above, and hence to strong limitations on the occurrence of variables in the words generated by the expanding automaton.

The relationship between this notion of an expanding automaton and the usual (synchronous or asynchronous) automata is somewhat analogous to the ways of adding another layer of implicitness discussed in Chapter 16. To make this more precise, we should introduce a bit of terminology. Let us think of the execution of an ordinary automaton as proceeding in a series of "steps", where an individual step corresponds to the reading of a single letter. For an expanding automaton E_A, we might use the phrase "big step" to describe what happens in a single step in the execution of the ordinary automaton A. That is, a big step for E_A consists of the reading of a single letter by A, followed by the series of substitutions performed by the expanding automaton before the reading of the next letter by A.

A single big step for an expanding automaton E_A is of roughly the same kind of computational strength as the entire execution of an ordinary automaton. There are a number of ways that one might look at a big step for E_A, but let us leave that aside for the moment. An expanding automaton is roughly like starting with the notion of an ordinary automaton, and making a system in which automata are executed over and over again in a way that is controlled by another automaton. (There are a number of ways in which one might make this precise.)

18.4 Tapes that cross

Let us describe now a different extension of automata. Roughly speaking, we want to consider a system in which autonomous tapes can cross each other and diverge from each other, instead of remaining "parallel". Let us call these *crossing automata.*

In this extension, there are two main ingredients. The first is an oriented graph G that represents the possible transitions in the system. We ask that G be a *marked* graph in the sense of Definition 15.1, so that G comes equipped with a designated initial vertex $q_0(G)$ and a collection $F(G)$ of final vertices. The second main ingredient is another oriented graph X together with a mapping

$$\xi : X \to G \tag{18.21}$$

It is X which controls the behavior of the tapes; in particular, X can carry labels which would correspond to the formation of words.

One can imagine X as lying directly "above" G, with ξ representing a projection from X to G. For a standard automaton, one would take X to simply be the same as G. In general, X could be more complicated, in such a way that as one follows a path in G, there may be several strands in X, which may come together, split apart, or remain parallel (or some of each).

We shall require that ξ be orientation-preserving and a local +-surjection (Definition 10.12). This ensures that at each vertex in X one can always go "forward" with the transitions of the automaton. With these conditions, there can still be branching within X that does *not* come from branching in G, however, since we do not ask that ξ be a local +-injection.

We also ask that a (nonempty) collection of vertices $Q_0(X)$ in X be specified, all of which are mapped to $q_0(G)$ by ξ. (We do not require that $Q_0(X)$ contain *all* of the vertices in X which are mapped to $q_0(G)$, however.) These will be our starting vertices in X, and there may be several of them. As we follow along a path in G, we can "read off" a graph from X, starting at the elements of $Q_0(X)$. This is easy to imagine visually, but let us formulate it more precisely.

Let L_n denote the *linear graph* of length n. Specifically, let us think of L_n as having vertices $0, 1, 2, \ldots, n$, and exactly one edge from the vertex i to the vertex $i+1$ for $i \leq n-1$, and no other edges. Thus a *path* in G of length n is the same as a mapping λ from L_n into G, while an *oriented path* means a mapping λ which preserves orientations.

Suppose that we are given an oriented path in G which begins at the initial vertex $q_0(G)$ and has length n, and let $\lambda : L_n \to G$ be the mapping which represents this path. We want to associate to λ an oriented graph X_λ that reflects the portions of X above λ and follows λ in a natural way, starting with the "initial" vertices $Q_0(X)$ in X. If we are in the special case where λ does not cross itself, then X_λ will be a subgraph of X, but this will not be true in general. Even in this case, X_λ may not contain all of the parts of X that lie above λ (and even if $Q_0(X)$ contains all of the vertices in X that lie above $q_0(G)$).

Let us begin by choosing some sets of vertices $V_0, V_1, \ldots, V_n$ in X. We shall do this in such a way that

$$\xi(v) = \lambda(j) \qquad \text{for every } v \in V_j, \tag{18.22}$$

but we also want to be careful to make certain that no such vertices appear spontaneously, without connections to the past.

For V_0 we take $Q_0(X)$, the set of initial vertices in X. If V_j has been chosen for some $j < n$, then we take V_{j+1} to be the vertices v in X such that $\xi(v) = \lambda(j+1)$, and for which there is an edge e in X such that e goes from a vertex in V_j to v, and the image of e under $\xi : X \to G$ is the same as the edge in G that the path λ traverses in going from the vertex $\lambda(j)$ to $\lambda(j+1)$. In this manner, we can define V_j for all j, $0 \leq j \leq n$.

Note that the V_j's may not be pairwise disjoint, since λ could cross itself (when G contains nontrivial oriented cycles). It is convenient to define $\widetilde{V}_j$ to be the set of ordered pairs (v, j) with $v \in V_j$, so that the $\widetilde{V}_j$'s are pairwise disjoint by construction.

Let V_λ be the union of the $\widetilde{V}_j$'s. We use V_λ for the set of vertices in X_λ. For the edges, we attach an edge from a vertex (v, j) in V_λ to a vertex $(w, j+1)$ for each edge f in X such that f goes from v to w, and the image of f under ξ is the same as the edge in G that is traversed by the path λ as it goes from the vertex $\lambda(j)$ to $\lambda(j+1)$. Thus there may be multiple edges in X_λ between a given pair of vertices, when there are multiple edges in X. We do not add any other edges. This defines our graph X_λ.

Note that every vertex in $\widetilde{V}_j$ has at least one outgoing edge in X_λ when $j < n$. This uses our original assumption that $\xi : X \to G$ be a local +-surjection. On the other hand, the vertices $\widetilde{V}_n$ do not have any edges in X_λ coming out of them.

For a simple example, suppose that X is the disjoint union of several "parallel" copies of G. In this event, X_λ looks like the disjoint union of the same number of parallel copies of L_n itself. In general, there can be branching in X_λ, inherited from branching in X.

The construction of X_λ gives rise to a pair of natural mappings $\alpha_\lambda : X_\lambda \to L_n$ and $\beta_\lambda : X_\lambda \to X$. These mappings are defined on vertices by

$$\alpha_\lambda(v, j) = j, \quad \beta_\lambda(v, j) = v, \tag{18.23}$$

and they are extended to edges in the obvious manner, following the definitions above. These mappings α_λ and β_λ preserve orientations, and are compatible with the mappings $\xi : X \to G$ and $\lambda : L_n \to G$, in the sense that

$$\lambda \circ \alpha_\lambda = \xi \circ \beta_\lambda \tag{18.24}$$

(as can be verified from the definitions). This is all very similar to the notion of *fiber products* from Section 15.6, and in fact one can simply view X_λ as being a subgraph of the fiber product of $\xi : X \to G$ and $\lambda : L_n \to G$. Notice that X_λ can be a *proper* subgraph of the fiber product in general, because of the way that the introduction of vertices in X_λ is restricted. It is easy to see that X_λ is always a +-complete subgraph of the fiber product (in the sense of Definition 10.102), by construction.

The graph X_λ never contains a nontrivial oriented cycle, because any such cycle in X_λ would be mapped by α_λ to a nontrivial oriented cycle in L_n, in which there are none. On the other hand, X and G certainly may contain nontrivial oriented cycles, and the presence of these cycles is manifested in the fact that λ and X_λ can be arbitrarily large (at least when there are cycles in G which can be reached by oriented paths beginning at the initial vertex $q_0(G)$, and cycles in X which can be reached by oriented paths that begin in $Q_0(X)$).

In the usual manner, one can be concerned in the end with the graphs X_λ corresponding to paths λ in G which begin at the initial vertex $q_0(G)$ and end at a final vertex in G (an element of $F(G)$). As one might wish, one can use these graphs to represent words over a given alphabet, or other objects. Here is a basic situation. Let Σ be a finite set, which serves as an alphabet. The graph X can be equipped with extra data of labellings of the edges of X by elements of Σ, just as for ordinary automata. One can allow edges to not have labels as well, which would be like ϵ-moves for ordinary automata. When one "reads" a graph X_λ from X as above, similar labels for X_λ are inherited from the ones on X in a natural way (i.e., following the mapping $\beta_\lambda : X_\lambda \to X$). Using these labels, one can interpret X_λ as a *feasibility graph* for constructing words over Σ, as in Section 7.1. That is, focusing branch points in X_λ would represent concatenation of words, while defocusing branch points would induce duplications of words constructed so far. In this regard, the order in which the words are combined at the focusing branch points should be specified, and for this one can ask that the specifications be inherited from similar ones given at the vertices of X (which would be additional data for X, like the labellings for the edges before).

The interpretation of X_λ as a feasibility graph provides constructions of words over Σ, through the notion of *normalized value functions* from Section 7.1. In particular, one can use the values of the normalized value function on X_λ at the elements of $\widetilde{V}_n$. (The elements of $\widetilde{V}_n$ are exactly the vertices in X_λ which correspond to the endpoint of λ.)

19

GEOMETRY OF SCALES IN METRIC SPACES

In this chapter, we elaborate on the dichotomy between *paths* and *scales* from Sections 17.7, 17.8, and 17.9. We shall review a well-known mechanism for taking a metric space and producing a graph whose paths reflect movement by scales and locations in the original space. This helps to clarify the role of automatic structures in geometry and their relationship to ideas about "patterns".

For this chapter, we drop our usual convention that the word "graph" always mean "finite graph".

19.1 Metric spaces and length spaces

Let $(M, d(\cdot,\cdot))$ be a metric space. That is, M is a nonempty set, and $d(\cdot,\cdot)$ is a nonnegative function on $M \times M$ which vanishes exactly on the diagonal, is symmetric, and satisfies the triangle inequality. A basic class of examples is provided by the set of vertices in a connected graph G, in which the distance between two points is given by the *path metric*, i.e., the minimum of the lengths of the paths in G connecting the points. (The length of a path in G is simply the number of edges which it traverses.)

One can think of metric spaces as being transcendental versions of *graphs*, but with more flexibility about what constitutes a single step. This analogy works better for *length spaces*, in which one asks that for every pair of points x, y in M there be a path γ in M which connects them and which has length equal to $d(x, y)$. (The *length* of γ can be defined as in Euclidean spaces, as the supremum of sums of successive distances over a finite sequence of points along γ.) Under this condition, the space is connected, and the metric is approximately like the path metric on a connected graph.

Note that the length of a path is always greater than or equal to the distance between its endpoints, because of the triangle inequality for $d(u, v)$ (which says that distances are less than or equal to sums of intermediate distances).

Actually, one often calls a metric space a length space if it satisfies the following slightly weaker condition: for each $x, y \in M$ and each $\epsilon > 0$, there is a path γ in M which goes from x to y that has length less than $d(x, y) + \epsilon$. Under modest hypotheses (that every closed and bounded subset of M be compact), it can be shown that this condition implies the previous one.

Here is a simpler condition which contains similar information. Instead of asking for a path γ between x and y, one asks that for each $\epsilon > 0$ there be a finite sequence of points $z_1, z_2, \ldots, z_n$ such that $z_1 = x$, $z_n = y$, $d(z_i, z_{i+1}) < \epsilon$ for all $i < n$, and

$$\sum_{i=1}^{n-1} d(z_i, z_{i+1}) < d(x,y) + \epsilon. \tag{19.1}$$

This is like asking for arbitrarily fine discrete approximations to paths as above. If closed and bounded subsets of M are compact, then the existence of these sequences also implies the existence of a path from x to y with length $d(x,y)$.

The Euclidean space $\mathbf{R}^n$ with the standard metric $|x-y|$ is a length space. For each $x, y \in \mathbf{R}^n$, the line segment between x and y has length $|x-y|$, and it is the unique path with this property. (In general, one need not have uniqueness.)

If E is a nonempty subset of $\mathbf{R}^n$, then E can be viewed as a metric space in its own right, using the same Euclidean distance. This metric space will be a length space in the first sense if and only if E is convex. Each of the conditions above implies that the closure of E is convex, and the converse is true for the third condition, involving finite sequences. These assertions are not difficult to verify.

Convexity is a rather strong restriction, and one sometimes considers a weaker property of being *quasiconvex.*. If E is a subset of $\mathbf{R}^n$ and C is a positive real number, then E is said to be C-*quasiconvex* (as in [Gro81b, G$^+$99]) if for every $x, y \in E$ there is a path in E joining x and y which has length $\le C\,|x-y|$.

If every pair of points in a set E can be connected by a curve of finite length, then one can define a new distance function $d_1(x,y)$ on E by taking $d_1(x,y)$ to be the infimum of the lengths of all paths in E which connect x and y. This is analogous to the path metric on a graph, and it makes sense for general metric spaces (and not just subsets of $\mathbf{R}^n$). It is always true that $d_1(x,y)$ is greater than or equal to the usual distance between x and y. If E is C-quasiconvex, then $d_1(x,y)$ is less than or equal to C times the usual distance. This distance function $d_1(x,y)$ will automatically satisfy the second of the length-space conditions defined above, as one can verify.

Reasonably-nice domains or submanifolds of Euclidean spaces are quasiconvex. Some other examples include the Sierpinski gasket and fractal tree discussed in Section 17.6, and the Sierpinski carpet shown in Section 16.11.

The continuous versions of the Heisenberg groups have similar features. They are length spaces if one chooses the metrics properly, and otherwise one still has curves of length at most a constant times the distance between their endpoints, if one measures distances as in Section 17.3. This is closely related to the fact (discussed in Section 17.3) that the word metrics on the discrete Heisenberg groups are bounded by constants times the distances defined in Section 17.3.

It is by no means true that all metric spaces behave well in terms of conditions like these. Connectedness and local connectedness properties are involved, and also more than that. A basic example of this is given by $\mathbf{R}^n$ with a "snowflake" metric of the form $|x-y|^\alpha$, where $0 < \alpha < 1$. With α in this range, the triangle inequality still holds for this distance function, but there are no nontrivial paths of finite length. The analogous condition in terms of finite sequences fails in the same way, as one can show. (I.e., even for finite sequences, the lengths become too large, when the size of the steps is small.) On the other hand, these spaces

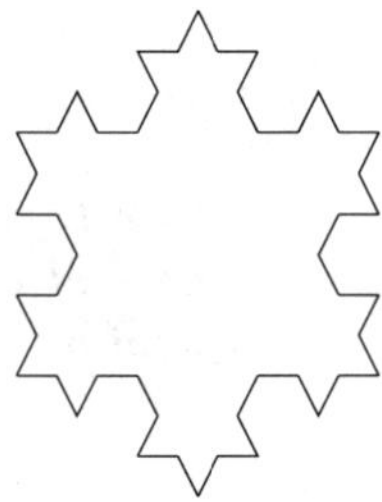

FIG. 19.1. A snowflake curve (after three generations)

are connected, and they have the property that arbitrary pairs of points can be connected by paths whose *diameter* (instead of length) is equal to the distance between the points.

The more classical versions of "snowflakes" were fractal curves in the plane constructed through explicit procedures, and with roughly the same kind of geometry. (Bilipschitz mappings are relevant for the latter.) See Fig. 19.1 for an illustration of the Von Koch snowflake, after three generations of the construction.

Let us say that a metric space $(M, d(x, y))$ is a *quasi-length space* if there is a constant $C > 0$ so that every pair of points $x, y \in M$ can be joined by a curve with length $\leq C\, d(x, y)$. One can also consider a version of this defined in terms of finite sequences of points, as we did earlier in this section. This will be come up explicitly in the hypotheses of Lemma 19.2 in the next section.

19.2 Discretizations of metric spaces

Let $(M, d(x, y))$ be a metric space. We would like to look at approximations of M which are graphs (perhaps infinite).

For this chapter, we shall make the following standing assumption:

If B is any ball in M and $r > 0$, then B can be covered by a finite number of balls of radius r. (19.2)

This condition ensures that bounded regions of M should admit *finite* approximations of arbitrary precision.

Fix $t > 0$. This parameter determines the *scale* at which we want to make our approximation.

Lemma 19.1 *Notations and assumptions as above. There is a subset A_t of M which is t-separated, which means that*

$$d(x, y) \geq t \quad \textit{whenever } x, y \in A_t,\ x \neq y, \tag{19.3}$$

and which practically exhausts M at the scale of t, in the sense that

$$M \subseteq \bigcup_{x \in A_t} B(x, t) \tag{19.4}$$

holds. Here $B(x,t)$ denotes the open ball in M with center x and radius t, i.e., $B(x,t) = \{z \in M : d(x,z) < t\}$.

Moreover, given any t-separated subset E of M and a ball B in M, there can be at most a finite number of elements of E inside B. This number can be bounded by a constant that does not depend on E (but which may depend on B, M, and t).

Proof This is quite standard (in some quarters). To prove the *existence* of such a set A_t, we shall want to find a *maximal* t-separated subset of M. To do this, it will be helpful to establish the last part of the lemma first.

Given a ball B in M and $t > 0$, there is a number $N = N(B,t)$ such that any t-separated subset E of M has at most N elements that lie in B. Indeed, our finite covering assumption (19.2) ensures that we can cover B by a finite number N of open balls of radius $t/2$. Any t-separated subset of M cannot have more than one element which lies in one of these smaller balls, and hence cannot contain more than N points in all of B.

This uniform bound permits us to do the following. Given any ball B and a t-separated subset E of B (which may be empty), there is a *maximal* t-separated subset E_1 of B such that $E_1 \supseteq E$. (Here "maximal" means with respect to set-theoretic inclusion.) One can simply take E_1 to be a t-separated subset of B that contains E and which has a maximal number of elements. Such a subset exists and is finite, because of our uniform bound on the possible number of elements of a t-separated subset of B. (Imagine simply adding points to E, one-by-one, until one has to stop.)

Next, we want to show that there is a *maximal* t-separated subset A of M. To do this, we begin by fixing an increasing sequence of balls $\{B^j\}_{j=1}^{\infty}$ whose union is all of M. (For instance, take $B^j = B(p,j)$ for some $p \in M$.) For each j, there is a subset A^j of B^j which is t-separated, contains A^{j-1} (when $j > 1$), and is maximal. This follows from the observation of the preceding paragraph. Once we have this sequence $\{A^j\}$, we take A to be the union of all the A^j's. It is easy to see that A is t-separated, since the A^j's are t-separated, and since $A^j \subseteq A^{j+1}$ for each j. As for maximality, if $z \notin A$, then $A \cup \{z\}$ cannot be t-separated, because $A^j \cup \{z\}$ fails to be t-separated as soon as j is large enough that $z \in B^j$, by the maximality of the A^j's.

Let us check that the maximality of A implies (19.4). If z were a point in M which did not manage to lie inside $B(x,t)$ for some $x \in A$, then we would be able to add z to A and maintain t-separatedness. This would violate maximality. This completes the proof of Lemma 19.1. □

Let us convert this discrete approximation A_t of M into an (unoriented) *graph* $\mathcal{A}_t$. Given $t > 0$, we take A_t for the set of vertices in the graph $\mathcal{A}_t$, we attach an edge between a pair of distinct points x and y in A_t when

$$d(x,y) \leq 4\,t, \tag{19.5}$$

and not otherwise. This fits well with the idea of A_t being an approximation to M at the scale of t. Notice that (19.5) holds when

$$B(x, 2t) \cap B(y, 2t) \neq \emptyset. \tag{19.6}$$

Also, the part of $\mathcal{A}_t$ that comes from a bounded region in M is always finite, because of the last assertion in Lemma 19.1.

The next lemma says that if M is a length space (or approximately like a length space), then the path metric on $\mathcal{A}_t$ provides a good approximation to the original metric on M.

Lemma 19.2 *Notations and assumptions as above. Assume also that there is a constant $C > 0$ so that if x and y are distinct points in M and $\epsilon > 0$, then there exists a finite sequence of points $z_1, z_2, \ldots, z_n$ in M such that $z_1 = x$, $z_n = y$, $d(z_i, z_{i+1}) < \epsilon$ when $1 \leq i < n$, and*

$$\sum_{i=1}^{n-1} d(z_i, z_{i+1}) \leq C\, d(x, y). \tag{19.7}$$

(This is a simplified version of the "quasi-length space" property mentioned at the end of Section 19.1.)

Fix $t > 0$, let A_t be as in Lemma 19.1, and let $\mathcal{A}_t$ be the graph obtained from A_t as above. Let $\delta(x, y)$ be the metric on A_t defined by paths in $\mathcal{A}_t$, so that $\delta(x, y)$ is the minimal number of edges in $\mathcal{A}_t$ traversed by a path connecting x and y when $x, y \in A_t$. Then $\delta(x, y)$ is finite for all $x, y \in A_t$ (which is the same as saying that $\mathcal{A}_t$ is connected), and

$$C^{-1}\, t\, \delta(x, y) \leq d(x, y) \leq 4\, t\, \delta(x, y) \tag{19.8}$$

for all $x, y \in A_t$.

The factor of t in (19.8) reflects the fact that each edge in the graph corresponds to a step of about size t in M.

Proof This is also quite standard (in some quarters).

The upper bound for $d(x, y)$ in (19.8) is automatic. This follows from the triangle inequality for the metric $d(\cdot, \cdot)$ and the fact that adjacent elements of A_t satisfy (19.5).

The lower bound in (19.8) is *not* automatic, but requires our hypothesis concerning the quasi-length space property. Indeed, Lemma 19.2 would not work for metric spaces like the Cantor set or a snowflake.

Fix $t > 0$ and $x, y \in A_t$. We may as well assume that $x \neq y$. By assumption, there exists a finite sequence of points $z_1, z_2, \ldots, z_n$ in M such that $z_1 = x$, $z_n = y$, $d(z_j, z_{j+1}) < t$ when $1 \leq j < n$, and such that (19.7) holds. We would like to convert this sequence of points into one which lies in A_t and enjoys similar properties.

We define a new sequence $w_1, \ldots, w_{k+1} \in A_t$ as follows. Set $w_1 = x$. Let $j_1 \leq n$ be the *smallest* positive integer such that $z_{j_1} \notin B(x, t)$. Such a j_1 exists, because $y = z_n$ does not lie in $B(x,t)$ (since $x \neq y$, $x, y \in A_t$, and A_t is t-separated). We choose $w_2 \in A_t$ so that $z_{j_1} \in B(w_2, t)$, which we can do, by (19.4).

Now choose j_2, if possible, so that j_2 is the smallest integer which satisfies $j_1 < j_2 \leq n$ and $z_{j_2} \notin B(w_2, 2t)$. If this is not possible, so that $z_j \in B(w_2, 2t)$ for all $j > j_1$, $j \leq n$, then we simply stop the construction. (Note that we use $B(w_2, 2t)$ here, rather than $B(w_2, t)$, as in the first step. This reflects the fact that we do not have $z_{j_1} = w_2$, but $z_{j_1} \in B(w_2, t)$, while in the first step we had $z_1 = x = w_1$.) We again choose $w_3 \in A_t$ so that $z_{j_2} \in B(w_3, t)$.

We repeat this process for as long as we can. This gives a finite sequence of integers $1 < j_1 < j_2 < \cdots < j_k \leq n$ and points $w_1, \ldots, w_{k+1} \in A_t$ with the following properties: (a) $w_1 = x$; (b) $z_{j_i} \in B(w_{i+1}, t)$ for $i = 1, \ldots, k$; (c) $z_j \in B(w_{i+1}, 2t)$ whenever $j_i < j < j_{i+1}$; (d) $z_{j_i} \notin B(w_i, 2t)$, $1 \leq i \leq k$. For $i = k$, we have $z_j \in B(w_{k+1}, 2t)$ when $j_k \leq j \leq n$ instead of (c). In particular,

$$d(y, w_{k+1}) < 2t, \tag{19.9}$$

since $y = z_n$.

Let us check that w_i is adjacent to w_{i+1} in $\mathcal{A}_t$ for all $1 \leq i \leq k$. To do this, it is enough to show that

$$B(w_i, 2t) \cap B(w_{i+1}, 2t) \neq \emptyset \tag{19.10}$$

for all i, as in (19.6). For each i, we have that $z_{j_i - 1}$ lies in $B(w_i, 2t)$, by construction (as in (c) above). Let us check that it also lies in $B(w_{i+1}, 2t)$. This is true because $z_{j_i} \in B(w_{i+1}, t)$, as in (b), while

$$d(z_{j_{i+1}-1}, z_{j_{i+1}}) < t \tag{19.11}$$

because of the properties of the z_j's at the beginning of the argument. This proves that (19.10) holds for all i, $1 \leq i \leq k$.

Let us define w_{k+2} to be equal to y. Then w_{k+2} is adjacent to w_{k+1} in $\mathcal{A}_t$, because of (19.9) and (19.5). Thus the sequence $w_1, \ldots, w_{k+2} \in A_t$ defines a path in $\mathcal{A}_t$ which connects x to y. It has length $k+1$ by definition, so that

$$\delta(x, y) \leq k + 1. \tag{19.12}$$

We need to get a bound for $k+1$ in terms of $d(x,y)$. For this purpose, we look for an estimate in terms of the sum of the distances between the successive z_j's.

A key point is that

$$t \leq \sum_{j=j_i}^{j_{i+1}-1} d(z_j, z_{j+1}) \tag{19.13}$$

when $1 \leq i < k$. To see this, observe that

$$d(z_{j_i}, z_{j_{i+1}}) \leq \sum_{j=j_i}^{j_{i+1}-1} d(z_j, z_{j+1}) \tag{19.14}$$

for all $i < k$, by the triangle inequality. We also have that

$$t \leq d(z_{j_i}, z_{j_{i+1}}), \tag{19.15}$$

because $z_{j_i} \in B(w_{i+1}, t)$ and $z_{j_{i+1}} \notin B(w_{i+1}, 2t)$, by construction. This proves (19.13). Similarly,

$$t \leq \sum_{j=1}^{j_1-1} d(z_j, z_{j+1}), \tag{19.16}$$

because $z_1 = x$ and $z_{j_1} \notin B(x, t)$, by construction.

Summing over i in (19.13), and adding in (19.16), we get that

$$k\,t \leq \sum_{j=1}^{j_1-1} d(z_j, z_{j+1}) + \sum_{i=1}^{k-1} \sum_{j=j_i}^{j_{i+1}-1} d(z_j, z_{j+1}) \leq \sum_{j=1}^{j_k-1} d(z_j, z_{j+1}). \tag{19.17}$$

Let us assume for the moment that

$$t \leq d(z_{j_k}, y). \tag{19.18}$$

In this case, we have that

$$t \leq \sum_{j=j_k}^{n-1} d(z_j, z_{j+1}), \tag{19.19}$$

by the triangle inequality again (since $z_n = y$), and hence that

$$(k+1)\,t \leq \sum_{j=1}^{n-1} d(z_j, z_{j+1}), \tag{19.20}$$

by (19.17). From the choice of the z_j's, we have that

$$\sum_{j=1}^{n-1} d(z_j, z_{j+1}) \leq C\,d(x, y), \tag{19.21}$$

as in (19.7). Combining this with the previous inequality, we get that

$$(k+1)\,t \leq C\,d(x, y). \tag{19.22}$$

The lower bound for $d(x, y)$ in (19.8) now follows from this and (19.12), at least when (19.18) holds. If (19.18) does not hold, then we can go back adjust the sequence, taking $w_{k+1} = y$ and dropping the w_{k+2} term. We still have that $d(z_{j_k}, w_{k+1}) < t$, and the sequence $w_1, \ldots, w_{k+1}$ defines a path in $\mathcal{A}_t$ from x to y

(with w_k adjacent to w_{k+1} for the same reasons as before). This path has length k, so that $\delta(x,y) \le k$, instead of (19.12). On the other hand, (19.17) and (19.21) yield

$$k\,t \le C\,d(x,y), \tag{19.23}$$

and this again implies that $t\,\delta(x,y) \le C\,d(x,y)$, as desired. This completes the proof of Lemma 19.2. □

19.3 The scale–geometry graph

Assume that M satisfies the standing assumption (19.2) mentioned near the beginning of Section 19.2. For each $t > 0$, we can choose A_t as in Lemma 19.1, and we can make it into a graph $\mathcal{A}_t$ in the same way as before.

We want to define now a graph which takes into account *all* scales and locations, and not just the behavior of M at a fixed scale t. We shall call this graph the "scale–geometry graph", and denote it by U. For the set of vertices of U, we take the set of all ordered pairs (x,t) with $x \in A_t$ and t is an integer power of 2 (including *negative* powers of 2). (It is natural to restrict oneself to t's which are not too much larger than the diameter of M, if it is finite. One might go up to the first power of 2 which is greater than or equal to the diameter of M, for instance.) For the edges, we do the following. We attach an edge between (x,t) and (y,t) whenever we attached an edge between x and y in $\mathcal{A}_t$ before. These are called the *horizontal edges* in U. We also add some edges *between* scales, as follows. Given vertices (x,t) and $(z,2t)$, we attach an edge between them whenever

$$d(x,z) \le 2t. \tag{19.24}$$

These are called the *vertical edges* in U. Notice that for each $x \in A_t$, there is a $z \in A_{2t}$ which satisfies (19.24), and that for each $z \in A_{2t}$, there is an $x \in A_t$ which satisfies (19.24). This follows from the exhaustion property (19.4) of the A_t's, as in Lemma 19.1.

This defines the graph U. Its vertices represent all possible locations and scales in M, to within modest errors, while its edges represent transitions between scales and locations of approximately "unit" size, where the units reflect the scale t at which one is working.

This is like the coding of triangles in the Sierpinski gasket by words over a finite alphabet (in Section 17.6), where each new letter represented exactly one step in the chain of scales. We had a similar phenomenon for the coding of the vertices in the linear graphs L_n by binary sequences in Section 17.6. Similar phenomena occurred in Section 5.5, but in the opposite order; we started with a tree, and then defined a geometry on the limiting set, in such a way that the relationship between the two would be like the relationship between U and M here.

If M is just the usual Euclidean space $\mathbf{R}^n$, then the graph U has approximately the same large-scale geometry as the upper-half space $\mathbf{R}^n \times (0,\infty)$ equipped with the standard hyperbolic metric. There is a similar statement for

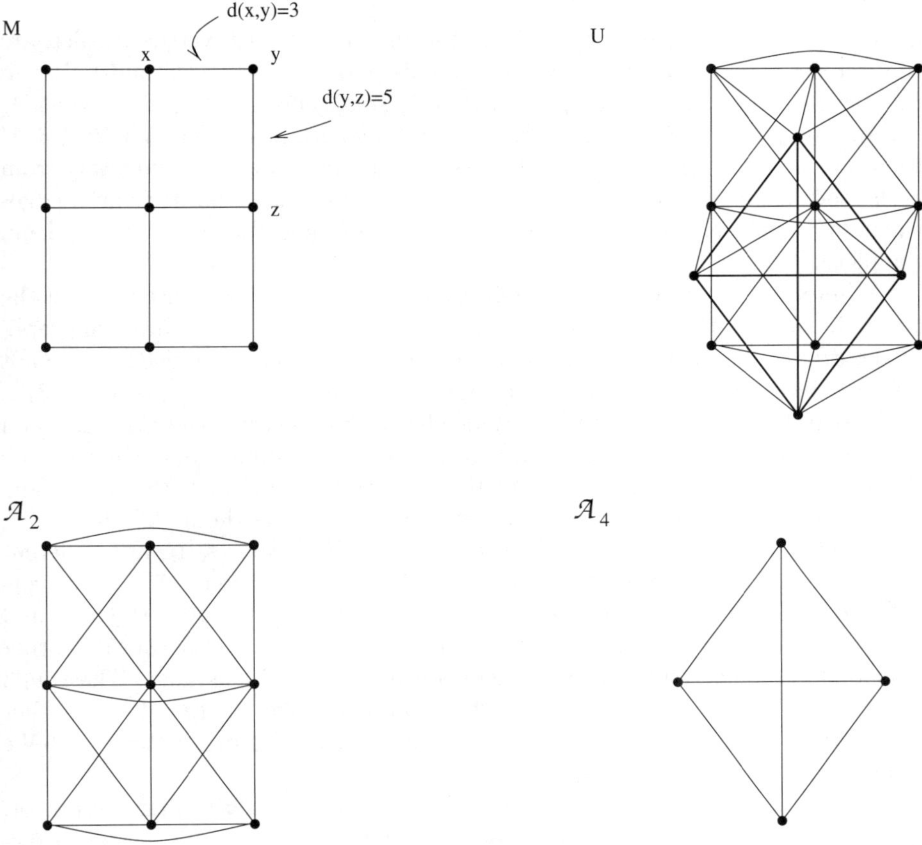

FIG. 19.2. The construction of U

the Heisenberg group with its distance function (17.15) and "complex hyperbolic space". If M is a Cantor set (with some geometric conditions), then we get an approximation to the usual kind of infinite tree which has the Cantor set as its natural limit. In general, this is a standard construction in geometry and analysis.

A concrete illustration of the construction of the graph U is shown in Fig. 19.2. In this picture, the metric space M is represented by the set of vertices in a graph. The edges provide a way to measure distances between these points, through the lengths of paths. For this we take the horizontal edges to have length 3, and the vertical edges to have length 5, rather than the usual custom of assigning unit length to every edge.

The diagram for U does not show all of the scales involved, but it does show the most interesting ones. For t large enough, the set A_t will consist of only a single element. In fact, we can choose the A_t's so that this will happen for $t > 8$ (and also for $t = 8$, if we reduce the distances in M slightly). This leads to an infinite (linear) string of vertices in U for the large scales, vertices which are connected successively by vertical edges. This part of U has been omitted from the picture. The opposite happens when t is small enough ($t < 3$), in that A_t necessarily consists of all points in M. For t somewhat smaller still ($t < 3/4$) there are no edges in the graph $\mathcal{A}_t$, because the points are too far away from each other compared to t. Thus the copies of A_t in U have no horizontal edges when t is small, only vertical edges. This part of U has also been omitted from the picture.

At intermediate levels there is more structure. The set A_2 contains all of the points in M, but the graph $\mathcal{A}_2$ contains more edges than in the original graph. In the graph $\mathcal{A}_4$, we did not use all of the vertices in M, and indeed we cannot do this, since the elements of A_4 are supposed to be 4-separated (as in (19.3)).

Notice that there are several possible choices for the set A_4 in this case, and hence several possible choices for the graph $\mathcal{A}_4$. For instance, one could choose A_4 to consist of the four vertices in the corners together with the one in the middle, or so that it contains just the three vertices along the middle line.

In general, there is no special recipe for choosing the sets A_t, and different choices can lead to different graphs U. Similarly, the restriction to integer powers of 2 for the scale parameter t in the definition of U is somewhat arbitrary, and at times other choices may be more convenient. One can also consider other adjustments, concerning the way edges are attached, for instance. The rough geometry of U is not affected much by modest changes in the A_t's or other parameters, but precise patterns can be made more obscure in this way. We shall return to these matters in Section 19.6.

In other words, for a particular metric space M, like a self-similar Cantor set, there might be a natural way to choose U so that it has special properties. However, this choice may depend on knowing about M in a fairly precise way. Here we make a general construction which does not involve any special information or features of M.

Remark 19.3 The scale–geometry graph U reflects the behavior of M in a good way whether or not M satisfies anything like a quasi-length space condition, as in Lemma 19.2. For that matter, M can be disconnected, as in the case of a Cantor set. The main point is that one can always get from one location and scale to another by going up to a sufficiently large scale (compared to the two scales and the distance between the two locations with which one starts), and then down again along a different route (to the other scale and location). We shall discuss some matters related to this in Section 19.7.

In the next section, we discuss conditions under which these approximations have locally bounded complexity.

19.4 Conditions of bounded geometry

Let us be a bit more careful about the nature of the graphs $\mathcal{A}_t$ and U associated to a metric space as in the preceding sections.

We made a point of assuming (in (19.2)) that every ball in our metric space could be covered by finitely many balls of arbitrarily small radius. This ensures that $\mathcal{A}_t$ and U are always locally finite, but how finite are they? It would be better if they did not have too many edges coming out of their vertices.

Let us start with the graphs $\mathcal{A}_t$ which approximate a given metric space M as in Section 19.2.

Definition 19.4 *A metric space* $(M, d(x,y))$ *has* locally bounded geometry *if, for each choice of positive real numbers* r, R *with* $r < R$*, there is a constant* $K = K(r, R) > 0$ *so that any ball of radius* R *can be covered by at most* K *balls of radius* r*.*

For example, this holds when M is compact, or when closed and bounded subsets of M are compact, and M admits a transitive group of isometries. In particular, it is true for the Euclidean spaces $\mathbf{R}^n$, and for the standard hyperbolic spaces. If M has locally bounded geometry, then the same is true of any subset of M, viewed as a metric space in its own right (and with the same distance function as on M). If G is an infinite connected graph, then the set of vertices of G equipped with the path metric has locally finite geometry if and only if the number of edges attached to any vertex in G is bounded. This is not hard to verify. Notice that the Cayley graph of any finitely-generated group satisfies this criterion, by construction.

Lemma 19.5 *If* $(M, d(x,y))$ *is a metric space with locally bounded geometry, then for each* $t > 0$ *there is a uniform upper bound (which may depend on* t*) for the number of edges in* $\mathcal{A}_t$ *attached to any given vertex in* $\mathcal{A}_t$*.*

Proof Let us first check that if B is any closed ball in M of radius $4t$, then B can only contain a bounded number of elements of A_t, where the bound may depend on t but not on B. This follows from the same argument as the one given at the beginning of the proof of Lemma 19.1 in Section 19.2. Specifically, the assumption of locally bounded geometry implies that we can cover B by a bounded number of open balls of radius $t/2$, and none of these smaller balls can contain more than 1 element of A_t, because of the t-separated property (19.3).

Thus the ball B can only contain a bounded number of elements of A_t. Using this, it is easy to control the number of edges attached to any fixed vertex in $\mathcal{A}_t$, because of the criterion for attaching edges to $\mathcal{A}_t$ given in (19.5). This proves Lemma 19.5. □

The following is condition on M prevents the bounds from depending on t.

Definition 19.6 *A metric space* $(M, d(x,y))$ *is said to be* doubling *if there is a constant* $C > 0$ *so that every ball* B *in* M *can be covered by at most* C *balls with half the radius of* B*.*

This condition is satisfied by the Euclidean space $\mathbf{R}^n$, for example. One can verify that all subsets of a metric space which is doubling are doubling in their own right (using the same metric). Also, the doubling condition is preserved by bilipschitz equivalence (defined in Section 17.2). For both of these assertions, the next lemma can be useful.

Lemma 19.7 *If $(M, d(x, y))$ is doubling, then there exist constants $C_1, d > 0$ such that each ball B in M of radius R can be covered by at most $C_1(R/r)^d$ balls of radius r, where R, r are arbitrary positive numbers with $R \geq r$.*

Proof This follows easily by iterating the doubling condition. That is, B can be covered by at most C balls of radius $R/2$, at most C^2 balls of radius $R/4$, C^3 balls of radius $R/8$, etc. □

This shows that metric spaces which are doubling satisfy a kind of "polynomial growth condition". This is similar to the considerations of volume growth that we have encountered before, as in Section 4.11 and Chapter 17, especially Sections 17.4, 17.8, and 17.9. Spaces roughly like the standard hyperbolic spaces, or (nontrivial) homogeneous trees, or Cayley graphs of hyperbolic groups, do not behave in this way, but instead have *exponential* growth. They do have locally bounded geometry, but the constant $K(r, R)$ from Definition 19.4 does not remain bounded when R gets large.

The continuous version of the Heisenberg groups do satisfy the doubling property, with respect to the geometry discussed in Section 17.3. This can be shown using the translations and dilations on a Heisenberg group to reduce the question for an arbitrary ball to that of a single ball, e.g., the ball centered at the identity element and having radius 1. For that ball, one can make coverings directly. This implies the analogous statement for the discrete version of the Heisenberg group, equipped with the word metric, because the word metric is bounded from above and below by the one in Section 17.3, as discussed there.

Lemma 19.8 *Suppose that $(M, d(x, y))$ is doubling. Then the number of edges attached to any vertex in $\mathcal{A}_t$ can be bounded from above by a constant which depends neither on the vertex in $\mathcal{A}_t$ nor on $t > 0$. There is also a uniform bound for the number of edges attached to any vertex in the graph U from Section 19.3.*

This can be checked in essentially the same manner as for Lemma 19.5.

We shall discuss the doubling condition further in Section 19.8. For the moment, we would like to return to the concept of automatic structures from Chapter 17.

19.5 Automatic structures

We have already discussed notions of automatic structures for graphs (beginning in Section 17.6), but what about analogues for metric spaces in general?

For spaces which are *length spaces* (or roughly like that), one might approximate the space by graphs in the manner discussed in Section 19.2. If the metric space is *doubling* (Definition 19.6), then these discrete approximations have only polynomial growth, and this suggests that it is better to make codings through *scales and locations* instead of *paths*. This follows the discussion in Section 17.8.

To do this, one can use the graph U defined in Section 19.3, in which moving around by paths corresponds to moving around by scales and locations in M. We know from Lemma 19.8 in the previous section that U will also have only a bounded number of edges coming out of each vertex, which is important if we want to have anything like an automatic structure.

What kind of automatic structure might we look for on U? We discussed several variations on this theme in Chapter 17. It would be reasonable to consider automatic structures which move around U by *paths* (and this is a basic point about working with U, or graphs like it).

There is some special structure in the present setting, including the division of edges in U into *horizontal* and *vertical*. We have a very natural subgraph T of U, which consists of the same set of vertices, but only *vertical* edges. For an automatic structure for U, one might like to have a regular language which represents T in a good way. Each new letter in a word could correspond to exactly one step "down" a vertical edge (at least if one starts at some fixed scale). One can also forget about words and languages here, and ask for a realization of T as the image of a visibility graph, or a finite union of visibility graphs. (Compare with Section 17.10.)

This would only be part of an automatic structure, since one should also have an good way to represent adjacency by horizontal edges. Of course, many standard fractals come with simple descriptions of this nature already, by construction. Indeed, this kind of general approach follows closely some of the particular examples discussed in Section 17.6.

We should be more careful about some of the foundational issues here, however, as in the next section.

19.6 Making choices

There is a flaw in the idea of looking for automatic structures for the scale–geometry graph U associated to a metric space M as above, which lies in the choice of the graph U. Although U does indeed reflect the approximate geometry of M in a good way, the details of its construction are somewhat ad hoc, and they might obscure some of the patterns or symmetry in M.

Consider the *Sierpinski gasket* and *fractal tree* from Section 17.6, for instance. Each has a natural division by scales, but these divisions proceed at different rates, by factors of 2 for the gasket and factors of 3 for the fractal tree. In the definition of U, we used 2 for the scale factor, but this was merely a simple selection to make in the absence of any other information. It would not be so good for the fractal tree.

There are similar problems with the construction of the graph $\mathcal{A}_t$ in Section 19.2. It was based on ad hoc choices for which there was a priori no clear preference. For the Sierpinski gasket and fractal tree, there are particular graphs which have the same basic role as $\mathcal{A}_t$, and which are very regular in their behavior (as in the discussions of Section 17.6), but it is not necessarily clear how to obtain graphs like these through a general procedure.

To formulate good notions of *automatic structures* for metric spaces, it makes sense to use graphs like U, but also to allow some freedom in the specific choice of the graph. We shall now define a class of "admissible graphs" for this purpose. We start with an auxiliary definition which will be useful.

Definition 19.9 *Let $(M, d(x, y))$ be a metric space. We say that two points (x, t) and (y, s) in $M \times (0, \infty)$ lie within a single "step" of each other if $d(x, y) \leq s + t$ and $t/2 \leq s \leq 2t$.*

In effect, this notion of a single step defines a "quasi-hyperbolic" geometry for $M \times (0, \infty)$ (which is quite standard). The idea is that the size of a step should be adapted to the scale on which one is working, as though one is looking at the space through an adjustable microscope.

This normalization of a "step" in $M \times (0, \infty)$ is a bit arbitrary, but this does not matter too much. Any other reasonable notion like this can be simulated by a bounded number of these steps, and vice-versa.

The graphs that we are concerned with here are ones whose internal geometry approximates the geometry by steps on $M \times (0, \infty)$. The next definition provides a precise version of this. Let $\mathrm{diam} M$ denote the diameter of M, i.e., the supremum of distances between pairs of points in M.

Definition 19.10 *Let $(M, d(x, y))$ be a metric space. We say that a graph W is* admissible *if it satisfies the following four conditions.*

First, every vertex in W should be an element of $M \times (0, \infty)$, and every element of $M \times (0, \mathrm{diam} M)$ should lie within a bounded number of steps of a vertex in W.

Second, if two vertices in W are adjacent to each other, then they should lie within a bounded number of steps from each other. Also, we do not allow there to be more than one edge between any given pair of vertices.

Third, the distance between any pair of distinct vertices in W should be bounded by a constant multiple of the minimal number of steps between the vertices in $M \times (0, \infty)$. (As usual, we measure distance between vertices in W by the minimal number of edges traversed by a path between the given vertices.)

Fourth, we ask that there be no more than a bounded number of vertices in W which lie within a single step of any other fixed vertex. (This prevents overcrowding in W.)

If the metric space M is doubling, then the graph U defined before satisfies all of these conditions. This is not hard to check from the definition of U. (The considerations of Lemmas 19.5 and 19.8 and their proofs are relevant here.)

Note that for any admissible graph W there is always a bound on the number of edges attached to a given vertex when the underlying metric space M is doubling. This is not hard to verify, using the prohibition against overcrowding and the requirement that adjacent vertices lie within a bounded number of steps of each other.

The third requirement in the definition above is easier to check than it might appear to be initially. The point is that if the graph W is reasonably well-connected locally, between vertices which lie within a bounded number of steps of each other, then the analogous property for vertices which are far apart will follow from a thickness condition like the first requirement.

For particular self-similar fractals, it is often very easy to find such graphs W which are admissible for the given space and which are very regular. This includes self-similar Cantor sets, the Sierpinski gasket, the fractal tree from Section 17.6, and the Sierpinski carpet (Section 16.11), for instance. In some situations, it may not be very easy to pass from self-similarity properties of a space M to an especially well-behaved choice of W, though. One might know that M enjoys strong self-similarity properties without knowing a particular procedure which lead to these properties. This is a basic problem for the class of "BPI spaces" in [DS97]. ("BPI" stands for "big pieces of itself".)

19.7 A geometric caveat

Let M be a metric space, which we assume to be doubling, and let W be an admissible graph, in the sense of Definition 19.10. To what extent can we recapture the geometry of M from the geometry of W?

Roughly speaking, the answer is that the "bilipschitz" geometry of W determines the "quasisymmetric" geometry of M, and conversely, but that the bilipschitz geometry of W does *not* determine the bilipschitz geometry of M (or anything too close to that) without additional information, or except in certain situations. We shall not pursue these phenomena in detail here, but we want to give at least some indications about them. (These matters are very well-known in some contexts of geometry and analysis. Compare with [Ahl87, Pan89a], for instance.)

In speaking of "bilipschitz" geometry, we mean to allow measurements of distance which are known to within a bounded factor. This is natural for our admissible graphs W, whose main properties are only defined to within bounded distortions. In "quasisymmetric" geometry, one does not really know distances at all, a priori. One knows only *relative* distances, e.g., z is much closer to x than y is, or p is much further from x than y is, without saying how far away p actually is. From this kind of information, one can often deduce approximate sizes of *logarithms* of distances, but not the distances themselves.

For quasisymmetric geometry, one knows roughly what a ball looks like, and what it means to expand a ball by a bounded factor, but one does not know anything a priori about the *diameter* of a ball. This is part of the idea of *relative* distances. It is not hard to see how things like this can correspond to the geometry of "steps" on $M \times (0, \infty)$ as in Definition 19.9.

Here is a basic class of examples. Let $(M, d(x, y))$ be a metric space, and let $(M, d(x, y)^s)$ be a of M, $0 < s < 1$. These two spaces automatically have the same quasisymmetric geometry, even if they can be very different in more precise ways. (For instance, $(M, d(x, y))$ might contain nontrivial curves of finite length, as in the case of ordinary Euclidean spaces, but $(M, d(x,y)^s)$ never does when $s < 1$. Similar matters arose in Section 19.1.) Notice that these two spaces have exactly the same classes of balls, even if they can assign substantially different values to the radii of the balls. The idea of "expansion of a ball by a bounded factor" is essentially the same for both spaces, in the sense that increasing the radius of a ball in $(M, d(x,y))$ by a factor of $\alpha > 1$ corresponds exactly to increasing the radius of a ball in $(M, d(x,y)^s)$ by a factor of α^s.

As a simpler situation, start with a metric space $(M, d(x,y))$, and consider modifying it by multiplying the metric $d(x,y)$ by some positive constant factor λ. Then distances in M change, perhaps by a lot, but relative distances do not change. If one makes the change of variables in $M \times (0,\infty)$ given by $(x,t) \mapsto (x, \lambda t)$, then the notion of a "step" in $M \times (0,\infty)$ from Definition 19.9 for the original metric on M is equivalent to the same notion of a step in $M \times (0,\infty)$ for the new metric, after the change of variables. An analogous assertion holds for the snowflake transform described in the previous paragraph, and the change of variables $(x,t) \mapsto (x, t^s)$, except that the "steps" do not quite match up exactly. They do match up within simple bounds, and one could modify the definition of a step slightly so that they would match up exactly.

19.8 The doubling condition

Let us briefly mention some other facts about metric spaces which are doubling (Definition 19.6).

If $(M, d(x,y))$ is a metric space which is doubling, then the corresponding "snowflake" spaces $(M, d(x,y)^s)$ are also doubling $(0 < s < 1)$, and conversely. This can be checked using Lemma 19.7.

It is natural to ask whether every metric space which is doubling is actually bilipschitz equivalent to a subset of a Euclidean space. In some sense, the doubling condition provides the exactly the right "size" condition for this to work, but there can be additional subtleties to the geometry, and the answer to this question turns out to be *no*. Counterexamples are provided by the Heisenberg groups, as in Proposition 17.1. There is a positive result, however, due to P. Assouad [Ass77, Ass79, Ass83].

Theorem 19.11 (Assouad's embedding theorem) *If $(M, d(x,y))$ is a metric space which is doubling, and if s is a real number with $0 < s < 1$, then there is a bilipschitz embedding of $(M, d(x,y)^s)$ into some $\mathbf{R}^n$, where n and the bilipschitz constant for the embedding depend on s and the doubling constant for M (from Definition 19.6).*

Note that the doubling condition on M is *necessary* for such an embedding to exist for any s in $(0,1)$. This uses the statements mentioned just after Definition 19.6.

Assouad's theorem has the useful feature that it often permits one to work inside a finite-dimensional Euclidean space, even if one starts with spaces which are more abstract.

20

THE CORONA DECOMPOSITION REVISITED

We have discussed some notions of symmetry and compact descriptions for graphs in earlier chapters. In particular, one might say that the visibility of an oriented graph G enjoys a lot of symmetry when the visibility is much larger than G itself, even if this symmetry is not reflected in a perfect kind of periodicity (especially when the visibility is finite).

The patterns that do exist are brought out in part by the Calderón–Zygmund and Corona decompositions, as in Sections 8.3 and 8.4. A key point is provided by Lemma 8.4, which says that the subgraphs $F(s)$ of the visibility are isomorphic when the corresponding s's project down to the same vertex in G. This will occur with some frequency when the visibility is large compared to G.

In this chapter, we present an alternate view of the Corona decomposition which employs notions about mappings and weak mappings between graphs in a stronger way.

20.1 Interesting paths

The following provides a refinement of the notion of *focal pairs* (from Section 4.14) which will be useful for us.

Definition 20.1 (Interesting paths) *Let G be an oriented graph, and let p be an oriented path in G, with initial vertex u and final vertex w. We call p an* interesting path *if (u, w) is a focal pair, and if p does not reach a vertex z such that (u, z) is a focal pair before it reaches the end of the whole path (i.e, the final occurrence of w).*

This concept was implicit in our earlier constructions in Sections 8.3 and 8.4. Now we want to bring it out explicitly, and use it to make a slightly different geometric construction.

In this chapter, we shall often restrict ourselves to situations where the paths of interest cannot reach an oriented cycle. This will prevent the possibility of interesting paths which are degenerate.

Let us make another definition, and then explain how oriented paths can be decomposed into interesting ones.

Definition 20.2 (Dull paths) *Let G be an oriented graph, and let p be an oriented path in G. We call p a* dull path *if it does not contain a focal pair of vertices.*

Fig. 20.1 shows an example of an interesting path p and a dull path q.

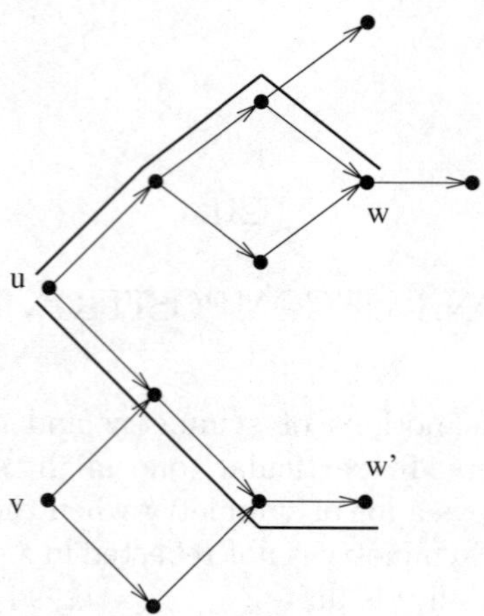

FIG. 20.1. An interesting path (in the top half of the diagram) and a dull path (in the bottom half)

Lemma 20.3 (Checking for dull paths) *If G is an oriented graph, and p is an oriented path in G which begins at a vertex u, then p is dull as soon as there is no vertex z lying on p such that (u,z) is a focal pair.*

Proof Indeed, if p is not dull then it means that there are vertices x and y lying on p such that (x,y) is a focal pair. This implies that (u,y) is a focal pair. This is easy to check from the definitions, since we know that there is an oriented path from u to x, namely a subpath of p. This proves the lemma. □

Note that this lemma is valid when G contains oriented cycles. We simply have to allow the vertices to coincide, or for y to come before x along p. Neither of these possibilities would disturb the proof.

Lemma 20.4 (Decomposition into interesting subpaths) *Let G be an oriented graph without nontrivial oriented cycles, and let α be a (finite) oriented path in G. Then there is nonnegative integer k and a sequence of paths $\alpha_1, \alpha_2, \ldots, \alpha_{k+1}$ with the following properties:*

(a) the initial point of α_j is the endpoint of α_{j-1} for each $j \geq 2$;

(b) each α_j is a subpath of α, and in fact α is exactly the path obtained by starting at the initial point of α_1, following α_1 to the end, then following α_2 to the end, and so forth, until we reach the end of α_{k+1};

(c) α_j is both interesting and nondegenerate for $1 \leq j \leq k$;

(d) α_{k+1} is a dull path (and perhaps degenerate as well).

The assumption that there be no nontrivial oriented cycles can be weakened to the condition that there be no oriented cycles which have a vertex in common with α.

Proof This is pretty straightforward. It may be that α is itself dull, in which case we take $k = 0$ and $\alpha_1 = \alpha$. Otherwise, we start at the initial vertex u of α,

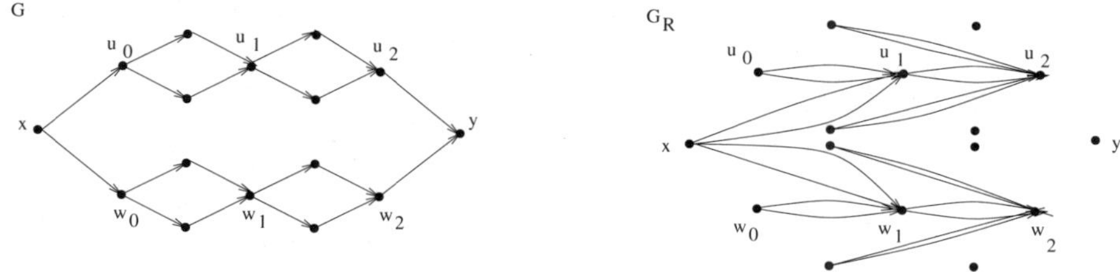

FIG. 20.2. An example of the reduced graph

and follow α until the first time that we reach a vertex w_1 such that (u, w_1) is a focal pair. We must reach such a vertex w_1 if α is not dull, because of Lemma 20.3. Our assumption that G be free of nontrivial oriented cycles implies that $w_1 \neq u$. We take α_1 to be the subpath of α which goes from u to this first focal vertex w_1, so that α_1 is both interesting and nondegenerate by construction. We denote by γ_1 the remaining part of α.

Now we repeat the process. If γ_1 is dull then we stop. Otherwise, we follow it up to the first time that we reach a vertex w_2 such that (w_1, w_2) is a focal pair. Again $w_2 \neq w_1$, since there are no oriented cycles. We take α_2 to be the subpath of α which goes from w_1 to w_2, and α_2 is both interesting and nondegenerate. We denote by γ_2 the part of α which remains.

We keep repeating this process until we have to stop. We have to stop eventually, since α is finite, and since each α_j has positive length. In the end, we get a decomposition of α as described in the lemma. This is easy to check. □

20.2 Reduced graphs

Definition 20.5 *Let G be an oriented graph. We define the* reduced graph G_R *associated to G (as an oriented graph) as follows. For the vertices of G_R, we use the same vertices as for G. Given a pair of such vertices u, w, we attach an edge going from u to w in G_R for each nondegenerate interesting path in G that goes from u to w. These are all the edges that we attach.*

Notice that the reduced graph contains no edges at all when G does not contain any focal pairs (and hence no interesting paths). A more interesting situation is shown in Fig. 20.2. In this example, we have vertices x and y in G such that (x, y) defines a focal pair in G, but y is not connected to x in G_R, and in fact y is an isolated point in G_R.

Distinct (non-isomorphic) oriented graphs can have reduced graphs that are the same, up to isomorphism. This is illustrated by the examples in Fig. 20.3.

A different type of example is given in Fig. 20.4. Unlike the previous examples, the reduced graph has exactly one edge going from u to v.

In a reduced graph G_R, one can have focal pairs that do not come about because of two edges between a single pair of vertices. An example of this is given in Fig. 20.5 (in the next section). Specifically, in the reduced graph G_R shown there, (u, z_2) is a focal pair, and this results from the oriented paths that

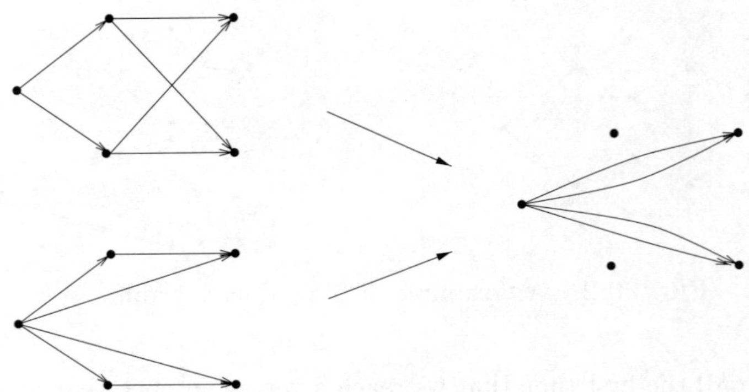

FIG. 20.3. Two oriented graphs and their common reduced graph (up to isomorphism)

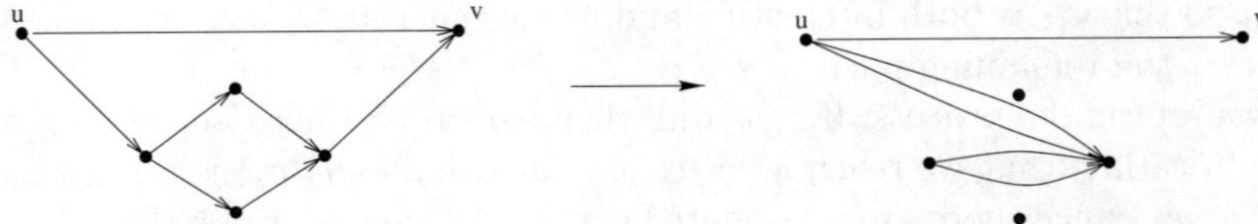

FIG. 20.4. Another example of a graph and its reduced graph

start at u and pass through z_1, x_1, z_2 and w_1, y_1, z_2, respectively. Similarly, (u, w_2) is a focal pair in this reduced graph, because of the oriented paths that start at u and pass through z_1, x_1, w_2 and w_1, y_1, w_2, respectively.

As in these examples, edges in the reduced graph often come in pairs, but not always. The next lemma provides a general statement related to this.

Lemma 20.6 *Let G be an oriented graph which has no nontrivial oriented cycles, and let u, w be vertices in G such that (u, w) is a focal pair in G. Then there is a vertex z in G such that (a) there are (at least) two interesting paths in G that begin at u and arrive at z along different edges, and (b) there is an oriented path in G which goes from z to w.*

Condition (a) ensures that there are at least two edges in the reduced graph G_R that go from u to z.

This lemma does not work when G is allowed to contain nontrivial oriented cycles. For instance, u and w might be the only vertices in G, with an edge from u to w and one from w to itself, and no others. In this case, the path from u to w which consists only of a single step is an interesting path, but there are no others that start at u, because they are not allowed to go around the loop.

Note that Lemma 20.6 does work in the case shown in Fig. 20.4 (where there is no nontrivial oriented cycle). The lemma does not rule out the possibility of a pair of vertices with a single edge from one to the other in the reduced graph (as in Fig. 20.4), but it does lead to the existence of another vertex with two edges going to it in the reduced graph, and coming from the first of the previous two vertices. These features are clearly present in the example shown in Fig. 20.4.

Proof Let u and w be given as in the lemma. Let us say that a pair of oriented paths α, β in G is *admissible* if it satisfies the following conditions: α and β both begin at u; they both end at the same vertex x; they arrive at x along different edges; and there is an oriented path in G which goes from x to w.

Our assumption that (u, w) be a focal pair implies that an admissible pair of paths exists, with $x = w$.

If x is any vertex in G which admits an oriented path to w, let $\lambda(x)$ denote the length of the *longest* such path. This number is always finite, because of our assumption that G be free of nontrivial oriented cycles.

Let α_0, β_0 be an admissible pair of paths which end at a vertex x_0 such that $\lambda(x_0)$ is as *large* as possible. Such a pair of paths exists, since there are only finitely many possible values for the quantity $\lambda(x)$.

Claim 20.7 *α_0 and β_0 are interesting paths.*

Indeed, suppose that this is not the case, so that one of the paths passes through a vertex y such that (u, y) is a focal pair, and so that the path in question does not reach y before coming to its end at x_0. If this happens, then we can find oriented paths α_1, β_1 which go from u to y and which arrive at y along different edges, since (u, y) is a focal pair. From the construction, we also know that there is a nontrivial oriented path in G which goes from y to x_0, namely a subpath of α or β, whichever contains y. This implies that there is an oriented path going from y to w, since there is already an oriented path which goes from x_0 to w (coming from the assumed admissibility of the pair α_0, β_0).

This shows that α_1, β_1 is an admissible pair of paths. On the other hand, the existence of a nontrivial oriented path going from y to x_0 ensures that

$$\lambda(y) > \lambda(x_0). \tag{20.1}$$

This contradicts the choice of α_0, β_0, and the claim follows.

To finish the proof of Lemma 20.6, we simply take $z = x_0$. This satisfies the required properties, since α_0 and β_0 are interesting paths which go from u to z, and because the existence of an oriented path from z to w comes from the admissibility of α_0, β_0. □

Corollary 20.8 *Suppose that G is an oriented graph which contains no nontrivial cycles, and let G_R denote the reduced graph. If G contains a chain of focal pairs of length n which begins at some vertex v, then the same is true of G_R.*

Notice that we cannot necessarily use the same vertices for the chain of focal pairs in G_R as we used in G. We saw this already in the example pictured in Fig. 20.2, for the single focal pair (x, y).

Proof To say that G contains a chain of focal pairs of length n starting at v means that there is a sequence of vertices $w_0, w_1, \ldots, w_n$ in G such that $w_0 = v$

and (w_i, w_{i+1}) is a focal pair for each $i < n$. (See Definition 4.17.) Let us define a new sequence $z_0, z_1, \ldots, z_n$ as follows. Set $z_0 = v$. Since (z_0, w_1) is a focal pair in G, we can apply Lemma 20.6 to get a vertex z_1 such that there exist two distinct interesting paths going from z_0 to z_1 and an oriented path going from z_1 to w_1. The latter ensures that (z_1, w_2) is a focal pair, since (w_1, w_2) is, and so we can repeat the process to find a vertex z_2 such that there are two interesting paths in G that go from z_1 to z_2 and an oriented path from z_2 to w_2.

Continuing in this manner, we end up with a sequence $z_0, z_1, \ldots, z_n$ such that $z_0 = v$ and there are (at least) two interesting paths from z_i to z_{i+1} for each $i < n$. Thus there are at least two edges in G_R going from z_i to z_{i+1} for each $i < n$, so that $z_0, z_1, \ldots, z_n$ provides a chain of focal pairs in G_R which begins at v and has length n. This proves Corollary 20.8. □

On the other hand, arbitrary oriented paths in G_R give rise to chains of focal pairs in G, as in the following lemma.

Lemma 20.9 *Let G be an oriented graph and let G_R be its reduced graph. Suppose that p is an oriented path in G_R which traverses the vertices $u_0, u_1, \ldots, u_n$ (in that order). Then $(u_0, u_1), (u_1, u_2), \ldots, (u_{n-1}, u_n)$ is a chain of focal pairs in G.*

Again, the converse to this is not true, as in the example shown in Fig. 20.2.

Proof This is an easy consequence of the definitions. □

Corollary 20.10 *Let G be an oriented graph which contains no nontrivial oriented cycles, and let G_R be the corresponding reduced graph. If v is any vertex in G, then the following three numbers are equal to each other: the length A of the longest chain of focal pairs in G which begins at v; the length B of the longest chain of focal pairs in G_R which begins at v; the length C of the longest oriented path in G_R which begins at v.*

Proof Indeed, $A \leq B$ by Lemma 20.6, $B \leq C$ automatically, and $C \leq A$ by Lemma 20.9. □

The length of the longest oriented path in an oriented graph (without nontrivial oriented cycles) can normally be much longer than the length of the longest chain of focal pairs. This is one of the main reasons for considering chains of focal pairs, since they can be used to obtain better estimates for the visibility than the naive exponential bounds in terms of the length of the longest oriented path (as in Lemma 4.9). With the reduced graph, we eliminate this gap, while retaining much of the important structure of the original graph G (in a way that is made precise in the next sections).

20.3 Crisp paths

We want to be able to compare the visibilities $\mathcal{V}_+(v, G)$ and $\mathcal{V}_+(v, G_R)$. To do this, we first make an identification between some of their vertices.

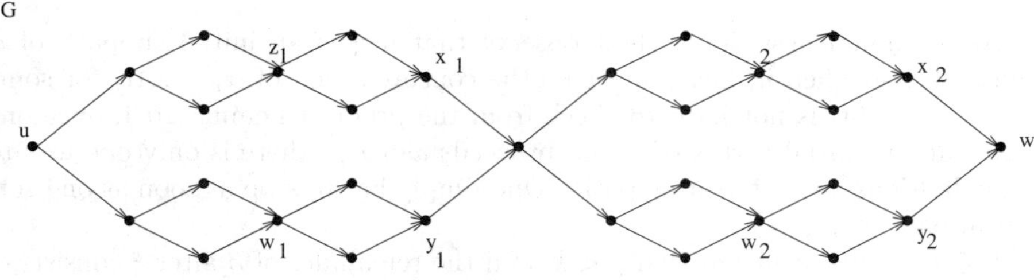

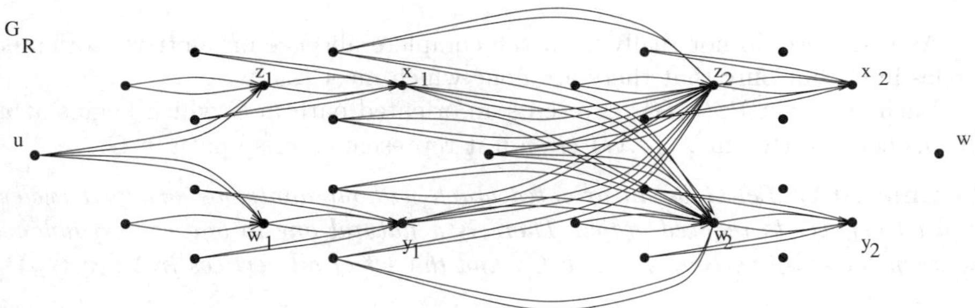

FIG. 20.5. A graph and its reduced graph

Each oriented path in G_R determines an oriented path in G. That is, each edge in G_R itself represents a path in G, and the paths corresponding to the edges in an oriented path can be connected together to make a path in G. Let us call an oriented path in G *crisp* if it comes from a path in G_R in this manner. (We allow degenerate paths in both G and G_R here.) With this terminology, we have the following.

Lemma 20.11 *Suppose that G is an oriented graph with no nontrivial oriented cycles. Then every oriented path α in G can be realized as the combination of a crisp path β followed by a dull path δ, and this realization is unique. (Either the crisp part or the dull part might be degenerate.) If ξ is any other initial subpath of α which is crisp, then ξ must be an initial subpath of β. If $\xi \neq \beta$, then the remainder of β after ξ must also be crisp.*

For some examples, consider the graph G shown in Fig. 20.5. If α is an oriented path in G which goes from u to w, then α cannot be crisp, but its penultimate subpath which reaches x_2 or y_2 is always crisp (no matter the specific choice of α). The last part of α, going from x_2 or y_2 to w, is always dull.

Proof Let G, α be given as above. Apply Lemma 20.4 to get subpaths $\alpha_1, \ldots,$ α_{k+1} of α as described there. We can take β to be the combination of $\alpha_1, \ldots, \alpha_k$, and δ to be α_{k+1}, to get our desired decomposition.

As for uniqueness, let us first observe that if ξ is an initial subpath of α which is crisp, then in fact ξ must be the concatenation of $\alpha_1, \ldots, \alpha_j$ for some $0 \leq j \leq k$. This is not hard to check from the proof of Lemma 20.4; once one begins at the initial vertex of α and proceeds along α, there is only one way to make divisions into interesting paths. One simply has to stop as soon as one gets a focal pair, etc.

If $\xi \neq \beta$, then it means that $j < k$, and the remainder of β after ξ consists of $\alpha_{j+1}, \ldots, \alpha_k$. In particular, this remainder is crisp.

If ξ is a crisp initial subpath of α, and if the remainder of α after ξ is dull, then we must have $j = k$, i.e., ξ must be the combination of $\alpha_1, \ldots, \alpha_k$. Otherwise, the remainder would contain a focal pair of vertices. From here uniqueness follows easily. □

As usual, we do not really need the complete absence of nontrivial oriented cycles in G, but only that there are none which meet α.

Each vertex in $\mathcal{V}_+(v, G)$ represents an oriented path in G which begins at v. Let us call a vertex in $\mathcal{V}_+(v, G)$ *crisp* if it represents a crisp path in G.

Lemma 20.12 *Let G be an oriented graph without nontrivial oriented cycles, and let G_R be its reduced graph. There is a natural one-to-one correspondence between the crisp vertices in $\mathcal{V}_+(v, G)$ and the set of all vertices in $\mathcal{V}_+(v, G_R)$.*

This would also work if we assumed instead that G contain no nontrivial oriented cycles which are accessible from v by an oriented path.

Proof Indeed, the crisp paths in G are in a natural one-to-one correspondence with all oriented paths in G_R, just because of the definition of a crisp path (and the observation of uniqueness in Lemma 20.11). This induces a one-to-one correspondence between the crisp vertices in $\mathcal{V}_+(v, G)$ and the set of all vertices in $\mathcal{V}_+(v, G_R)$, as desired. □

Given a vertex s in $\mathcal{V}_+(v, G)$, let us write $\mathcal{D}(s)$ for the set of vertices t in $\mathcal{V}_+(v, G)$ with the following property. Both t and s represent oriented paths in G which begin at v, and we ask that t be obtained from s by the addition of a dull path at the end. We allow a degenerate path here, so that s itself lies in $\mathcal{D}(s)$ (when G contains no nontrivial oriented cycles).

Lemma 20.13 *Suppose that G is an oriented graph with no nontrivial oriented cycles, and let v be a vertex in G. Then every vertex t in the visibility $\mathcal{V}_+(v, G)$ lies in $\mathcal{D}(s)$ for exactly one crisp vertex s in $\mathcal{V}_+(v, G)$.*

This also holds if we assume that there are no nontrivial oriented cycles in G which are accessible by an oriented path from v.

Proof This follows from Lemma 20.11. □

Thus we have a decomposition of the vertices in $\mathcal{V}_+(v, G)$ which is parameterized by the "crisp" vertices, which can themselves be identified with vertices in the visibility $\mathcal{V}_+(v, G)$. Let us look at what happens for edges.

Lemma 20.14 *Let G be an oriented graph without nontrivial oriented cycles, and let v be a vertex in G. Let t and t' be vertices in $\mathcal{V}_+(v, G)$ which lie in $\mathcal{D}(s)$ and $\mathcal{D}(s')$, respectively, where s and s' are crisp vertices in $\mathcal{V}_+(v, G)$. Suppose that there is an edge in $\mathcal{V}_+(v, G)$ that goes from t to t', and that $s \neq s'$. Then $t' = s'$, and the path in G represented by s' is the same as the path in G represented by s followed by an interesting path in G. In particular, there is an edge in G_R which goes from the endpoint of s to the endpoint of s'.*

Again, we could replace the complete absence of nontrivial oriented cycles here by the requirement that there be none which are accessible by an oriented path that begins at v.

Proof By definitions, t, t', s, and s' all represent oriented paths in G which begin at v, and in fact the paths t, s, and s' are all initial subpaths of t'.

From Lemma 20.11, we know that s' is the unique *maximal* crisp initial subpath of the path t'. In particular, the path s must be an initial subpath of the one represented by s', since it is both crisp and an initial subpath of t'. It must also be a *proper* subpath of s', since $s \neq s'$, by assumption. The remainder of s' after s is crisp, as in Lemma 20.11.

Since the paths t and s' are both initial subpaths of t', one of them has to be an initial subpath of the other. If s' is an initial subpath of t, then we get into trouble, because it would mean that t contains the remainder of s' after s, a crisp path, while the remainder of t after s is supposed to be dull (since t lies in $\mathcal{D}(s)$). Thus t must be an initial subpath of s', and in fact a proper subpath.

We have now that t is a proper initial subpath of s', that s' is an initial subpath of t', and that t' is obtained from t by adding just one extra edge. We conclude that $t' = s'$, as desired. We already know that the remainder of s' after s is a crisp path, but it must be an interesting path, since it consists of a dull path (the remainder of t after s) followed by a single edge. This proves Lemma 20.14. □

20.4 A weak mapping between visibilities

We can reformulate the decomposition of the visibility $\mathcal{V}_+(v, G)$ indicated in Lemma 20.13 as a "weak mapping" (Definition 10.2) from $\mathcal{V}_+(v, G)$ to $\mathcal{V}_+(v, G_R)$, in the following manner. Given a vertex t in $\mathcal{V}_+(v, G)$, we look to see which crisp vertex s in $\mathcal{V}_+(v, G)$ has $\mathcal{D}(s)$ containing t, and then we associate t to the vertex in $\mathcal{V}_+(v, G_R)$ that corresponds to s as in Lemma 20.12. This defines a mapping from vertices in $\mathcal{V}_+(v, G)$ to vertices in $\mathcal{V}_+(v, G_R)$.

Suppose now that we have an edge from a vertex t to a vertex t' in $\mathcal{V}_+(v, G)$. If t and t' lie in the same $\mathcal{D}(s)$, then they are mapped to the same vertex in $\mathcal{V}_+(v, G_R)$, and we do not assign an edge in $\mathcal{V}_+(v, G_R)$ to the edge in $\mathcal{V}_+(v, G)$ that goes from t to t'. If we have $t \in \mathcal{D}(s)$, $t' \in \mathcal{D}(s')$ for different crisp vertices s, s' in $\mathcal{V}_+(v, G)$, then we have to assign an edge in $\mathcal{V}_+(v, G_R)$ to the edge from t to t'. In this situation, we know that there is an edge in G_R which goes from the endpoint of s to the endpoint of s', because of Lemma 20.14. This says

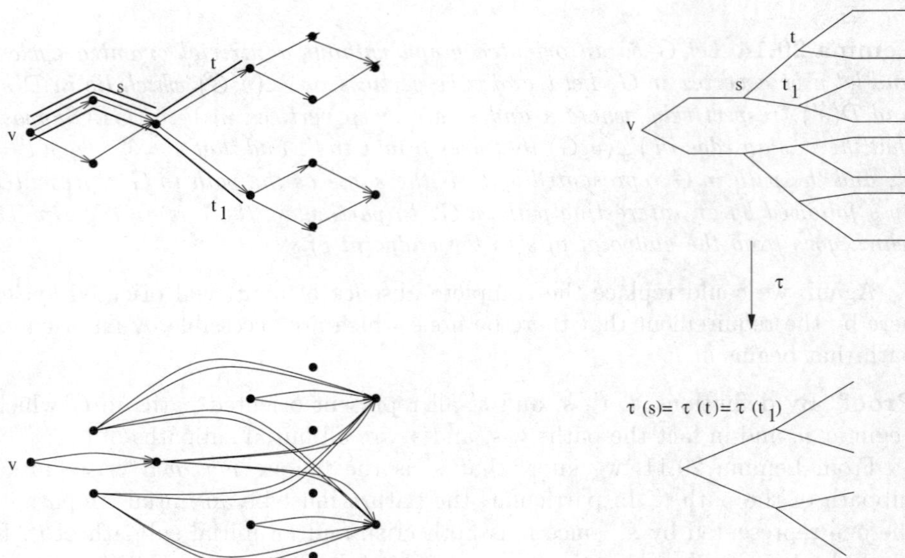

FIG. 20.6. An illustration of a weak mapping $\tau : \mathcal{V}_+(v, G) \to \mathcal{V}_+(v, G_R)$

exactly that there is an edge in $\mathcal{V}_+(v, G_R)$ which goes from the counterpart of s in $\mathcal{V}_+(v, G_R)$ to the counterpart of s' in $\mathcal{V}_+(v, G_R)$, and it is this edge that we associate to the one in $\mathcal{V}_+(v, G)$ which goes from t to t'.

This defines a weak mapping $\tau : \mathcal{V}_+(v, G) \to \mathcal{V}_+(v, G_R)$. See Fig. 20.6 for an illustration, and in particular for the way that τ can collapse several vertices in $\mathcal{V}_+(v, G)$ down to a single vertex in $\mathcal{V}_+(v, G_R)$. Notice that τ preserves orientations, as one can easily check from the definitions.

We should emphasize that in this definition of $\tau : \mathcal{V}_+(v, G) \to \mathcal{V}_+(v, G_R)$ we have continued to use the assumption that G be free of nontrivial oriented cycles. As usual, we really only need to know that there be no nontrivial oriented cycles in G which are accessible by an oriented path from v. One could make similar constructions when there are nontrivial oriented cycles, but some adjustments would be needed, and the exposition is simpler in the present circumstances. (The basic ideas are all the same anyway. Related issues are addressed in Remark 8.7.)

The weak mapping $\tau : \mathcal{V}_+(v, G) \to \mathcal{V}_+(v, G_R)$ is really just a repackaging of the information given in the Corona decomposition before, in Section 8.4. The "tree" which arises implicitly in the Corona decomposition (i.e., the tree of $\mathcal{W}$ subgraphs) is the same as $\mathcal{V}_+(v, G_R)$. Indeed, if one unwinds the definitions, one finds that the set of crisp vertices in $\mathcal{V}_+(v, G)$ is just the union of all the $\mathcal{B}$-sets generated in the Corona decomposition, together with the basepoint of $\mathcal{V}_+(v, G)$. The $\mathcal{W}$ subgraphs of the visibility produced in the Corona decomposition are really the same as the sets $\mathcal{D}(s)$ above (with s a crisp vertex), together with the edges between the vertices in each fixed $\mathcal{D}(s)$. The edges between different $\mathcal{D}(s)$'s correspond to the $\mathcal{E}$ sets before.

To make this more precise, let us look at the Calderón–Zygmund decomposition.

Lemma 20.15 *Let G be an oriented graph (without nontrivial oriented cycles), and let v be a vertex in G. Let b denote the basepoint of $\mathcal{V}_+(v,G)$, i.e., the degenerate path which traverses v alone. Then $\mathcal{D}(b)$ is the same as the set of vertices in the subgraph $\mathcal{W}$ obtained in Proposition 8.5 and its proof (associated to this choice of vertex v). A vertex in $\mathcal{V}_+(v,G)$ lies in the set $\mathcal{B}$ provided by Proposition 8.5 exactly when it represents an interesting path in G which begins at v.*

Proof This is a matter of unwinding definitions. Let us first recall the definition of $\mathcal{W}$ from the beginning of the proof of Proposition 8.5. Before we took $\mathcal{I}$ to be the set of vertices t in $\mathcal{V}_+(v,G)$ such that $(v,\pi(t))$ is a focal pair, and we defined $\mathcal{W}$ so that its vertices are the ones in $\mathcal{V}_+(v,G)$ which can be reached by an oriented path from the basepoint which does not touch $\mathcal{I}$.

Now t lies in $\mathcal{D}(b)$ exactly when t represents an oriented path in G that begins at v and which is dull. If this is the case, then t lies in $\mathcal{W}$. Indeed, we can take the *lifting* of the path represented by t in G (as in Section 4.6) to get a path in the visibility $\mathcal{V}_+(v,G)$ which goes from b to t without touching $\mathcal{I}$.

Conversely, if t lies in $\mathcal{W}$, then there is an oriented path in the visibility $\mathcal{V}_+(v,G)$ which goes from b to t and does not intersect $\mathcal{I}$. From Corollary 4.6, the canonical projection of this path in $\mathcal{V}_+(v,G)$ to a path in G is the same as the path represented by t. It must be dull precisely because the path up in the visibility did not meet $\mathcal{I}$.

This proves that the set of vertices in $\mathcal{W}$ is the same as $\mathcal{D}(b)$.

Now let us consider $\mathcal{B}$. By definition, this is the set of vertices s in $\mathcal{V}_+(v,G)$ which do not lie in $\mathcal{W}$, but for which there is an edge going from a vertex t in $\mathcal{W}$ to s. This means that the path in G represented by s consists of the path in G represented by t together with one additional edge added at the end. Since t is dull but s is not, it is easy to see that s actually represents an interesting path. If s represents an interesting path in G, then s is nondegenerate (by the assumed absence of oriented cycles), and we can reverse the process, by choosing t in $\mathcal{V}_+(v,G)$ so that it represents the entire initial subpath of s except for the last vertex and edge. This path is necessarily dull, since the one represented by s is interesting, so that t lies in $\mathcal{W}$. We also have that there is an edge in $\mathcal{V}_+(v,G)$ which goes from t to s, and hence that s lies in $\mathcal{B}$. This proves Lemma 20.15. □

There is a version of this lemma for any vertex in $\mathcal{V}_+(v,G)$, and not just the basepoint b. Indeed, let u be a vertex in $\mathcal{V}_+(v,G)$, and let $F(u)$ be the part of $\mathcal{V}_+(v,G)$ that "follows" u, as in Lemma 8.1. Then $F(u)$ is "naturally" isomorphic to the visibility $\mathcal{V}_+(\pi(u),G)$, as in Lemma 8.4. One can apply Lemma 20.15 inside of this visibility, with the role of v played now by $\pi(u)$, and the same analysis applies.

This fits with the recursion used to make the Corona decomposition. There we started with v and $\mathcal{V}_+(v, G)$, and then we applied the Calderón–Zygmund decomposition to get $\mathcal{W}$ and $\mathcal{B}$. We then applied Lemma 8.4 to say that each $F(s)$, $s \in \mathcal{B}$, is isomorphic to the visibility $\mathcal{V}_+(\pi(s), G)$ in its own right, so that the same argument could be repeated there. By repeating the argument over and over again, we obtained a certain decomposition of the original visibility, and that decomposition is the same as the one given by Lemma 20.13. Lemma 20.15 represents the "induction step" in making this link between the present construction and the one given before.

Instead of pursuing this link in precise and explicit terms, we can work directly with the version that we have here, and derive conclusions analogous to the ones before.

20.5 Injectivity properties of the weak mapping

Lemma 20.16 *Let G be an oriented graph which is free of nontrivial oriented cycles, let v be a vertex of G, and let s be a crisp vertex of $\mathcal{V}_+(v, G)$. Let $\pi : \mathcal{V}_+(v, G) \to G$ be the canonical projection, as in Section 4.5. Then π is injective on the vertices in $\mathcal{D}(s)$.*

As always, it is only the cycles in G which are accessible from v which can cause trouble here.

Proof This corresponds to (i) in Proposition 8.5. Rather than unwinding all the definitions to reduce to that case, we sketch another proof, for which the language is now a bit simpler.

So suppose that we have vertices $t, t' \in \mathcal{D}(s)$ such that $\pi(t) = \pi(t')$, and let us show that $t = t'$. As usual, t and t' actually represent oriented paths in G which begin at v, and $\pi(t) = \pi(t')$ means that they have the same endpoint.

Our assumption that t and t' both belong to $\mathcal{D}(s)$ implies that t, t' can each be obtained from the path represented by s by adding dull paths γ, γ' at the end. All we have to do is to show that γ, γ' are the same path.

Suppose that this is not the case. Both of γ and γ' begin at the endpoint of s and end at $\pi(t) = \pi(t')$. If γ and γ' are not the same path, then either one is a proper subpath of the other, or they have to diverge apart at some moment. The first possibility would imply the existence of a nontrivial oriented cycle in G, since γ and γ' have the same endpoint, and this is ruled out by the hypotheses of the lemma. Thus we assume that γ and γ' diverge apart at some moment. This contradicts the assumption that γ and γ' be dull, since they have to meet again after diverging from each other, which leads to a focal pair in G though which γ and γ' both pass. This completes the proof of Lemma 20.16. □

Corollary 20.17 (Joint injectivity) *Let G be an oriented graph with no oriented cycles, and let v be a vertex in G. Let t and t' be two vertices in the visibility $\mathcal{V}_+(v, G)$, and suppose that they have the same images under the canonical projection $\pi : \mathcal{V}_+(v, G) \to G$, and also under the weak mapping $\tau : \mathcal{V}_+(v, G) \to \mathcal{V}_+(v, G_R)$ defined in Section 20.4. Then $t = t'$.*

Proof Indeed, to say that $\tau(t) = \tau(t')$ says exactly that there is a crisp vertex s in $\mathcal{V}_+(v, G)$ such that $t, t' \in \mathcal{D}(s)$. (Compare with Lemma 20.13 and the first paragraph in Section 20.4.) From Lemma 20.16 we conclude that $t = t'$ if also $\pi(t) = \pi(t')$. □

Thus we get a nice picture here. Neither of the mappings π or τ is normally injective in its own right, but together they are. They also serve different roles; π sees local structure better, but sends points in $\mathcal{V}_+(v, G)$ which may be far away from each other to points which might be the same or close together, while τ respects the large-scale structure, but may collapse many vertices which are relatively close together into a single vertex.

Corollary 20.18 *Let G be an oriented graph with no oriented cycles, and let v be a vertex in G. Then the number of vertices in the visibility $\mathcal{V}_+(v, G)$ is no greater than the product of the number of vertices in G and the number of vertices in $\mathcal{V}_+(v, G_R)$.*

Proof This is an immediate consequence of Corollary 20.17. □

One might recall here that different oriented graphs can have the same reduced graph, up to isomorphism, as in Fig. 20.3.

20.6 Bounds

Lemma 20.19 (Bounds for G_R) *Let G be an oriented graph which contains no nontrivial oriented cycles, and let v be a vertex in G.*

(a) Suppose that k is a positive integer such that no vertex in G has more than k edges arriving into it. Then for each pair of vertices x, y in G_R, there are no more than k edges that go from x to y.

(b) If ℓ is the length of the longest chain of focal pairs in G which begins at v, then G_R does not contain any oriented path which begins at v and has length greater than ℓ.

Note that a vertex in G_R may have more incoming edges than it has in G, as in the example shown in Fig. 20.5.

Proof Part (b) is an immediate consequence of Lemma 20.9.

As for (a), by definition of G_R, the conclusion is equivalent to the assertion that there are no more than k distinct nondegenerate interesting paths which go from x to y. Suppose that this is not the case, that there are more than k such paths. Then two of these paths must arrive at y along the same edge going into y. Thus we have two distinct paths α, β which both begin at x and come together at least one step before they end at y. The absence of nontrivial oriented cycles in G implies that α and β must diverge from each other before reaching y, since otherwise one of α and β would be a proper subpath of the other, and G would contain a nontrivial oriented cycle. From here it is not hard to show that there must be a vertex z in G which occurs in these paths before their endpoints and such that (x, z) is a focal pair. Specifically, one can take z to be the first vertex at which α and β meet after diverging from each other. This contradicts the assumption that α and β be *interesting*, since that would imply that α and β do not reach such a vertex z before arriving to y at the end. This proves the lemma. □

Using the same kind of observations as in Section 8.7, we can convert this into the following bound for the total number of vertices in $\mathcal{V}_+(v, G_R)$.

Corollary 20.20 *Let G be an oriented graph without nontrivial oriented cycles, and let v be a vertex in G. Let N be the number of vertices in G. Assume that no vertex in G has more than k edges arriving into it, where $k \geq 2$, and that ℓ is the length of the longest chain of focal pairs in G which begins at v. Then the number of vertices in $\mathcal{V}_+(v, G_R)$ is at most*

$$2 \cdot k^{\ell} \cdot \frac{(N-1)^{\ell}}{\ell!}. \tag{20.2}$$

The case where $k = 1$ is not interesting here, because then there are no interesting paths in G, and hence no edges in G_R. In the degenerate case where $N = 1$ and $\ell = 0$, we interpret $(N-1)^{\ell}$ as being 1. Also, the extra factor of 2 in (20.2) is not needed when $\ell = 0$.

Proof The number of vertices in $\mathcal{V}_+(v, G_R)$ is the same as the total number of oriented paths in G_R which begin at v. From Lemma 20.19, we know that every such path has length at most ℓ, and in particular that G_R does not contain any nontrivial oriented cycles which can be reached by an oriented path that begins at v. (There are no nontrivial oriented cycles in G_R at all, since there are none in G, but we do not really need this here.)

Claim 20.21 *Let A be an (unordered) set of exactly j vertices from G_R, all of which are distinct from v, and let $\mathcal{A}$ denote the set of oriented paths in G_R which begin at v and pass through each of the vertices in A but do not pass through any vertices besides the ones in $A \cup \{v\}$. Then $\mathcal{A}$ contains at most k^j elements.*

To show this, we may as well assume that $\mathcal{A}$ is nonempty, so that there is at least one path α which begins at v and passes exactly through the set of vertices in $A \cup \{v\}$. Note that α cannot pass through any vertex more than once, since G_R contains no nontrivial oriented cycles which can be reached by an oriented path beginning at v. If α' is any other element of $\mathcal{A}$, then it must pass through the elements of A in exactly the same order as α does, also because of the absence of cycles in G_R which are accessible from v. Thus α and α' can differ from each other only in the choices of edges from one element of $A \cup \{v\}$ to the next. The number of such transitions is bounded by k at each step, because of Lemma 20.19, and this implies a bound of k^j for the total number of elements of $\mathcal{A}$. This proves Claim 20.21.

From the claim it follows that there are at most

$$k^j \cdot \binom{N}{j} \tag{20.3}$$

oriented paths in G_R which begin at v and have length equal to j. Hence there are at most

$$\sum_{j=0}^{\ell} k^j \cdot \binom{N}{j} \tag{20.4}$$

oriented paths in G_R which begin at v all told, since no such path has length longer than ℓ. From here the rest of the proof of Corollary 20.20 is the same as for Theorem 8.11 in Section 8.7, starting with Claim 8.15. □

APPENDIX A

FORMAL PROOFS: A BRIEF REVIEW

For the convenience of the reader, we mention a few basic facts about formal proofs. See [CS97] for a more thorough introduction, and see [Gir87c, Tak87] for more detailed information.

A.1 Sequent calculus

We shall assume that the reader is familiar with the idea of *formulae* in mathematical logic. For the record, let us just recall that one can build formulae with the logical connectives $\wedge$, $\vee$, $\neg$, and $\supset$, which represent conjunction, disjunction, negation, and implication. In *predicate logic* one also permits universal and existential quantifiers $\forall, \exists$. These are the connectives, but we need to be a bit careful about the underlying building blocks. In propositional logic, one simply uses *propositional variables* p, q, etc., which would represent single statements that might be true or false. In predicate logic, one has *relation symbols* (or *predicates*), which can take any (fixed positive) number of arguments. For the arguments of the relation symbols one uses *terms*, which are expressions made from *constants*, *variables*, and *function symbols*. In the interpretation ("semantics"), the *relation symbols* represent some property which might be either true or false for the particular *terms* in question. For example, one might have a binary predicate Q, and $Q(\phi(x), \theta(x, \psi(y), z))$ would then be a logical formula, where $\phi(x)$ and $\theta(x, \psi(y), z)$ are terms built out of the variables x, y, and z and the function symbols ϕ, θ, and ψ.

Note that we are restricting ourselves to *first-order languages* in this discussion. (Roughly speaking, this means that we do not try to define predicates which take relation symbols as arguments, nor do we try to quantify over predicates, i.e., to say "for all unary predicates R, ...")

Logical formulae which do not involve connectives are called *atomic formulae.* For propositional logic, this simply means propositional variables, while in predicate logic, it means a relation symbol with particular terms as its arguments.

A *sequent* is an expression of the form

$$A_1, A_2, \ldots, A_m \to B_1, B_2, \ldots, B_n$$

where the A_i's and the B_j's are formulae. The interpretation of this sequent is "from A_1 and A_2 and ... and A_m follows B_1 or B_2 or ... or B_n". In particular, if the sequent above is valid, then so is any other ("weaker") sequent of the form

$$A_1, A_2, \ldots, A_m, C_1, \ldots, C_k \to B_1, B_2, \ldots, B_n, D_1, \ldots, D_l.$$

We should emphasize that the sequent arrow $\to$ is *not* a connective, nor is a sequent a formula. The notion of a sequent provides a convenient combinatorial formalism in which we can work with *collections* of formulae as individual objects. It is because we use the symbol $\to$ to divide the two sides of a sequent that we use $\supset$ to denote the connective of implication. (See [Carc] for a purely combinatorial "model" for sequents and their logical calculi.)

In a sequent as above, the formulae A_i and B_j are permitted to have repetitions, and this turns out to be important. We do not care about the ordering, however. We also permit empty sets of formulae, so that

$$A_1, A_2, \ldots, A_m \to \qquad \text{and} \qquad \to B_1, B_2, \ldots, B_n$$

are permissible sequents. As a matter of notation, we shall typically use upper-case Roman letters $A, B, C, \ldots$ to denote formulae, and upper-case Greek letters like $\Gamma, \Delta, \Lambda, \ldots$ to denote collections of formulae.

To make precise the notion of a formal proof in classical logic, we shall use the *sequent calculus LK* [Gen34, Gir87c, Tak87]. This is a certain collection of *axioms* and *rules of inference*, for which a *proof* of a particular sequent is a derivation of that sequent from the axioms which follows the rules of inference. The *axioms* in LK are the sequents of the form

$$A, \Gamma \to \Delta, A$$

where A is any formula and Γ, Δ are any collections of formulae. The formulae in Γ, Δ are called *weak formulae.* The *rules of inference* come in two types, namely the *logical rules* and *structural rules.* To describe these rules, we write Γ, Γ_1, Γ_2, etc., for collections of formulae, and we write $\Gamma_{1,2}$ as a shorthand for the combination of Γ_1 and Γ_2 (counting multiplicities). (The same convention applies to Δ_1, Δ_2, $\Delta_{1,2}$.)

The *logical rules* are used to introduce connectives, and are given as follows:

$\neg$: *left* $\quad \dfrac{\Gamma \to \Delta, A}{\neg A, \Gamma \to \Delta}$ $\qquad$ $\neg$: *right* $\quad \dfrac{A, \Gamma \to \Delta}{\Gamma \to \Delta, \neg A}$

$\wedge$: *right* $\quad \dfrac{\Gamma_1 \to \Delta_1, A \quad \Gamma_2 \to \Delta_2, B}{\Gamma_{1,2} \to \Delta_{1,2}, A \wedge B}$

$\wedge$: *left* $\quad \dfrac{A, B, \Gamma \to \Delta}{A \wedge B, \Gamma \to \Delta}$

$\vee$: *left* $\quad \dfrac{A, \Gamma_1 \to \Delta_1 \quad B, \Gamma_2 \to \Delta_2}{A \vee B, \Gamma_{1,2} \to \Delta_{1,2}}$

$\vee$: *right* $\quad \dfrac{\Gamma \to \Delta, A, B}{\Gamma \to \Delta, A \vee B}$

$\supset$: *left* $\dfrac{\Gamma_1 \to \Delta_1, A \quad B, \Gamma_2 \to \Delta_2}{A \supset B, \Gamma_{1,2} \to \Delta_{1,2}}$

$\supset$: *right* $\dfrac{A, \Gamma \to \Delta, B}{\Gamma \to \Delta, A \supset B}$

$\exists$: *left* $\dfrac{A(b), \Gamma \to \Delta}{(\exists x)A(x), \Gamma \to \Delta}$ $\exists$: *right* $\dfrac{\Gamma \to \Delta, A(t)}{\Gamma \to \Delta, (\exists x)A(x)}$

$\forall$: *left* $\dfrac{A(t), \Gamma \to \Delta}{(\forall x)A(x), \Gamma \to \Delta}$ $\forall$: *right* $\dfrac{\Gamma \to \Delta, A(b)}{\Gamma \to \Delta, (\forall x)A(x)}$

The *structural rules* do not involve connectives, and are the following:

Cut $\dfrac{\Gamma_1 \to \Delta_1, A \quad A, \Gamma_2 \to \Delta_2}{\Gamma_{1,2} \to \Delta_{1,2}}$

Contraction $\dfrac{\Gamma \to \Delta, A, A}{\Gamma \to \Delta, A}$ $\dfrac{A, A, \Gamma \to \Delta}{A, \Gamma \to \Delta}$

One should be a little careful about the rules for the quantifiers. In $\exists : right$ and $\forall : left$, any term t is allowed which does not include a variable which lies already within the scope of a quantifier in the given formula A. In $\exists : left$ and $\forall : right$ one has the "eigenvariable" b which should not occur free (not bound by a quantifier) in Γ, Δ. (Let us emphasize that b should simply be a single variable, and not a more complicated term with function symbols.)

This is the system LK for classical predicate logic. For propositional logic, it is the same, except that one drops the quantifier rules.

This system has very nice combinatorial properties. The formulae never simplify in the course of a proof; they can disappear through the cut rule, and their repetitions can be reduced through the contraction rule, but they cannot be "decomposed" directly. The effect of simplifying a formula can be achieved indirectly, however, through the use of the cut rule.

Note that LK represents "pure" logic, without any special rules or axioms. In particular mathematical contexts, one would employ additional axioms or rules of inference to reflect the structure at hand. Thus, in arithmetic for instance, one would accept as an axiom a sequent of the form

$$\Gamma \to \Delta, x = x,$$

and one would allow special rules of inference which include the basic mechanisms for manipulating equations (e.g., sums of equations give equations, etc.), and an additional rule representing induction. In the context of feasible numbers, one has special rules of inference which encode the idea that feasibility is preserved by sums, products, and successor, such as the F : *times* rule

$$\frac{\Gamma_1 \to \Delta_1, F(s) \quad \Gamma_2 \to \Delta_2, F(t)}{\Gamma_{1,2} \to \Delta_{1,2}, F(s \cdot t)}$$

where s and t are arbitrary terms (in arithmetic). One also has the special axioms

$$\Gamma \to \Delta, F(0)$$

which express the feasibility of 0. For these axioms, one has only one "distinguished occurrence" $F(0)$, as opposed to the two distinguished occurrences of A in the ordinary axiom $A, \Gamma \to \Delta, A$. A similar point arises above, with the axiom containing $x = x$.

A.2 Cut elimination

Theorem A.1 (Gentzen [Gen34, Gir87c, Tak87]) *Any proof in LK can be effectively transformed into a proof which never uses the cut rule. This works for both propositional and predicate logic.*

Note that this result may *not* work if one has special axioms or rules of inference. In the context of arithmetic, for instance, the use of induction can force one to use *infinite* proofs in order to eliminate cuts. (See [Gir87c, Tak87] for more information concerning arithmetic and the sequent calculus.)

Theorem A.1 is very striking, particularly in view of its combinatorial consequences for formal proofs. In a proof without cuts, there is no way to simplify formulae, even indirectly as before, and no way to make a formula disappear completely. This leads to the famous *subformula property*, which says that every formula which appears in a proof without cuts also occurs as a subformula of a formula in the endsequent. (One should be a little careful about the notion of a *subformula* in the presence of quantifiers, which can have the effect of changing the terms within.)

One of the results of cut elimination is the simplification of the dynamical processes which can occur within proofs. Predicate logic gives a way to code substitutions which may not be expressed explicitly, and an effect of cut elimination is to make all the substitutions explicit. A related point is that proofs with cuts can have *(oriented) cycles* in their logical flow graphs (discussed in Section A.3), while proofs without cuts cannot. See [Car97, Car00, Cara, Car99] for more information and examples. Note that [Car00, Car99] provide an alternative proof system in which there are no (oriented) cycles, and in which it can be easier to make explicit the construction of terms which lies below a proof.

The "price" of cut elimination is that the cut-free proof may have to be much larger than proofs without cuts. There are propositional tautologies for which cut-free proofs must be exponentially larger than proofs with cuts, and in predicate logic the expansion can be non-elementary. See [Ore82, Ore93, Sta74, Sta78, Sta79, Tse68].

The *contraction rule* plays a key role in this expansion. While it may seem harmless enough, it can in fact be very powerful in connection with the cut rule. Imagine having a piece of information which is represented by a formula A that is known to be provable, and then needing to use this piece of information twice in the course of an argument. By employing the contraction rule on the left hand side of a sequent, the argument can be arranged in such a way that A only has to be verified once. A cut-free proof represents "direct" reasoning, without lemmas, and this can force one to duplicate the proof of A. (See [CS97] for more information.)

Thus the cut rule provides a mechanism by which the contraction rule can have the effect of a "duplication" rule. It is this point for which we have often aimed in the geometric models in the main part of the text.

A.3 The logical flow graph

Girard [Gir87a] has introduced a notion of *proof nets*, which provides a way to study proofs and their *global* aspects, and the interaction between formulae in them. This was an important new step, for looking at the manner in which components of a proof fit together.

Another type of graph associated to proofs was introduced subsequently by Buss [Bus91], called the *logical flow graph*. This graph traces the flow of formula occurrences in a proof. For the present purposes, we shall modify the original definition slightly by restricting ourselves to atomic formulae (as in [Car97]).

The logical flow graph of a proof Π of a sequent S is defined as follows. For the set of *vertices* in the graph we take the set of all atomic occurrences in Π. We add edges between these occurrences in the following manner. If we have an axiom

$$A, \Gamma \to \Delta, A$$

then we attach an edge between the vertices which correspond to the "same" atomic formula in A. That is, each atomic subformula of A has an occurrence within each of the occurrences of A in the axiom above, and we connect the corresponding pair of vertices by an edge. We leave undisturbed the vertices which come from Γ and Δ.

When we apply a rule from LK to one or two sequents in the proof Π, we have that every atomic occurrence in the upper sequents has a counterpart in the lower sequent, except for those occurrences in the cut formulae for the cut rule. In all but the latter case, we simply attach an edge between the occurrences in the upper sequents and their counterparts in the lower sequent. In the case of atomic occurrences within cut formulae, we attach edges between the occurrences which occupy the same position within the cut formula.

Note that there is also a subtlety with the contraction rule. Every atomic occurrence in the contraction formula in the lower sequent will be connected by an edge to an atomic occurrence in *each* of the formulae being contracted in the upper sequent. In all other cases, there is exactly one counterpart in the upper sequents of an atomic occurrence in the lower sequent of a rule.

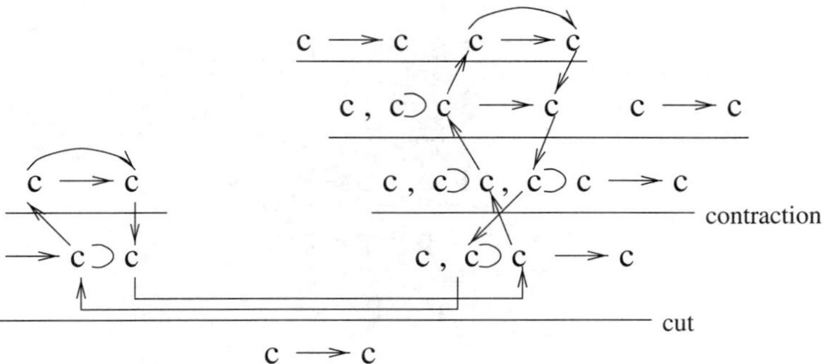

FIG. A.1. A proof whose logical flow graph contains a nontrivial oriented cycle

Thus, in all cases *except* axioms, cuts, and contractions, the logical flow graph simply makes a kind of "prolongation" which does not affect the topology of the graph.

This defines the set of edges for the logical flow graph. We can define an *orientation* in the following manner. We first define the notion of the *sign* of an atomic occurrence within a sequent. One can do this inductively, but the following is a bit more direct. If P is an *atomic* subformula of a formula A, then we say that P occurs *positively* in A if it lies within the scope of an *even* number of negations, and we say that it occurs *negatively* if it occurs an odd number of times within the scope of a negation. Suppose now that A is a formula which appears in a given sequent $\Gamma \to \Delta$, i.e., A appears as one of the elements of Γ or Δ. If A appears within Δ, then we say that P occurs *positively* in the sequent if it occurs positively in A, and we say that P occurs *negatively* in the sequent if it occurs negatively in A. If A appears within Γ, then we do the opposite, saying that P occurs positively in the sequent if it is negative as a subformula of A, and that P occurs negatively in the sequent if it is positive in A. This takes into account the negation which is implicit in the sequent.

With this notion of *sign* of an atomic formula, we can define the *orientation* for the logical flow graph as follows. If an edge in the logical flow graph comes from an axiom as above, then we orient the edge *from* negative occurrences *to* positive occurrences. If the edge comes from a cut formula, then we do the opposite, and orient it *from* positive occurrences *to* negative occurrences. Otherwise, the edge goes between "upper" and "lower" occurrences of an atomic formula in the application of a rule. If the occurrence is *negative* then we orient the edge *from* the lower occurrence *to* the upper occurrence, and for positive occurrences we go the other way around. (Note that atomic occurrences never change sign in the transition from an upper sequent to a lower sequent in a rule of inference. This is not hard to check.)

As in [Car97], logical flow graphs of proofs with cuts can contain nontrivial oriented cycles, but this cannot occur for proofs without cuts. A simple example of a formal proof in which a cycle is present is shown in Fig. A.1. In proofs

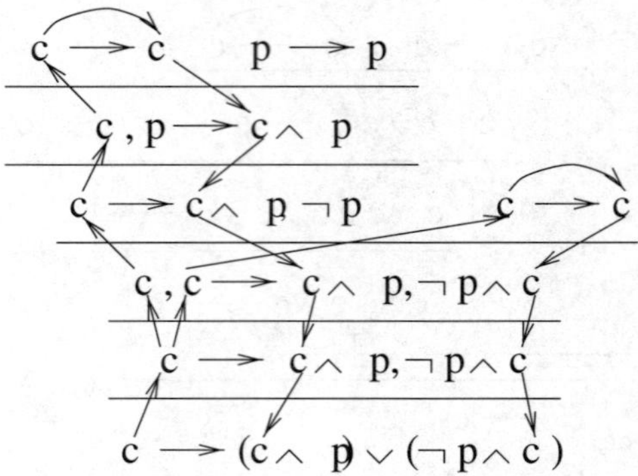

FIG. A.2. A formal proof without cuts, and a pair of oriented paths in the logical flow graph of the proof

without cuts, oriented paths can go up and over an axiom, but once they start going down, they have no chance to turn up again. This is illustrated by the example shown in Fig. A.2. By contrast, one can see how the presence of a cut in the proof in Fig. A.1 permits the path to turn up again. Figure A.2 also indicates the way that oriented paths can split apart at contractions.

See Sections 6.12 and 6.18 for more information about the structure of logical flow graphs when the corresponding proofs contain no cuts, or only very simple cuts. The evolution of the logical flow graph under Gentzen's method of cut elimination is also discussed extensively in Chapter 6.

One can think of the orientation on the logical flow graph as showing the natural flow of information in the proof. Individual oriented paths reflect the way that the "information" in a given atomic occurrence is being used or transmitted. This is especially relevant for the *substitutions* which can occur within a proof in predicate logic. This point emerges clearly in [Car00], where one sees how the absence of *oriented* cycles can enable one to track more easily the substitutions involved in the construction of terms in a proof, in such a way that one can obtain estimates on the complexity of terms as a function of the size of the proof. (See also Section 16.16.)

Notice that only the use of the contraction rule leads to branching in the logical flow graph. This need not be true if one allows additional rules of inference, as in the setting of *feasible numbers*, for instance. If one extends the notion of logical flow graph to the context of feasible numbers in the natural way (as in [Car00]), then the special rules of inference that govern the feasibility of sums and products of arbitrary terms also give rise to (focusing) branch points. (Note that the F : *times* rule was given near the end of Section A.1.) This reflects the fact that two different pieces of information are being combined into one, just as in the contraction rule.

REFERENCES

[Ahl87] L. Ahlfors. *Lectures on Quasiconformal Mappings.* Wadsworth & Brooks/Cole, Monterey, CA, 1987.

[Ajt88] M. Ajtai. The complexity of the pigeonhole principle. In *Proceedings of the IEEE 29th Annual Symposium on the Foundation of Computer Science*, pages 346–355, 1988.

[Ajt90] M. Ajtai. Parity and the pigeonhole principle. In S. Buss and P. Scott, editors, *Feasible Mathematics: Proceedings of the Workshop held at Cornell University, 1989*, volume 9 of *Progress in Computer Science and Applied Logic*, pages 1–24. Birkhäuser, Boston, 1990.

[AS60] L. Ahlfors and L. Sario. *Riemann Surfaces*, volume 26 of *Princeton Mathematical Series.* Princeton University Press, Princeton, NJ, 1960.

[Ash65] R. Ash. *Information Theory.* Dover, New York, 1965.

[Ass77] P. Assouad. *Espaces Métriques, Plongements, Facteurs.* Thèse de Doctorat, Université de Paris XI, 91405 Orsay, France, January 1977.

[Ass79] P. Assouad. Étude d'une dimension métrique liée à la possibilité de plongement dans $\mathbf{R}^n$. *Comptes Rendus de l'Académie des Sciences Paris, Sér. A*, 288:731–734, 1979.

[Ass83] P. Assouad. Plongements lipschitziens dans $\mathbf{R}^n$. *Bulletin de la Société Mathématique de France*, 111:429–448, 1983.

[Bar96] M. Barnesly. Fractal image compression. *Notices of the American Mathematical Society*, 43:657–662, 1996.

[Ber91] C. Berge. *Graphs*, volume 6 of *North-Holland Mathematical Library.* North-Holland, Amsterdam, 1991. Third revised edition.

[Bes78] A. Besse. *Manifolds all of whose Geodesics are Closed.* Ergibnisse der Mathematik und ihrer Grenzgebeite. Springer-Verlag, Berlin, New York, 1978.

[BIK+92] P. Beame, R. Impagliazzo, J. Krajíček, T. Pitassi, P. Pudlák, and A. Woods. Exponential lower bounds for the pigeonhole principle. In *Proceedings of the 24th Annual ACM Symposium on Theory of Computing*, pages 200–221. Association for Computing Machinery, 1992.

[BIK+96] P. Beame, R. Impagliazzo, J. Krajíček, T. Pitassi, and P. Pudlák. Lower bounds on Hilbert's Nullstellensatz and propositional proofs. *Proceedings of the London Mathematical Society*, 73:1–26, 1996.

[Boo75] W. Boothby. *An Introduction to Differentiable Manifolds and Riemannian Geometry*, volume 63 of *Pure and Applied Mathematics.* Academic Press, New York, London, 1975.

[Bou97] M. Bourdon. Immeubles hyperbolique, dimension conforme et rigidité de Mostow. *Geometric and Functional Analysis*, 7:245–268, 1997.

[BP99] M. Bourdon and H. Pajot. Poincaré inequalities and quasiconformal structure on the boundary of some hyperbolic buildings. *Proceedings of the American Mathematical Society*, 127:2315–2324, 1999.

[Bur96] J. Burillo. Lower bounds of isoperimetric functions for nilpotent groups. In G. Baumslag et al., editors, *Geometric and Computational Perspectives on Infinite Groups*, volume 25 of *DIMACS Series in Discrete Mathematics and Theoretical Computer Science*, pages 1–8. American Mathematical Society, Providence, RI, 1996.

[Bus87] S. Buss. Polynomial size proofs of the propositional pigeonhole principle. *Journal of Symbolic Logic*, 52:916–927, 1987.

[Bus88] S. Buss. Weak formal systems and connections to computational complexity. Course notes, University of California at Berkeley, 1988.

[Bus91] S. Buss. The undecidability of k-provability. *Annals of Pure and Applied Logic*, 53:75–102, 1991.

[Can84] J. Cannon. The combinatorial structure of cocompact discrete hyperbolic groups. *Geometriae Dedicata*, 16:123–148, 1984.

[Cara] A. Carbone. Asymptotic cyclic expansion and bridge groups of formal proofs. Submitted.

[Carb] A. Carbone. The cost of a cycle is a square. *Journal of Symbolic Logic*. To appear.

[Carc] A. Carbone. Some combinatorics behind proofs. Submitted.

[Car62] L. Carleson. Interpolations by bounded analytic functions and the corona problem. *Annals of Mathematics*, 76:547–559, 1962.

[Car97] A. Carbone. Interpolants, cut elimination and flow graphs for the propositional calculus. *Annals of Pure and Applied Logic*, 83:249–299, 1997.

[Car99] A. Carbone. Turning cycles into spirals. *Annals of Pure and Applied Logic*, 96:57–73, 1999.

[Car00] A. Carbone. Cycling in proofs and feasibility. *Transactions of the American Mathematical Society*, 352:2049–2075, 2000.

[CE75] J. Cheeger and D. Ebin. *Comparison Theorems in Riemannian Geometry*, volume 9 of *North-Holland Mathematical Library*. North-Holland, Amsterdam, 1975.

[Cha87] G. Chaitin. *Information, Randomness & Incompleteness — Papers on Algorithmic Information Theory*, volume 8 of *Series in Computer Science*. World Scientific, River Edge, NJ, 1987.

[Cha92] G. Chaitin. *Information-Theoretic Incompleteness*, volume 35 of *Series in Computer Science*. World Scientific, River Edge, NJ, 1992.

[Che99] J. Cheeger. Differentiability of Lipschitz functions on metric measure spaces. *Geometric and Functional Analysis*, 9:428–517, 1999.

[Coi91] R. Coifman. Adapted multiresolution analysis, computation, signal processing, and operator theory. In *Proceedings of the International Congress of Mathematicians, Kyoto/Japan, 1990*, volume II, pages 879–887. Mathematical Society of Japan, Tokyo, 1991.

[CP93] M. Coornaert and A. Papadopoulos. *Symbolic Dynamics and Hyperbolic*

Groups, volume 1539 of *Lecture Notes in Mathematics*. Springer-Verlag, Berlin, New York, 1993.

[CR79] S. Cook and R. Reckhow. The relative efficiency of propositional proof systems. *Journal of Symbolic Logic*, 44:36–50, 1979.

[CS] A. Carbone and S. Semmes. Looking from the inside and from the outside. *Synthèse*. To appear.

[CS97] A. Carbone and S. Semmes. Making proofs without modus ponens: An introduction to the combinatorics and complexity of cut elimination. *Bulletin of the American Mathematical Society*, 34:131–159, 1997.

[CS99] A. Carbone and S. Semmes. Propositional proofs via combinatorial geometry and the search for symmetry. In *Collegium Logicum, Annals of the Kurt Gödel Society*, volume 3, pages 85–98. Institute of Computer Science AS CR Prague, 1999.

[CZ52] A. Calderón and A. Zygmund. On the existence of certain singular integrals. *Acta Mathematica*, 88:85–139, 1952.

[Dau93] I. Daubechies, editor. *Different Perspectives on Wavelets*, volume 47 of *Proceedings of Symposia in Applied Mathematics*. American Mathematical Society, Providence, RI, 1993.

[DJS97] V. Danos, J.-B. Joinet, and H. Schellinx. A new deconstructive logic: Linear logic. *Journal of Symbolic Logic*, 62:755–807, 1997.

[Dra85] A. Dragalin. Correctness of inconsistent theories with notions of feasibility. In A. Skowron, editor, *Computation Theory: Proceedings of the Fifth Symposium held in Zaborŏw, 1984*, volume 208 of *Lecture Notes in Computer Science*, pages 58–79. Springer-Verlag, Berlin, New York, 1985.

[DS93] G. David and S. Semmes. *Analysis of and on Uniformly Rectifiable Sets*, volume 38 of *Mathematical Surveys and Monographs*. American Mathematical Society, Providence, RI, 1993.

[DS97] G. David and S. Semmes. *Fractured Fractals and Broken Dreams: Self-Similar Geometry through Metric and Measure*, volume 7 of *Oxford Lecture Series in Mathematics and its Applications*. The Clarendon Press, Oxford University Press, Oxford, 1997.

[ECH+92] D. Epstein, J. Cannon, D. Holt, S. Levy, M. Paterson, and W. Thurston. *Word Processing in Groups*. A K Peters, Natick, MA, 1992.

[Eil74] S. Eilenberg. *Automata, Languages, and Machines, Volume A*, volume 59-A of *Pure and Applied Mathematics*. Academic Press, New York, London, 1974.

[Fal90] K. Falconer. *Fractal Geometry: Mathematical Foundations and Applications*. John Wiley & Sons, Chichester UK, New York, 1990.

[Far92] B. Farb. Automatic groups: A guided tour. *L'Enseignement Mathématique*, 38:291–313, 1992.

[Fed69] H. Federer. *Geometric Measure Theory*. Springer-Verlag, Berlin, New York, 1969.

[FM98] B. Farb and L. Mosher. A rigidity theorem for the solvable Baumslag–Solitar groups. *Inventiones Mathematicae*, 131:419–451, 1998.

[G+99] M. Gromov et al. *Metric Structures for Riemannian and Non-Riemannian Spaces*, volume 152 of *Progress in Mathematics*. Birkhäuser, Boston, 1999.

[Gar81] J. Garnett. *Bounded Analytic Functions*, volume 96 of *Pure and Applied Mathematics*. Academic Press, New York, London, 1981.

[Gd90] E. Ghys and P. de la Harpe, editors. *Sur les Groupes Hyperboliques d'aprés Mikhael Gromov*, volume 83 of *Progress in Mathematics*. Birkhäuser, Boston, 1990.

[Gen34] G. Gentzen. Untersuchungen über das logische Schließen I-II. *Mathematische Zeitschrift*, 39:176–210, 405–431, 1934.

[Gen69] G. Gentzen. *The Collected Papers of Gerhard Gentzen*. Studies in Logic and the Foundations of Mathematics. North-Holland, Amsterdam, 1969. M. E. Szabo, editor.

[Ger92] S. Gersten. Bounded cocycles and combings of groups. *International Journal of Algebra and Computation*, 2:307–326, 1992.

[Gir87a] J.-Y. Girard. Linear Logic. *Theoretical Computer Science*, 50:1–102, 1987.

[Gir87b] J.-Y. Girard. Linear logic and parallelism. In M. Venturini Zilli, editor, *Mathematical Models for the Semantics of Parallelism: Proceedings of the Advanced School held in Rome, 1986*, volume 280 of *Lecture Notes in Computer Science*, pages 166–182. Springer-Verlag, Berlin, New York, 1987.

[Gir87c] J.-Y. Girard. *Proof Theory and Logical Complexity*, volume 1 of *Studies in Proof Theory. Monographs*. Bibliopolis, via Arangio Ruiz, 83, Napoli, Italy, 1987.

[Gir88] J.-Y. Girard. Normal functors, power series and λ-calculus. *Annals of Pure and Applied Logic*, 37:129–177, 1988.

[Gir89a] J.-Y. Girard. Geometry of interaction I: Interpretation of system F. In R. Ferro et al., editors, *Logic Colloquium '88: Proceedings of the Colloquium held at the University of Padova, 1988*, volume 127 of *Studies in Logic and the Foundations of Mathematics*, pages 221–260. North Holland, Amsterdam, 1989.

[Gir89b] J.-Y. Girard. Towards a geometry of interaction. In J. Gray and A. Scedrov, editors, *Categories in Computer Science and Logic: Proceedings of the AMS-IMS-SIAM Joint Summer Research Conference held at the University of Colorado, 1987*, volume 92 of *Contemporary Mathematics*, pages 69–108. American Mathematical Society, Providence, RI, 1989.

[Gir90] J.-Y. Girard. Geometry of interaction II: Deadlock-free algorithms. In P. Martin-Löf and G. Mints, editors, *COLOG-88: Proceedings of the International Conference on Computer Logic held in Tallinn, 1988*, volume 417 of *Lecture Notes in Computer Science*, pages 76–93. Springer-Verlag, Berlin, New York, 1990.

[Gir92] J.-Y. Girard. Logic and exceptions: A few remarks. *Journal of Logic and*

Computing, 2:111–118, 1992.

[Gir93] J.-Y. Girard. On the unity of logic. *Annals of Pure and Applied Logic*, 59:201–217, 1993.

[Gir95a] J.-Y. Girard. Geometry of interaction III: Accommodating the additives. In J.-Y. Girard, Y. Lafont, and L. Regnier, editors, *Advances in Linear Logic: Proceedings of the Workshop held at Cornell University, 1993*, volume 222 of *London Mathematical Society Lecture Note Series*, pages 329–389. Cambridge University Press, Cambridge, 1995.

[Gir95b] J.-Y. Girard. Light linear logic. In D. Leivant, editor, *Logic and Computational Complexity: Papers from the International Workshop (LCC '94) held in Indianapolis, 1994*, volume 960 of *Lecture Notes in Computer Science*, pages 145–176. Springer-Verlag, Berlin, New York, 1995.

[GJ79] M. Garey and D. Johnson. *Computers and Intractability : A Guide to the Theory of NP-Completeness.* W. H. Freeman, San Francisco, 1979.

[GL87] J.-Y. Girard and Y. Lafont. Linear logic and lazy computation. In H. Ehrig et al., editors, *TAPSOFT '87, Volume 2: Proceedings of the Second International Joint Conference on Theory and Practice of Software Development held in Pisa, 1987*, volume 250 of *Lecture Notes in Computer Science*, pages 52–66. Springer-Verlag, Berlin, New York, 1987.

[GLT89] J.-Y. Girard, Y. Lafont, and P. Taylor. *Proofs and Types*, volume 7 of *Cambridge Tracts in Theoretical Computer Science.* Cambridge University Press, Cambridge, New York, Sydney, 1989.

[GP91] M. Gromov and P. Pansu. Rigidity of lattices: An introduction. In P. De Bartolomeis and F. Tricerri, editors, *Geometric Topology: Recent Developments*, volume 1504 of *Lecture Notes in Mathematics*, pages 39–137. Springer-Verlag, Berlin, New York, 1991.

[Gro81a] M. Gromov. Groups of polynomial growth and expanding maps. *Publications Mathématiques IHES*, 53:183–215, 1981.

[Gro81b] M. Gromov. *Structures Métriques pour les Variétés Riemanniennes.* Cedic/Fernand Nathan, Paris, 1981. J. Lafontaine, P. Pansu, editors.

[Gro84] M. Gromov. Infinite groups as geometric objects. In *Proceedings of the International Congress of Mathematicians, Warsaw, 1983*, pages 385–392. PWN, Warsaw, 1984.

[Gro87] M. Gromov. Hyperbolic groups. In S. Gersten, editor, *Essays in Group Theory*, volume 8 of *Mathematical Sciences Research Institute Publications*, pages 75–263. Springer-Verlag, Berlin, New York, 1987.

[Gro90] M. Gromov. Cell division and hyperbolic geometry. IHES Preprint M/90/54, Bures-sur-Yvette, France, 1990.

[Gro93] M. Gromov. Asymptotic invariants of infinite groups. In G. Niblo and M. Roller, editors, *Geometric Group Theory*, volume 182 of *London Mathematical Society Lecture Note Series.* Cambridge University Press, Cambridge, 1993.

[Gro94] M. Gromov. Sign and geometric meaning of curvature. *Rendiconti del Seminario*

Matematico e Fisico di Milano, 61:9–123, 1994.

[GSS92] J.-Y. Girard, A. Scedrov, and P. Scott. Bounded linear logic: A modular approach to polynomial-time computability. *Theoretical Computer Science*, 97:1–66, 92.

[Haj61] G. Hajós. Über eine konstruktion night n-färberer graphen. *Wissenshaftliche Zeitschrift der Martin–Luther–Universität Halle–Wittenberg*, 10:116–117, 1961.

[Hak85] A. Haken. The intractability of resolution. *Theoretical Computer Science*, 39:297–308, 1985.

[HE73] S. Hawking and G. Ellis. *The Large-Scale Structure of Space-Time.* Cambridge Monographs of Mathematical Physics. Cambridge University Press, Cambridge, 1973.

[Hei95] J. Heinonen. Calculus on Carnot groups. In *Fall School in Analysis (Jyväskylä, 1994)*, pages 1–31. Universität Jyväskylä, Jyväskylä, Finland, 1995.

[HK96] J. Heinonen and P. Koskela. From local to global in quasiconformal structures. *Proceedings of the National Academy of Sciences (U.S.A.)*, 93:554–556, 1996.

[HK98] J. Heinonen and P. Koskela. Quasiconformal maps in metric spaces with controlled geometry. *Acta Mathematica*, 181:1–61, 1998.

[HS97] J. Heinonen and S. Semmes. Thirty-three yes-or-no questions about mappings, measures, and metrics. *Conformal Geometry and Dynamics (an Electronic Journal of the American Mathematical Society)*, 1:1–12, 1997.

[HU79] J. Hopcroft and J. Ullman. *Introduction to Automata Theory, Languages, and Computation.* Addison-Wesley, Reading, MA, 1979.

[HW41] W. Hurewicz and H. Wallman. *Dimension Theory*, volume 4 of *Princeton Mathematical Series.* Princeton University Press, Princeton, NJ, 1941.

[HW74] J. Hopcroft and J. Wong. Linear time algorithm for isomorphism of planar graphs (preliminary report). In *ACM Symposium on Theory of Computing*, volume 6, pages 172–184. Association for Computing Machinery, New York, 1974.

[IM76] D. Isaacson and R. Madsen. *Markov Chains, Theory and Applications.* Wiley Series in Probability and Mathematical Statistics. John Wiley & Sons, New York, 1976.

[JN61] F. John and L. Nirenberg. On functions of bounded mean oscillation. *Communications on Pure and Applied Mathematics*, 14:415–426, 1961.

[Jou83] J.-L. Journé. *Calderón–Zygmund Operators, Pseudo-Differential Operators, and the Cauchy Integral of Calderón*, volume 994 of *Lecture Notes in Mathematics.* Springer-Verlag, Berlin, New York, 1983.

[KN69] S. Kobayashi and K. Nomizu. *Foundations of Differential Geometry.* Number 15 in Interscience Tracts in Pure and Applied Mathematics. Interscience Publishers, New York, 1963, 1969. Two volumes.

[Kol68] A. Kolmogorov. Logical basis for information theory and probability theory. *IEEE Transactions on Information Theory*, 14:662–664, 1968.

[KR85] A. Korányi and M. Reimann. Quasiconformal mappings on the Heisenberg group. *Inventiones Mathematicae*, 80:309–338, 1985.

[Kra95] J. Krajíček. *Bounded Arithmetic, Propositional Logic, and Complexity Theory*, volume 60 of *Encyclopedia of Mathematics and its Applications*. Cambridge University Press, Cambridge, 1995.

[Kre77] G. Kreisel. From foundations to science: Justifying and unwinding proofs. In *Set Theory, Foundations of Mathematics (Proceedings of Symposium, Belgrade, 1977)*, number 2 (10) in Zbornik Radova, Matematički Institut, Beograd, Nova Serija, pages 63–72, 1977.

[Kre81a] G. Kreisel. Extraction of bounds: Interpreting some tricks of the trade. In P. Suppes, editor, *University-Level Computer-Assisted Instruction at Stanford: 1968–1980*, pages 149–163. Institute for Mathematical Studies in the Social Sciences, Stanford University, 1981.

[Kre81b] G. Kreisel. Neglected possibilities of processing assertions and proofs mechanically: Choice of problems and data. In P. Suppes, editor, *University-Level Computer-Assisted Instruction at Stanford: 1968–1980*, pages 131–147. Institute for Mathematical Studies in the Social Sciences, Stanford University, 1981.

[KS86] W. Kuich and A. Salomaa. *Semirings, Automata, Languages*, volume 5 of *EATCS Monographs on Theoretical Computer Science*. Springer-Verlag, Berlin, New York, 1986.

[Laa98] T. Laakso. Ahlfors q-regular spaces with arbitrary q admitting weak Poincaré inequality. March 1998. Preprint 180, Reports of the Department of Mathematics, University of Helsinki; to appear, Geometric and Functional Analysis.

[LM95] D. Lind and B. Marcus. *An Introduction to Symbolic Dynamics and Coding*. Cambridge University Press, Cambridge, 1995.

[LV90] M. Li and P. Vitányi. Kolmogorov complexity and its applications. In J. van Leeuwen, editor, *Handbook of Theoretical Computer Science Volume A: Algorithms and Complexity*, pages 187–254. The MIT Press/Elsevier, Cambridge, MA, Amsterdam, 1990.

[Man71] A. Manning. Axiom A diffeomorphisms have rational zeta functions. *Bulletin of the London Mathematical Society*, 3:215–220, 1971.

[Man77] Yu. Manin. *A Course in Mathematical Logic*, volume 53 of *Graduate Texts in Mathematics*. Springer-Verlag, Berlin, New York, 1977.

[Mañ87] R. Mañé. *Ergodic Theory and Differentiable Dynamics*. Ergebnisse der Mathematik und ihrer Grenzgebiete. Springer-Verlag, Berlin, New York, 1987.

[Mar82] D. Marr. *Vision: A Computational Investigation into the Human Representation and Processing of Visual Information*. W.H. Freeman, San Francisco, 1982.

[Mas91] W. Massey. *A Basic Course in Algebraic Topology*, volume 127 of *Graduate Texts in Mathematics*. Springer-Verlag, Berlin, New York, 1991.

[MS95] J.-M. Morel and S. Solimini. *Variational Methods in Image Segmentation*, volume 14 of *Progress in Nonlinear Differential Equations and their Ap-*

plications. Birkhäuser, Boston, 1995.

[O'N83] B. O'Neill. *Semi-Riemannian Geometry. With Applications to Relativity*, volume 103 of *Pure and Applied Mathematics*. Academic Press, New York, London, 1983.

[Ore82] V. Orevkov. Lower bounds for increasing complexity of derivations after cut elimination. *Journal of Soviet Mathematics*, 20(4):2337–2350, 1982.

[Ore93] V. Orevkov. *Complexity of Proofs and their Transformations in Axiomatic Theories*, volume 128 of *Translations of Mathematical Monographs*. American Mathematical Society, Providence, RI, 1993.

[Pan89a] P. Pansu. Dimension conforme et sphère à l'infiniti des varieétés à courbure négative. *Annales Academiae Scientiarum Fennicae Series A. I. Mathematica*, 14:177–212, 1989.

[Pan89b] P. Pansu. Métriques de Carnot–Carathéodory et quasiisométries des espaces symétriques de rang un. *Annals of Mathematics*, 129:1–60, 1989.

[Pap94] C. Papadimitriou. *Computational Complexity*. Addison-Wesley, Reading, MA, 1994.

[Par71] R. Parikh. Existence and feasibility in arithmetic. *Journal of Symbolic Logic*, 36:494–508, 1971.

[Par73] R. Parikh. Some results on the length of proofs. *Transactions of the American Mathematical Society*, 177:29–36, 1973.

[PL96] P. Prusinkiewicz and A. Lindenmayer. *The Algorithmic Beauty of Plants*. The Virtual Laboratory. Springer-Verlag, Berlin, New York, 1996.

[Pra75] V. Pratt. Every prime has a succinct certificate. *SIAM Journal on Computing*, 4:214–220, 1975.

[Pud91] P. Pudlák. Ramsey's theorem in bounded arithmetic. In E. Börger et al., editors, *Computer Science Logic: Proceedings of the Fourth Workshop (CSL '90) held in Heidelberg, 1990*, volume 533 of *Lecture Notes in Computer Science*, pages 308–317. Springer-Verlag, Berlin, New York, 1991.

[Pud98] P. Pudlák. The lengths of proofs. In S. Buss, editor, *Handbook of Proof Theory*, volume 137 of *Studies in Logic and the Foundations of Mathematics*, pages 547–637. North Holland, Amsterdam, 1998.

[R^+92] M. Ruskai et al., editors. *Wavelets and Their Applications*. Jones and Bartlett, Boston, 1992.

[Rag72] M. Raghunathan. *Discrete Subgroups of Lie Groups*. Ergebnisse der Mathematik und ihrer Grenzgebiete. Springer-Verlag, Berlin, New York, 1972.

[Rii93] S. Riis. *Independence in Bounded Arithmetic*. PhD thesis, Oxford University, 1993.

[Rii97] S. Riis. Count(q) does not imply Count(p). *Annals of Pure and Applied Logic*, 90:1–56, 1997.

[RS80] G. Rozenberg and A. Salomaa. *The Mathematical Theory of L Systems*, volume 90 of *Pure and Applied Mathematics*. Academic Press, New York, London, 1980.

[Rue94] D. Ruelle. *Dynamical Zeta Functions for Piecewise Monotone Maps of the Interval*, volume 4 of *CRM Monograph Series*. American Mathematical Society, Providence, RI, 1994.

[Sem96] S. Semmes. On the nonexistence of bilipschitz parameterizations and geometric problems about strong A_∞ weights. *Revista Matemática Iberoamericana*, 12:337–410, 1996.

[Sem99a] S. Semmes. Analysis on metric spaces. In M. Christ, C. Kenig, and C. Sadosky, editors, *Harmonic Analysis and Partial Differential Equations: Essays in Honor of Alberto P. Calderón*. University of Chicago Press, Chicago, 1999.

[Sem99b] S. Semmes. Metric spaces and mappings seen at many scales. In *Metric Structures for Riemannian and Non-Riemannian Spaces,* M. Gromov et al., volume 152 of *Progress in Mathematics*. Birkhäuser, Boston, 1999.

[Sin76] Y. Sinai. *Introduction to Ergodic Theory,* volume 18 of *Mathematical Notes.* Princeton University Press, Princeton, NJ, 1976.

[Sin94] Y. Sinai. *Topics in Ergodic Theory,* volume 44 of *Princeton Mathematical Series.* Princeton University Press, Princeton, NJ, 1994.

[Spi79] M. Spivak. *A Comprehensive Introduction to Differential Geometry.* Publish or Perish, Houston, 1979. Five volumes.

[Sta74] R. Statman. *Structural Complexity of Proofs.* PhD thesis, Stanford University, 1974.

[Sta78] R. Statman. Bounds for proof-search and speed-up in predicate calculus. *Annals of Mathematical Logic*, 15:225–287, 1978.

[Sta79] R. Statman. Lower bounds on Herbrand's Theorem. *Proceedings of the American Mathematical Society*, 75:104–107, 1979.

[Ste70] E. Stein. *Singular Integrals and Differentiability Properties of Functions*, volume 30 of *Princeton Mathematical Series.* Princeton University Press, Princeton, NJ, 1970.

[Ste93] E. Stein. *Harmonic Analysis: Real-Variable Methods, Orthogonality, and Oscillatory Integrals*, volume 43 of *Princeton Mathematical Series.* Princeton University Press, Princeton, NJ, 1993.

[Tak87] G. Takeuti. *Proof Theory,* volume 81 of *Studies in Logic.* North-Holland, Amsterdam, second edition, 1987.

[Tse68] G. Tseitin. Complexity of a derivation in the propositional calculus. *Zap. Nauchn. Sem. Leningrad Otd. Mat. Inst. Akad. Nauk SSSR*, 8:234–259, 1968.

[vL90a] J. van Leeuwen, editor. *Handbook of Theoretical Computer Science Volume A: Algorithms and Complexity.* The MIT Press/Elsevier, Cambridge, MA, Amsterdam, 1990.

[vL90b] J. van Leeuwen, editor. *Handbook of Theoretical Computer Science Volume B: Formal Models and Semantics.* The MIT Press/Elsevier, Cambridge, MA, Amsterdam, 1990.

INDEX